Earth Surface: Structure and Processes

Earth Surface: Structure and Processes

Editor: Joe Carry

RCALLISTO REFERENCE

www.callistoreference.com

Callisto Reference,
118-35 Queens Blvd., Suite 400,
Forest Hills, NY 11375, USA

Visit us on the World Wide Web at:
www.callistoreference.com

ISBN: 978-1-64116-050-6 (Hardback)

Cataloging-in-Publication Data

Earth surface : structure and processes / edited by Joe Carry.
    p. cm.
Includes bibliographical references and index.
ISBN 978-1-64116-050-6
1. Earth (Planet)--Surface. 2. Earth sciences. I. Carry, Joe.
QE511 .E27 2019
551.13--dc23

Table of Contents

Preface

The study of the Earth surface, its structure and processes, falls under the umbrella of Earth science. The lithosphere forms the outer surface of the Earth. It consists of the upper mantle and crust. Lithosphere is divided into several tectonic plates. This book delves into a detailed account of the structure of the Earth and the varied processes occurring on it. It includes some of the vital pieces of work being conducted across the world, on various topics related to the surface of the Earth. This book consists of contributions made by international experts. A number of latest researches have been included herein to keep the readers updated with the global concepts in this area of study.

Various studies have approached the subject by analyzing it with a single perspective, but the present book provides diverse methodologies and techniques to address this field. This book contains theories and applications needed for understanding the subject from different perspectives. The aim is to keep the readers informed about the progresses in the field; therefore, the contributions were carefully examined to compile novel researches by specialists from across the globe.

Indeed, the job of the editor is the most crucial and challenging in compiling all chapters into a single book. In the end, I would extend my sincere thanks to the chapter authors for their profound work. I am also thankful for the support provided by my family and colleagues during the compilation of this book.

Editor

Deformation in cemented mudrock (Callovo–Oxfordian Clay) by microcracking, granular flow and phyllosilicate plasticity: insights from triaxial deformation, broad ion beam polishing and scanning electron microscopy

Guillaume Desbois[1], Nadine Höhne[1], Janos L. Urai[1], Pierre Bésuelle[2], and Gioacchino Viggiani[2]

[1] Structural Geology, Tectonics and Geomechanics, RWTH Aachen University, Lochnerstrasse 4–20, 52056 Aachen, Germany
[2] Univ. Grenoble Alpes, CNRS, Grenoble INP, 3SR, 1270 Rue de la Piscine, 38610 Gières, France

Correspondence to: Guillaume Desbois (guillaume.desbois@emr.rwth-aachen.de)

Abstract. The macroscopic description of deformation and fluid flow in mudrocks can be improved by a better understanding of microphysical deformation mechanisms. Here we use a combination of scanning electron microscopy (SEM) and broad ion beam (BIB) polishing to study the evolution of microstructure in samples of triaxially deformed Callovo–Oxfordian Clay. Digital image correlation (DIC) was used to measure strain field in the samples and as a guide to select regions of interest in the sample for BIB–SEM analysis. Microstructures show evidence for dominantly cataclastic and minor crystal plastic mechanisms (intergranular, transgranular, intragranular cracking, grain rotation, clay particle bending) down to the nanometre scale. At low strain, the dilatant fabric contains individually recognisable open fractures, while at high strain the reworked clay gouge also contains broken non-clay grains and smaller pores than the undeformed material, resealing the initial fracture porosity.

1 Introduction

Mudrocks constitute up to 80 % of the Earth's sedimentary rocks (Stow, 1981). Due to their low permeability and self-sealing properties (Boisson, 2005; Bernier et al., 2007), claystones are considered for nuclear waste disposal and seals for storage in deep geological formations (Salters and Verhoef, 1980; Shapira, 1989; Neerdael and Booyazis, 1997; Bonin, 1998; Ingram and Urai, 1999; ONDRAF/NIRAS, 2001; NA-GRA, 2002; NEA, 2004; ANDRA 2005; IAEA, 2008). Predictions of mechanical and transport properties over long timescales are essential for the evaluation of subsurface integrity. For this, it is generally agreed that a multiscale experimental approach that combines measurement of bulk mechanical and transport properties with microstructural study to identify deformation mechanisms is required to develop microphysics-based constitutive equations, which can be extrapolated to timescales not available in the laboratory, after comparison with naturally deformed specimens (Morgenstern and Tchalenko, 1967; Tchalenko, 1968; Lupini et al., 1981; Rutter et al., 1986; Logan et al., 1979, 1987, 1992; Marone and Scholz, 1989; Evans and Wong, 1992; Katz and Reches, 2004; Niemeijer and Spiers, 2006; Colletini et al., 2009; Haines et al., 2009, 2013; French et al., 2015; Crider, 2015; Ishi, 2016).

In the field of rock mechanics and rock engineering, experiments are performed to low strain and over a relatively short time in order to predict damage and deformation in tunnelling and mining, for example. Here, a macroscopic and phenomenological approach is common to characterise mechanical and transport properties and to establish the constitutive laws. Microstructures are rarely studied because the strained regions are difficult to find (except for macroscopic fractures) and because microstructures below micrometre scales are elusive. However, it is well established that for long-term predictions a microphysics-based understanding of mechanical and fluid flow properties in mudrocks provides a better basis for extrapolating constitutive equations beyond the timescales accessible in the laboratory. This requires integration of measurement of the mechanical and transport properties with microstructures in order to obtain a multiscale description of deformation in mudrocks at low strain.

The microstructural geology community studied microstructures in deformed mudrocks to infer deformation mechanisms (Dehandschutter et al., 2004; Gratier et al., 2004; Klinkenberg et al; 2009; Renard, 2012; Robinet et al., 2012; Richard et al., 2015; Kaufhold et al., 2016), but this was limited by problems with sample preparation for high-resolution electron microscopy. Conversely, the mechanical properties and related microstructures of natural and experimental high-strain fault rocks have been studied extensively (Bos and Spiers, 2001; Faulkner et al., 2003; Marone and Scholz, 1989). For Opalinus Clay (OPA) deformed in laboratory, Nüesch (1991) and Jordan and Nüesch (1989) concluded that cataclastic flow was the main deformation mechanism, with kinking and shearing on R and P surfaces at the micro-scale; however, this was only based on observations with optical microscopy, so that grain-scale processes were not resolved. Klinkerberg et al. (2009) demonstrated a correlation between compressive strength and carbonate content of two claystones; this correlation is positive for OPA but negative for Callovo–Oxfordian Clay (COX). This was explained by the differences in grain size, shape, and spatial distribution of the carbonate (Klinkerberg et al., 2009; Bauer-Plaindoux et al., 1998). Microstructural investigations using BIB–SEM (broad ion beam and scanning electron microscopy) and FIB–TEM (focussed ion beam and transmitted electron microscopy) milling tools in OPA from the main fault in the Mont Terri underground research laboratory (Laurich et al., 2014, 2017) showed that inter- and transgranular microcracking, pressure solution, clay neoformation, phyllosilicate crystal plasticity and grain boundary sliding all play an important role during the early stages of faulting in OPA. However, simple cataclastic microstructures are rare due to the high shear strain and there was an almost complete loss of porosity in micro-shear zones.

Digital image correlation (DIC) applied to images acquired during experimental deformation provides a method to directly measure the local displacement fields (in 2-D or 3-D depending on the imaging method) and locally quantifies strain over time (Lenoir et al., 2007 (claystone, 3-D, X-ray tomography); Bornert et al., 2010 (claystone, 2-D, optical microscopy); Bésuelle and Hall, 2011 (claystone, 2-D, optical microscopy); Dautriat et al., 2011 (carbonates, 2-D, optical microscopy and SEM); Wang et al., 2013, 2015 (claystone, 2-D, environmental SEM); Fauchille et al., 2015 (claystone, 2-D, optical microscopy); Sone et al., 2015 (shale, 2-D, SEM)). For samples with grain sizes above micrometres, this approach allows the study of processes that occur at grain scale with high resolution (Hall et al., 2010 (sand, 3-D, X-ray tomography); Andò et al., 2012 (sand, 3-D, X-ray tomography); Bourcier et al., 2012, 2013 (rock salt, 2-D, optical microscopy and environmental SEM); Wang et al., 2015 (claystone, 2-D, environmental SEM)). On claystones, DIC was used to study swelling in environmental SEM (Wang et al., 2013, 2015) to measure strain between the clay matrix and non-clay minerals.

Microstructural studies in naturally compacted mudrocks are currently in rapid development, enabled by the development of ion beam milling tools (e.g. FIB and BIB), which allow imaging of mineral fabrics and porosity down to the nanometre scale in very high-quality cross sections with SEM and TEM (Lee et al., 2003; Desbois et al., 2009, 2011, 2013, 2016; Loucks et al., 2009; Curtis et al., 2010; Heath et al., 2011; Klaver et al., 2012; Keller et al., 2011, 2013; Houben et al., 2013, 2014; Hemes et al., 2013, 2015; Laurich et al., 2014; Warr et al., 2014; Song et al., 2016). Serial sectioning allows the reconstruction of microstructure in 3-D (Keller et al., 2011, 2013; Milliken et al., 2013; Hemes et al., 2015), and cryogenic techniques can image the pore fluid in the samples and avoid artefacts produced by drying (Desbois et al., 2013, 2014; Schmatz et al., 2015).

Previous work has shown that the mechanical properties of COX do not only depend on the fraction and mineralogy of the clay but also on water content and texture (Bauer-Plaindoux et al., 1998). Chiarelli et al. (2000) showed that COX is more brittle with increasing calcite content and more ductile with increasing clay content, and they proposed two deformation mechanisms: plasticity induced by slip of clay sheets and induced anisotropic damage as indicated by microcracks at the interface between grains and matrix; however, they provided little microstructural evidence to support this. Gasc-Barbier et al. (2004), Fabre and Pellet (2006), Chiarelli et al. (2003) and Fouché et al. (2004) reported that the COX has an unconfined compressive strength of 20 to 30 MPa and a Young's modulus of 2 to 5 GPa. In the context of underground storage of radioactive waste, these papers try to predict the mechanical evolution of COX over the period of thousands of years. The effects studied included creep, pore-pressure dissipation, swelling, contraction, chemical effects, pressure solution and force of crystallisation. Although these papers develop elaborate constitutive laws, they provide very limited microstructural observations. The need for micromechanical observations was already recognised by Yang et al. (2012) and Wang et al. (2013, 2015). From DIC applied to optical and environmental scanning electron microscope (ESEM) images, these authors showed how heterogeneous strain fields correlate with microstructure and recognised shear bands and tensile microcracks.

For highly overconsolidated claystones from the Variscan foreland thrust belt in the Ardennes and Eifel, Holland et al. (2006) proposed an evolutionary model starting with mechanical fragmentation of the original fabric. In this model, the initial loss of cohesion is driven by kinking, folding and microfracturing processes, with an increasing porosity and permeability. Abrasion during progressive deformation increases the amount of clay gouge, and resealing occurs by decrease in pore size of the clay gouge.

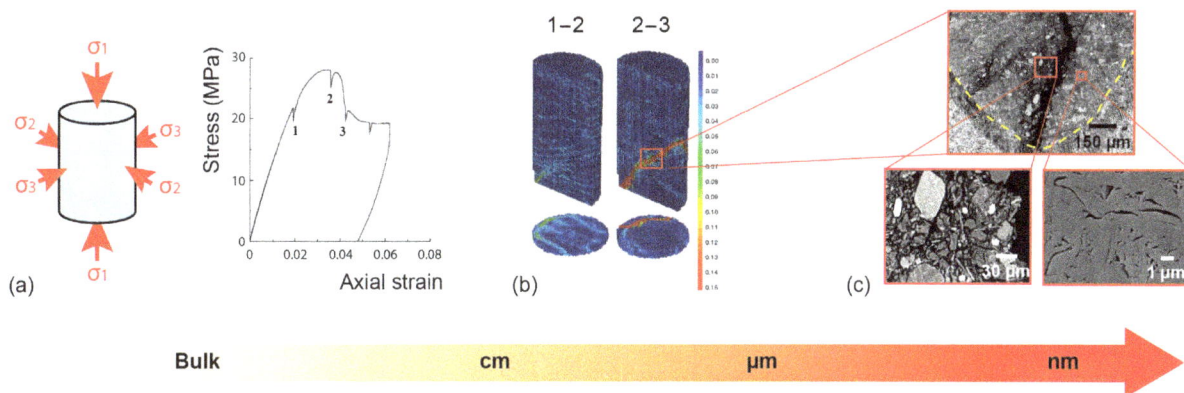

Figure 1. Drawing of the experimental concept used for the investigation of experimentally deformed fine-grained mudrocks from bulk scale to nanometre scale. The example is based on **(a)** a triaxial deformation test (10 MPa confining pressure) performed on a cylindrical Callovo–Oxfordian Clay, **(b)** volumetric DIC on X-ray microtomography images to follow displacement fields (after Lenoir et al., 2007), and **(c)** SEM imaging on high-quality cross sections prepared by BIB. **(a)** Steps when X-ray microtomography images were acquired are indicated by 1, 2 and 3. **(b)** Deformation increments between steps 1 and 2 and steps 2 and 3 are indicated by 1–2 and 2–3 respectively.

In summary, deformation mechanisms in mudrocks are poorly understood, especially at low strain. Although as a first approximation the plasticity of cemented and uncemented mudrocks can be described by effective pressure-dependent constitutive models, the full description of their complex deformation and transport properties would be much improved by better understanding of the microscale deformation mechanisms. There is a wide range of possible mechanisms: intra- and intergranular fracturing, cataclasis, grain boundary sliding, grain rotation and granular flow, plasticity of phyllosilicates, and the poorly known plasticity of nanoclay aggregates with the strong role of clay-bound water, cementation, fracture sealing and solution precipitation.

This contribution combines stress-strain data and measurement of displacement fields by DIC with microstructural investigations in areas selected based on the DIC results. For this, we prepared millimetre-sized high quality cross sections by (BIB) milling followed by SEM to infer microphysical processes of deformation with submicron resolution (Fig. 1). The two samples used are from the COX (a cemented claystone): one deformed in plane strain compression at 2 MPa confining pressure (COX-2MPa; Bésuelle and Hall, 2011) and another in triaxial compression at 10 MPa confining pressure (COX-10MPa; Lenoir et al., 2007). Specimens were taken from the Bure site in Meuse-Haute Marne in France and belong to the clay-rich facies of the COX.

2   Material studied and DIC-derived strain fields

Triaxial experiments were performed on two COX samples collected at the ANDRA Underground Research Laboratory located at Bure (Meuse-Haute Marne, eastern France) at approximately 550 m below ground surface (Boisson, 2005). The clay fraction (illite–smectite, illite, chlorite) is 40–45 %,

carbonate (mostly calcite) and quartz are 25–35 and 30 % respectively and the samples contain minor feldspar, mica and pyrite (Gaucher et al., 2004).

The details of these experiments, including instrumentation, boundary conditions and DIC interpretations are comprehensively described in Bésuelle and Hall (2011) and Lenoir et al. (2007). This contribution mostly presents the microstructural analysis performed on these previously deformed two samples.

The first sample considered in this study (COX-2MPa, sample reference: EST32896) was tested in plane strain compression at 2 MPa confining pressure. Two-dimensional DIC was performed on consecutive photographs of one side of the specimen (in the plane of deformation) throughout the test. Further details are given in Bésuelle and Hall (2011). The second sample (COX-10MPa) was tested in triaxial compression at 10 MPa confining pressure. Three-dimensional DIC was performed on consecutive X-ray images of the specimen obtained in a synchrotron throughout the test. Further details are given in Lenoir et al. (2007). Please note that in this publication this sample is referred to as ESTSYN01 with drilling reference EST261.

In the following paragraphs, the relevant findings in Bésuelle and Hall (2011) and Lenoir et al. (2007) are summarised.

The prismatic sample COX-2MPa was tested in plane strain compression in a true triaxial apparatus at a constant value of $\sigma_3 = 2$ MPa. The size of the specimen is 50 mm in the vertical direction, which is the direction of major principal stress (σ_1), 30 mm in the direction of intermediate principal stress (σ_2), and 25 mm in the direction of minor principal stress (σ_3). The test was displacement controlled, with a constant rate of displacement (in direction 1) of 1.25 μm s^{-1}, i.e. a strain rate of 2.5×10^{-5} s^{-1} (see Bésuelle and Hall, 2011 for further details). Figure 2a shows the evolution of the dif-

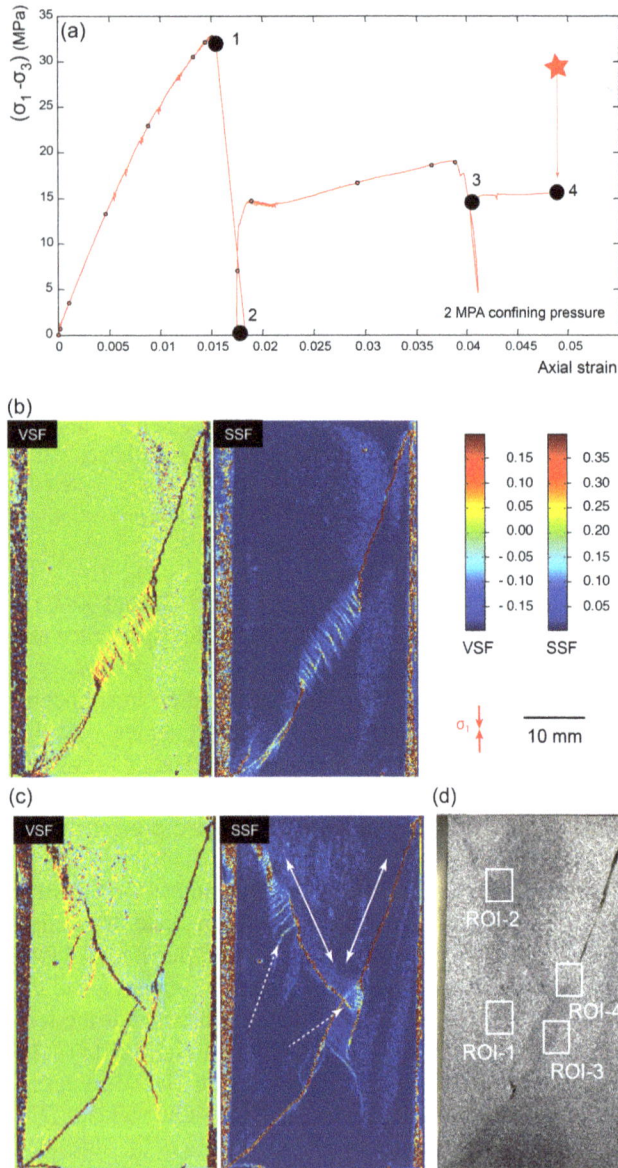

Figure 2. Results of deformation test done on sample COX-2MPa. **(a)** Deviator stress vs. axial strain response. The red star indicates the state of the sample when BIB–SEM microstructural analyses are done. **(b, c)** Incremental volumetric strain fields (VSF) and maximum shear strain fields (SSF) for deformation increments 1–2 **(b)** and 3–4 **(c)** indicated in **(a)** after DIC. Arrows with solid lines indicate the set of two conjugated synthetic fractures, whereas the arrows with dashed lines show antithetic fractures oblique to the conjugated fractures. **(d)** Selection of differently strained areas (region of interest (ROI)) highlighted from DIC analysis for BIB–SEM microstructural analyses. Four ROI were analysed: three at conjugate synthetic fractures in areas with a different amount of diffuse strain and antithetic fractures (ROI-2, ROI-3 and ROI-4) and one in a region without measurable strain (ROI-1). The bedding is perpendicular to the mean principal stress σ_1. Performed according to Bésuelle et al. (2011).

ferential stress (σ_1–σ_3) vs. axial strain. The curve shows a first stress peak at 0.02 axial strain, followed by a strong stress drop. Then, a slow stress increase is observed, followed by a second stress drop at 0.42 axial strain. Afterwards, the stress is quite constant. As shown in Fig. 2b and c (gage length of 180 µm), these two stress drops are associated with major faulting in the specimen. The crack that appeared during the second drop is conjugate to the first crack set, which appeared at the first drop. This set of conjugate fractures, at an angle of 20 to 45° about direction 1, will be referred to as "main synthetic fractures" in the following sections. The DIC-derived strain fields in Fig. 2b and c also show that the development of each single conjugate fracture is accompanied by relay zones with a set of antithetic fractures. Moreover, the fracture appearing during the second stress drop (Fig. 2c) also reactivates the first fracture and its associated antithetic fractures. At this resolution (pixel size is $10 \times 10\,\mu m^2$), the set of conjugate fractures and the associated antithetic fractures are the major features of localised deformation: they represent zones where the sample was sheared, with damaged zones having a thickness of about 60 µm. Dilatancy was also measured in the damaged zones mentioned above (see volumetric strain fields, Fig. 2b and c).

The cylindrical sample COX-10MPa (10 mm in diameter and 20 mm in height) was deformed in triaxial compression at a confining pressure of 10 MPa. The test was carried out under tomographic monitoring at the European Synchrotron Radiation Facility (ESRF) in Grenoble, France, using an original experimental set-up developed at Laboratoire 3SR at the University of Grenoble Alpes (France). Complete 3-D images of the specimens were recorded throughout the test using X-ray microtomography (voxel size was $14 \times 14 \times 14\,\mu m^3$). The test was displacement controlled, with a displacement rate of $0.05\,\mu m\,s^{-1}$, i.e. an axial strain rate of $2.5 \times 10^{-6}\,s^{-1}$. The stress-strain curve (Fig. 3a) shows only one stress peak at an axial strain of 0.04. The peak stress is followed by a major stress drop corresponding to the formation of a shear fracture (referred to as main synthetic fracture in the following sections) oriented at an angle of 30–40° about the direction of the principal stress σ_1 (the DIC-based maximum shear strain fields are given in Fig. 3b, gage length of 280 µm). The DIC-derived volumetric strain fields (not shown here; see Lenoir et al., 2007) indicate that the shear fracture is accompanied by some slight dilatancy.

3 Methods: BIB–SEM imaging of deformed microstructures

After the experiments of Lenoir et al. (2007) and Bésuelle and Hall (2011), deformed samples were stored at low vacuum and room temperature in a desiccator, where they dried slowly. From these deformed samples, subsamples were selected to represent areas with different strain history based on the DIC analysis. For COX-2MPa, three BIB cross sections

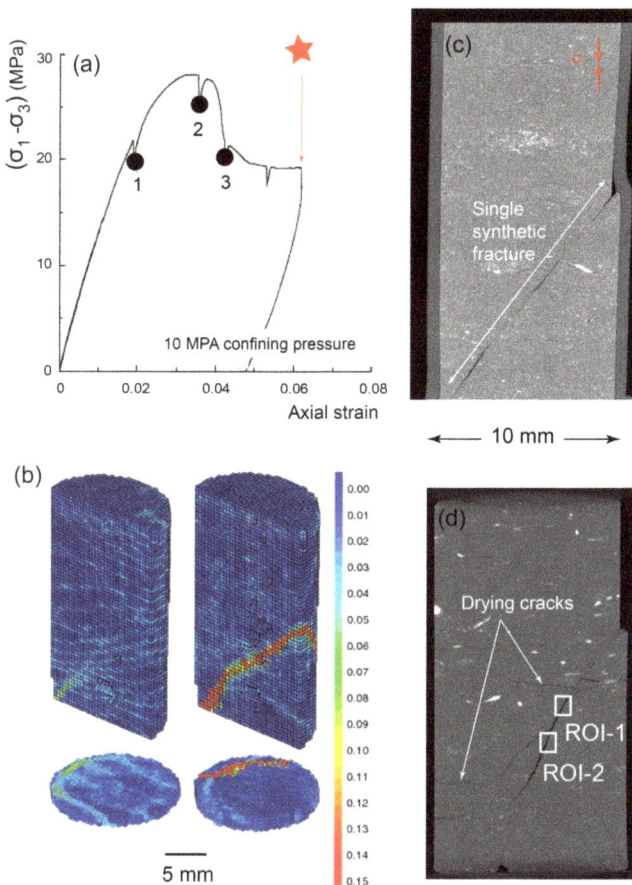

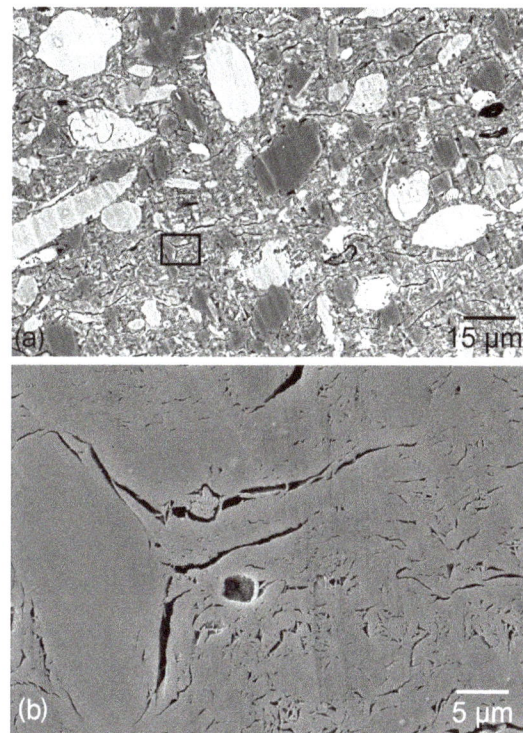

Figure 4. (a) BSE–SEM micrograph of the typical mineral fabric in undeformed COX. (b) SE2 SEM micrograph of a detail indicated by the black box in (a) showing the typical pore fabric in undeformed COX. In both micrographs, the bedding is horizontal.

Figure 3. Results of deformation test done on sample COX-10MPa. (a) Deviator stress vs. axial strain response. The red star indicates the state of sample when BIB–SEM microstructural analyses are done. (b) Incremental maximum shear strain fields for deformation increments 1–2 and 2–3 indicated in (a) interpreted after DIC. (c) Shows the X-ray radiography of the sample taken directly at the end of the deformation test, whereas (d) shows the X-ray radiography of the same sample but taken about 10 years after the end of the deformation: drying cracks developed following the bedding, and the aperture of the single shear fracture became larger. (d) Also indicates that two ROI were analysed, both around the single synthetic shear fracture. In (c) and (d) the bedding is perpendicular to the principal stress σ_1 indicated in (c). Performed according to Lenoir et al. (2007).

1.10^{-4} Pa, 6 kV, 150 µA) to remove a 100 µ m thick layer of material interpreted to be the layer of damage after polishing with SiC papers. BIB cross sections were all prepared parallel to the σ_1 and direction and perpendicular to the shear fracture. The BIB cross sections of about 1.5 mm^2 (Figs. 5 and 6) were imaged with a Supra 55-Zeiss SEM (SE2 and BSE detectors at 20 kV and working distance (WD) = 8 mm). Further details of the method are given in Klaver et al., 2012, 2015; Houben et al., 2013, 2014; Hemes et al., 2013, 2015; and Desbois et al., 2016.

4 Results

4.1 Overview of microstructures

were prepared around the conjugate fractures in areas with different amounts of diffuse strain (at the resolution of DIC) and antithetic fractures (ROI-2, ROI-3 and ROI-4; Figs. 2d, 5b, c, d and 6) and a fourth BIB cross section is from a region without measurable strain (ROI-1; Figs. 2d and 5a). For COX-10MPa, two BIB–SEM analyses were done around the single shear fracture (Figs. 3d and 5e, f).

Subsamples were first embedded in epoxy, extracted with a low speed diamond saw in dry conditions, pre-polished dry using SiC papers (down to P4000 grade) and BIB polished in a JEOL SM-09010 cross section polisher (for 8 h, 1.10^{-3}–

The subsample without measurable strain (i.e. ROI-1_COX-2MPa, Fig. 5a) shows non-clay minerals in a clay matrix with a weak-shape-preferred orientation parallel to bedding (perpendicular to the experimental σ_1). The clay matrix contains submicron pores typical of compaction and diagenesis, with a power law distribution of pore sizes. Pores commonly have very high aspect ratio, with the long axis oriented subparallel to the bedding. Mineral fabric is very similar to those in the undeformed COX sample (Fig. 4; see Robinet et al., 2012).

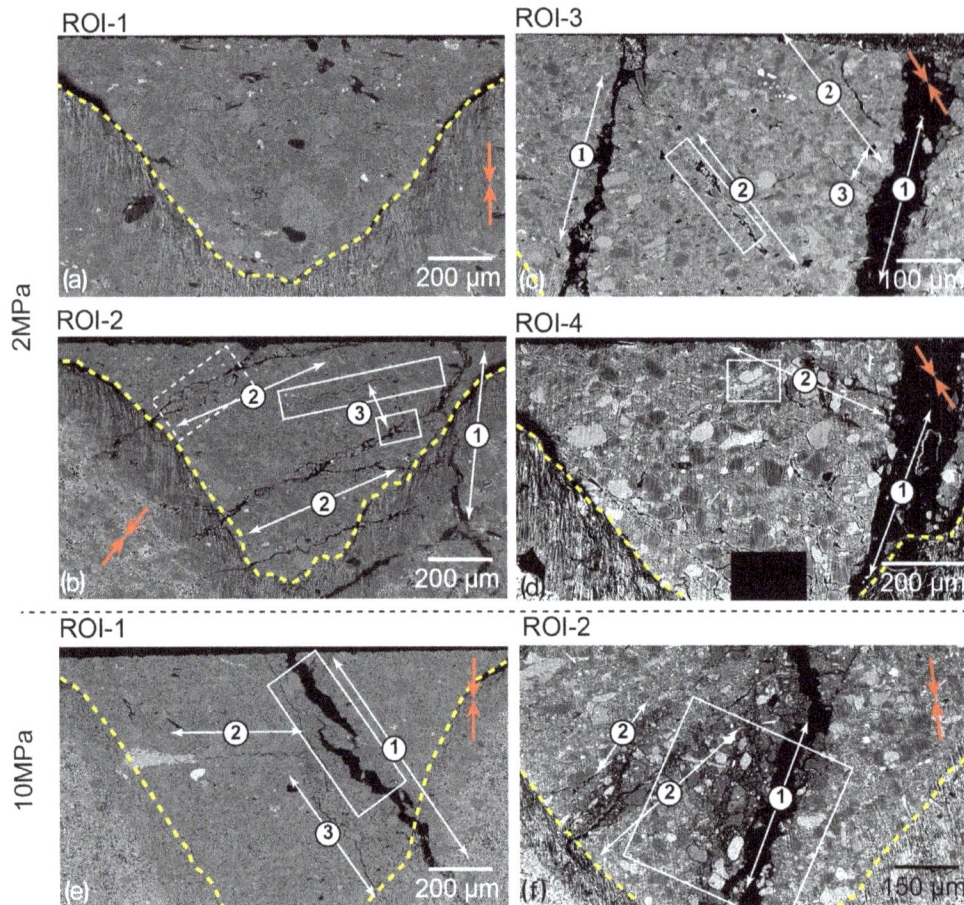

Figure 5. BSE–SEM micrographs of the BIB cross section overviews of COX-2MPa **(a–d)** and COX-10MPa **(e–f)** at differently strained areas (ROI) highlighted from DIC analysis in Figs. 2 and 3. Highly strained ROI **(b–f)** display damaged microstructures where three different types of fracture are identified: (1) the main synthetic fracture, (2) antithetic fractures oriented about 60° to the main fracture and (3) joints subparallel to the main synthetic fracture. These fractures are respectively indicated by numbers 1, 2 and 3 in the figure. In **(b)** the white box in dashed lines refers to Fig. 7f; the upper white box in solid lines refers to Fig. 7a and the lower white box in solid lines refers to Fig. 7e. In **(c)** the white box in solid lines refers to Fig. 7g. In **(d)** the white box in solid lines refers to Fig. 7c. In **(e)** the white box in solid lines refers to Fig. 9. In **(c)** the white box in solid lines refers to Fig. 11. In all micrographs, the orientation of the principal stress (σ_1) is indicated by red arrows. The bedding is perpendicular to σ_1. Dashed yellow lines indicate the boundaries of the BIB-polished areas.

In all other BIB cross sections (Figs. 5c–f and 6), both samples show damaged microstructures. At the sample scale, three different types of fracture are identified: (i) the main synthetic fracture (Sect. 2), (ii) antithetic fractures (Fig. 5) and (iii) joints subparallel to the main fracture. The material between the fracture zones has very similar microstructure to undeformed COX.

4.2 Detailed description of microstructures

4.2.1 Arrays of antithetic fractures

In COX-2MPa, the antithetic fractures (Fig. 6) are of two different types. Type I is located only in the clay matrix (Fig. 7a), with apertures up to a few micrometres, with boundaries closely matching, suggesting that these are opening mode fractures (Mode I). Type II fractures consist of a

damage zone with a thickness of up to 25 µm (Fig. 7e, f, g, h, i), containing angular fragments of non-clay minerals and clay aggregates (Fig. 7h), sometimes with preferred orientation parallel to the fracture. The transition between the damage zone and the undeformed host rock is sharp (Fig. 7f, g, h, i). In relay zones the fracturing becomes so intense that the clay matrix is fragmented into submicron-size fragments (Fig. 7i). Porosity in these relay zones is locally much higher and pores are much larger than in undeformed COX. Fracture boundaries usually do not match (Fig. 7h). Figure 7e shows examples where parts of broken non-clay minerals can be matched.

In COX-10MPa, we observed the two types of antithetic fractures mentioned above. Antithetic fractures of Type I are very similar (indicated in Fig. 5f) to those in COX-2MPa but they are rare, whereas antithetic fractures of Type II contain a

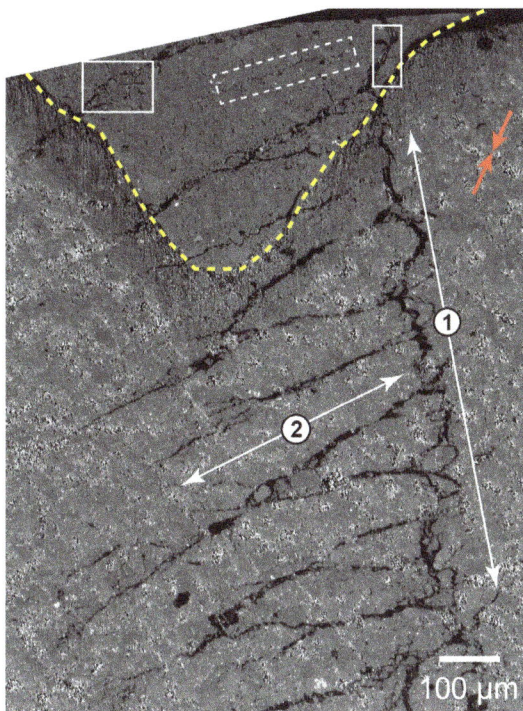

Figure 6. Larger field of BSE–SEM micrograph of the BIB cross section overview at ROI-1 in COX-2MPa sample. It shows the network of antithetic fractures (indicated by number 1) oblique to the principle main synthetic fracture (indicated by number 2). The left white box in solid lines refers to Fig. 7f; the white box in dashed lines refers to Fig. 7a and the right white box in solid lines refers to Fig. 8. Orientation of the principal stress (σ_1) is indicated by red arrows. The bedding is perpendicular to σ_1. Dashed yellow lines indicate the boundaries of the BIB-polished areas.

wider damage zone in comparison to those in COX-2MPa, in which the average grain size and the pore size is significantly smaller, consistent with stronger cataclasis at high confining pressure. In parts of the damage zones interpreted to be restraining sections, pores in the reworked clay aggregates cannot be resolved in the SEM.

In both samples, the fragments between the arrays of antithetic fractures show only minor deformation indicated by fractured grains of organic matter (Fig. 7b), calcite (Fig. 7d, c) or quartz (Fig. 7d). Visible relative rotation of parts of fractured grain is rare (Fig. 7d).

4.2.2 Synthetic fractures

The synthetic fractures are the regions that localised most of the strain and have the thickest damage zone (Figs. 2 and 3). Here, COX-2MPa and COX-10MPa show very similar microstructures. The grain (fragment) size of non-clay minerals is significantly smaller than in the host rock and their sizes are poorly sorted. In comparison to undeformed samples (Fig. 4a), non-clay minerals also have dominant angular and/or chipped edges (Figs. 8, 9 and 11). Locally, grains

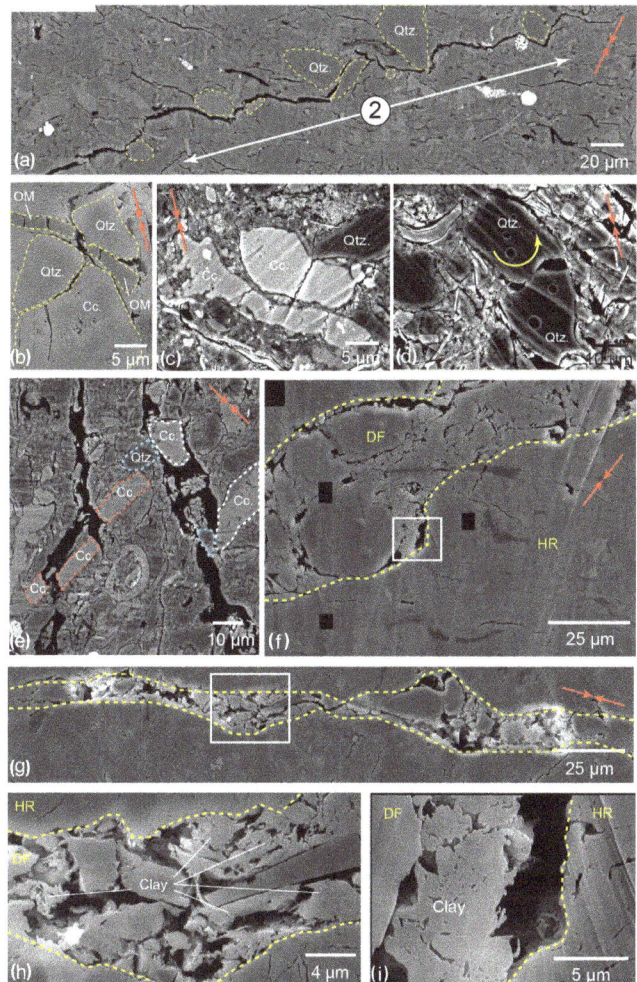

Figure 7. Detailed microstructures in sample COX-2MPa. **(a)** A fracture running parallel to the antithetic fractures and at the interfaces between non-clay mineral and clay matrix. In **(b)** intragranular fractures are in calcite grain (Cc.) and transgranular fractures are in organic matter (OM), whereas in **(c)** transgranular fractures are in Cc. at impingement with quartz grain (Qtz.). **(d)** A broken quartz grain showing evidence for rotation of its broken fragments. **(e)** Incipient of flow of broken non-clay minerals within the antithetic fractures indicated by red (Cc.), blue (Qtz.) and white (Cc.) dashed lines. **(f)** and **(g)** show parts of antithetic fractures displaying thick damaged fabrics made of broken grains and clay matrix fragments. **(h)** Details of the white box indicated in **(g)**. **(i)** Details of the white box indicated in **(f)**, showing the denser and deformed fabric of a part of the clay matrix squeezed between a quartz grain located in the damaged fabric and the boundary with the host rock. In **(f)**–**(i)**, the damaged fabric is related to a higher porosity in comparison to the host rock. In all micrographs, the orientation of the principal stress (σ_1) is indicated by red arrows. The bedding is perpendicular to σ_1. Dashed yellow lines indicate the boundaries between the damaged fabric (DF) and the host rock (HR) and also some grain boundaries in **(a)**. Black squares in **(f)** are missing pictures.

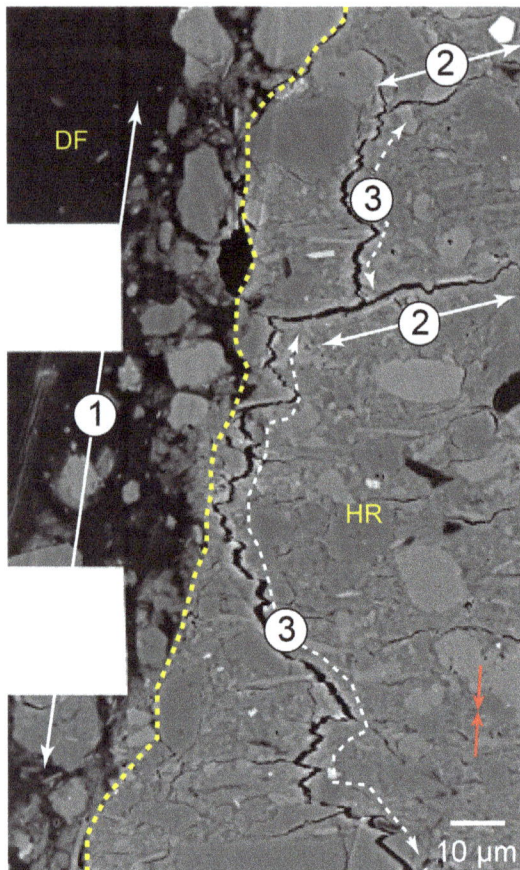

Figure 8. Detailed microstructure close the main fracture (indicated by number 1) in sample COX-2MPa. The main fracture displays internal damaged fabric made of fragments of broken non-clay minerals and clay matrix. Close to the main synthetic fracture, the host rock displays jagged joints subparallel to the main synthetic fracture (indicated by number 3) starting and ending at the antithetic fracture (indicated by number 2). In all micrographs, the orientation of the principal stress (σ_1) is indicated by red arrows. The bedding is perpendicular to σ_1. The dashed yellow line indicates the boundary between the damaged fabric (DF) and the host rock (HR).

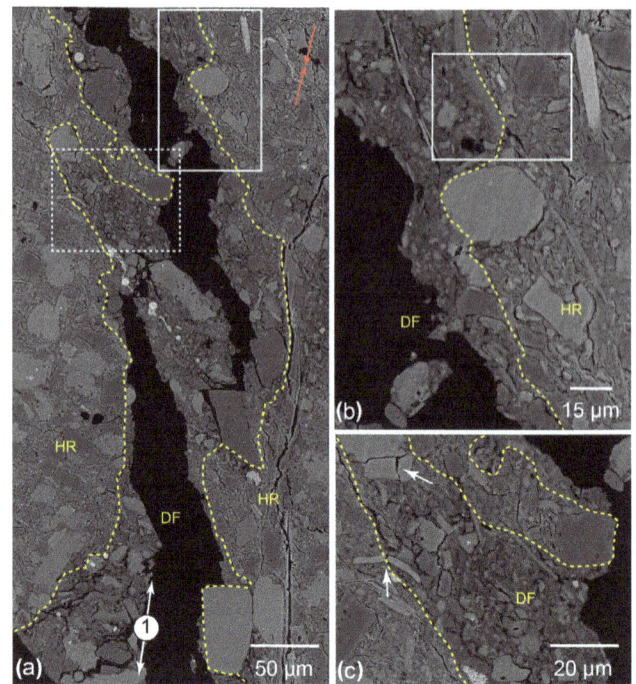

Figure 9. Microstructures of ROI-1 in sample COX-10MPa. (**a–c**) The damaged fabric (DF) within the main fracture (1) is made of fragments of non-clay minerals derived from the dense, tight clay matrix. (**a**) The large open fracture in the middle of the main fracture (black) is interpreted to develop after the experiment by unloading and/or drying (see Sect. 5.1 for details). In (**a**) the white box in dashed lines refers to (**c**), whereas the one in solid lines refers to (**b**). (**b**) Details of difference in mineral fabric between DF and the host rock (HR). The white box refers to Fig. 10. (**c**) Some grains within the damaged fabric, but close to the boundary between the damaged fabric and the host rock, show transgranular fracturing (white arrows). In all micrographs, the orientation of the principal stress (σ_1) is indicated by red arrows. The bedding is perpendicular to σ_1. The dashed yellow lines indicate the boundaries between DF and HR.

in the damaged zone show transgranular fractures (Figs. 9c and 11a). In parts of the damage zone, dilatancy and a strong increase in connected porosity (ROI-4_COX-2MPa, Fig. 8) are indicated by epoxy impregnation. In other parts, (ROI-1_COX-10MPa, Figs. 9 and 10) strongly reworked clay matrix is not impregnated and shows no pores visible at the resolution of image (83.8 nm pixel size in Fig. 10b, c).

For COX-2MPa, the DIC analysis shows that the conjugated synthetic fractures form a complex network of fracture branches in the region where they both intersect (Fig. 2c). The ROI-3_COX-2MPa subsample (Fig. 2d) covers two of these branches. Microstructural analysis of these two branches in ROI-3_COX-2MPa shows similar microstructures, with only the fracture apertures being different (Fig. 5c).

In both COX-2MPa and COX-10MPa, the damage zone of the synthetic fractures contains an open fracture (Figs. 8, 9 and 11), with apertures of 50–70 µm. These large open fractures are filled with epoxy, have matching boundaries and never crosscut the non-clay minerals in the damage zone. Similar fractures are found in COX-2MPa but parallel to the antithetic fractures, with jagged morphologies and matching walls never crossing the non-clay minerals (Fig. 7b, c, e). These fractures are not resolved by DIC at the resolution of the X-ray images and at the strain gage length used in this contribution.

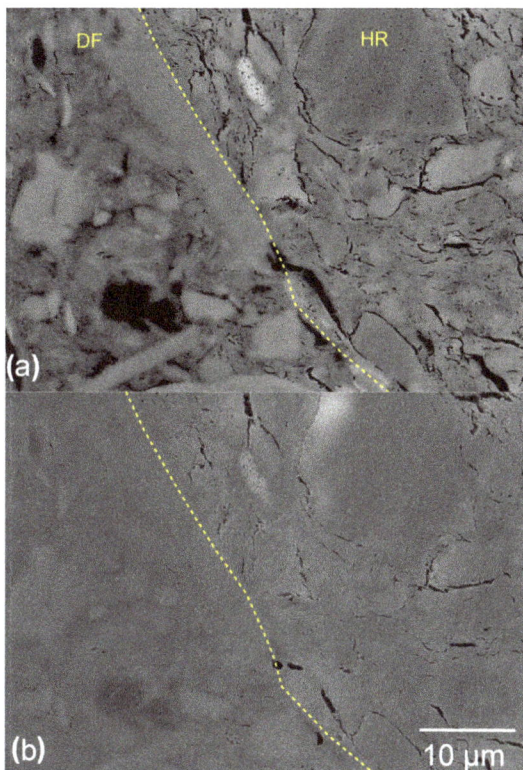

Figure 10. Details of Fig. 9b. Microstructures (ROI-1_COX-10MPa) showing details of porosity in BSE–SEM micrograph (a) and SE2 SEM micrograph (b). At the resolution of the SEM micrograph, the damaged fabric appears to be very low porous in comparison to the host rock. The dashed yellow line indicates the boundary between the damaged fabric (DF) and the host rock (HR).

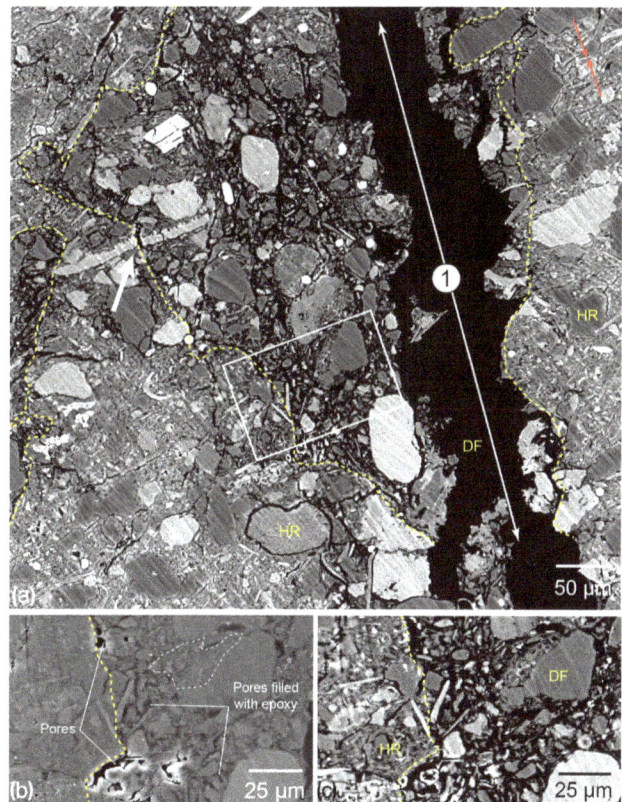

Figure 11. Detailed microstructures at ROI-2 in sample COX-10MPa. (a–c) The damaged fabric (DF) within the main synthetic fracture (indicated by number 1) is made of fragments of non-clay minerals and clay matrix derived from the host rock (HR). (a) Some grains within the damaged fabric, but close to the boundary between the damaged fabric and the host rock, show transgranular fracturing (white arrow). The white box refers to Fig. 11a and b. Detailed observations in (b) and c (SE2 SEM and BSE–SEM micrographs of the same sub-area, respectively) show that parts of the damaged fabric display (i) porous island, where pores are between the fragments of non-clay and clay matrix, whereas other parts display (2) low porous islands made of fragments of non-clay minerals embedded in a dense, tight clay matrix (within the region bounded by the dashed white line). Pores within the porous island can either be filled with epoxy (in deep black pixel values) or not. The orientation of the principal stress (σ_1) is indicated by red arrows. The bedding is perpendicular to σ_1. The dashed yellow lines indicate the boundaries between the damaged fabric and the host rock.

5 Discussion

5.1 Artefacts caused by drying and unloading

Claystones are sensitive to changes in hydric conditions that can lead to the shrinkage or the swelling of the clay matrix (Galle, 2001; Kang et al., 2003; Soe et al., 2007; Gasc-Barbier and Tessier, 2007; Cosenza et al., 2007; Pineda et al., 2010; Hedan et al., 2012; Renard, 2012; Wang et al., 2013, 2015; Desbois et al., 2014).

The DIC analysis is not affected by this because the images were acquired during deformation of preserved (wet) samples. SEM analysis is done on samples that have been deformed and unloaded, followed by slow drying in a low vacuum and further dehydration in the high vacuum of the BIB and SEM. In COX-10MPa, this is illustrated by Fig. 3c and d. Figure 3c shows the sample at the end of the deformation experiment, whereas Fig. 3d shows the same sample but about 10 years later, both X-ray imaged. The comparison of Fig. 3c and d shows that cracks developed parallel to the bedding and that the apertures of fractures developed during the deformation became larger. These are interpreted to result

from unloading and shrinkage during drying of specimens. Though the second sample was not scanned with X-ray in the dry condition, we infer that similar changes also occurred in COX-2MPa: by analogy, there is no reason that the clay matrix in COX-2MPa behaves differently that in COX-10MPa.

The considerations above indicate that some fractures developed during deformation but drying damage overprinted them. Unfortunately, BIB–SEM images (performed on dried samples) do not provide direct information to distinguish if the visible fractures and cracks developed during deformation (and subsequently overprinted by drying) or only by drying. However, as presented in the following paragraphs, indirect evidence suggests that the fractures in the fragments between the arrays of antithetic fractures and the antithetic fractures of Type I and Type II developed during deformation.

The fractures in the fragments between the arrays of antithetic fractures (Fig. 7b, c, d) are not present in the low-strain ROI-1_COX-2MPa, and they are subparallel to σ_1 and crosscut the bedding, suggesting strongly that they are formed by experimental deformation.

Antithetic fractures of Type II (Figs. 5, 6 and 7e–i) are interpreted to develop during deformation because (i) the internal microstructures and fabrics are damaged and (ii) DIC recorded a clear localisation of strain in these. Though the antithetic fractures of Type I are not clearly recognised at the resolution of DIC, most of these in COX-2MPa (Fig. 7a) are interpreted to develop during deformation because they are oblique to the bedding and parallel to the antithetic fractures of Type II (Figs. 5, 6 and 7f–g). One exception is the antithetic fractures of Type I observed in ROI-1_COX-10MPa (Fig. 5e), which are parallel to bedding. Mode I fractures subparallel to the main synthetic fractures are less easy to interpret: they may be related to the rotation of blocks between the antithetic fractures (Kim et al., 2004). Cryogenic techniques to preserve wet fabrics combined with ion beam milling and cryo-SEM (Desbois et al., 2008, 2009, 2013, 2014) are the dedicated techniques for addressing this question in the future.

5.2 Deformation mechanisms

In our experiments, differential stresses exceed the confining pressure by a factor of 3–15, which would suggest that dilatant fracturing prevails over other mechanisms (e.g. Kohlstedt et al., 1995). This is partly corroborated from the stress-strain measurements that show major stress drops after peaks of stress (Figs. 2 and 3). In agreement with this, at micro-scale the first conclusion based on the microstructural observations above is the dominantly cataclastic deformation in Callovo–Oxfordian Clay at confining pressures up to 10 MPa. Microfracturing, which produces fragments at a range of scales and reworks them into a phyllosilicate-rich cataclastic gouge during frictional flow, is the main process in both samples. This is accompanied by dilatancy and by microfracturing of the original fabric but also by progressive decrease in porosity and pore size in the gouge with the non-clay particles embedded in reworked clay. The structure of macro-scale fracture in the samples compares well with Ishii et al. (2011, 2016).

Although in many cases the initial fractures propagate around the hard non-clay grains, there is also significant fracturing of the hard non-clay minerals (e.g. Fig. 7b–d). This can be due to local stress concentrations at contacts between adjacent non-clay minerals or because the clay matrix is so strongly cemented that it can transmit stresses sufficient to fracture calcite and quartz grains. Broken non-clay minerals can displace or rotate with respect to each other (Fig. 7d) with local dilatancy during deformation (Fig. 2b), in agreement with the interpretation of DIC measurements in Bésuelle and Hall (2011) and Lenoir et al. (2007).

In COX-2MPa, the propagation of antithetic fractures of Type I (Fig. 7a) is predominantly in the clay matrix. This is in agreement with the smaller strain in comparison to antithetic fractures of Type II. Antithetic fractures of Type II contain angular non-clay grains with smaller size than those in the host rock. We interpret these as evidence for comminution by grain fracturing. Matching broken grains (Fig. 7e) are rare and in agreement with high-strain cataclastic flow. Fragments of clay aggregates in the antithetic fractures of Type II are much less coherent (Fig. 7h) and more porous than the undeformed COX (Fig. 7i), indicating strong remolding by cataclastic flow and perhaps also plastic deformation of phyllosilicates. Here, because pore morphologies do not show typical shapes that originate from drying, we interpret this to mean that these developed during deformation.

Microstructures in the main synthetic fractures, both in COX-2MPa (Fig. 8) and COX-10 MPa (Figs. 9 and 11), are similar. Angular non-clay minerals in the reworked clay matrix have a wide range of grain sizes, smaller than those in the host rock. These characteristics are typical for cataclasis (Passchier and Trouw, 2005). In COX-2MPa, the cataclastic gouge seems to be more porous than in COX-10 MPa; this is as expected for the lower mean stress, but firm conclusions require further study to exclude that this is an unloading and drying effect. For COX-10MPa, the porosity in the clay matrix is clearly reduced in comparison to the one in the host rock: most pores, if present, are below the resolution of SEM (Figs. 9 and 10). The mechanism of this compaction during shearing is interpreted to be a combination of cataclasis of the cemented clay matrix and shear-induced rearrangement of clay particles around the fragments of non-clay particles.

5.3 Conceptual model of microstructure development in triaxially deformed COX

Based on BIB–SEM microstructural observations, we propose the following sequence of micromechanisms in the Callovo–Oxfordian Clay (Fig. 12):

(1) & (2) Microfracturing

Incipient deformation occurs by intergranular microfractures propagating in the clay matrix and transgranular and intragranular microfractures propagating in non-clay minerals, both resulting in the fragmentation of the original fabric and in agreement with the high compressive strength of this cemented mudstone. Intergranular microfractures are interpreted to be initiated from pores, propagating along weak contacts at non-clay mineral–clay matrix interfaces or along

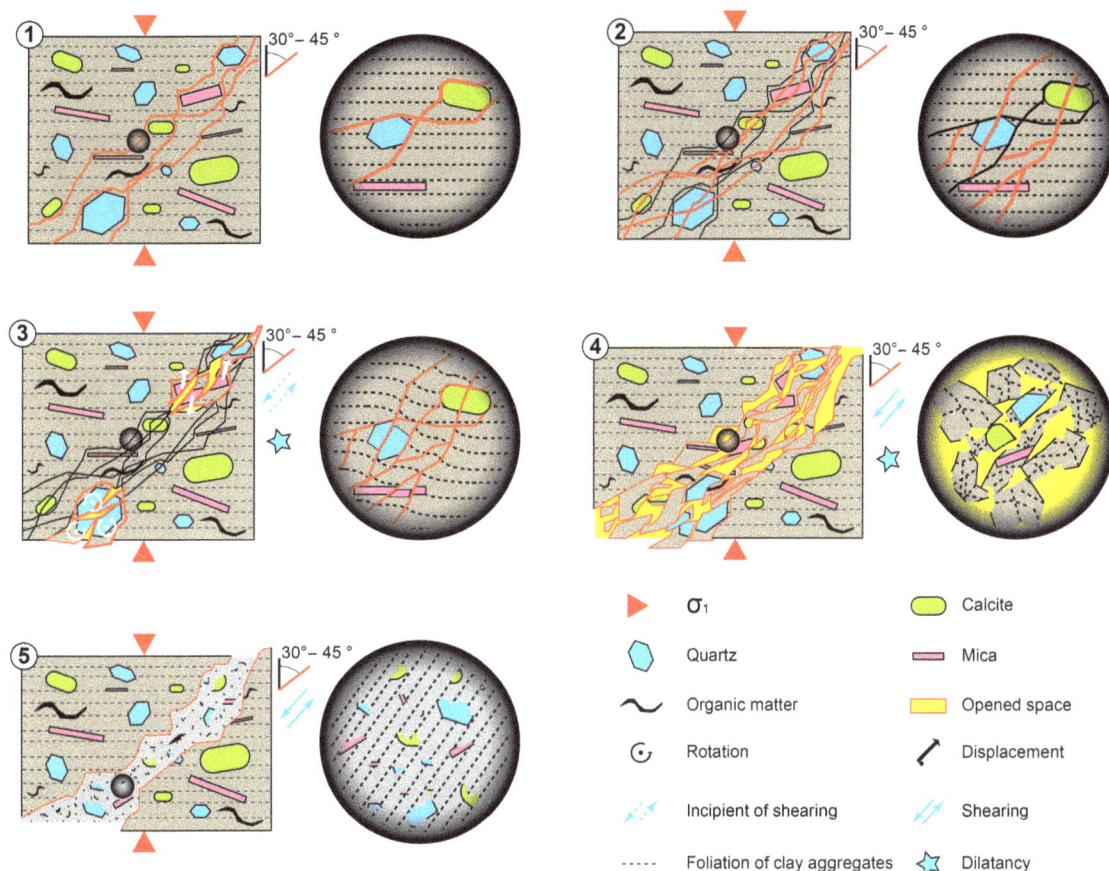

Figure 12. Conceptual model of microstructure development in triaxially deformed COX. (1) and (2) show microfracturing. In (1) intergranular microcracking initiating at non-clay minerals and clay minerals (NCM–CM) interfaces and propagating within CM. In (2) fragmentation of original fabric by transgranular and intragranular microfracturing of NCM. (3, 4) Cataclastic shearing with plasticity of phyllosilicates, macroscopic failure. In (3) incipient of shearing enhanced by plasticity of phyllosilicates at microfracture boundaries initiates cataclastic flow of original fabric's fragments. In (4) ongoing shearing drives cataclastic flow, and reworking of CM in original fabric's fragments. (5) Resealing of the damage zone by shear and pore collapse, evolution of clay gouge. See text for details. CM: clay matrix; NCM: non-clay minerals.

(001) cleavage planes of phyllosilicates (Chiarelli et al., 2000; Klinkenberg et al., 2009; Den Hartog and Spiers, 2014, Jessel et al., 2009). Note here that probably the biggest unknown at present in the micromechanisms of deformation in claystones is the nature of cement bonds between grains; further work in this project is aimed at understanding this better.

(3 & 4) Cataclastic shearing with plasticity of phyllosilicates, macroscopic failure

Further deformation occurs by frictional sliding affecting the process zone at microfracture boundaries and in relays between fractures. Mechanisms are abrasion and bending of phyllosilicates by cataclastic and crystal plastic mechanisms. This is accompanied by rotation of fragments and cataclastic flow. This stage is interpreted to start at the peak stress in the stress-strain curve, accompanied by local dilatancy. At the specimen scale, fractures link up, resulting in loss of cohe-

sion. In restraining sections along the fractures, reworking of the clay matrix reduces porosity and eliminates large pores, changing the pore size distributions. The specimen suffers from a major loss of cohesion accompanied by dilatancy and stress drop after peak stress.

(5) Resealing of the damage zone by shear and pore collapse, evolution of clay gouge

Ongoing abrasion of the fragments and comminution develop a cataclastic fabric. A full understanding of the deformation mechanisms in cataclastic clay aggregates requires more work, but the grain sliding (Chiarelli et al., 2000) and grain rotation between low-friction clay particles together with collapsing of porosity is inferred because (i) slip on the (001) basal planes of clay particles is much easier than shearing related to grain breakage (see Haines et al., 2013 and Crider, 2015) and (ii) residual strength observed after specimen failure argues for sliding between low frictional clay particles (Lupini et al., 1981). At sufficiently high strain, this stage would correspond to the residual strength, resulting in the resealing of initial fracture porosity by filling the fractures with clay gouge. In this stage, cataclasis of non-clay particles is expected to become less important because they are embedded in reworked clay.

The conceptual model above for microstructure evolution in triaxially deformed COX is a first look based on direct grain-scale observation of microstructures. Our ongoing studies focus on the nature of the cement and microstructures of the damage zone at fracture tips to better understand the localisation mechanisms.

6   Conclusions

The integration of bulk stress-strain data and the analysis of displacement fields by 3-D and 2-D digital image correlation (DIC) with broad ion beam cutting and scanning electron microscopy (BIB–SEM) is a powerful multi-scale method to study the deformation behaviour of mudstones.

We studied samples of Callovo–Oxfordian Clay (COX) subjected to triaxial compression at 2 and 10 MPa confining pressure. DIC was used to locate regions deformed to different states of strain and BIB–SEM allows microstructural investigations of mineral and porosity fabrics down to the nanometre scale.

Microstructures show evidence for dominantly cataclastic mechanisms (intergranular, transgranular, intragranular cracking, grain rotation, clay particle bending) down to the nanometre scale.

At low strain, the dilatant fabric contains individually recognisable open fractures, while at high strain in shear fractures the reworked clay gouge evolves towards smaller pores than the undeformed material and corresponding resealing of initial fracture porosity. This shear-induced resealing is more important at the higher confining pressure.

This study provides a first step towards a microphysical basis for constitutive models of deformation and fluid flow in cemented mudstones, with an improved extrapolation of these models for long timescales.

In the future, the microstructures in experimentally deformed specimens need to be compared with the microstructures in naturally deformed claystones (Laurich et al., 2014) in order to help extrapolate the constitutive models to long timescales.

Competing interests. The authors declare that they have no conflict of interest.

Acknowledgements. We thank ANDRA for providing samples. We are very grateful to the reviewers G. Dresen and A. Dimanov for their constructive and valuable comments.

Edited by: R. Heilbronner

References

Andò, E., Hall, S. A., Viggiani, G., Desrues, J., and Besuelle, P.: Grain-scale experimental investigation of localised deformation in sand: a discrete particle tracking approach, Acta Geotechnica, 7, 1–13, 2012.

ANDRA: Evaluation of the feasibility of a geological repository in an argillaceous formation, Meuse/Haute Marne site, Dossier 2005, Argiles, Report Series, ANDRA, 2005.

Bauer-Plaindoux, C., Tessier, D., and Ghoreychi, M.: Propriétés mécaniques des roches argileuses carbonateés: importance de la relation calcite-argile, C. R. Acad. Sci. Paris, Sciences de la Terre et des planètes, Earth Planet. Sci., 326, 231–237, 1998.

Bernier F., Li, X. L., Bastien,s W., Ortiz, L., Van Geet, M., Wouters, L., Frieg, B., Blümling, P., Desrues, J., Viaggiani, G., Coll, C., Chancole, S., De greef, V., Hamza, R., Malinsky, L., Vervoort, A., Vanbrabant, Y., Debeker, B., Verstraelen, J., Govaerts, A., Wevers, M., Labiouse, V., Escoffier, S. Mathier, J.-F., Gastaldo, L., and Bühler, C.: Fractures and Self-healing within the Excavation Disturbed Zone in Clays (SELFRAC), Final report, European Commission, CORDIS Web Site, EUR 22585, 56 pp., 2007.

Bésuelle, P. and Hall, S. A.: Characterization of the Strain Localization in a Porous Rock in Plane Strain Condition Using a New True-Triaxial Apparatus, Advances in bifurcation and degradation in geomaterials, Springer Series in geomechanics and geo-engineering, 11, 345–352, 2011.

Boisson, J. Y.: Clay Club Catalogue of Characteristics of Argillaceous Rocks, OECD/NEA/RWMC/IGSC (Working Group on measurement and Physical understanding of Groundwater flow through argillaceous media) august 2005 Report NEA no. 4436 (Brochure and CD-Rom including data base), OECD/NEA Paris, France, 72 pp., 2005.

Bonin, B.: Deep geological disposal in argillaceous formations: studies at the tournemire test site, J. Contam. Hydrogeol., 35, 315–330, 1998.

Bornert, M., Vales, F., Gharbi, H., and Nguyen Minh, D.: Multiscale full-field strain measurements for micromechanical investigations of the hydromechanical behaviour of clayey rocks, Strain, 46, 33–46, 2010.

Bos, B. and Spiers, C. J.: Experimental investigation into the microstructural and mechanical evolution of phyllosilicate-bearing fault rock under conditions favouring pressure solution, J. Struct. Geol., 23, 1187–1202, 2001.

Bourcier, M., Dimanov, A., Héripré, E., Raphanel, J. L., Bornert, M., and Desbois, G.: Full field investigation of salt deformation at room temperature: cooperation of crystal plasticity and grain sliding, in: mechanical behavior of salt VII, Berest, edited by:

Ghoreychi, M., Hadj-Hassen, F., and Tijani, M., Taylor & Francis group, London, 37–43, 2012.

Bourcier, M., Bornert, M, Dimanov, A., heripré, E., and Raphanel, J. L.: Multiscale experimental investigation of crystal plasticity and grain boundary sliding in synthetic halite using digital image correlation, J. Geophys. Res.-Sol. Ea., 118, 511–526, 2013.

Chiarelli, A. S., Ledesert, B., Sibal, M., Karami, M., and Hoteit, N.: Influence of mineralogy and moisture content on plasticity and induced anisotropic damage of a claystone: application to nuclear waste disposal, Bull. Soc. Géol. France, 171, 621–627, 2000.

Chiarelli, A. S., Shao, J. F., and Hoteit, N.: Modeling of elastoplastic damage behavior of a claystone, Int. J. Plasticity, 19, 23–45, 2003.

Collettini, C., Niemeijer, A., Viti, C., and Marone, C.: Fault zone fabric and fault weakness, Nature, 462, 907–910, 2009.

Cosenza, P., Ghorbani, A., Florsch, N., and Revil, A.: Effects of Drying on the Low-Frequency Electrical Properties of Tournemire Argillites, Pure Appl. Geophys., 164, 2043–2066, 2007.

Crider, J. G.: The initiation of brittle faults in crystalline rock, J. Struct. Geol., 77, 159–174, 2015.

Curtis, M. E., Ambrose, R. J., Sondergeld, C. H., and Rai, C. S.: Structural Characterization of Gas Shales on the Micro- and Nano-Scales, Canadian Unconventional Resources and International Petroleum Conference, Society of Petroleum Engineers, Calgary, Alberta, Canada, 15 pp., 2010.

Dautriat, J., Bornert, M., Gland, N., Dimanov, A., and Raphanel, J.: Localized deformation induced by heterogeneities in porous carbonate analysed by multi-scale digital image correlation, Tectonophysics, 503, 100–116, 2011.

Dehandschutter, B., Vandycke, S., Sintubin, M., Vandenberghe, N., Gaviglio, P., Sizun, J.-P., and Wouters, L.: Microfabric of fractured Boom Clay at depth: a case study of brittle-ductile transitional clay behaviour, Appl. Clay Sci., 26, 389–401, 2004.

Den Hartog, S. A. M. and Spiers, C.: A microphysical model for fault gouge friction applied to subduction megathrusts, J. Geophys. Res., 119, 1510–1529, 2014.

Desbois, G., Urai, J. L., Burkhardt, C., Drury, M. R., Hayles, M., and Humbel, B.: Cryogenic vitrification and 3-D serial sectioning using high resolution cryo-FIB SEM technology for brine-filled grain boundaries in halite: first results, Geofluids, 8, 60–72, 2008.

Desbois, G., Urai, J. L., and Kukla, P. A.: Morphology of the pore space in claystones – evidence from BIB/FIB ion beam sectioning and cryo-SEM observations, E-Earth, 4, 15–22, 2009.

Desbois, G., Urai, J. L., Kukla, P. A., Konstanty, J., and Baerle, C.: High-resolution 3-D fabric and porosity model in a tight gas sandstone reservoir: a new approach to investigate microstructures from mm- to nm-scale combining argon beam cross-sectioning and SEM imaging, J. Petrol. Sci. Eng., 78, 243–257, 2011.

Desbois G., Urai, J. L., Pérez-Willard, F., Radi, Z., van Offern, S., Burkart, I., Kukla, P. A., and Wollenberg, U.: Argon broad ion beam tomography in cryogenic scanning electron microscope: a novel tool for the investigation of preserved representative microstructures, Application to rock salt and other sedimentary rocks, J. Microsc.-Oxford, 249, 215–235, 2013.

Desbois, G., Urai, J. L., Hemes, S., Brassinnes, S., De Craen, M., and Sillen, X.: Nanometer-scale pore fluid distribution and drying damage in preserved clay cores from Belgian clay formations inferred by BIB-cryo-SEM, Eng. Geol., 170, 117–131, 2014.

Desbois, G., Urai, J. L, Hemes, S., Schröppel, B., Schwarz, J.-O., Mac, M., and Weiel, D.: Multiscale analysis of porosity in diagenetically altered reservoir sandstone from the Permian Rotliegend (Germany), J. Petrol. Sci. Eng., 140, 128–148, 2016.

Evans, B. and Wong, T.-F.: Fault mechanics and transport properties of rocks, Academic Press, International Geophysics, 51, 524 pp., 1992.

Fabre, G. and Pellet, F.: Creep and time-dependent damage in argillaceous rocks, Int. J. Rock Mech. Min., 43, 950–960, 2006.

Fauchille, A.-L., Hedan, S., Prêt, D., Cosenza, P., Valle, V., and Cabrera, J.: Relationships between desiccation cracking behavior and microstructure of the Tournemire clay-rock by coupling DIC and SEM methods, in: Geomechanics from micro to macro, edited by: Soga, K., Kumar, K., Biscontin, G., and Kuo, M., 1421–1424, 2015.

Faulkner, D. R., Lewis, A. C., and Rutter, E. H.: On the internal structure and mechanics of large strike-slip fault zones: field observations of the Carboneras fault in southeastern Spain, Tectonophysics, 367, 235–251, 2003.

Fouché, O., Wright, H., Le Cléac'h, J.-M., and Pellenard, P.: Fabric control on strain and rupture of heterogeneous shale samples by using a non-conventional mechanical test, Appl. Clay Sci., 26, 367–387, 2004.

French, M. E., Chester, F. M., and Schester, J. S.: Micromechanisms of creep in clay-rich gouge from the Central Deforming Zone of the San Andreas Fault, J. Geophys. Res.-Sol. Ea., 120, 827–849, 2015.

Galle, C.: Effect of drying on cement-based materials pore structure as identified by mercury intrusion porosimetry: A comparative study between oven-, vacuum-, and freeze-drying, Cement Concrete Res., 31, 1467–1477, 2001.

Gasc-Barbier, M. and Tessier, D.: Structural Modifications of a Hard Deep Clayey Rock due to Hygro-Mechanical Solicitations, Int. J. Geomech., 7, 227–235, 2007.

Gasc-Barbier, M., Chanchole, S., and Bérest, P.: Creep behavior of Bure clayey rock, Appl. Clay Sci., 26, 449–458, 2004.

Gaucher, E., Robelin, C., Matray, J. M., Negrel, G., Gros, Y., Heitz, J. F., Vinsot, A., Rebours, H., Cassagnabere, A., and Bouchet, A.: ANDRA underground research laboratory: interpretation of the mineralogical and geochemical data acquired in the Callovian-oxfordian Formation by investigative drilling, Phys. Chem. Earth, 29, 55–77, 2004.

Gratier, J. P., Jenatton, L., Tisserand, D., and Guiguet, R. Indenter studies of the swelling, creep and pressure solution of Bure argillite, Appl. Clay Sci., 26, 459–472, 2004.

Haines, S. H., Van der Pluijm, B., Ikari, M. J., Saffer, D. M., and Marone, C.: Clay fabric intensity in natural and artificial fault gouges: Implications for brittle fault zone processes and sedimentary basin clay fabric evolution, J. Geophys. Res., 114, B05406, doi:10.1029/2008JB005866, 2009.

Haines, S. H., Kaproth, B., Marone, C., Saffer, D., and Van der Pluijm, B.: Shear zones in clay-rich fault gouge: A laboratory study of fabric development and evolution, J. Struct. Geol., 51, 206–225, 2013.

Hall, S., Bornert, M., Desrues, J., Pannier, Y., Lenoir, N., Viggiani, G., and Bésuelle, P.: Discrete and Continuum analysis of localised deformation in sand using X-ray CT and Volumetric Digital Image Correlation, Géotechnique, 60, 315–322, 2010.

Heath, J. E., Dewers, T. A., McPherson, B. J. O. L., Petrusak, R., Chidsey, T. C., Rinehart, A. J., and Mozley, P. S.: Pore networks in continental and marine mudstones: Characteristics and controls on sealing behavior, Geosphere, 7, 429–454, 2011.

Hedan, S., Cozensa, P., Valle, V., Dudoignon, P., Fauchille, A.-L., and Cabrera, J.: Investigation of the damage induced by desiccation and heating of Tournemire argillite using digital image correlation, Int. J. Rock Mech. Min., 51, 64–75, 2012.

Hemes, S., Desbois, G., Urai, J. L., De Craen, M., and Honty, M.: Variations in the morphology of porosity in the Boom Clay Formation: insights from 2-D high resolution, BIB-SEM imaging and Mercury injection Porosimetry, Neth. J. Geosci., 92, 275–300, 2013.

Hemes, S., Desbois, G., Urai, J. L., Schröppel, B., and Schwarz, J.-O.: Multi-scale characterization of porosity in Boom Clay (HADES, Mol, Belgium) using a combination of μ-CT, BIB-SEM and serial FIB-SEM techniques, Micropor. Mesopor. Mat., 208, 1–20, 2015.

Holland, M., Urai, J. L., van der Zee, W., Stanjek, H., and Konstanty, J.: Fault gouge evolution in highly overconsolidated claystones, J. Struct. Geol., 28, 323–332, 2006.

Houben, M. A., Desbois, G., and Urai, J. L.: Pore morphology and distribution in the shaly facies of Opalinus clay (Mont Terri, Switzerland): insights from representative 2-D BIB-SEM investigations on mm- to nm- scales, Appl. Clay Sci., 71, 82–97, 2013.

Houben, M. A., Desbois, G., and Urai, J. L.: A comparative study of representative 2-D microstructures in Shaly and Sandy facies of Opalinus Clay (Mont Terri, Switzerland) inferred form BIB-SEM and MIP methods, Mar. Petrol. Geol., 49, 143–161, 2014.

IAEA: The safety case and safety assessment for radioactive waste disposal, Draft safety guide, International atomic energy agency, report no. DS 355, Vienna, 2008.

Ingram, G. M. and Urai, J. L.: Top-seal leakage through faults and fractures, the role of mudrock properties, Geol. Soc. Sp., 158, 125–135, 1999.

Ishii, E.: Far-field stress dependency of the failure mode of damage-zone fractures in fault zones: Results from laboratory tests and field observations of siliceous mudstone, J. Geophys. Res.-Sol. Ea., 121, 70–91, doi:10.1002/2015JB012238, 2016.

Ishii, E., Sanada, H., Funaki, H., Sugita, Y., and Kurikami, H.: The relationships among brittleness, deformation behavior, and transport properties in mudstones: An example from the Horonobe Underground Research Laboratory, Japan, J. Geophys. Res., 116, B09206, doi:10.1029/2011JB008279, 2011.

Jessell, M. W., Bons, P. D., Griera, A., Evans, L., and Wilson, C. J. L.: A tale of two viscosities, J. Struct. Geol., 31, 719–736, 2009.

Jordan, P. and Nüesch, R.: Deformation behavior of shale interlayers in evaporite detachment horizons, Jura overthrust, Switzerland, J. Struct. Geol., 11, 859–871, 1989.

Kang, M.-S., Watabe, Y., and Tsuchida, T.: Effect of Drying Process on the Evaluation of Microstructure of Clays using Scanning Electron Microscope (SEM) and Mercury Intrusion Porosimetry (MIP), Proceedings of The Thirteenth International Offshore and Polar Engineering Conference Honolulu, Hawaii, USA, 25–30 May, 2003.

Katz, O. and Reches, Z.: Microfracturing, damage, and failure of brittle granites, J. Geophys. Res., 109, B01206, doi:10.1029/2002JB001961, 2004.

Kaufhold, A., Halisch, M., Zacher, G., and Kaufhold, S.: X-ray computed tomography investigation of structures in Opalinus Clay from large-scale to small-scale after mechanical testing, Solid Earth, 7, 1171–1183, doi:10.5194/se-7-1171-2016, 2016.

Keller, L., Schuetz, P., Erni, R., Rossell, M. D., Lucas, F., Gasser, P., and Holzer, L.: Characterization of multi-scale microstructural features in Opalinus Clay, Micropor. Mesopor. Mat., 170, 83–94,2013.

Keller, L. M., Holzer, L., Wepf, R., and Gasser, P.: 3-D geometry and topology of pore pathways in Opalinus clay: Implications for mass transport, Appl. Clay Sci., 52, 85–95, 2011.

Kim, Y.-S., Peacock, D. C. P., and Sanderson, D. J.: Fault damage zones, J. Struct. Geol., 26, 503–517, 2004.

Klaver, J., Desbois, G., Urai, J. L., and Littke, R.: BIB-SEM study of porosity of immature Posidonia shale from the Hils area, Germany, Int. J. Coal Geol., 103, 12–25, 2012.

Klaver, J., Desbois, G., Littke, R., and Urai, J. L.: BIB-SEM characterization of pore space morphology and distribution in postmature to overmature samples from the Haynesville and Bossier Shales, Mar. Petrol. Geol., 59, 451–466, 2015.

Klinkenberg, M., Kaufhold, S., Dohrmann, R., and Siegesmund, S.: Influence of carbonate microfabrics on the failure strength of claystones, Eng. Geol., 107, 42–54, 2009.

Kohlstedt, D. L., Evans, B., and Mackwell, S. J.: Strength of the lithosphere: constraints imposed by laboratory experiments, J. Geophys. Res., 100, 17587–17602, 1995.

Laurich, B., Urai, J. L., Desbois, G., Vollmer, C., and Nussbaum, C.: Microstructural evolution of an incipient fault zone in Opalinus Clay: Insights from an optical and electron microscopic study of ion-beam polished samples from the Main Fault in the Mont Terri underground research laboratory, J. Struct. Geol., 67, 107–128, 2014.

Laurich, B., Urai, J. L., and Nussbaum, C.: Microstructures and deformation mechanisms in Opalinus Clay: insights from scaly clay from the Main Fault in the Mont Terri Rock Laboratory (CH), Solid Earth, 8, 27–44, doi:10.5194/se-8-27-2017, 2017.

Lee, M. R., Bland, P. A., and Graham, G.: Preparation of TEM samples by focused ion beam (FIB) techniques: applications to the study of clays and phyllosilicates in meteorites, Mineral. Mag., 67, 581–592, 2003.

Lenoir, N., Bornert, M., Desrues, J., Besuelle, P., and Viggiani, G.: Volumetric digital image correlation applied to X-ray microtomography images from triaxial compression tests on argillaceous rock, Strain, 43, 193–205, 2007.

Logan, J. M., Friedman, M., Higgs, N., Dengo, C., and Shimamoto, T.: Experimental studies of simulated gouge and their application to studies of natural fault zones, in: Proceedings of Conference VIII on Analysis of Actual Fault Zones in Bedrock, US Geological Survey, Open File Report, 79–1239, 1979.

Logan, J. M., Dengo, C. A., Higgs, N. G., and Wang, Z. Z.: Fabrics of Experimental Fault Zones: Their Development and Relationship to Mechanical Behavior, in: Fault Mechanics and Transport Properties of Rocks – A Festschrift in Honor of W. F. Brace, edited by: Evans, B. and Wong, T., Academic Press, 33–67, 1992.

Logan, J. M. and Rauenzahn, K. A.: Frictional dependence of gouge mixtures of quartz and montmorillonite on velocity, composition and fabric, Tectonophysics, 144, 87–108, 1987.

Loucks, R. G., Reed, R. M., Ruppel, S. C., and Jarvie, D. M.: Morphology, Genesis, and Distribution of Nanometer-Scale Pores in Siliceous Mudstones of the Mississippian Barnett Shale, J. Sediment. Res., 79, 848–861, 2009.

Lupini, J. F., Skinner, A. E., and Vaughan, P. R.: The drained residual strength of cohesive soils, Géotechnique 31, 181–213, 1981.

Marone, C. and Scholz, C. H.: Particle-size distribution and microstructures within simulated fault gouge, J. Struct. Geol., 11, 799–814, 1989.

Milliken, K. L., Rudnicki, M., Awwiller, D. N., and Zhang, T.: Organic matter-hosted pore system, Marcellus formation (Devonian), Pennsylvania, AAPG bulletin, 97, 177–200, 2013.

Morgenstern, N. R. and Tchalenko, J. S.: Microscopic structures in kaolin subjected to direct shear, Geotechnique, 17, 309–328, 1967.

Nagra: Technischer Bericht 02-03, Projekt Opalinuston, Synthese der geowissenschaftlichen Untersuchungsergebnisse, 2002.

NEA: Post-closure safety case for geological repositories. Nature and purpose, OECD/NEA, no. 3679, Paris, France, 2004.

Neerdael, B. and Boyazis J. P.: The Belgium underground research facility: status on the demonstration issues for radioactive waste disposal in clay, Nucl. Eng. Des., 176, 89–96, 1997.

Niemeijer, A. R. and Spiers, C. J.: Velocity dependence of strength and healing behaviour in simulated phyllosilicate-bearing fault gouge, Tectonophysics, 427, 231–253, 2006.

Nüesch, R.: Das mechanische Verhalten von Opalinuston, PhD Thesis, ETH Zürich, 244 pp., 1991.

ONDRAF/NIRAS: SAFIR 2, Safety Assessment and Feasibility Interim Report 2, NIROND 2001-06, 2001.

Passchier, C. W. and Trouw R. A. J.: Microtectonics, Springer, 366 pp., 2005.

Pineda, J., Romero, E., Gómez, S., and Alonso, E.: Degradation effects at microstructural scale and their consequences on macroscopic behaviour of a slightly weathered siltstone, in: Geomechanics and Geotechnics, From Micro to Macro, Two Volume Set, edited by: Bolton, M., CRC Press 2010, 73–78, 2010.

Renard, F.: Microfracturation in rocks: from microtomography images to processes, Eur. Phys. J. Appl. Phys., 60, 24203, 2012.

Richard, J., Gratier, J. P., Doan, M.-L., Boullier, A.-M., and Renard, F.: Rock and mineral transformations in a fault zone leading to permanent creep: Interactions between brittle and viscous mechanisms in the San Andreas Fault, J. Geophys. Res.-Sol. Ea., 119, 8132–8153, 2015.

Robinet, J. C., Sardini, P., Coelho, D., Parneix, J.-C., Dimitri, P., Sammartino, S., Boller, E., and Altmann, S.: Effects of mineral distribution at mesoscopic scale on solute diffusion in a clay-rich rock: Example of the Callovo-Oxfordian mudstone (Bure, France), Water Resour. Res., 48, W05554, doi:10.1029/2011WR011352, 2012.

Rutter, E. H., Maddock, R. H., Hall, S. H., and White S. H.: Comparative microstructures of natural and experimentally produced clay-bearing fault gouges, Pure Appl. Geophys., 124, 3–30, 1986.

Salters, V. J. M. and Verhoef, P. N. W. (Eds.): Geology and nuclear waste disposal, Geologica Ultraiectina Special Publication, Instituut voor Aardwetenschappen der Rijksuniversiteit te Utrecht, Institute of Earth Sciences, Utrecht, 1, 399 pp., 1980.

Schmatz, J., Berg, S., Urai J., and Ott, H.: Nano-scale imaging of pore-scale fluid-fluid-solid contacts in sandstone, Geophys. Res. Lett., 42, 2189–2195, 2015.

Shapira, J. P.: Long-term waste management: present status and alternatives, Nucl. Instrum. Methods, A280, 568–582, 1989.

Soe, A. K. K., Osada, M., Takahashi, M., and Sasaki, T.: Characterization of drying-induced deformation behaviour of Opalinus Clay and tuff in no-stress regime, Environ. Geol., 58, 1215–1225, 2009.

Sone, H., Morales, L. F., and Dresen, G.: Microscopic observations of shale deformation from in-situ deformation experiments conducted under a scanning electron microscope, ARMA, 15–27, 2015.

Song, Y., Davy, C. A., Bertier, P., and Troadec, D.: Understanding fluid transport through claystones from their 3-D nanoscopic pore network, Micropor. Mesopor. Mat., 228, 64–85, 2016.

Stow, D. A. V.: Fine-grained sediments: Terminology, Q. J. Eng. Geol., 14, 243–244, 1981.

Tchalenko, J. S.: The evolution of kink-bands and the development of compression textures in sheared clays, Tectonophysics, 6, 159–174, 1968.

Wang, L. L., Bornert, M., Chancole, S., Yang, S., Heripré, E., Tanguy, A., and Caldemaison, D.: Micro-scale experimental investigation of the swelling anisotropy of the Callovo- Oxfordian argillaceous rock, Clay Miner., 48, 391–402, 2013.

Wang, L. L., Bornert, M., Chancole, S., Heripré, E., and Yang, S.: Micromechanical experimental investigation of mudstones, Géotechnique letters, 4, 306–309, 2015.

Warr, L. N., Wojatschke, J., Carpenter, B. M., Marone, C., Schleicher, A. M., and van der Pluijm, B. A.: A "slice-and-view" (FIB-SEM) study of clay gouge from the SAFOD creeping section of the San AndreasFault at 2.7 km depth, J. Struct. Geol., 69, 234–244, 2014.

Yang, D. S., Bornert, M., Chanchole, S., Gharbi, H., Valli, P., and Gatmiri, B.: Dependence of elastic properties of argillaceous rocks on moisture content investigated with optical full-field strain measurement techniques, Int. J. Rock Mech. Min., 53, 45–55, 2012.

Switching deformation mode and mechanisms during subduction of continental crust: a case study from Alpine Corsica

Giancarlo Molli[1], Luca Menegon[2], and Alessandro Malasoma[3]

[1]Dipartimento di Scienze della Terra, Università di Pisa, Via S. Maria, 53, Pisa 56126, Italy
[2]School of Geography, Earth and Environmental Sciences, Plymouth University, Plymouth, UK
[3]TS Lab and Geoservices, Via Vecchia Fiorentina, 10, Cascina 56023, Pisa, Italy

Correspondence to: Giancarlo Molli (giancarlo.molli@unipi.it)

Abstract. The switching in deformation mode (from distributed to localized) and mechanisms (viscous versus frictional) represent a relevant issue in the frame of crustal deformation, being also connected with the concept of the brittle–"ductile" transition and seismogenesis. In a subduction environment, switching in deformation mode and mechanisms and scale of localization may be inferred along the subduction interface, in a transition zone between the highly coupled (seismogenic zone) and decoupled deeper aseismic domain (stable slip). However, the role of brittle precursors in nucleating crystal-plastic shear zones has received more and more consideration being now recognized as fundamental in some cases for the localization of deformation and shear zone development, thus representing a case in which switching deformation mechanisms and scale and style of localization (deformation mode) interact and relate to each other. This contribution analyses an example of a millimetre-scale shear zone localized by brittle precursor formed within a host granitic protomylonite. The studied structures, developed in ambient pressure–temperature (P–T) conditions of low-grade blueschist facies (temperature T of ca. 300 °C and pressure $P \geq 0.70$ GPa) during involvement of Corsican continental crust in the Alpine subduction. We used a multidisciplinary approach by combining detailed microstructural and petrographic analyses, crystallographic preferred orientation by electron backscatter diffraction (EBSD), and palaeopiezometric studies on a selected sample to support an evolutionary model and deformation path for subducted continental crust. We infer that the studied structures, possibly formed by transient instability associated with fluctuations of pore fluid pressure and episodic strain rate variations, may be considered as a small-scale example of fault behaviour associated with a cycle of interseismic creep and coseismic rupture or a new analogue for episodic tremors and slow-slip structures. Our case study represents, therefore, a fossil example of association of fault structures related to stick-slip strain accommodation during subduction of continental crust.

1 Introduction

The study of deformation fabric of fault rocks has been crucial for the development of a general conceptual model for crustal-scale fault zones (Sibson, 1977, 1983; Scholz, 1988, 2002; Handy et al., 2007; Cooper et al. 2010; Platt and Behr, 2011). In this model, the increasing PT conditions determine the transition from a seismogenic frictional regime, dominated by pressure-sensitive deformation and involving cataclasis and frictional sliding, to a viscous regime (Rutter, 1986; Schmid and Handy, 1991; Montési and Hirth, 2003; Handy and Brun, 2004), where dominantly aseismic, mainly crystal-plastic and continuous shearing is localized within mylonitic shear zones. In quartz-feldspathic rocks, this transition is primarily determined by the temperature-related quartz response to change of deviatoric stress, with dislocation creep becoming the principal deformation mechanism at $T > 270$ °C (e.g. White, 1971, 1973; Stipp et al., 2002, and references therein).

Rock deformation experiments (Rutter, 1986; Shimamoto and Logan, 1986; Bos and Spiers 2002; Scholz, 2002) have shown that the shear strength of simulated faults at the brittle–viscous transition may depend on normal stress (as with faulting dominated by cataclastic mechanisms) although strain is achieved through crystal plasticity and/or solution transfer. These results support the observations that some mylonitic shear zones are produced by coupled frictional and viscous mechanisms under semi-brittle conditions (Shimamoto and Logan, 1986; Shimamoto, 1989; Chester, 1989; Scholz, 2002; Pec et al., 2012, and references therein). The experimental approach on semi-brittle behaviour, however, still leaves several questions open for natural fault zones, such as the mechanisms controlling bulk-rock deformation at microscopic scale, the degree of interdependence of active deformation mechanisms, their cyclicity, and the associated bulk rock style of deformation (Sibson, 1980; White and White, 1983; Viola et al., 2006; Takagi et al., 2000; Handy and Brun, 2004; Pec et al., 2012). A complex transitional behaviour involving mixed continuous and discontinuous, distributed vs. localized, and cyclic switching in deformation mechanisms over large variations in strain rates is inferred at the transition between frictional and viscous domains, a depth interval in the crust which contains the typical hypocentres and rupture depths of large earthquakes in continental crust (Sibson, 1983; Kohlstedt et al., 1995; Scholz, 2002; Handy and Brun, 2004; Pennacchioni et al., 2006).

In subduction settings, this transition zone is located between 10 and 35 km depth depending on slab dip and thermal structure (i.e. between temperatures of 150 and 350–450 °C) and along the subduction interface is recognized as the site of megathrust earthquake nucleation and concentrated post-seismic afterslip, as well as the focus site of episodic tremors and slow-slip events (Rogers and Dragert, 2003; Liu and Rice, 2007; Hacker et al., 2003; Vannucchi et al., 2008; Meneghini et al., 2010; Angiboust et al., 2014, 2015; Andersen et al., 2014; Hayman and Lavier, 2014; Fagereng et al., 2014).

The feedback between brittle and viscous deformation mechanisms is also relevant for the mode of shear zone nucleation, and fracturing has been proposed to be a prerequisite for the initiation of ductile shear zones in the lithosphere (e.g. Handy and Stünitz, 2002; Pennacchioni and Mancktelow, 2007; Fusseis and Handy, 2008). Shear zones and style of strain accommodation and localization processes in granitoids have been described by different authors (Ramsay and Graham, 1970; Burg and Laurent, 1985, 1986; Gapais et al., 1987; Goncalves et al., 2016, and references therein). In this context, the role of brittle precursors in nucleating shear zones has received increasing consideration as having in some cases a fundamental role in the localization process (e.g. Segall and Simpson, 1986; Mancktelow and Pennacchioni, 2005; Pennacchioni, 2005; Pennacchioni and Mancktelow, 2007; Menegon and Pennacchioni, 2010; Pennacchioni and Zucchi, 2013).

These studies, however, mostly deal with brittle precursors consisting of inherited structures, such as discontinuities already existing at the beginning of the viscous deformation history, as in the case of cooling joints, cataclasites, and veins (e.g. Guermani and Pennacchioni, 1998; Pennacchioni and Mancktelow 2007); most of these studies consider deformation at shallow to intermediate crustal depths (Fusseis and Handy, 2008; Mazzoli et al., 2009; Molli et al., 2011), although case studies under conditions of upper amphibolite to granulite (White, 1996; Pennacchioni and Cesare, 1997; Pittarello et al., 2013; Altenberger et al., 2013) and eclogite facies (Austrheim and Boundy, 1994; Austrheim, 2013, and references therein) were also investigated.

The deformation of crustal units and granitoids during subduction has only recently been analysed in terms of preserved rock records of the palaeo-seismic cycle and/or slow-slip phenomena (Angiboust et al., 2015; Goncalves et al., 2016). A well-documented study from the Dent Blanche Thrust in the western Alps, for example, is framed in the hanging wall of an ancient subduction interface zone (Angiboust et al., 2014, 2015). Our contribution integrates the existing literature by analysing a meso- to microscale example of a brittle precursor to a shear zone derived from the footwall of a subduction interface in blueschist-facies conditions thus representing a hitherto undocumented or less thoroughly studied case (Molli et al., 2005; Molli, 2007).

We used a multidisciplinary approach that combined detailed microstructural, petrographic, and electron backscatter diffraction (EBSD) analysis in order to derive a model of the deformation sequence experienced by our sample during subduction of the continental crust in Corsica. EBSD data have been used to identify the dominant deformation and recrystallization mechanisms in quartz, and to derive quantitative grain size data that were used to evaluate the differential stresses experienced by the rock during dislocation creep deformation by means of the recrystallized grain size piezometry.

By virtue of their deep origin, the analysed structures are ideal to contribute to the ongoing discussion on the deformation style and mechanisms associated with the broad spectrum of fault-slip behaviour (from seismic slip to stable aseismic creep to episodic slow-slip events and nonvolcanic tremors) recorded by seismic and geodetic observations at active plate boundaries (e.g. Peng and Gomberg, 2010; Beroza and Ide, 2011, and references therein). Slow-slip events (i.e. fault slip events with slip rates in between coseismic rate and aseismic creep, and generating equivalent seismic moments similar to large earthquakes) in subduction zones have been recorded in a depth interval that experiences temperatures between 250 and 650 °C and pressures between 0.6 and 1.2 GPa (Beroza and Ide, 2011), typically in areas of high V_p/V_s ratios suggestive of local high fluid pressures. This places slow-slip events at the lower end of the seismogenic zone, under metamorphic conditions where the rheology is expected to be viscous or at the frictional–viscous transition. Accordingly, recent studies of shear zones exhumed from similar conditions along the subduction interface have suggested that coupled fracture and viscous flow, possibly associated with fluctuations in fluid pressure, can originate tremors and slow slips (e.g. Peng and Gomberg, 2010; Beroza and Ide, 2011; White, 2012; Fagereng et al., 2014; Hayman and Lavier, 2014; Angiboust et al., 2015; Malatesta et al., 2017). Here we show that the brittle–viscous transition preserved in the Popolasca granitoids of northern Corsica can also be explained by transient high fluid pressures triggering brittle deformation in an otherwise viscous regime, and discuss the related implications for fault-slip behaviours in subduction zones.

2   Regional background and geological setting of studied sample

In Alpine-type orogens the study of meso- and microstructural record of exhumed subduction-related thrust zones and its interpretation in terms of subduction zone rheology and seismicity have received increasing attention since the end of the 1990s (Stöckhert et al.,1999; Küster and Stöckhert, 1999). This subject has been well developed and explored (Austrheim and Andersen, 2004; Andersen and Austrheim, 2006; Healy et al., 2009; Andersen et al., 2014; Deseta et al., 2014a, b; Magott et al., 2016a, b) in northeast Corsica (so-called "Alpine Corsica"), which is formed by relicts of subduction orogen related to the Cretaceous–Middle Tertiary Europe–Adria convergence history (Lahondère, 1988; Fournier et al., 1991; Handy et al., 2010; Molli and Malavieille, 2011).

These studies focused on oceanic units made up of peridotite, serpentinite, gabbro, basalt, calcareous and siliceous schist, and marble exposed as remnants of the lithosphere of the Jurassic Piemonte–Liguria oceanic basin and its pelagic sedimentary cover (Mattauer et al., 1981; Bezert and Caby, 1988; Jolivet et al., 1990; Molli, 2008; Vitale Brovarone et al., 2013, and references therein).

The widespread occurrence and preservation of relicts of exhumed seismogenic structures are mainly due to the peculiar geologic history of Alpine Corsica, which is connected with the development of the Alps–Apennine orogenic system (Molli and Malavieille, 2011; Guyedan et al., 2017; Beaudoin et al., 2017). The latter did not develop continent–continent "hard collision"-related structures and lacks the thermal reworking observable instead in the Alps (e.g. Polino et al., 1995; Schmid et al., 1996; Berger and Busquet, 2008; Butler, 2013; Rosenberg and Kissling, 2013; Carminati and Doglioni, 2014), thus resulting in the better preservation of the early stages of subduction-related structures and fabrics.

Corsica, therefore, represents an exceptional natural laboratory for the investigation of subduction-related processes in oceanic and continental crust, as firstly suggested by Mattauer et al. (1981) and Gibson and Horak (1984).

Continental-derived units in Corsica sourced from the footwall of an ancient subduction interface zone can be observed in three different structural positions (Fig. 1), each of them corresponding to different peak metamorphic con-

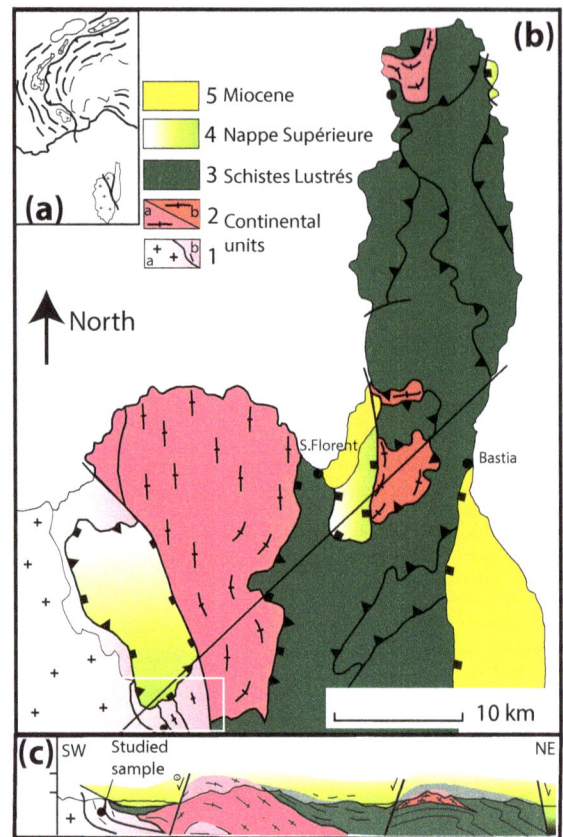

Figure 1. (a) Alpine Corsica within the Alps–Apennine framework. **(b)** Tectonic map of north Corsica showing the main tectonic units and the area of studied sample. (1, 2) Corsican continental crust, mainly Carboniferous–Permian granitoids, their host pre-Carboniferous basement, and a Mesozoic to Eocene cover: (1, **a**) "autochthonous" Hercynian Corsica and (1, **b**) greenschist/lower-blueschist external continental units (Corte and Popolasca units); (2) inner continental units: (2, **a**) upper-blueschist units (Tenda Massif; Centuri) and (2, **b**) eclogite slices (Serra di Pigno-Farinole); (3) Schistes Lustrés nappe (undifferentiated); (4) Nappe Supérieure i.e. upper non-metamorphic units (Balagne, Nebbio, and Macinaggio units); and (5) Miocene sediments. **(c)** Regional sketch cross section with location of the studied sample from the Popolasca area.

ditions (Tribuzio and Giacomini, 2004; Molli, 2008; Vitale Brovarone et al., 2013):

1. The innermost slices (Serra di Pigno/Farinole units) interleaved with oceanic units show eclogite peak conditions at 1.5–1.8 GPa; $500 \pm 50\,°C$.

2. The intermediate units (e.g. the Tenda and Centuri units) are instead characterized by upper blueschist facies peak conditions at 0.9–1.1 GPa, $450 \pm 50\,°C$.

3. The most external units (e.g. Corte/Popolasca) contain high-pressure greenschist and/or blueschist facies peak assemblages developed at $T = 325–370\,°C$, $P = 0.75–0.85$ GPa (Malasoma et al., 2006; Di Rosa et al., 2016).

These occurrences document the progressive underthrusting of these units down to different depths and their in-sequence contractional exhumation within an Alpine-age, east-dipping subduction system (Mattauer et al., 1981; Bezert and Caby, 1988; Jolivet et al., 1990; Garfagnoli et al., 2009; Molli and Malavieille, 2011; Maggi et al., 2012; Di Rosa et al., 2016; Guyedan et al., 2017; Beaudoin et al., 2017).

The studied samples come from the external continental units (Molli, 2008; Molli and Malavieille, 2011; Vitale Brovarone et al., 2013) of Alpine Corsica, more specifically from the Popolasca unit (Bezert and Caby, 1988; Malasoma et al., 2006; Di Rosa et al., 2016). This unit is characterized by a pre-Mesozoic basement mainly formed by granitoids, a Permo-Mesozoic metasedimentary sequence and an early Eocene flysch (Fig. 1).

Blueschist assemblages in the unit have been described by regional studies (Bezert and Caby, 1988; Malasoma et al., 2006; Molli, 2008; Di Rosa et al., 2016) and are observed in all suitable rock types, such as in metapelites from cover rocks, metabasic dykes, and in some of the granitoid suites from the basement. Peak metamorphism has been initially constrained to 250–350 °C and 0.4–0.55 GPa (Bezert and Caby, 1988) or more recently more closely bracketed at 325–370 °C and 0.75–0.85 GPa (Malasoma et al., 2006; Di Rosa et al., 2016). The prograde to peak pressure assemblage has been recently dated with $^{40}Ar/^{39}Ar$ at 45–36 Ma by Di Vincenzo et al. (2016).

The contributions of Malasoma et al. (2006) and Di Rosa et al. (2016) analysed the cartographic to mesoscopic-scale structural geometries of deformation in the area of the studied sample. Superimposed foliations and fold structures are typical of the metasedimentary cover, with blueschist assemblages as relict fabrics. In contrast, in basement rocks a main continuous foliation can be observed wrapping around undeformed granitoids. The main foliation is associated with blueschist or greenschist facies assemblages (Malasoma et al., 2005; Di Rosa et al., 2016; Di Vincenzo et al., 2016). Anastomosing networks of fault zones associated with sub-greenschist facies assemblages overprints all previous structures and may be related to the activity of a major N–S-trending transcurrent fault zone (Central Corsica Fault Zone) reworked by normal kinematics active during Oligocene–Miocene in the frame of rototranslation of Corsica–Sardinia microblock and upper-plate extension associated with Apenninic subduction (Faccenna et al., 2004; Molli, 2008; Molli and Malavieille, 2011; Carminati and Doglioni, 2014; Guyedan et al., 2017; Beaudoin et al., 2017).

3 Geometry of deformation and microstructures

The analysed sample (Fig. 2) is a deformed granitoid of calc-alkaline affinity (K-feldspar, plagioclase, quartz, biotite) common to Hercynian Corsica (Rossi et al., 2001). It con-

Figure 2. The analysed sample of a protomylonitic metagranite with millimetre-scale ultramylonite (compare **a** and **b**). The dashed lines in (**b**) are the trace of foliation in the protomylonitic host and the sharp shear zone boundary of the ultramylonite developed after a brittle precursor. All observable deformation structures were developed under high-pressure, low-temperature (HP/LT) metamorphic conditions.

tains a continuous foliation (Figs. 2, 3a, b) mainly defined by the shape preferred orientation of quartz, feldspars, and biotite grains.

Shape anisotropy of quartz in the host protomylonite shows an aspect ratio around 0.37 (mean X/Z of 2.7), which indicates a shear strain γ of ca. 1 assuming homogeneous simple shear deformation.

Quartz displays features of typical low-temperature plasticity (Hirth and Tullis, 1992; Stipp et al., 2002; Vernooij et al., 2006; Trepmann et al., 2007; Derez et al., 2015; Kjøll et al., 2015) (Fig. 3), such as undulatory extinction, localized extinction bands (LEBs, following the terminology of Derez et al., 2015), typically forming conjugate sets, and up to 100 μm thick intracrystalline bands of recrystallized grains (5–10 μm in size). Bands of recrystallized grains occur parallel to the main foliation as well as in conjugate sets intersecting at ca. 90° (see below) and parallel to the conjugate sets of LEBs.

Feldspars show local evidence of bulging recrystallization and of grain size reduction by microcracking and microfaulting (Fig. 3), often associated with K-feldspar breakdown to

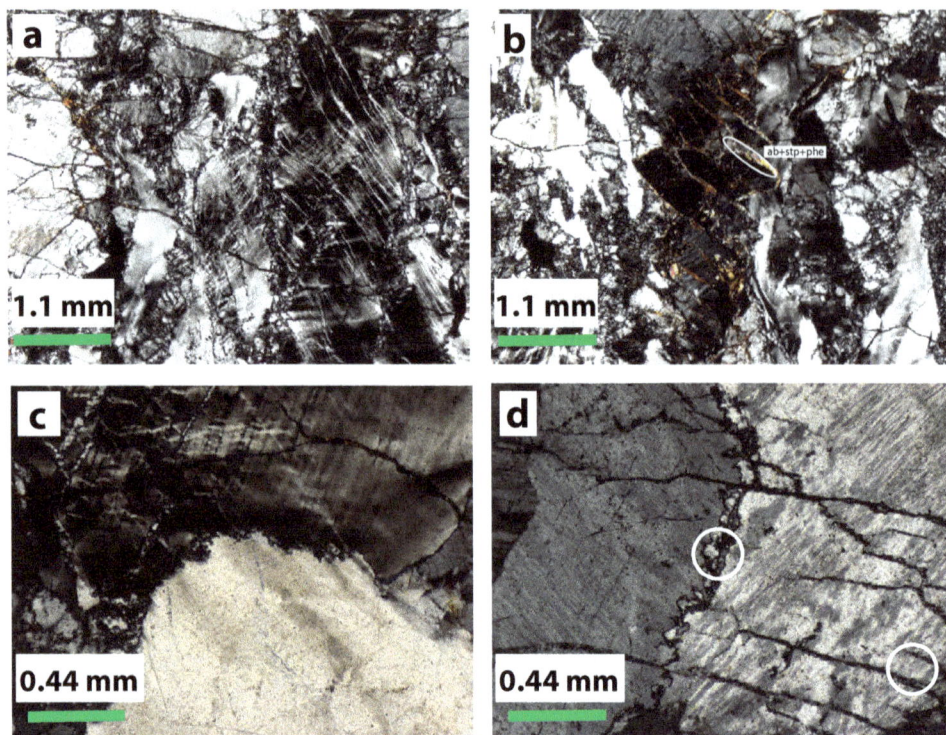

Figure 3. Microphotographs of host-protomylonite showing quartz microstructures typical of low-temperature plasticity: **(a)** intracrystalline deformation (undulatory extinction, deformation lamellae, deformation bands, localized extinction bands, LEBs) associated with recrystallization along intragranular conjugate shear bands sets; **(b)** domino-style brittle fracturing of K-feldspar and associated K-feldspar breakdown to albite, stilpnomelane, and phengite; **(c)** quartz deformation lamellae and intergranular recrystallization by bulging; and **(d)** K-feldspar intergranular recrystallization and albite neocrystallization and recrystallization also observable along later cracks and microfractures (horizontal to sub-horizontal).

albite, stilpnomelane, and phengite (Fig. 3). Thin needles of Na amphibole (Fig. 4) attest the development of this fabric in high-pressure, low-temperature (HP/LT) conditions as described below.

The main foliation is cross-cut at a high angle by a millimetre-thick localized zone of deformation (Figs. 2, 5a). This shows a sharp boundary, truncating flattened quartz and feldspar grains (Figs. 6, 7) – a feature suggesting its development as an unstable fracture (Schmid and Handy, 1991; Passchier and Trouw, 2011).

Three compositionally controlled domains can be recognized within the thin localized shear zone (Figs. 2, 5):

Domain 1 is feldspar- and quartz-dominated and shows microstructural features typical of a cataclastic rock (Fig. 5a, c). Clasts are embedded in a fine-grained matrix (< 10–20 µm in size) present in a patchy variable amounts in different portions of the cataclasite (from less than 10 % to more than 50 %, Fig. 5b, c, f). The clasts show angular shapes and a polymodal grain size distribution (Fig. 5d–g).

Feldspar clasts and quartz grains reveal both displacive intragranular fractures and intergranular "stable" cracks (Atkinson, 1982; Schmid and Handy, 1991). The clasts show relict microstructural features similar to those found in the host protomylonite, such as intracrystalline recrystallized shear bands, which in some cases (where the clasts were not substantially rotated, Fig. 5c–f) have the same geometry and orientation as those observed in the host protomylonite (Fig. 5d–g).

Domain 2 is "phyllosilicate"-dominated (stilpnomelane and phengite) and shows microstructural features of a foliated-cataclasite/phyllonite (Figs. 5a, 6). Asymmetric porphyroclasts and shear band systems characterize the phyllosilicate-rich parts of the shear zone (Fig. 6a–d).

Domain 3 is a quartz–albite-rich domain showing microstructural features typical of an ultramylonite (Figs. 5a, 7). This is characterized by very fine recrystallized albite and quartz grains (5–10 µm in size) (Fig. 7) with strong crystallographic preferred orientation (see below). Quartz porphyroclasts show a mean aspect ratio around 0.09 (mean X/Z of 11), which corresponds to a calculated shear strain γ of ca. 3. At this shear strain the corresponding angle between the foliation and the shear zone boundary in simple shear is around 15°, which agrees with the mean orientation of the quartz porphyroclasts shape preferred orientation (SPO) defining the ultramylonitic foliation (Fig. 7a–c). Syn- to post-

Table 1. Representative microprobe analyses (wt%) of coexisting minerals within the metamorphic assemblage. Na-amphiboles structural formulae were calculated assuming 23 oxygens per anhydrous formula unit.

Amphibole	Am1	Am3	Am4	Am5	Am6N	Am6B	Am7	Am8	Am9
Wt%									
SiO$_2$	51.58	51.30	50.62	50.97	50.28	50.61	52.62	51.26	51.10
TiO$_2$	0.35	0.32	0.52	0.24	0.20	0.39	0.16	0.20	0.14
Al$_2$O$_3$	1.67	1.89	1.97	1.73	2.22	1.85	2.59	1.74	1.87
Cr$_2$O$_3$	0.00	0.00	0.00	0.00	0.00	0.00	0.00	0.00	0.00
FeO	28.13	29.65	28.80	28.77	30.25	30.18	27.94	30.06	29.57
MnO	0.00	0.15	0.00	0.20	0.00	0.00	0.00	0.10	0.00
MgO	2.24	1.86	1.97	2.16	1.64	1.98	1.90	1.89	2.24
CaO	0.06	0.05	0.10	0.09	0.13	0.14	0.06	0.08	0.05
Na$_2$O	6.44	6.63	6.66	6.79	6.02	6.42	6.88	6.73	6.97
K$_2$O	0.41	0.39	0.46	0.38	0.35	0.43	0.71	0.29	0.29
TOT	90.88	92.24	91.10	91.33	91.09	92.00	92.86	92.35	92.23
Cations									
Si	8.16	8.09	8.09	8.11	8.07	8.05	8.17	8.09	8.08
Al IV	0.00	0.00	0.00	0.00	0.00	0.00	0.00	0.00	0.00
Sum T	8.16	8.09	8.09	8.11	8.07	8.05	8.17	8.09	8.08
Al VI	0.31	0.35	0.37	0.32	0.42	0.35	0.47	0.32	0.35
Fe$_3$+	1.70	1.56	1.42	1.54	1.46	1.50	1.45	1.59	1.50
Ti	0.04	0.04	0.06	0.03	0.02	0.05	0.02	0.02	0.02
Cr	0.00	0.00	0.00	0.00	0.00	0.00	0.00	0.00	0.00
Mg	0.53	0.44	0.47	0.51	0.39	0.47	0.44	0.44	0.53
Fe$_2$+	2.02	2.35	2.43	2.28	2.61	2.51	2.18	2.38	2.41
Mn	0.00	0.02	0.00	0.03	0.00	0.00	0.00	0.01	0.00
Sum C	4.60	4.76	4.75	4.71	4.90	4.88	4.56	4.76	4.81
Mg	0.00	0.00	0.00	0.00	0.00	0.00	0.00	0.00	0.00
Mn	0.00	0.00	0.00	0.00	0.00	0.00	0.00	0.00	0.00
Fe$_2^+$	0.00	0.00	0.00	0.00	0.00	0.00	0.00	0.00	0.00
Ca	0.01	0.01	0.02	0.02	0.02	0.02	0.01	0.01	0.01
Na	1.97	1.99	1.98	1.98	1.87	1.98	1.99	1.99	1.99
Sum B	1.98	2.00	2.00	2.00	1.89	2.00	2.00	2.00	2.00
Na	0.00	0.04	0.08	0.11	0.00	0.00	0.08	0.07	0.14
K	0.08	0.08	0.09	0.08	0.07	0.09	0.14	0.06	0.06
Sum A	0.08	0.12	0.17	0.19	0.07	0.09	0.22	0.13	0.20
TOT	14.82	14.97	15.01	15.01	14.93	15.02	14.95	14.98	15.09

–: below detection limits

kinematic Na amphibole (Fig. 7c, d) documents shearing in HP/LT metamorphic conditions, as illustrated below.

It is worth noting that no fractures or cataclastic fabrics overprint the ultramylonite. This demonstrates that the ultramylonites postdate and rework a former cataclastic band (and exploits it as a nucleation site), which in turn developed after the protomylonitic fabric in the host rock.

4 Mineral chemistry and estimate of metamorphic pressure–temperature (P-T) conditions

Chemical analyses of coexisting minerals within the metamorphic assemblage (Tables 1, 2, 3) were obtained using a JEOL JXA-8600 electron microprobe, equipped with four wavelength-dispersive spectrometers, at the CNR – Istituto di Geoscienze e Georisorse, Firenze, Italy. Running conditions were 15 kV accelerating voltage and 10 nA beam current on a Faraday cage. Counting time for the determined elements

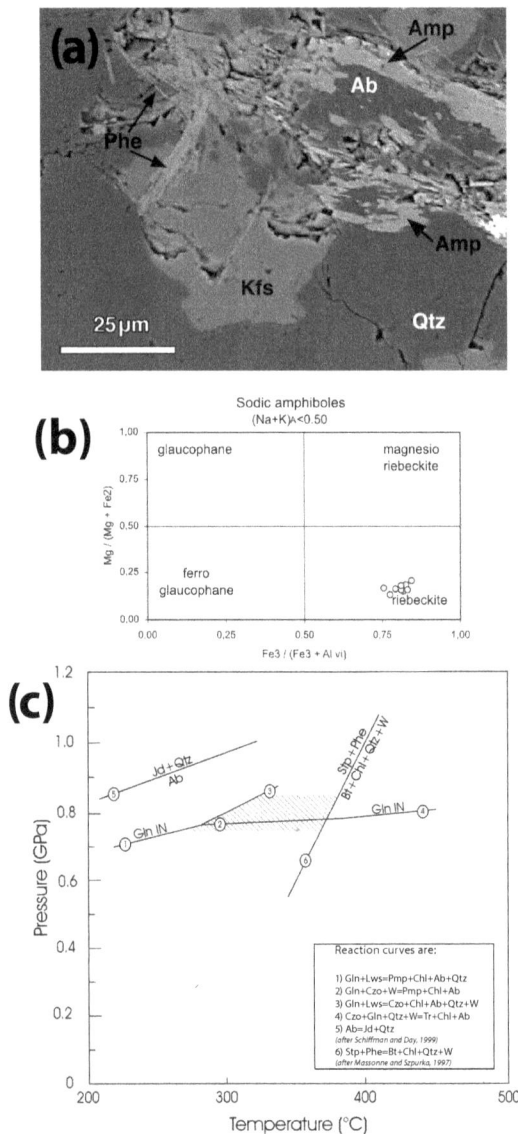

Table 2. Representative microprobe analyses of Stilpnomelane whose structural formulae were calculated assuming 24 oxygens and all Fe as divalent (Fe_2^+).

Stilpnomelane	phyll 1	phyll 2	phyll 3	phyll 4
Wt %				
SiO$_2$	45.20	48.73	40.85	42.13
TiO$_2$	0.13	0.07	0.30	0.18
Al$_2$O$_3$	21.19	12.22	17.08	20.28
Cr$_2$O$_3$	–	0.11	–	–
FeO	13.54	20.45	19.60	16.13
MnO	–	0.16	–	–
MgO	2.39	0.95	1.94	1.92
CaO	0.16	0.20	0.29	0.32
Na$_2$O	0.16	1.00	0.37	0.26
K$_2$O	4.68	3.00	5.05	4.89
TOT	87.45	86.89	85.48	86.11
Cations				
Si	7.418	8.292	7.235	7.200
Al vi	1.582	0.708	1.765	1.800
Sum Z	9.000	9.000	9.000	9.000
Al iv	2.516	1.743	1.800	2.284
Ti	0.016	0.009	0.040	0.023
Cr	0.000	0.015	0.000	0.000
Fe$_3^+$	0.000	0.000	0.000	0.000
Fe$_2^+$	1.858	2.909	2.902	2.304
Mn	0.000	0.023	0.000	0.000
Mg	0.584	0.241	0.512	0.489
Sum Y	4.974	4.939	5.254	5.100
Ca	0.028	0.036	0.055	0.059
Na	0.051	0.330	0.127	0.086
K	0.979	0.651	1.141	1.066
Sum X	1.058	1.017	1.323	1.210
TOT	15.032	14.956	15.576	15.311

– : below detection limit

Figure 4. (a) Backscattered electron image of the metamorphic mineral assemblage in the host granitoids of the studied sample; Ab, albite; Amp, Na amphibole; Kfs, K-feldspar; Phe, phengite; and Qtz, quartz. (b) Composition of sodic amphiboles, using the classification of Leake et al. (1997). (c) Estimated peak-metamorphic pressure and temperature conditions for the studied sample (crosshatched area) constrained by the reactions indicated.

ranged from 10 to 60 s at both peak and background. The Bence and Albee (1968) method was employed for the correction of all data. A number of synthetic and mineral standards were used for instrumental calibration.

Structural formulae of amphiboles were calculated assuming 23 oxygens, and the classification of Leake et al. (1997) was adopted. Site assignment and ferric iron contents were calculated using the scheme proposed by Schumacher in Leake et al. (1997). Because of the small sizes of the crystals (widths ca. 10 μm), it was not possible to make compositional traverses across individual crystals to detect intracrystalline variations in chemical composition, such as core-to-rim zonation. Thus, each analysis reported in Table 1 is from a different crystal. In the studied sample the Na amphiboles are mostly riebeckite (Table 1 and Fig. 4b) with low Mg / (Mg + Fe^{2+}) ratio (0.13–0.21) and are characterized by Si contents close to the maximum of 8.0 apfu.

Albites were calculated assuming 8 oxygens and all have composition close to the pure sodic end-member. Stilpnomelane structural formulae were calculated assuming 24 oxygens and all Fe as divalent (Fe^{2+}). Stilpnomelanes have Fe amounts ranging from 1.86 to 2.91 apfu and Mg amounts ranging from 0.24 to 0.58 apfu (Table 2). For K-feldspar and biotite porphyroclasts, structural formulae were calculated

Table 3. Representative microprobe analyses for K-Feldpars and albites which were calculated assuming 8 oxygens.

Feldspar	KFeld	Kfeld	Kfeld	Kfeld	Kfeld	Albite	Kfeld	Kfeld	Albite	Kfeld	Kfeld
Wt %											
SiO_2	64.47	68.51	63.50	64.25	68.06	67.82	64.35	68.12	67.75	63.98	68.33
TiO_2	–	–	–	–	–	–	0.02	–	–	–	0.07
Al_2O_3	18.75	20.33	18.32	18.66	20.50	19.84	18.28	19.43	20.20	18.91	20.43
Cr_2O_3	–	–	–	–	–	–	–	–	–	–	–
FeO	–	0.15	-	–	0.13	–	–	0.27	0.09	0.11	–
MnO	–	–	–	–	–	–	0.03	–	–	–	0.07
MgO	–	–	0.04	–	0.04	–	–	–	–	–	–
CaO	–	–	–	–	–	0.03	–	–	–	–	–
Na_2O	2.44	0.18	0.17	0.19	11.03	11.44	–	11.31	11.56	0.20	11.17
K_2O	13.48	9.05	16.32	14.72	0.28	0.05	15.74	0.16	0.04	14.39	0.56
TOT	99.14	98.22	98.35	97.82	100.04	99.18	98.42	99.29	99.64	97.59	100.63
Cations											
Si	11.92	12.25	11.95	12.02	11.88	11.93	12.03	11.99	11.88	11.98	11.88
Al	4.09	4.28	4.06	4.11	4.22	4.11	4.03	4.03	4.17	4.17	4.18
Sum Z	16.01	16.53	16.02	16.13	16.09	16.05	16.06	16.01	16.05	16.16	16.06
Fe_3^+	0.00	0.00	0.00	0.00	0.00	0.00	0.00	0.00	0.00	0.00	0.00
Fe_2^+	0.00	0.02	0.00	0.00	0.02	0.00	0.00	0.04	0.01	0.02	0.00
Mg	0.00	0.00	0.01	0.00	0.01	0.00	0.00	0.00	0.00	0.00	0.00
Ca	0.00	0.00	0.00	0.00	0.00	0.01	0.00	0.00	0.00	0.00	0.00
Na	0.87	0.06	0.06	0.07	3.73	3.90	0.00	3.86	3.93	0.07	3.76
K	3.18	2.06	3.92	3.51	0.06	0.01	3.75	0.04	0.01	3.44	0.12
Sum X	4.05	2.15	3.99	3.58	3.82	3.92	3.75	3.93	3.95	3.53	3.89
TOT	20.06	18.68	20.01	19.71	19.91	19.97	19.82	19.95	20.00	19.68	19.95
% Ab	21.58	2.93	1.56	1.92	98.36	99.57	0.00	99.08	99.77	2.07	96.81
% An	0.00	0.00	0.00	0.00	0.00	0.14	0.00	0.00	0.00	0.00	0.00
% Or	78.42	97.07	98.44	98.08	1.64	0.29	100.00	0.92	0.23	97.93	3.19

assuming 8 and 22 oxygens respectively. The analysed K-feldspars have composition close to the orthoclase pure end-member, with minor amounts of Na (Table 3).

The peak metamorphic mineral assemblage is defined by Na amphibole + phengite + quartz + albite + stilpnomelane. Na amphibole is a typical mineral of the blueschist facies and is indicative of HP/LT gradient metamorphism (e.g. Evans, 1990; Schiffman and Day, 1999, and references therein). The minimum temperature (T) and pressure (P) conditions of this mineral assemblage can be estimated using the reaction curves proposed by Schiffman and Day (1999) for the appearance of Na amphibole (stability field of the blueschist facies); the maximum temperature (T) conditions were instead constrained by the presence of stilpnomelane, as shown by the following equilibrium: Stp + Phe = Bt + Chl + Qtz + W (Massonne and Szpurka, 1997). Summing up all the available thermobarometric information, the metamorphic conditions can be estimated as temperature (T) around $320 \pm 50\,°C$ and pressure (P) greater than 0.70 GPa (Fig. 4c), consistent with those reported by

Malasoma et al. (2006) in metagranitic rocks from the area of our studied sample.

5 EBSD data

EBSD analysis of quartz was conducted with a Jeol 7001 FEG-SEM equipped with a NordlysMax EBSD detector (Oxford Instruments) at the Plymouth University Electron Microscopy Centre. Working conditions during acquisition of the EBSD patterns were 20 kV, 70° sample tilt, and high vacuum. Thin sections were chemically polished with colloidal silica and carbon-coated before the analysis. EBSD patterns were acquired on rectangular grids with step size of 1 and 2 µm. EBSD patterns were acquired and indexed with the AZtec software and processed with Channel 5 software (Oxford Instruments). Raw EBSD data were processed to reduce data noise following the procedure tested by Prior et al. (2002) and Bestmann and Prior (2003). EBSD results are shown in form of an inverse pole figure map, pole figures (equal angle, lower hemisphere) of crystallographic axes and

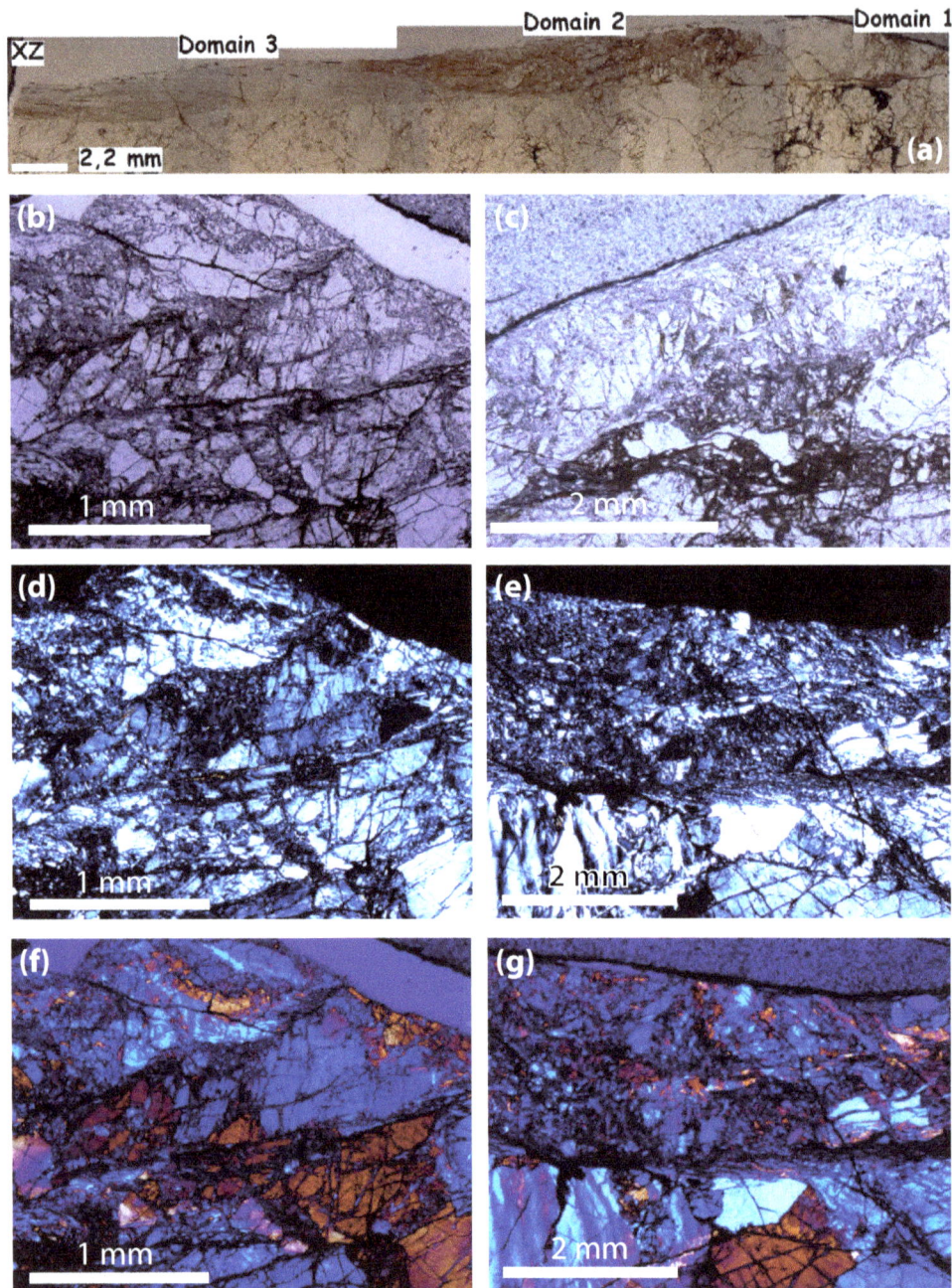

Figure 5. **(a)** Microscopic view of millimetre-scale shear zone nucleated after brittle precursor, with indication of the structural–compositional domains within the shear zone. Domain 1: relict domain of cataclasite; Domain 2: foliated-cataclasite/phyllonite; Domain 3: ultramylonite. **(b–g)** Micrographs showing details of cataclastic relicts in Domain 1; **(b, c, d)** crush microbreccia showing angular to subangular quartz and feldspars clasts with intergranular and intragranular microfractures. Stilpnomelane is observable in **(c)** as filling microfractures. In **(d)** clasts of quartz and feldspars show relict internal deformations similar to those of the host protomylonite. **(e)** Protocataclasite with submillimetre-sized quartz and feldspar clasts within a fine-grained matrix. **(f, g)** Cataclasite with quartz and feldspar clasts in fine-grained matrix. The angular fragments and their grain size distribution are indicative of brittle comminution. **(a, b, e)** Images with plane-polarized light, **(d, f)** cross-polarized light, and **(d, g)** cross-polarized with a lambda plate.

planes ($\langle 0001 \rangle$ c axis, $\langle 11-20 \rangle$ a axis, $\{10\text{–}10\}$ prism $\{m\}$, $\{10\text{–}11\}$ positive rhomb $\{r\}$, $\{01\text{–}11\}$ negative rhomb $\{z\}$), misorientation profiles, and plots of misorientation axis in sample coordinates.

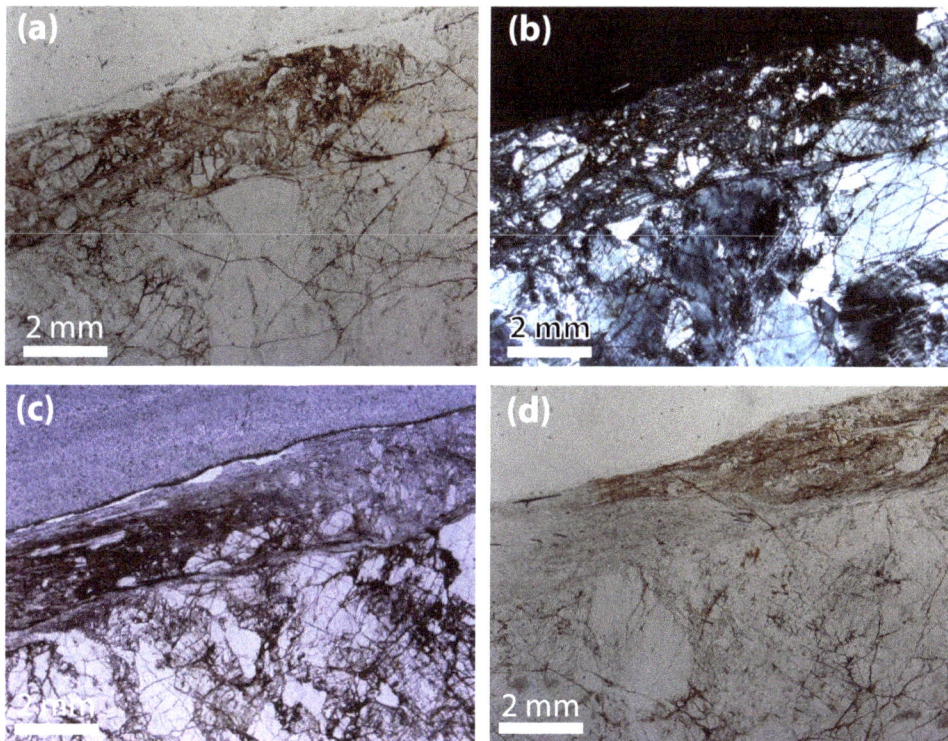

Figure 6. Micrographs showing details of foliated cataclasite/phyllonite of Domain 2 (a); (b) quartz and feldspar clasts are embedded in phyllosilicate-rich matrix defining sub-millimetre-spaced bands in Riedel (R'/C') orientation with respect to shear zone boundary. In (a) and (b) the subangular to subrounded shape of clasts and their grain size distribution due to brittle comminution are evident; (c) cataclasite to phyllonite transition and (d) foliated cataclasite to ultramylonite transition. (a, c, d) Plane-polarized light; (b) cross-polarized light.

5.1 Quartz in the host rock

We analysed a monocrystalline quartz ribbon elongated parallel to the host-rock foliation and sharply cut by the thin localized shear zone. The quartz ribbon contains two nearly orthogonal sets of intracrystalline shear bands of recrystallized grains (Fig. 8a). These are defined by the presence of a SPO oblique to the shear bands boundaries and by the c-axis orientation of the grains within the shear bands. SPO and crystallographic preferred orientation (CPO) in the bands are consistent with the kinematic framework of the proto-mylonitic host rock (see below). One set of shear band is oriented at a high angle ($90° \pm 20$, set 1) and one at a low angle ($\leq 20°$, set 2) to the host rock foliation (vertical in Fig. 8a). The average grain size of the recrystallized grains is $5 \pm 2\,\mu\text{m}$. The host grain also contains fine localized extinction bands (up to $20\,\mu\text{m}$ thick) (LEBs: Derez et al., 2015) subparallel to the bands of recrystallized grains. Low-angle boundaries are ubiquitous in the ribbon; on EBSD maps they are typically straight, poorly connected, and subparallel to the bands of recrystallized grains (Fig. 8b). Some low-angle boundaries are connected to form subgrains of approximately the same size of the recrystallized grains. Subgrains occur with a higher frequency at the intersection between two sets of recrystallized bands, and in regions sandwiched between

closely spaced ($\leq 100\,\mu\text{m}$) subparallel bands of recrystallized grains (Fig. 8b).

The c axis of the host ribbon grain is oriented near the pole to the host-rock foliation, i.e. in a position suitably oriented for the activation of the basal $\langle a \rangle$ slip system of quartz (Fig. 9a). The c-axis orientation of the recrystallized grains in the intracrystalline bands is mostly spread out along the periphery of the pole figure, although some scattered grains have their c axis in intermediate positions between the X and Y directions of the pole figures (Fig. 9b). Such an orientation suggests that the recrystallized grains have experienced a rotation around the Y direction of finite strain (i.e. centre of the pole figure) from the host-grain orientation (e.g. Van Daalen et al., 1999; Menegon et al., 2011).

This is confirmed by the boundary trace analysis (Prior et al., 2002; Menegon et al., 2010) of the two main sets of straight low-angle boundaries defining localized extinction bands (parallel to bands of recrystallized grains), one running ENE–WSW (subset 1) and one running ca. N–S (subset 2) in Fig. 8b. The dispersion paths of crystallographic directions in the pole figures of subset 1 (Fig. 9c) identifies $\{m\}$ as the rotation axis, which lies very close to the centre of the pole figure. The pole to the prismatic plane $\{m\}$ is the rotation axis associated with the basal $\{a\}$ and with the $\{a\} \langle c \rangle$

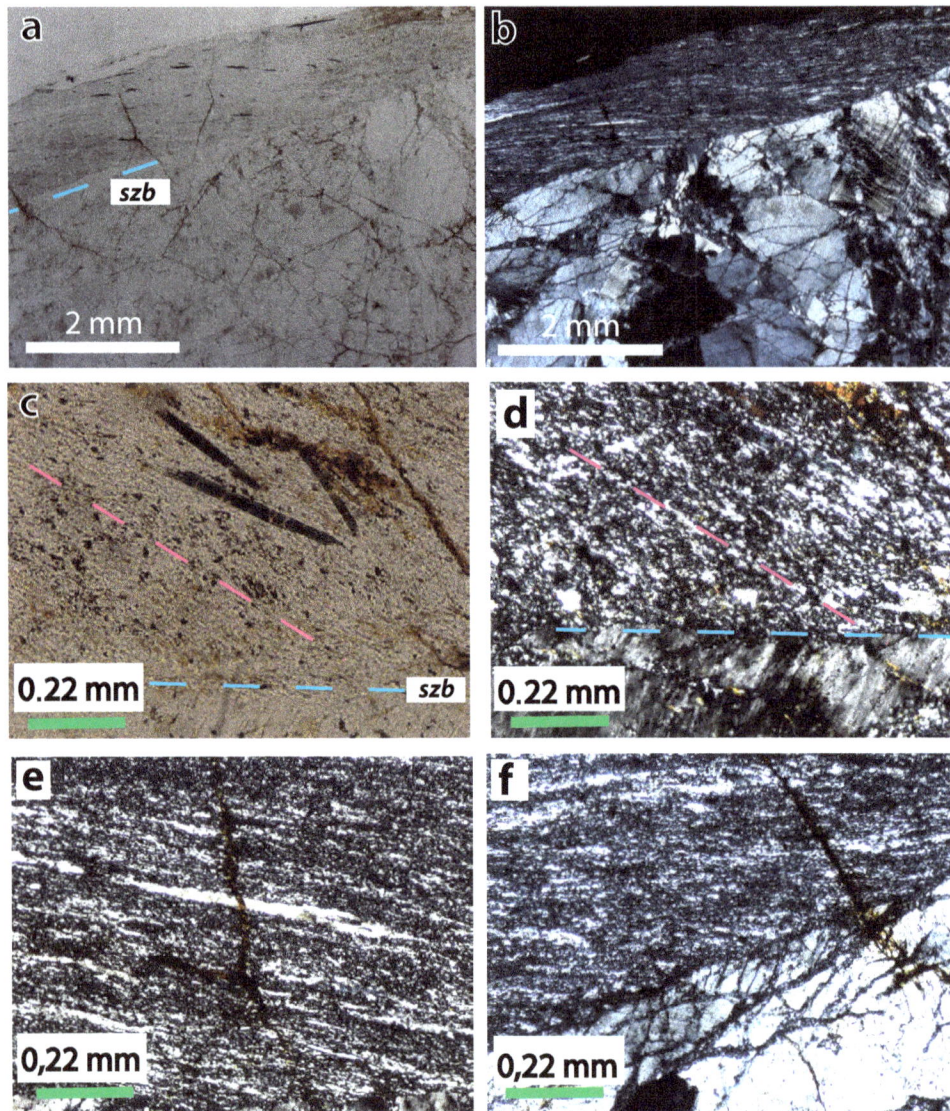

Figure 7. Microscopic view of ultramylonite. **(a, b)** Detail of ultramylonite and host protomylonite with the sharp shear zone boundary (szb) heritage of former host-fracture contact: **(c)** synkinematic Na amphibole, boudinaged within the quartz and albite ultramylonite matrix. Red dashed line (see also in **d**) indicates the shape preferred orientation of the quartz and albite matrix, whereas in blue the shear zone boundary (szb) is indicated; **(d)** quartz and albite recrystallized matrix, in the lower part the boundary (blue dashed line) of microscale shear zone to a host quartz porphyroclast. **(e)** Quartz ribbon with a strong elongation (1 : 15, X/Z ratio) defining a shape preferred orientation oblique (15°) to the shear zone boundary (see **b** for location of quartz ribbon within the shear zone). **(f)** Detail of shear zone boundary showing intragranular fractures in the host protomylonite showing Riedel geometry. Along the Riedel shear, former cataclastic fragments recrystallized to produce the fine (5–10 μm in size) new grains.

slip system in quartz (e.g. Neumann, 2000), and, accordingly, subset 1 can be interpreted as a tilt boundary plane produced by the activity of the slip system basal $\langle a \rangle$ (Fig. 9c) and containing the boundary trace of subset 1 and the rotation axis.

Subset 2 has a similar dispersion path as subset 1, with $\{m\}$ as the identified rotation axis. However, in this case the boundary trace analysis is not consistent with a tilt boundary produced by the activity of the slip system basal $\{a\}$, but could indicate the activity of the $\{a\} \langle c \rangle$ slip system

(Fig. 9d). However, activity of c slip in quartz typically requires temperature in excess of 600 °C (Kruhl, 1996; Zibra et al., 2010) and, therefore, appears unlikely in our samples. Moreover, misorientation profiles across low-angle boundaries with a subset 2 orientation show abrupt misorientation jumps of up to 6° (profiles c–d in Figs. 8b and 9e), as opposed to a gradual accumulation of misorientation towards low-angle boundaries with a subset 1 orientation (profile a–b in Figs. 8b and 9c). Thus, the low-angle boundaries with a

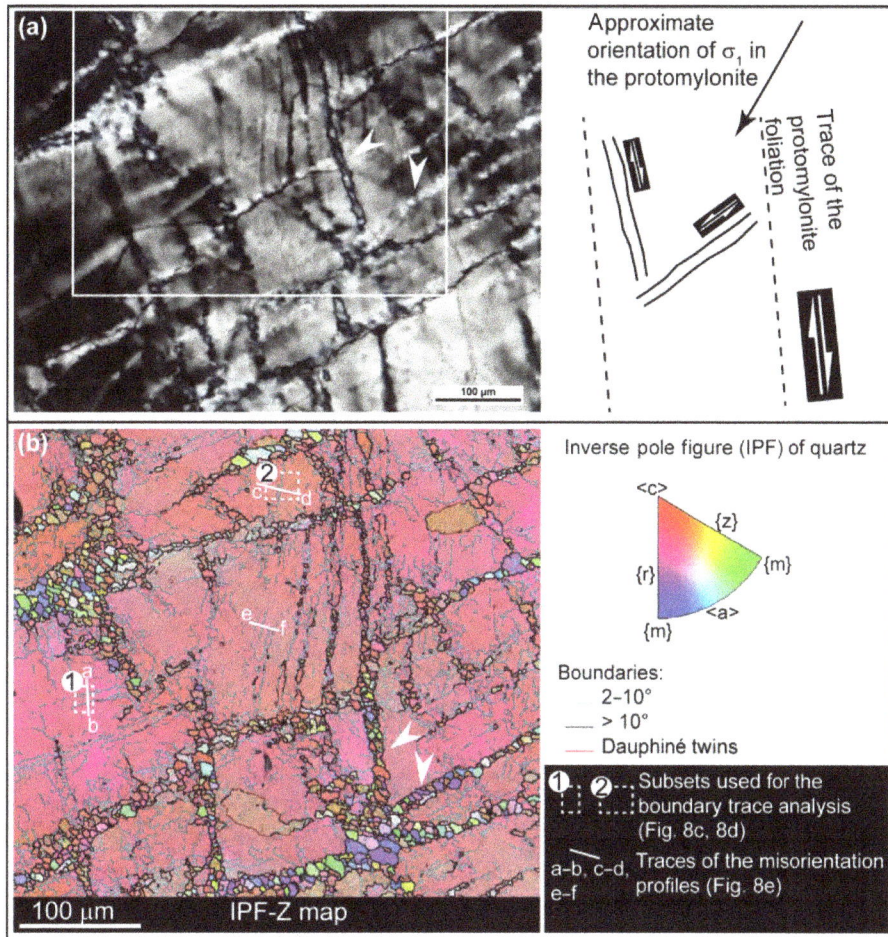

Figure 8. EBSD analysis of quartz in the host protomylonite. **(a)** Microstructure of the analysed site. The white arrows in **(a)** and **(b)** indicate the same pair of conjugate shear. **(b)** Inverse pole figure map with the respect to the protomylonitic foliation in the host rock (vertical in the figure). Location of subset 1 and 2 and trace of misorientation profiles **(a)–(b)**, **(c)–(d)**, and **(e)–(f)** are shown.

subset 2 orientation could represent microcracks subparallel to the basal planes (e.g. Kjøll et al., 2015) that localized rigid body rotation of fragments around the Y direction (e.g. Trepmann et al., 2007; Menegon et al., 2013). The rotation around Y of (1) the crystallographic directions of the host grain and (2) the recrystallized grains (in this case for large misorientation > 10°) is confirmed by the plots of the misorientation axis in sample coordinates (Fig. 9f).

5.2 Quartz in the ultramylonite

We analysed polycrystalline ribbons of recrystallized grains from domain 3 of the localized shear zone (Fig. 10a). The c-axis orientation of the recrystallized grains defines an inclined type I crossed girdle synthetically oriented with respect to the bulk shear sense of the shear zone (Fig. 10b). The c-axes are preferentially clustered near the pole to the shear zone boundary, i.e. in an orientation suitably oriented for the activity of the basal $\langle a \rangle$ slip system. The average grain size of the recrystallized grains in domain 3 is $6 \pm 2 \, \mu m$.

6 Ultramylonite: palaeopiezometry, flow stress, and strain rate

The microstructure and the crystallographic preferred orientation of quartz in the ultramylonite indicate that quartz deformed by dislocation creep and recrystallized to a fine-grained aggregate. Thus, the rheology and the flow stress in the ultramylonite can be evaluated extrapolating experimentally calibrated flow laws of quartz to the deformation conditions.

The rheology of quartz deforming by dislocation creep is generally described in terms of a power-law equation:

$$\dot{\varepsilon} = A f_{H_2O}^m e^{(-Q/RT)} \sigma^n, \tag{1}$$

where $\dot{\varepsilon}$ is the strain rate, f_{H_2O} is the water fugacity (raised to the power of m), Q is the activation energy, R is the universal gas constant, T is the temperature, σ is the differential stress, and n is the stress exponent. We used the theoretical dislocation creep flow law of Hirth et al. (2001), which has derived a linear dependence of strain rate on the water fu-

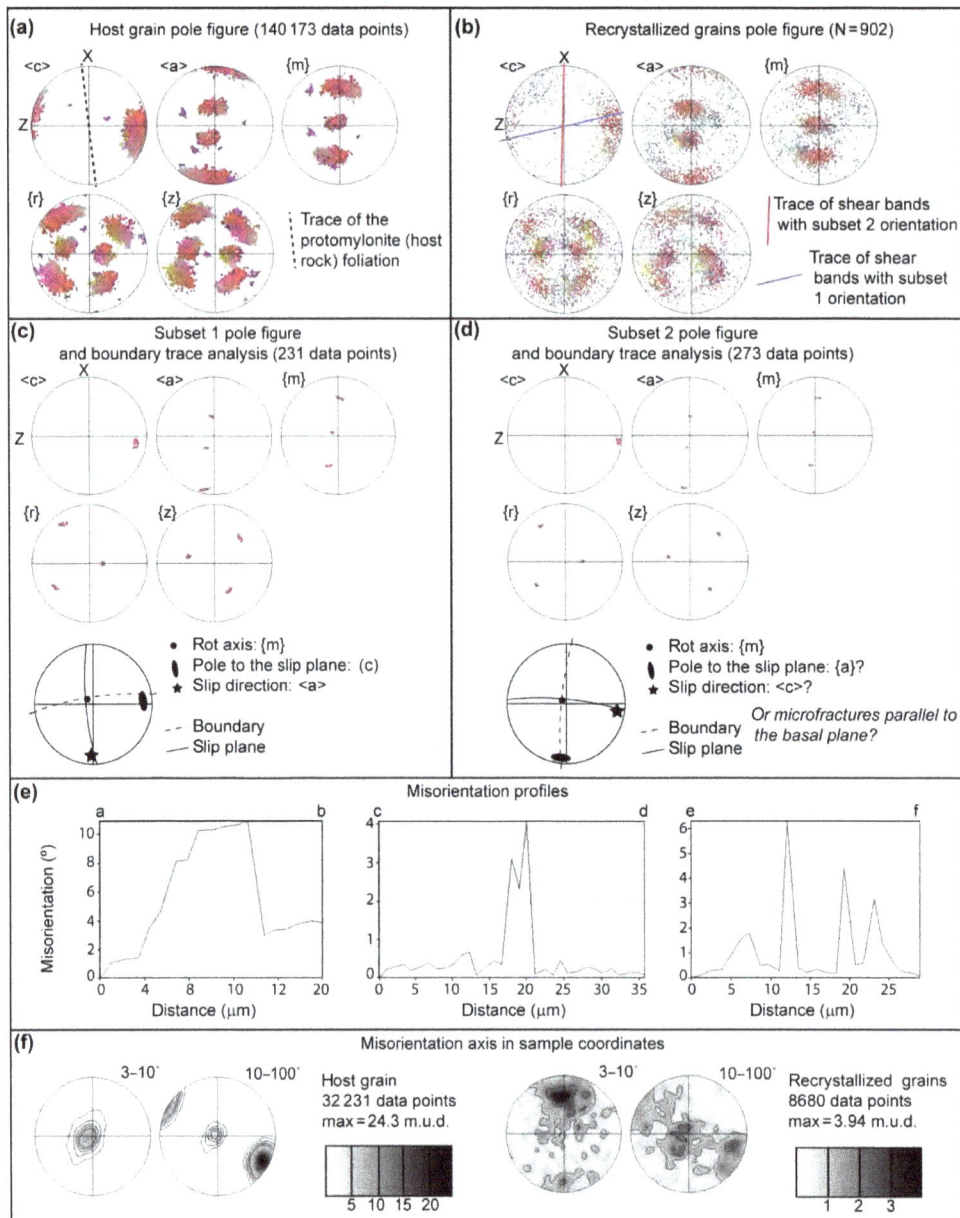

Figure 9. EBSD analysis of quartz in the host protomylonite. (**a**) Pole figure of the host grain, colour-coded per the quartz inverse pole figure with respect to the protomylonitic foliation shown in Fig. 8b. (**b**) Pole figure of recrystallized quartz in the intracrystalline bands. The blue and red line indicates the average orientation of the trace of the intracrystalline bands with a subset 1 and 2 orientation, respectively. (**c**) Boundary trace analysis of the localized extinction band of subset 1. (**d**) Boundary trace analysis of the localized extinction band of subset 2. (**e**) Point-to-point misorientation profiles. See Fig. 8b for location of the traces of the profiles. (**f**) Misorientation axis of the host grain and of the recrystallized grains in sample coordinates.

gacity ($m = 1$) and a stress exponent of 4. A water fugacity of 172 MPa is calculated from the water fugacity coefficient reported in Tödheide (1972) at $T = 350\,°C$, $P = 0.8\,GPa$.

The palaeostress (assuming steady-state flow at the time of viscous deformation) can be determined by means of a recrystallized quartz grain size palaeopiezometer (e.g. Stipp and Tullis, 2003), which has been calibrated in the form $\Delta\sigma = BD^{-x}$, where $\Delta\sigma$ is the steady-state differential stress

($\sigma_1 - \sigma_3$), D is the recrystallized grain size, and B and X are empirical constants. Using the recrystallized grain size piezometer of quartz calibrated by Stipp and Tullis (2003), a recrystallized grain size of 5–10 μm indicates differential stress in the range of 110–190 MPa. Extrapolation of the flow law of Hirth et al. (2001) yields a strain rate in the range of 7.6×10^{-13}–$6.2 \times 10^{-12}\,s^{-1}$ in the ultramylonite at the estimated deformation temperature of ca. 300 °C.

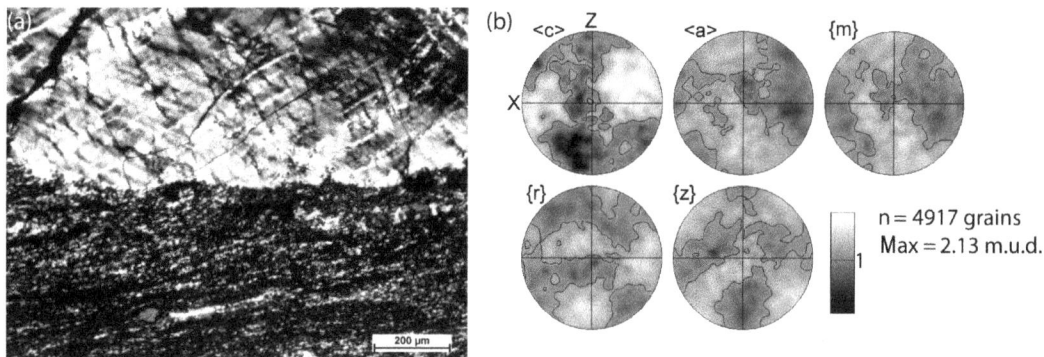

Figure 10. EBSD analysis of quartz in the ultramylonite. **(a)** Microstructure of the recrystallized quartz from domain 3 in the ultramylonite. **(b)** Pole figure of recrystallized quartz grains from **(a)**. Only grains from recrystallized aggregates in the ultramylonite were analysed.

7 Discussion and conclusion

7.1 Kinematic framework and deformation mechanisms of the host-protomylonite

On the basis of overprinting relationships we may infer that the oldest deformation fabric of our analysed sample is represented by the subvertical protomylonite foliation (Fig. 2), which shows microstructural features described in the text. This fabric type and the associated deformation microstructures represent a ubiquitous regional feature in the Popolasca granitoids and is not restricted to our studied sample only.

The c-axis maximum of the quartz ribbon in the host protomylonite is oriented near the pole of the protomylonite foliation and rotated clockwise to indicate a dextral sense of shear in the protomylonite (Fig. 9a). The dextral sense of shear corresponds to the regional top-to-the-west sense of shear observed in the field. The c-axis orientations of grains in the bands differ from the orientations of the host grains, as shown by our EBSD analysis (Fig. 9a, b). This difference is consistent with the relative rotation of the c axes of grains in the conjugate bands, reflecting their respective sense of shear assuming a shortening direction oriented approximately NE–SW in Fig. 8a, and consistent with the bulk dextral sense of shear of the protomylonite. In this kinematic framework, the respective sense of shear is dextral in bands with a subset 2 orientation and sinistral in bands with a subset 1 orientation (Fig. 8a). The SPO in the bands is also consistent with their respective sense of shear. Thus, we conclude that (1) the intracrystalline bands of recrystallized grains are conjugate shear bands; (2) they formed during the development of the regional proto-mylonitic foliation of the unit, and not in response to the localized brittle event investigated in the present study, and can therefore be used to evaluate the stress conditions in the protomylonite prior to the brittle event; and (3) they were overprinted by the cataclastic event forming the brittle precursor of the ultramylonite in our sample (Fig. 5f, g).

The microstructure and the trace analysis of low-angle boundaries indicates that deformation mechanisms in quartz in the host-protomylonite result from a combination of microfracturing and low-temperature dislocation activity (Fig. 9c, d). Although factors like the initial aperture of the (micro)cracks and the starting grain size of potential micro-gauge fragments can control the initial grain size in the quartz bands, the final microstructure of quartz in the bands is not cataclastic, but typical of a dynamic recrystallization aggregate with a clear SPO and CPO. Accordingly, our EBSD analysis suggests that the quartz grains in the intracrystalline bands deformed by dislocation creep on the basal $\langle a \rangle$ slip system. This is consistent with a microstructural evolution where dislocation creep overprinted fracturing and neocrystallization along cracks (e.g. Trepmann et al., 2007, 2017). Along cracks, new grains grew by strain-induced grain boundary migration from fractured fragments with a low dislocation density (e.g. Trepmann et al., 2007) and deformed by dislocation creep on the dominant basal $\langle a \rangle$ slip system, as shown by their CPO and SPO.

In conclusion, our observations suggest that the final grain size in the intracrystalline bands represents an equilibrium grain size with the flow stress during crystal-plastic flow of quartz in the bands. Thus, they can be used to evaluate the differential stress during dislocation creep deformation in the protomylonite before the development of the brittle precursor as developed furthermore below.

7.2 Viscous–brittle–viscous deformation under HP/LT conditions

The analysed example documents a switch in deformation mode and mechanisms within an HP/LT fault zone. Over-printing relationships and microstructural features suggest the following deformation sequence: stage (1) consists of distributed deformation (at a regional to sample scale) and development of a protomylonitic foliation in the granitoid by quartz low-temperature plasticity, microfracturing, and albite neo-crystallization from K-feldspar porphyroclasts (Fitzgerald and Stünitz, 1993); this was followed by stage (2), in which localized deformation by brittle fracturing formed a millimetre-thick cataclasite, which acted as a precursor for stage (3) localization of viscous deformation and ultramylonite development. The synkinematic and post-kinematic growth of Na amphibole in the host rock foliation and in localized ultramylonite indicates that the entire deformation sequence occurred under HP/LT conditions (ca. 300 °C at ≥ 0.7 GPa), corresponding to a depth of 23–30 km in the subduction channel.

The estimated P–T conditions are consistent with a low-T plasticity regime in quartz (Stipp et al., 2002; Derez et al., 2015, and references therein). Accordingly, the deformation microstructures of quartz produced during stage (1) in the host rock appear to be the product of the competition between dislocation activity and fracturing. Localized extinction bands at a high angle to the host rock foliation (subset 1 in Fig. 8b) are consistent with the activity of the basal $\langle a \rangle$ slip system, whereas localized extinction bands subparallel to the host-rock foliation (subset 2 in Fig. 8b) are interpreted as fractures subparallel to the basal plane, as previously observed by Kjøll et al. (2015). Some localized extinction bands (especially those with a subset 2 orientation; see Fig. 8b) contain isolated small new grains that are considerably smaller than the average grain size in the recrystallized bands, and that are only slightly misoriented with respect to the host grain. Moreover, the high-angle boundaries in such localized extinction bands are not always fully connected to define entire new grains. Together with the abrupt misorientation jumps (Fig. 9e), these observations further suggest that localized extinction bands with a subset 2 orientation represent fractured domains in which the fragments have rotated passively and sealed together, as proposed by, for example, Derez et al. (2015).

In conjunction with the local fracturing of the host grain, deformation of the recrystallized grains in the conjugate intracrystalline bands involved dislocation activity, as indicated by the cluster of misorientation axis around the prism $\{m\}$ for low misorientations (compare Fig. 9f with 9a). This is consistent with the local activity of the basal $\{a\}$ slip system in the recrystallized bands, as also indicated by the CPO of the recrystallized grains in the shear bands (Fig. 9b).

The type-I crossed girdle c-axis CPO of quartz in the ultramylonite suggests the concomitant activity of basal $\langle a \rangle$, rhomb $\langle a \rangle$ and prims $\langle a \rangle$ slip systems in the recrystallized grains (Fig. 10). The grain size of recrystallized quartz in the ultramylonites is in the same range (5–10 µm) as in the intracrystalline bands in the host rock. This suggests that 5–10 µm represents the equilibrium grain size for the flow stress (estimated in the range of 110–190 MPa with recrystallized grain size piezometry) during viscous deformation before and after the transient brittle event of stage (2).

During stage (2) the development of cataclasite and related dilatancy resulted in an increase in permeability and thus facilitated fluid access and fluid mobility in the shear zone. This enhanced mineral reactions as testified by modal enrichment of stilpnomelane (by biotite breakdown) and Na amphibole in the ultramylonite.

Therefore, the observed structures witness a change in deformation style (from sample-scale distributed to sample-scale localized strain in brittle precursor), and a switch in the dominant deformation mechanism (from low T plasticity in the host rock to cataclasis and back to crystal plasticity in the ultramylonite), which occurred at the footwall of the subduction interface under temperature conditions (ca. 300 °C) typical of the brittle–viscous transition in quartz-feldspathic rocks.

Finally, the transition from stage (2) and stage (3) is consistent with the general observation that nucleation of localized ductile shear zones requires, in some cases, the presence of a planar compositional or structural precursor (e.g. Pennacchioni and Zucchi, 2012, and references therein). The example from the Popolasca granite demonstrates that (i) nucleation on brittle precursors also occurs under HP/LT conditions in the subduction channel, and (ii) the brittle precursors are not necessarily inherited from an earlier deformation event, but can be the manifestation of switches in deformation mode in the footwall of the subduction interface.

7.3 Significance of the switch in deformation mode and implications for fault-slip behaviours in subduction zones

We interpreted the inferred deformation sequence and structures as the result of transient instabilities (Sibson, 1980; White, 1996; Handy and Brun, 2004), possibly representative of mixed fault-slip behaviours at seismogenic depth in the subduction channel. We have estimated the conditions resulting in the transient brittle event during stage (2) under the following assumptions and approximations:

1. The coefficient of internal friction, μ_i, is generally between 0.5 and 1.0 in intact rocks (Sibson, 1985). We considered $\mu_i = 0.6$ in the failure envelope for the intact Popolasca granitoid.

2. We used a cohesive strength of 35 MPa as representative of granitoids (Amitrano and Schmittbuhl, 2002), with a resulting tensile strength of 17.5 MPa.

3. We assumed an Andersonian stress field and a thrusting regime. The effective vertical stress σ_v' corresponds to the effective minimum principal stress $\sigma_3' = \sigma_3 - P_f$, where P_f is the pore fluid pressure. We considered a hydrostatic pore pressure ($\lambda = 0.4$, $P_f = 320$ MPa for $\sigma_3 = 800$ MPa) during the formation of the protomylonite foliation in the host rock.

4. We assumed a differential stress of 110–190 MPa during viscous flow of the granitoid prior to brittle failure at $\sigma_v' = 480$ MPa, as derived from the recrystallized grain size piezometry in the protomylonite.

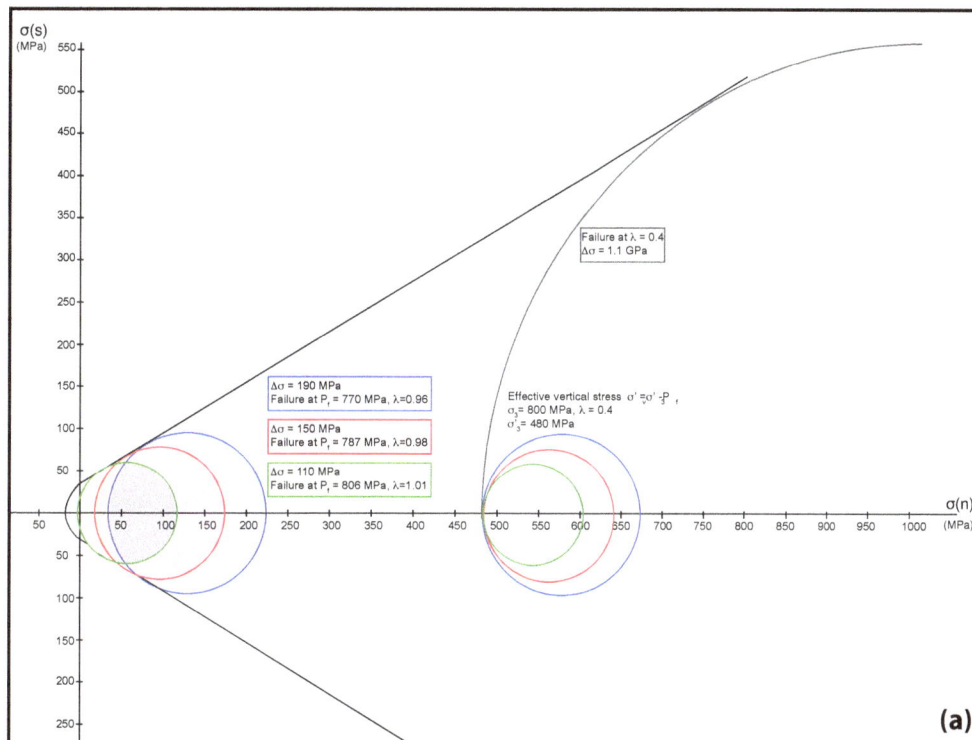

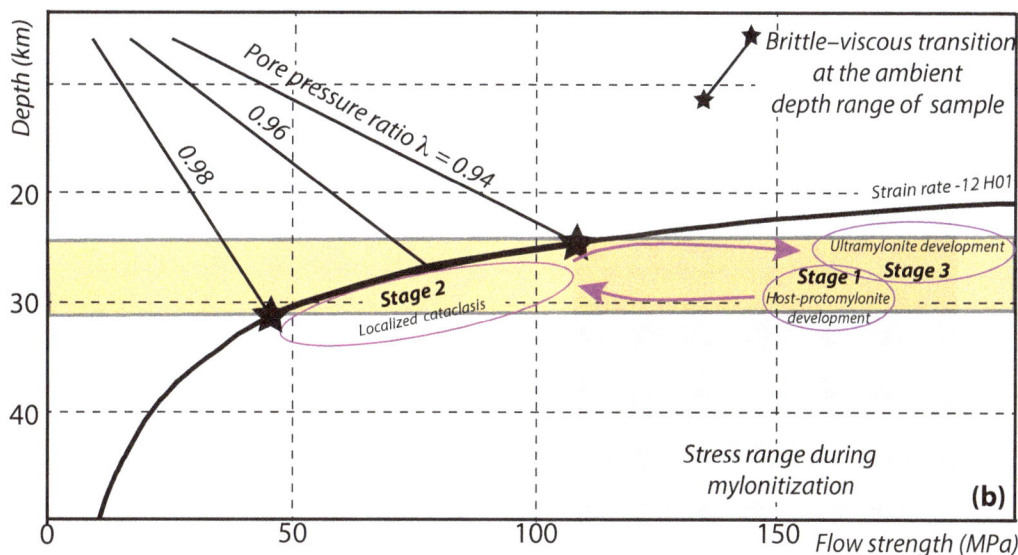

Figure 11. (a) Brittle failure analysis for our studied sample; see text for assumptions and approximations. For the differential stress range and the vertical stress considered, brittle failure requires (sub)lithostathic fluid pressure ($0.96 < \lambda < 1$), whereas viscous deformation at the estimated depth range (23–30 km) is only possible for pore fluid pressure ≤ 0.94. **(b)** Rheological profile calculated for a fixed strain rate of $10^{-12}\,s^{-1}$ (see text for calculation details) using the quartz flow law of Hirth et al. (2001). A cycle of switching in deformation style and mechanisms is suggested for the analysed sample in its ambient depth range. The frictional Byerlee envelope is calculated using an average friction coefficient of 0.7 for various pore fluid pressure ratio values (Fig. 11a and text). Stress estimates based on recrystallized quartz piezometry (grey shaded area) were calculated following Stipp and Tullis (2003).

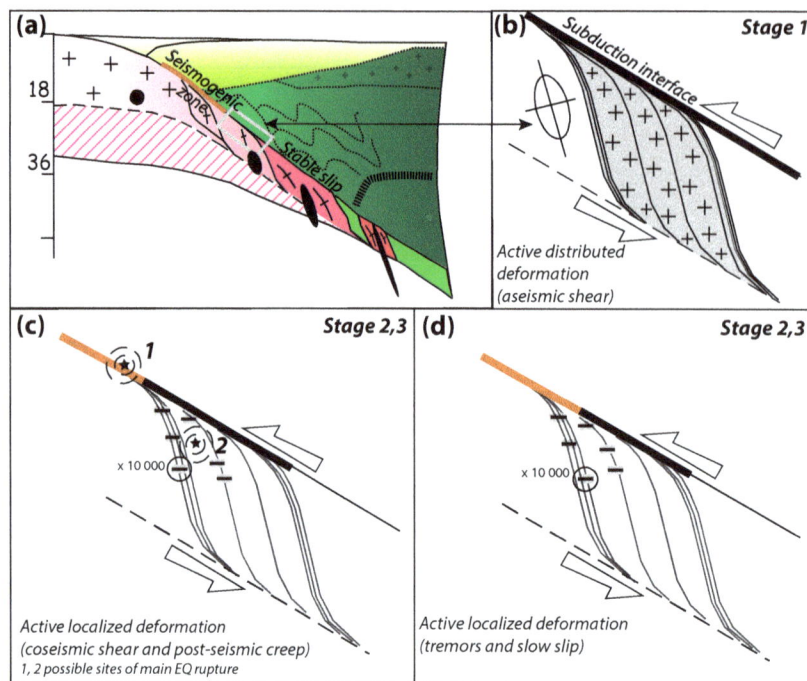

Figure 12. (a) Simplified model of the ancient subduction of the corsican continental crust, with indication where the mechanical coupling is the highest (seismogenic zone) and lowest (stable slip). The white rectangle corresponds to the inferred zone of the studied sample. The pink to red shading plots the lower to higher blueschist- to eclogite-facies peak metamorphism in continental-derived crustal units. **(a, b, c, d)** Multi-stage scenario for development of the association of granitic protomylonites and ultramylonites from brittle precursors in the Popolasca area: **(a)** general sketch showing the east-dipping ancient Alpine subduction of the corsican continental crust; **(b)** formation of a crustal-scale anastomosed network of shear zones within the host granitic crust and active distributed deformation with aseismic creep; **(c)** formation of brittle instabilities (stage 2) in the shear zone by seismic ruptures nucleated along (1) the subduction interface or (2) the core of granites and propagating across and beyond the granitic mylonites, followed by post-seismic creep, with localization of viscous shear zone after brittle precursor (stage 3); and **(d)** formation of brittle instabilities by seismic tremors and slow-slip events followed by post-seismic (post-tremors) creep.

The results of our analysis are shown in Fig. 11a. For brittle failure, it is necessary to invoke high differential stresses (of the order of 1.1 GPa assuming a hydrostatic pore fluid pressure), or higher fluid pressure (Fig. 11a). Differential stresses in excess of 1 GPa have been associated with intermediate depth (50–300 km) earthquakes in the subduction channel (e.g. John et al., 2009) and are expected to result in extensive development of pseudotachylytes in granitoid rocks (e.g. Austrheim, 2013, and references therein), which, however, have not been observed in the sample studied here.

For the differential stress range and the vertical stress considered here, brittle failure requires (sub)lithostathic fluid pressure ($0.96 < \lambda < 1$, Fig. 11). The strength envelope plotted for pore fluid pressure between 0.98 and 0.94 and for a strain rate of $10^{-12} \, s^{-1}$ (Fig. 11b) suggests that viscous deformation at the estimated depth range (23–30 km) is only possible for pore fluid pressure ≤ 0.94; otherwise brittle deformation is expected to occur. Thus, under the assumptions listed above, our analysis indicates that local fluctuations in pore fluid pressure can explain the cyclic viscous–brittle–viscous deformation switch. However, despite the synkine-

matic growth of hydrous minerals in the cataclasite and the high pore fluid pressure required at failure, there is no evidence of macroscopic veining or hybrid fractures in the samples. Our analysis is consistent with this observation, in that failure occurs entirely in the shear fractures field and not in the (hybrid) shear extension fractures field (Fig. 11a).

Drawing from the results shown in Fig. 11a, b and using concepts and inferences coming from modern studies of convergent subduction system (e.g. Ozowa et al., 2002; Fu and Freymuller, 2013; Bedford et al., 2013), the following deformation scenario (Fig. 12) is envisaged. A first stage of distributed deformation (stage 1) may be associated with development of map-scale anastomosing shear zones (protomylonitic foliation at sample scale) during aseismic creep; subsequently, following Angiboust et al. (2015), two possible deformation scenarios and slip patterns may be envisaged: (a) propagation at seismic rate of microrupture followed by afterslip or (b) slow-slip phenomena in an aseismically creeping crust. In the first case, a transient strain rate increase associated with brittle fracturing would represent the deep response of the transition zone to a specific stage of the seismic cycle taking place higher up along the seismogenic portion of the subduction interface (Fig. 12c, scenario 1) or in less deformed domains acting as local stress raisers nearby (Fig. 12c, scenario 2). Cataclastic deformation would then correlate with coseismic to post-seismic deformation higher up along the interface. In that case, the mylonitization of the brittle precursor would be the record of subsequent interseismic deformation at lower strain rates, taking place in the time frame between two earthquakes.

In the second case (Fig. 12d), the transient highs in strain rates expressed by brittle fracturing and cataclasis could be the record of deformation associated with slow-slip events (SSEs) or other transient slips, generally localized along the subduction interface at this depth range itself (e.g. Shelly et al., 2006; Fagereng et al., 2014). We are aware that the mechanism(s) of SSEs initiation is still poorly understood, and that stress/strain rate perturbations triggered by an earthquake nearby followed by post-seismic slip and interseismic creep is an equally feasible mechanism to explain the deformation sequence recorded in our sample. However, our preferred interpretation is that the transient brittle deformation recorded in our studied sample is the manifestation of a slow-slip event, for the following reasons: (1) subduction interface SSEs typically occur at the downdip transition from stick-slip behaviour to aseismic creep (e.g. Wallace and Beavan, 2010), and, in granitoid rocks, this transition is expected to occur at the T range of deformation of our case study, as witnessed by the deformation microstructures of quartz and feldspar in the host protomylonite, and (2) SSEs are often related to high pore fluid pressure (e.g. Liu and Rice, 2007).

Whatever the actual process triggering brittle deformation was, we want to emphasize that detailed microstructural studies of exhumed shear zones are a valuable complement to the geodetic, seismological, and experimental studies that aim to unravel the complex fault-slip behaviours at the subduction interface. In this context, our studied sample represents a new and still not yet documented case study of brittle–viscous transition zone and processes in subducted continental crust. Moreover, our study reinforces the concept that the external zones of Alpine Corsica represent a unique target to document and better understand footwall deformation structures and processes related to HP/LT continental subduction.

Competing interests. The authors declare that they have no conflict of interest.

Special issue statement. This article is part of the special issue "Analysis of deformation microstructures and mechanisms on all scales". It is a result of the EGU General Assembly 2016, Vienna, Austria, 17–22 April 2016.

Acknowledgements. This work was developed and funded by the Italian PRIN 2006 project. Luca Menegon acknowledges financial support from a FP7 Marie Curie Career Integration Grant (grant agreement PCIG13-GA-2013-618289). The staff at the Plymouth University Electron Microscopy Centre are thanked for support during EBSD analysis. Constructive reviews by Giulio Viola and of an anonymous reviewer significantly improved the manuscript and are greatly acknowledged, as well as the final anonymous peer reviewer. Editorial handling of the special issue by Renée Heilbronner, Rüdiger Kilian, Florian Fusseis, and Ilka Weikusat is also greatly acknowledged.

Edited by: Renée Heilbronner

References

Altenberger, U., Prosser, G., Grande, A., Günter, C., and Langone, A.: A seismogenic zone in the deep crust indicated by pseudotachylytes and ultramylonites in granulite-facies ricks of Calabria (Southern Italy), Contrib. Mineral. Petr., 166, 975–994, 2013.

Amitrano, D. and Schmittbuhl, J.: Fracture roughness and gouge distribution of a granite shear band, J. Geophys. Res., 107, 1–16, https://doi.org/10.1029/2002JB001761, 2002.

Andersen, T. B. and Austrheim, H.: Fossil earthquakes recorded by pseudotachylytes in mantle peridotites from the Alpine subduction complex of Corsica, Earth Planet. Sc. Lett., 242, 58–72, 2006.

Andersen, T. B., Austrheim, H., Deseta, N., Silkoset, P., and Ashwal, L. D.: Large subduction earthquakes along the fossil Moho in Alpine Corsica, Geology, 42, 395–398, https://doi.org/10.1130/G35345.1, 2014.

Angiboust, S., Glodny, J., Oncken, O., and Chopin, C.: In search of transient subduction interfaces in the Dent Blanche–Sesia Tectonic System (W. Alps), Lithos, 205, 298–321, 2014.

Angiboust, S., Kirsch, J., Oncken, O., Glodny, J., Monié, P., and Rybacki, E.: Probing the transition between seismically coupled and decoupled segments along an ancient subduction interface, Geochem. Geophys. Geosys., 16, 1905–1922, https://doi.org/10.1002/2015GC005776, 2015.

Atkinson, B. K.: Subcritical crack propagation in rocks: theory, experimental result and applications, J. Struct. Geol., 4, 41–56, 1982.

Austrheim, H.: Fluid and deformation induced metamorphic processes around Moho beneath continent collision zones: Examples from the exposed root zone of the Caledonian mountain belt, W-Norway, Tectonophysics, 609, 620–635, 2013.

Austrheim, H. and Andersen T. B.: Pseudotachylytes from Corsica: fossil earthquakes from a subduction complex, Terra Nova, 16, 193–197, 2004.

Austrheim, H. and Boundy, T. M.: Pseudotachylytes generated during seismic faulting and eclogitization of the deep crust, Science, 265, 82–83, 1994.

Beaudoin, A., Augier, R., Joilivet, L., Jourdon, A., Raimbourg, H., Scaillet, S., and Cardello, G. L.: Deformation behavior of continentala crust during subduction and exhumation: strain distribution over the Tenda massif (Alpine Corsica, France), Tectonophysics, 705, 12–32, 2017.

Bedford, J., Moreno, M., Baez, J. C., Lange, D., Tilmann, F., Rosenau, M., Heibach, O., Oncken, O., Bartsch, M., Rietbrock, A., Tassara, A., Bevis, M., and Vigny, C.: A high-resolution, time-variable afterslip model for the 2010 Maule Mw 8.8, Chile megathrust earthquake, Earth Planet. Sc. Lett., 383, 26–36, 2013.

Berger, A. and Bousquet, R.: Subduction-related metamorphism in the Alps: review of isotopic ages based on petrology and their geodynamic consequences, Geological Society London Special Publications, 298, 117–144, 2008.

Beroza, G. C. and Ide, S.: Slow Earthquakes and Non-volcanic Tremor, Annu. Rev. Earth Pl. Sc., 39, 271–296, 2011.

Bestmann, M. and Prior, D. J.: Intragranular dynamic recrystallization in naturally deformed calcite marble: diffusion accommodated grain boundary sliding as a result of subgrain rotation recrystallization, J. Struct. Geol., 25, 1597–1613, 2003.

Bezert, P. and Caby, R.: Sur l'age post-bartonien des événements tectono-métamorphiques alpines en bourdure orientale de la Corse cristalline, B. Soc. Geol. Fr., 8, 965–971, 1988.

Bos, B. and Spiers, C. J.: Frictional-viscous flow of phyllosilicate-bearung rocks: Microphysical model and implications for crustal strength profiles, J. Geophys. Res., 107, 1–13, https://doi.org/10.1029/2001JB000301, 2002.

Burg, J. P. and Laurent, P.: Strain analysis of a shear zone in a granodiorite, Tectonophysics, 47, 15–42, https://doi.org/10.1016/0040-1951(78)90149-X, 1978.

Butler, R. H. W.: Area balancing as a test of models for the deep structure of mountain belts, with specific reference to the Alps, J. Struct. Geol., 52, 2–16, 2013.

Carminati, E. and Doglioni, C.: Alps vs. Apennines: The paradigm of a tectonically asymmetric Earth, Earth-Sci. Rev., 112, 67–96, 2012.

Chester, F. M.: Dynamic recrystallization in semi-brittle faults, J. Struct. Geol., 11, 847–858, 1989.

Cooper, F. J., Platt, J. P., Platzman, E. S., Grove, M. J., and Seward, G.: Opposing shear senses in a subdetachment mylonite zone: Implications for core complex mechanics, Tectonics, 29, TC4019, https://doi.org/10.1029/2009TC002632, 2010.

Derez, T., Pennock, G., Drury, M., and Sintubin, M.: Low-temperature intracrystalline deformation microstructures in quartz, J. Struct. Geol., 71, 3–23, 2015.

Deseta, N., Andersen, T. B., and Ashwal, L. D.: A weakening mechanism for intermediate depth seismicity? Detailed petrographic and microstructural observation from blueschist facies pseudotachylytes, Cape Corse, Corsica, Tectonophysics 610, 138–149, 2014a.

Deseta, N., Ashwal, L. D., and Andersen, T. B.: Initiating intermediate-depth earthquakes: Insights from a HP-LT ophiolite from Corsica, Lithos, 206, 127–146, 2014b.

Di Rosa, M., De Giorgi, A., Marroni, M., and Vidal., O. Syn-convergent exhumation of continental crust: evidence from structural and metamorphic analysis of the Monte Cecu area, Alpine Corsica (Northern Corsica, France), Geol. J., https://doi.org/10.1002/gj.2857, 2016.

Di Vincenzo, G., Grande, A., Prosser, G., Cavazza, W., and De Celles, P. G.: ^{40}Ar–^{39}Ar laser dating of ductile shear zones from central Corsica (France): evidence of Alpine (middle to late Eocene) syn-burial shearing in Variscan granitoids, Lithos, 262, 369–383, https://doi.org/10.1016/j.lithos.2016.07.022, 2016.

Evans, B. W.: Phase relations of epidote-blueschists, Lithos, 25, 3–23, 1990.

Faccenna, C., Piromallo, C., Crespo-Blanc, A., Jolivet, L., and Rossetti, F.: Lateral slab deformation and the origin of the western Mediterranean arcs, Tectonics, 23, TC1012, https://doi.org/10.1029/2002, 2004.

Fagereng, Å, Hillary, G. W. B., and Diener, J. F. A.: Brittle-viscous deformation, slow slip, and tremor, Geophys. Res. Lett., 41, 4159–4167, https://doi.org/10.1002/2014GL060433, 2014.

Fitzgerald, J. D. and Stünitz, H.: Deformation of granitoids at low metamorphic grades, I. Reactions and grain size reduction, Tectonophysics, 221, 269–297, 1993.

Fournier, M., Jolivet, L., Goffé, B., and Dubois, R.: The Alpine Corsica metamorphic core complex, Tectonics, 10, 1173–1186, 1991.

Fu, Y. and Freymueller, J. T.: Repeated large slow slip events at the south central Alaska subduction zone, Earth Planet. Sci. Lett., 375, 303–311, 2013.

Fusseis, F. and Handy, M. R.: Micromechanisms of shear zone propagation at the brittle–viscous transition, J. Struct. Geol., 30, 1242–1253, 2008.

Gapais, D., Bale, P., Choukroune, P., Cobbold, P. R., Mahjoub, Y., and Marquer, D.: Bulk kinematics from shear zone patterns: some field examples, J. Struct. Geol., 9, 635–646, https://doi.org/10.1016/0191-8141(87)90148-9, 1987.

Garfagnoli, F., Menna, F., Pandeli, E., and Principi, G.: Alpine metamorphic and tectonic evolution of the Inzecca-Ghisoni area (southern Alpine Corsica, France), Geol. J., 44, 191–210, 2009.

Gibbons, W. and Horak, J.: Alpine metamorphism of Hercynian hornblende granodiorite beneath the blueschist facies schistes lustre's nappe of NE Corsica, J. Metamor. Geol., 2, 95–113, 1984.

Goncalves, P. Poilvet, J.-C., Oliot, E., Trap, P., and Marquer, D.: How does shear zone nucleate? An example from the Suretta nappe (Swiss Eastern Alps), J. Struct. Geol., 86, 166–180, 2016.

Guermani, A. and Pennacchioni, G: Brittle precursors of plastic deformation in a granite: An example from the Mont Blanc Massif (Helvetic, western Alps), J. Struct. Geol., 20, 135–148, 1998.

Gueydan, F., Brun, J. P., Philippon, M., and Noury, M.: Sequential extension as record of Corsica rotation during Apennines slab roll-back, Tectonophysics, 710/711, 149–161, https://doi.org/10.1016/j.tecto.2016.12.028, 2016.

Hacker, B. R., Peacock, S. M., Abers, G. A., and Holloway S. D.: Subduction factory, 2. Are intermediate-depth earthquakes in subducting slabs linked to metamorphic dehydration reactions?, J. Geophys. Res., 108, 2030, https://doi.org/10.1029/2001JB001129, 2003.

Handy M. R. and Brun, J. P.: Seismicity, structure and strength of the continental lithosphere, Earth Planet. Sc. Lett., 223, 427–441, 2004.

Handy, M. R. and Stünitz, H.: Strain localization by fracturing and reaction-weakening – a mechanism for initiating exhumation of subcontinental mantle beneath rifted margins, in: Deformation mechanisms, rheology and tectonics: current status and future perspectives, edited by: De Meer, S., Drury, M. R., De Bresser, J. H. P., and Pennock, G. M., Geological Society of London Special Publication, 200, 387–407, 2002.

Handy, M. R., Hirth, G., and Burgmann, R.: Continental fault structure and rheology from frictional-to-viscous transition downward, in: Tectonic Faults: Agents of Change on a Dynamic Earth, edited by: Handy, M. R., Hirth, G., and Hovious, N., Dahlem Workshop Reports, MIT Press, 139–181, 2007.

Hayman, N. W. and Lavier, L. L.: The geologic record of deep episodic tremor and slip, Geology, 42, 195–198, https://doi.org/10.1130/G34990.1, 2014.

Healy, D., Reddy, S. M., Timms, N. E., Gray, E. M., and Vitale Brovarone, A.: Trench parallel fast axes of seismic anisotropy due to fluid-filled cracks in subducting slabs, Earth Planet. Sc. Lett., 283, 75–86, 2009.

Hirth, G. and Tullis, J.: Dislocation creep regimes in quartz aggregates, J. Struct. Geol., 14, 145–159, 1992.

Hirth, G., Teyssier, C., and Dunlop, W. J.: An evaluation of quartzite flow laws based on comparisons between experimentally and naturally deformed rocks, Int. J. Earth Sci., 90, 77–87, 2001.

John, T., Medvedev, S., Rüpke, L. H., Andersen, T. B., Podladchikov, Y. Y., and Austrheim, Å.: Generation of intermediate-depth earthquakes by self-localizing thermal runaway, Nat. Geosci., 2, 137–140, https://doi.org/10.1038/NGEO419, 2009.

Jolivet, L., Dubois, R., Fournier, M., Goffé, B., Michard, A., and Jourdan, C.: Ductile extension in Alpine Corsica, Geology, 18, 1007–1010, 1990.

Kjøll, H. J., Viola, G., Menegon, L., and Sørensen, B. E.: Brittle–viscous deformation of vein quartz under fluid-rich lower greenschist facies conditions, Solid Earth, 6, 681–699, https://doi.org/10.5194/se-6-681-2015, 2015.

Kohlstedt, D. L., Evans, B., and Mackwell, S. J.: Strength of the lithosphere: Constraints imposed by laboratory experiments, J. Geophys. Res., 100, 17857–17602, 1995.

Kruhl, J. H.: Prism- and basal-plane parallel subgrain boundaries in quartz: a microstructural geothermobarometer, J. Metamorph. Geol., 14, 581–589, 1996.

Küster, M. and Stöckhert, B.: High differential stress and sublithostatic pore fluid pressure in the ductile regime – microstructural evidence for short-term post-seismic creep in the Sesia Zone, Western Alps, Tectonophysics, 303, 263–277, 1999.

Lahondére, D.: Les schistes blues et les eclogites a lawsonite des unités continentals et océanique de la Corse alpine: Nouvelles donnée e pétrologique et structurales (Corse), Documents du BRGM, 240 pp., 1996.

Lamb, S.: Shear stresses on megathrusts: Implications for mountain building behind subduction zones, J. Geophys. Res., 111, B07401, https://doi.org/10.1029/2005JB003916, 2006.

Liu, Y. and Rice, J. R.: Spontaneous and triggered aseismic deformation transients in a subduction fault model, J. Geophys. Res., 112, B09404, https://doi.org/10.1029/2007JB004930, 2007.

Maggi, M., Rossetti, F., Corfu, F., Theye, T., Andersen, T. B., and Faccenna, C.: Clinopyroxene-rutile phyllonites from the East Tenda Shear Zone (Alpine Corsica, France): pressure-temperature-time constraints to the Alpine reworking of Variscan Corsica, J. Geol. Soc., 169, 723–732, 2012.

Magott, R., Fabbri, O., and Fournier, M.: Polyphase ductile/brittle deformation along a major tectonic boundary in an ophiolitic nappe, Alpine Corsica: Insights on subduction zone intermediate-depth asperities, J. Struct. Geol., 87, 95–114, 2016.

Malatesta, C., Federico, L., Crispini, L., and Capponi, G.: Fluid-controlled deformation in blueschist-facies conditions: plastic vs brittle behaviour in a brecciated mylonite (Voltri Massif, Western Alps, Italy), Geol. Mag., https://doi.org/10.1017/S0016756816001163, 2017.

Malasoma, A., Marroni, M., Musumeci, G., and Pandolfi, L.: High pressure mineral assemblage in granitic rocks from continental units in Alpine Corsica, France, Geol. J., 41, 49–59, 2006

Mancktelow, N. S.: How ductile are ductile shear zones?, Geology, 34, 345–348, 2006

Mancktelow, N. S. and Pennacchioni, G.: The control of precursor brittle fracture and fluidrock interaction on the development of single and paired ductile shear zones, J. Struct. Geol., 4, 645–661, 2005.

Massonne, H. J. and Szpurka, Z.: Thermodynamic properties of white micas on the basis of high-pressure experiments in the systems $K_2OMgO-Al_2O_3-SiO_2-H_2O$ and $K_2O-FeO-Al_2O_3-SiO_2-H_2O$, Lithos, 41, 229–250, 1997.

Mattauer, M., Faure, M., and Malavieille, J.: Transverse lineation and large scale structures related to Alpine obduction in Corsica, J. Struct. Geol., 3, 401–409, 1981.

Mazzoli, S., Vitale, S., Del Monaco, G., Guerriero, G., Margottini, C., and Spizzichino, D.: "Diffuse faulting" in the Machu Pichu granitoid pluton, Eastern Cordillera, Perù, J. Struct. Geol., 31, 1395–1408, 2009.

Meneghini, F., Di Toro, G., Rowe, C. D., Moore, J. C., Tsutsumi, A., and Yamaguchi, A.: Record of mega-earthquakes in subduction thrusts: The black fault rocks of Pasagshak Point (Kodiak Island, Alaska), Geol. Soc. Am. Bull., 122, 1280–1297, 2010.

Menegon, L. and Pennacchioni, G.: Local shear zone pattern and bulk deformation in the Gran Paradiso metagranite (NW Italian Alps), Int. J. Earth Sci., 99, 1805–1825, 2010.

Menegon, L., Piazolo, S., and Pennacchioni, G.: The effect of Dauphiné twinning on plastic strain in quartz, Contrib. Mineral. Petr., 161, 635–652, 2011.

Menegon, L., Stünitz, H., Nasipuri, P., Heilbronner, R., and Svahnberg, H.: Transition from fracturing to viscous ow in granulite facies perthitic feldspar (Lofoten, Norway), J. Struct. Geol., 48, 95–112, 2013.

Molli, G.: Localizzazione di zone di taglio HP/LT su precursori fragili: Un esempio dalla Corsica Alpina, Rend. Soc. Geol. It., 4, 270–271, 2007.

Molli, G.: Northern Apennine-Corsica orogenic system: an updated review, in: Tectonic Aspects of the Alpine-Dinaride-Carpathian System, edited by: Siegesmund, S., Fugenschuh, B., and Froitzheim, N., Geol. Soc., 298, 413–442, 2008.

Molli, G. and Malavieille, J.: Orogenic processes and the Alps/Apennines geodynamic evolution: insights from Taiwan, Int. J. Earth Sci., 100, 1207–1224, https://doi.org/10.1007/s00531-010-0598-y, 2011.

Molli, G., Malasoma, A., and Meneghini, F.: Brittle precursors of HP/LT microscale shear zone: a case study from Alpine Corsica, 15th Conference of Deformation, Rheology and Tectonics, Zurich, Abstract Volume, 153 pp., 2005.

Molli, G., White, J. C., Kennedy, L., and Taini, V.: Low-temperature deformation of limestone, Isola Palmaria, Northern Apennine, Italy – The role of primary textures, precursory veins and intracrystalline deformation in localization, J. Struct. Geol., 33, 255–270, https://doi.org/10.1016/j.jsg.2010.11.015, 2011.

Montési, L. G. J. and Hirth, G.: Grain size evolution and the rheology of ductile shear zones: From laboratory experiments to postseismic creep, Earth Planet. Sc. Lett., 211, 97–110, https://doi.org/10.1016/S0012-821X(03)00196-1, 2003.

Neumann, B.: Texture development of recrystallised quartz polycrystals unravelled by orientation and misorientation characteristics, J. Struct. Geol., 22, 1695–1711, 2000.

Ozawa, S., Murakami, M., Kaidzu, M., Tada, T., Sagiya, T., Hatanaka, Y., Yarai, H., and Nishimura, T.: Detection and monitoring of ongoing aseismic slip in the Tokai region, central Japan, Science, 298, 1009–1012, 2002.

Passchier, C. W. and Trouw, R. A. J.: Microtectonics, Springer, 366 pp., 2011.

Pec, M., Stünitz, H., and Heilbronner, R.: Semi-brittle deformation of granitoid gouges in shear experiments at elevated pressures and temperatures, J. Struct. Geol., 38, 200–221, 2012.

Peng, Z. and Gomberg, J.: An integrated perspective of the continuum between earthquakes and slow slip phenomena, Nat. Geosci., 3, 599–607, 2010.

Pennacchioni, G. and Cesare, B. Ductile-brittle transition in pre-Alpine amphibolite facies mylonites during evolution from water-present to water-deficient conditions (Mont Mary, Italian Western Alps), J. Metamor. Geol., 15, 777–791, 1997.

Pennacchioni, G.: Control of the geometry of precursor brittle structures on the type of ductile shear zone in the Adamello tonalites, Southern Alps (Italy), J. Struct. Geol., 27, 627–644, 2005.

Pennacchioni, G. and Mancktelow, N. S.: Nucleation and initial growth of a shear zone network within compositionally and structurally heterogeneous granitoids under amphibolite facies conditions, J. Struct. Geol., 29, 1757–1780, 2007.

Pennacchioni, G. and Zucchi, E.: High temperature fracturing and ductile deformation during cooling of a pluton: The Lake Edison granodiorite (Sierra Nevada batholith, California), J. Struct. Geol., 50, 54–81, 2013.

Pennacchioni, G., Di Toro, G., Brack, P., Menegon, L., and Villa, I. M.: Brittle–ductile–brittle deformation during cooling of tonalite (Adamello, Southern Italian Alps), Tectonophysics, 427, 171–197, 2006.

Pittarello, L., Pennacchioni, G., and Di Toro, G.: Amphibolite-facies pseudotachylytes in Premosello metagabbro and felsic mylonites (Ivrea Zone, Italy), Tectonophysics, 580, 43–57, 2012.

Platt, J. P. and Behr, W. M.: Lithospheric shear zones as constant stress experiments, Geology, 39, 127–130, https://doi.org/10.1130/G31561.1, 2011.

Polino, R., Dal Piaz, G. V., and Gosso, G.: Tectonic erosion at the Adria margin and accretionary process for the Cretaceous orogeny of the Alps, in: Deep Structure of the Alps, edited by: Roure, F., Heitzman, P., and Polino, R., Volume Speciale Società Geologica Italiana, Roma, 1, 345–367, 1990.

Prior, D. J., Wheeler, J., Peruzzo, L., Spiess, R., and Storey, G.: Some garnet microstructures: an illustration of the potential of orientation maps and misorientation analysis in microstructural studies, J. Struct. Geol., 24, 999–1011, 2002.

Rogers, G. and Dragert, H.: Episodic tremor and slip on the Cascadia subduction zone: The chatter of silent slip, Science, 300, 1942–1943, 2003.

Ramsay, J. G. and Graham, R. H.: Strain variation in shear belts, Can. J. Earth Sci., 7, 786–813, 1970.

Rosenberg, C. and Kissling, E.: Three-dimensional insight into Central-Alpine collision: Lower-plate or upper-plate indentation?, Geology, 41, 1219–1222, https://doi.org/10.1130/G34584.1, 2013.

Rossi, P., Durand-Delga, M., Lahondére, J. C., and Lahondére, D.: Carte géologique de la France à 1/50 000, feuille Santo Pietro di Tenda, BRGM, 2003.

Rutter, E. H: On the nomenclature of mode of failure transitions in rocks, Tectonophysics, 122, 381–387, 1986.

Schiffman, P. and Day, H. W.: Petrologic methods for the study of Very Low-grade metabasites in Low-Grade metamorphism, edited by: Frey, M. and Robinson, D., Blackwell Publishing Ltd., Oxford, UK, Chap. 4, 108–144, https://doi.org/10.1002/9781444313345.ch4, 2009.

Schmid, S. M. and Handy, M. R.: Towards a genetic classification of fault rocks: Geological usage and tectonophysical implications, in: Controversies in modern geology, edited by: Muller, D. W., Mc Kenzie, J. A., and Weissert, H., London, Academic Press, 339–361, 1991.

Schmid, S. M., Pfiffer, O. A., Froitzheim, N., Schonborn, G., and Kissling, E.: Geophysical-geological transect and tectonic evolution of the Swiss-Italian Alps, Tectonics, 12, 1036–1064, 1996.

Scholz, C. H.: The brittle-plastic transition and the depth of seismic faulting, Geol. Rundsch., 77, 319–328, 1988.

Scholz, C. H.: The Mechanics of Earthquakes and Faulting, 2nd Edn., Cambridge University Press, 2002.

Segall, P. and Simpson, C.: Nucleation of ductile shear zones on dilatant fractures, Geology, 14, 56–59, 1986.

Shelly, D. R. G., Beroza, G. C., Ide, S., and Nakamura, S.: Low-frequency earthquakes in Shikouku, Japan, and their relationship to episodic tremor and slip. Nature, 442, 188–191, 2006.

Shimamoto, T.: The origin of S-C mylonites and a new fault-zone model, J. Struct. Geol., 11, 51–64, 1989.

Shimamoto, T. and Logan, J. M.: Velocity-dependent behaviours of simulated halite shear zones: an analog for silicates, Am. Geophys. Mongr., 37, 49–63, 1986.

Sibson, R. H.: Fault rocks and fault mechanisms, J. Geol. Soc. Lond., 133, 191–213, 1977.

Sibson, R. H.: Transient discontinuities in ductile shear zones, J. Struct. Geol., 2, 165–171, 1980.

Sibson, R. H.: Continental fault structure and the shallow earthquake source, J. Geol. Soc. Lond., 140, 741–767, 1983.

Sibson, R. H.: A note on fault reactivation, J. Struct. Geol., 7, 751–754, 1985.

Simpson, C.: Deformation of granitic rocks across the brittle-ductile transition, J. Struct. Geol., 7, 503–511, 1985.

Simpson, C.: Fabric development in brittle-to-ductile shear zones, Pure Appl. Geophys., 124, 269–288, 1986.

Stipp, M. and Tullis, J.: The recrystallized grain size piezometer for quartz, Geophys. Res. Lett., 30, 2088, https://doi.org/10.1029/2003GL018444, 2003.

Stipp, M., Stünitz, H., Heilbronner, R., and Schmid, S. M.: The eastern Tonale fault zone: a "natural laboratory" for crystal plastic deformation of quartz over a temperature range from 250 to 700 °C, J. Struct. Geol., 24, 1861–1884, 2002.

Stöckhert, B.: Stress and deformation in subduction zones – insight from the record of exhumed metamorphic rocks, in: Geological Society of London Special Publication, edited by: De Meer, S., Drury, M. R., De Bresser, J. H. P., and Pennock, G. M., Geological Society of London Special Publication, London, 200, 255–274, 2002.

Takagi, H., Goto, K., and Shigematsu, N.: Ultramylonite bands derived from cataclasite and pseudotachylyte in granites, northeast Japan, J. Struct. Geol., 22, 1325–1339, 2000.

Tödheide, K.: Water at high temperatures and pressures, in: Water: A Comprehensive Treatise, edited by: Franks, F., Springer, New York, vol. 1, chap. 13, 463–514, 1972.

Trepmann, C. A., Stockhert, B., Dorner, D., Moghadam, E. H.,

Kuster, M., and Roller, K.: Simulating coseismic deformation of quartz in the middle crust and fabric evolution during postseismic stress relaxation – an experimental study, Tectonophysics, 442, 83–104, 2007.

Trepmann, C. A., Hsu, C., Hentschel, F., Döhler, K., Schneider, C., and Wichmann, V.: Recrystallization of quartz after low-temperature plasticity – The record of stress relaxation below the seismogenic zone, J. Struct. Geol., 95, 77–92, 2017.

Tribuzio, R. and Giacomini, F., Blueschist facies metamorphism of peralkaline rhyolites from the Tenda crystalline massif (northern Corsica): evidence for involvement in the Alpine subduction event?, J. Metamorph. Geol., 20, 513–526, 2002.

Van Daalen, M., Heilbronner, R., and Kunze, K.: Orientation analysis of localized shear deformation in quartz fibres at the brittle–ductile transition, Tectonophysics, 303, 83–107, 1999.

Vannucchi, P., Remitti, F., and Bettelli, G.: Geological record of fluid flow and seismogenesis along an erosive subducting plate boundary, Nature, 451, 7179, 699–703, 2008.

Vernooij, M. G., den Brok, B., and Kunze, K.: Development of crystallographic preferred orientations by nucleation and growth of new grains in experimentally deformed quartz single crystals, Tectonophysics, 427, 35–53, 2006.

Viola, G., Mancktelow, N. S., and Miller, J. A.: Cyclic frictional–viscous slip oscillations along the base of an advancing nappe complex: insights into brittle-ductile nappe emplacement mechanisms from the Naukluft Nappe Complex, central Namibia, Tectonics, 25, TC3016, https://doi.org/10.1029/2005tc001939, 2006.

Vitale Brovarone, A., Beyssac, Ol., Malavieille, J., Molli, G., Beltrando, M., and Compagnoni, R.: Stacking and metamorphism of continuous segments of subducted lithosphere in a high-pressure wedge: The example of Alpine Corsica (France), Earth-Sci. Rev., 116, 35–56, 2013.

Wallace, L. M. and Beavan, J.: Diverse slow slip behavior at the Hikurangi subduction margin, New Zealand, J. Geophys. Res., 115, B12402, https://doi.org/10.1029/2010JB007717, 2010.

White, S. H.: Natural creep of quartzites, Nat. Phys. Sci., 234, 175–177, 1971.

White, S. H.: Syntectonic recrystallisation and texture development in quartz, Nature, 244, 276–278, 1973.

White, J. C.: Transient discontinuities revisited: Pseudotachylyte, plastic instability and the influence of low pore fluid pressure on deformation processes in the mid-crust, J. Struct. Geol., 18, 1471–1477, 1996.

White, J. C.: Paradoxical pseudotachylyte – Fault melt outside the seismogenic zone, J. Struct. Geol., 38, 11–20, 2012.

White, J. C. and White, S. H.: Semi-brittle deformation within the Alpine fault zone, New Zealand, J. Struct. Geol., 5, 579–589, 1982

Zibra, I., Kruhl, J. H., and Braga, R.: Late Palaeozoic deformation of post-Variscan lower crust: shear zone widening due to strain localization during retrograde shearing, Int. J. Earth Sci., 99, 973–991, 2010.

3

Seismic anisotropy inferred from direct *S*-wave-derived splitting measurements and its geodynamic implications beneath southeastern Tibetan Plateau

Ashwani Kant Tiwari[1], **Arun Singh**[1], **Tuna Eken**[2], **and Chandrani Singh**[1]

[1]Department of Geology and Geophysics, Indian Institute of Technology Kharagpur, Kharagpur, India
[2]Department of Geophysical Engineering, Istanbul Technical University, Istanbul, Turkey

Correspondence to: Arun Singh (arun@gg.iitkgp.ernet.in)

Abstract. The present study deals with detecting seismic anisotropy parameters beneath southeastern Tibet near Namcha Barwa Mountain using the splitting of direct *S* waves. We employ the reference station technique to remove the effects of source-side anisotropy. Seismic anisotropy parameters, splitting time delays, and fast polarization directions are estimated through analyses of a total of 501 splitting measurements obtained from direct *S* waves from 25 earthquakes (≥ 5.5 magnitude) that were recorded at 42 stations of the Namcha Barwa seismic network. We observe a large variation in time delays ranging from 0.64 to 1.68 s, but in most cases, it is more than 1 s, which suggests a highly anisotropic lithospheric mantle in the region. A comparison between direct *S*- and SKS-derived splitting parameters shows a close similarity, although some discrepancies exist where null or negligible anisotropy has been reported earlier using SKS. The seismic stations with hitherto null or negligible anisotropy are now supplemented with new measurements with clear anisotropic signatures. Our analyses indicate a sharp change in lateral variations of fast polarization directions (FPDs) from consistent SSW–ENE or W–E to NW–SE direction at the southeastern edge of Tibet. Comparison of the FPDs with Global Positioning System (GPS) measurements, absolute plate motion (APM) directions, and surface geological features indicates that the observed anisotropy and hence inferred deformation patterns are not only due to asthenospheric dynamics but are a combination of lithospheric deformation and sub-lithospheric (asthenospheric) mantle dynamics. Direct *S*-wave-based station-averaged splitting measurements with increased back-azimuths tend to fill the coverage gaps left in SKS measurements.

1 Introduction

The Tibetan Plateau has a long history of deformation within the last 50 million years (e.g. Rowley and Currie, 2006; Henderson et al., 2011). The reliability of seismic anisotropy measurements is a challenging issue as it is essential to identify the tectonics, coupling–decoupling of the crust-lithospheric mantle, multi-layered anisotropic modelling, and active seismicity in relation to the type of deformation and possible flow patterns, which are still a matter of debate in understanding the formation process and future challenges of this active region.

Lattice-preferred orientation (LPO) of olivine mineral in the mantle as a result of plate interactions is controlled by various geodynamic processes and is considered to be the main cause of the shear wave splitting observations on the teleseismic *S* and SKS waves. Deformation in the upper mantle generally takes place through two processes: diffusion and dislocation creep under favourable conditions. The dislocation creep process, which is the creeping motion of crystal dislocation, is considered to be the leading cause of mantle anisotropy (Karato, 1987; Nicolas and Christensen, 1987; Karato and Wu, 1993; Mainprice et al., 2000). It can be caused by either high-stress conditions or large grain size or both, but the nonlinear increase in the strain rate is independent of the grain size (Karato and Wu, 1993). This type of deformation is expected to occur at a depth range of less than 400 km (e.g. Karato, 1984, 1987) where olivine is the most common mineral, and hence LPO development and observed anisotropy mainly represents the upper 400 km of the mantle (Becker and Faccenna, 2011).

Several observations on seismic anisotropy have greatly contributed to elucidating these deformation patterns in relation to the past and present geodynamic activity of the region. Generally speaking, SKS splitting analyses are the most diagnostic, quick, and well-established way of detection and quantification of seismic anisotropy. The SKS phase does not propagate as an S wave in the liquid outer core and refracts from a P wave into an SV (radially polarized) wave when entering the receiver-side mantle. Hence a recorded SKS phase at the surface is not influenced by the source-side anisotropy. The main disadvantage of using the SKS phase in splitting measurements is that finding good-quality observations is restricted by several parameters, i.e. epicentral distance and propagation direction of the event, and therefore the measurements need to be supplemented with other phases (e.g. ScS and direct S) that can provide better azimuthal coverage. However, employing such additional phases may introduce contamination due to the source-side anisotropy. Splitting of shear waves is similar to the birefringence phenomena in optics. Shear waves split into fast and slow components when they pass through an anisotropic medium. In such a situation, we obtain particle motion (e.g. elliptical, cruciform) with different shapes that depend on the anisotropy along the ray path. If the anisotropy is the only cause of splitting, then the observed shear wave (fast or slow) can be rotated in such a way that two very similar phases are seen, apart from scaling and a time delay between them (Silver and Chan, 1991; Savage, 1999; Long and Silver, 2009). Resultant splitting parameters, ϕ and δt, indicate the rotation angle in relation to the flow direction and shearing or extension along the ray path under the assumption of LPO in the upper mantle (a.k.a. fast polarization direction or FPD), and to the strength and thickness of the anisotropic layer (a.k.a. delay time), respectively. Splitting measurements from the Himalaya–Tibet collision zone have long been explained by the presence of a single homogeneous layer with a horizontal axis of symmetry (e.g. McNamara et al., 1994; Chen et al., 2010; Sol et al., 2007; Herquel et al., 1995; Hirn et al., 1995; Lavé et al., 1996; Sandvol et al., 1997; Huang et al., 2000; Lev et al., 2006; Wang et al., 2008; Fu et al., 2008; Sato et al., 2012).

The use of direct S waves of earthquakes at teleseismic distances (30–90°) can provide complementary splitting measurements to SKS measurements as this helps in establishing a more complete database of anisotropy that will be inferred from good-quality S wave signals from an enhanced azimuthal distribution relative to SKS splitting only. This is crucial for the Indian subcontinent where SKS measurements are skewed towards eastern azimuths, and very few SKS measurements have been obtained due to temporary deployments (see Singh et al., 2015). However, the major problem in including direct S waves in splitting measurements is the contamination of the S wave signals due to the influence of anisotropic structures existing within the source-side region. Eken and Tilmann (2014) have recently shown that this problem can be overcome using an array-based ap-

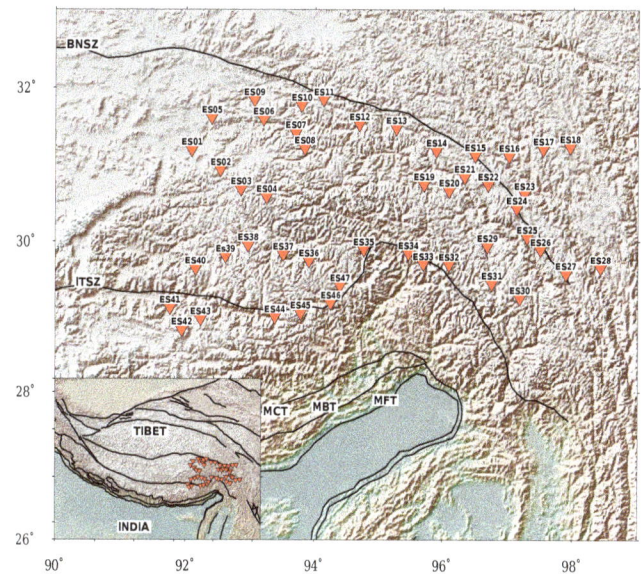

Figure 1. Tectonic and topographic map of the Himalayas and Tibet. Red triangles represent the broadband seismic stations of the XE network within the study region (MFT: Main Frontal Thrust; MBT: Main Boundary Thrust; MCT: Main Central Thrust; ITSZ: Indus–Tsangpo suture zone; BNSZ: Bangong–Nujiang suture zone).

proach, known as the reference station technique (RST). The method assumes an identical source-side anisotropy effect at two closely located stations (reference and target stations) with small differences in epicentral distances. In this case, optimum splitting parameters can be estimated by searching for receiver-side correction parameters for the target station that result in maximum similarity to the S wave signal corrected for previously known receiver-side anisotropy beneath the reference station in a grid search scheme. Signals used for that comparison are those of the reference station previously corrected for known reference receiver-side anisotropy and of the target station whose receiver-side splitting parameters are desired to be estimated. In this technique, we utilize seismic anisotropy parameters, which were previously inferred from the SKS measurements by Sol et al. (2007) as the reference knowledge of the receiver-side anisotropy beneath the reference station. The RST has been successfully tested through both synthetic and observed data collected along the northeastern and southwestern parts of the Tibetan Plateau and the Hellenic Trench in the eastern Mediterranean (e.g. Eken and Tilmann, 2014; Singh et al., 2016; Confal et al., 2016). The present study focuses on the southeastern part of Tibet near Namcha Barwa (Fig. 1). The study region is located between and around the Indus–Tsangpo suture zone (ITSZ) and Bangong–Nuijiang suture zone (BNSZ). Our major motivation is to calculate S-wave-derived seismic anisotropic parameters that may have a potential link to tectonic setting and deformation history with the help of a

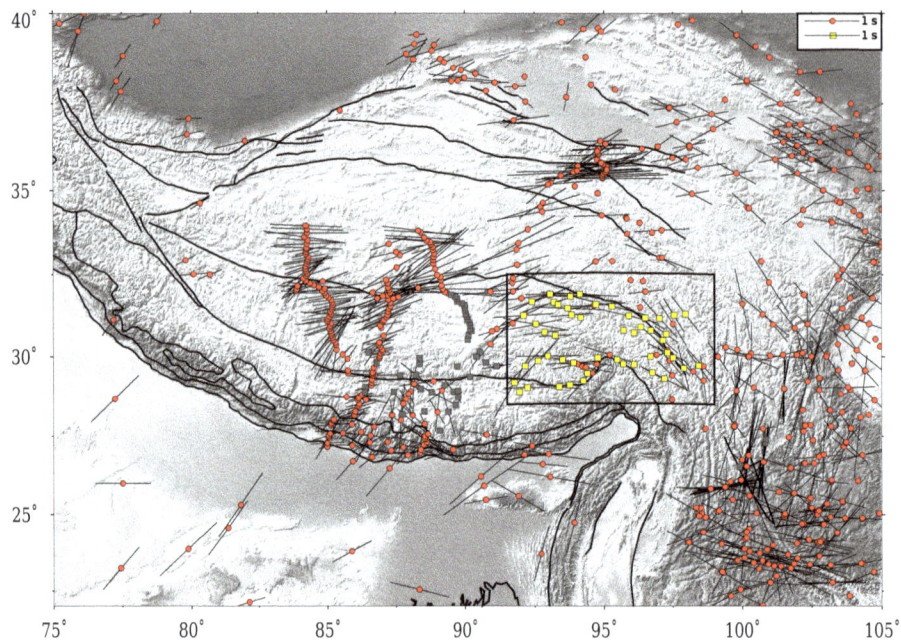

Figure 2. Earlier SKS and SKKS measurements in the study area (Sol et al., 2007; Wüstefeld et al., 2008). The length of solid bars shown for each seismic station is proportional to the splitting time delay (δt_S), and their orientation represents the fast polarization direction (ϕ_S). For clarity, the seismic stations used in the present study are shown by yellow-filled rectangles along with the SKS splitting measurements of Sol et al. (2007). Seismic stations where null or negligible anisotropy is reported in earlier studies (see Wüstefeld et al., 2008) are shown by grey-filled rectangles. All other stations are shown by red-filled circles.

correlative analysis of resultant anisotropy observations with absolute plate motion (APM) directions, GPS measurements, and the structural and topographic features. Our results contradict previous interpretations of an isotropic Indian lithospheric mantle (Chen and Ozalaybey, 1998; Barruol and Hoffmann, 1999; Chen et al., 2010) and add new constraints in understanding the types of deformation and their causes in the region.

2   Tectonics of the region

The formation of the Tibetan Plateau and the Himalayan mountain belt is due to collision and post-collision processes of the Indian and Eurasian plates starting at around 50 Ma (Argand, 1924; Garzanti and Van Haver, 1988; Molnar and Tapponnier, 1975; Yin and Harrison, 2000; Royden et al., 2008). Underthrusting of the Indian lithosphere beneath the Eurasian lithosphere has been proposed to be the main reason for the formation of the Himalayan and Karakorum ranges (Nelson et al., 1996; Kumar et al., 2006; Tseng et al., 2009) along with the formation of the central Tibetan region (Argand, 1924; Nelson et al., 1996; Li et al., 2008). The underlying reason for the development of the northern and eastern Tibetan Plateau, however, remains enigmatic (Karplus et al., 2011; Royden et al., 2008). McKenzie and Priestley (2008) discuss the development of the northern Tibetan lithosphere as an accreted one. Royden et al. (2008) argue that the Ti-

betan Plateau evolved due to the subduction of the Indian lithosphere beneath Eurasia, which is also responsible for the thickening of the Tibetan crust and afterwards the extrusion of the Tibetan lithosphere towards the east.

Various models have been developed regarding the deformation of Tibet (Royden et al., 1997; Molnar and Tapponnier, 1975; Houseman and England, 1986, 1993, 1996; Tapponnier et al., 1982, 2001; Shen et al., 2001; Holt et al., 1995, 2000; Replumaz and Tapponnier, 2003; Flesch et al., 2001), but no single model can explain all of it. The debate regarding the crust and mantle deformation patterns and ongoing geodynamics has not been settled. Recent work by Jagoutz et al. (2015) and Van Hinsbergen et al. (2012) suggests a model involving multistage subduction of the Tethys oceanic plate and the Indian Plate below the Eurasian Plate resulting in a highly heterogeneous and anisotropic lithosphere. The Tibetan and Himalayan region is mainly dominated by thrust and strike–slip faulting. Suture zones are extended in the E–W direction and take a sharp turn around the eastern Himalayan syntaxis (EHS; Fig. 1). Strike–slip faulting becomes more dominant to the east of the EHS. Figure 1 shows that the eastern portion of the subducting Indian Plate is found adjacent to the EHS (León Soto et al., 2012) where the structural and topographical features take a sharp trend from nearly W–E striking to N–S striking.

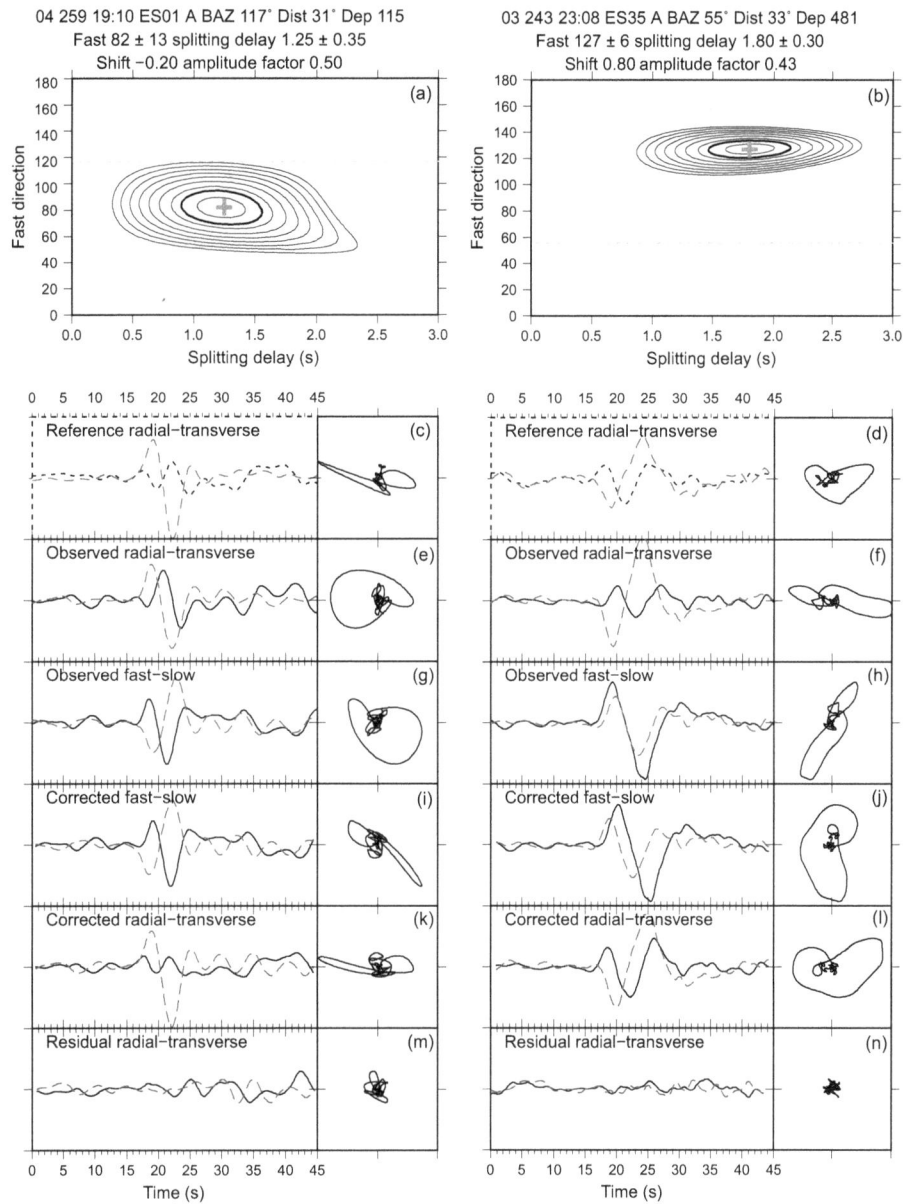

Figure 3. Examples of the direct S wave splitting measurements based on the reference station technique at stations ES01 and ES35. Figures on the left side represent the splitting measurement recorded at station pair ES03 (reference station)–ES01 (target station), and those on the right side represent the splitting measurement recorded at station pair ES16 (reference station)–ES35 (target station). **(a)** Misfit surface with splitting parameter $82° \pm 13°$ and 1.25 ± 0.35 s. **(c)** Signal at reference station (ES03) with receiver-side correction. **(e)** Signal at target station (ES01). **(g)** Fast and slow component after rotating signal at target station (ES01) using ϕ (82°). **(i)** Fast and slow component corrected for δt (1.25 s). **(k)** Corrected radial and transverse components at target station (ES01) using optimum ϕ and δt and isotropic delay (-0.2 s). **(m)** Residual trace. Panels on the right side follow the same order and explanation.

3   Data and method

In this study, we examine a total of 5285 waveforms with the direct S waves extracted from 161 teleseismic events with magnitudes ≥ 5.5 within an epicentral distance range from 30 to 90°. The teleseismic events used in this study are recorded at 47 seismic stations of the XE network, which operated between 2003 and 2004 (Sol et al., 2007). Of those

47 stations, we selected for use as reference stations only those 36 seismic stations where we have knowledge of seismic anisotropy inferred from SKS splitting measurements performed by Sol et al. (2007). Prior to the data analysis, we remove the instrument response from the original seismograms to overcome biases that can depend on the potential use of different stations (at reference and target sites). At the stage of the preprocessing, a band-pass filter between

0.03 and 0.2 Hz is performed to enhance S signals and re-sample the seismograms at 20 samples per second to avoid aliased signals and to reconstruct the waveforms in the appropriate frequency range. Signals with possible contamination with other phases, such as ScS, SKKS, and SKS, are omitted from the analysis. We select only those waveforms, which have ≥ 2.5 signal-to-noise ratio (SNR) on the transverse and radial components for further analysis. The selection of the waveforms is achieved by performing a manual visual inspection that allowed only 40 % of the direct S waveforms. We begin data analysis by determining station pairs over the entire area. We form the station pairs by selecting the same earthquake event recorded at both (reference and target) stations. Eken and Tilmann (2014) and Singh et al. (2016) successfully applied the RST to regional arrays with an interstation distance less than 300 km. By taking 300 km interstation spacing as the limit in a similar fashion, we have formed 22 816 station pairs with four horizontal components available at reference and target stations out of 35 649 possible station pairs; these are based on 161 teleseismic events prior to the application of the technique. To minimize the effects of the coda waves and converted phases, we use a 45 s time window starting 15 s before the theoretical onset of the direct S waves on the basis of the IASP91 1-D radial earth velocity model of Kennett (1991). This excludes the undesired effect of crustal S multiples in the thick Tibetan crust.

The approach used in the present study avoids the source-side anisotropy by minimizing the misfit function between the corrected seismic waveforms at the reference and target stations. At the first stage, an inverse splitting operator depending on a backward angular rotation with two horizontal components, a time shift, and the reversal of the back angular rotation are employed to correct the reference station for known receiver-side anisotropy (generally inferred from SKS splitting analyses) when estimating the direct-S-derived individual splitting parameter (Eken and Tilmann, 2014). Following the correction of the reference station, S signals are corrected for splitting parameters in a grid search manner at the target stations. Corrected S wave signals at reference and target stations are compared to each other for each pair of splitting parameters. Such comparison also allows for time shifts and amplitude corrections to account for the lateral heterogeneities and differences in site response between these stations by optimizing the time shift (Δt) and amplitude factor (a). First, we assign splitting parameters that minimized the misfit function simply representing the difference between the corrected reference and target station traces as optimum splitting parameters for the receiver-side beneath the target station at a given station pair. Later, taking the average of all optimum splitting parameters estimated at station pairs related to a given target station are considered representative of a given event. In the end, station-pair averaged splitting parameters are averaged over all events to estimate the final splitting parameters at each given target station.

Here we should note that our approach initially depends on the knowledge of receiver-side seismic anisotropy that can be most likely inferred from directionally averaged SKS splitting parameters at a given reference station. Conventional SKS splitting measurements are performed under the assumption of a single-layer anisotropic structure with a horizontal axis of symmetry. However, beneath the regions with complex anisotropic structures, for instance, in the case of a well-developed continental lithosphere with a dipping axis of symmetry (i.e. stable cratonic regions, see Plomerová et al., 2008) or an existing double-layer anisotropy (Silver and Savage, 1994), significant directional variation of apparent splitting parameters will likely be expected. In such regions, the average value of splitting parameters as reference knowledge of seismic anisotropy cannot be representative of the events from different directions, thus making resultant average S-derived splitting parameters misleading in our method since complicated anisotropic structures likely introduce a similar influence on both the SKS phase and direct S waves. However due to the fact that our approach certainly provides more splitting observations from an increased amount of back-azimuths, these new directionally enhanced apparent S-derived splitting parameters help in resolving the actual orientation of the anisotropic structure by using more sophisticated modelling strategies, which is not within the scope of the present work.

The RST relies on two important underlying assumptions: (i) the ray path at two stations can be considered equivalent in the deeper mantle part and near the source-side region due to the fact that the distance between receiver and target stations is small (< 300 km) compared to the epicentral distance; (ii) waveform differences between the receiver and target stations are only due to differences in anisotropic structure after correcting any waveform differences in time and amplitude presumably due to the lateral heterogeneities and differences in site response between these stations. Any potential difference between the thickness of the crust and sedimentary layers will also cause the timing and amplitude of converted phase, but Eken and Tilmann (2014) showed, numerically, its influence on expected splitting parameters would be negligible. During the application of the technique, we let ϕ and δt vary from 0 to $180°$ with an increment of $1°$ and from 0 to 3 s with an increment of 0.05 s, respectively. We perform an inverse F test error analysis for uncertainty estimates of obtained splitting parameters. In this process, we check the reliability of the individual splitting parameters by comparing variation in the residual energy distribution away from the minimum with the variation according to the preset confidence level of 95 %. At this stage, the number of degrees of freedom in the data and unknown model parameters becomes crucial. According to Silver and Chan (1991) 1 degree of freedom is set to 1 s and considering two horizontal components, the number of degrees of freedom becomes 2 times the data length, which could be considered a typical value for teleseismic data. In our case, however, the number

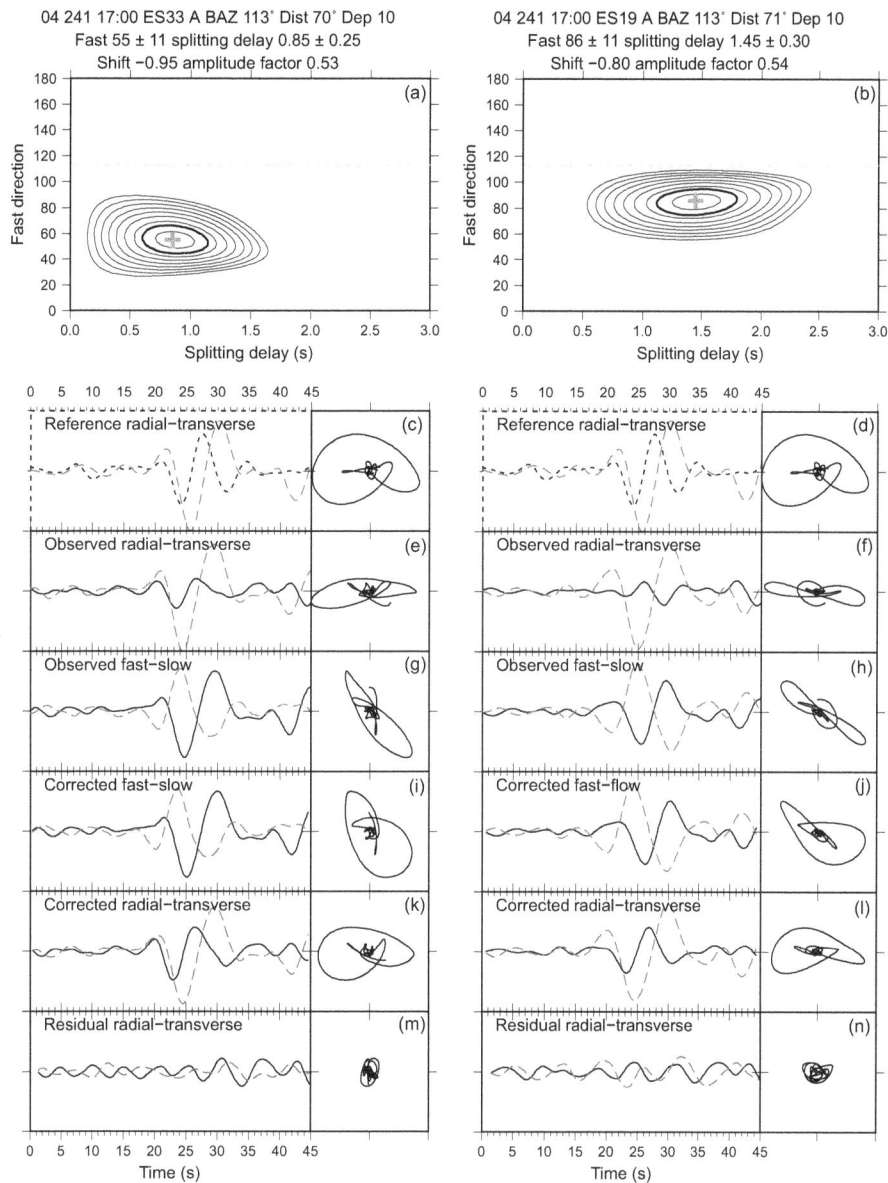

Figure 4. Examples of the direct *S* wave splitting measurements based on the reference station technique at stations ES33 and ES19, where previously null anisotropy was obtained using an SKS splitting measurement (Sol et al., 2007). Panels on the left side represent the splitting measurement observed at the station pair with ES12 (reference station) and ES33 (target station), and those on the right side represent the splitting measurement observed at the station pair with ES12 (reference station) and ES19 (target station). The explanation for each panel is the same as in Fig. 3.

of unknown model parameters is four at the minimum point (ϕ and δt, isotropic delay, amplitude correction factor) or two at any given splitting parameters tried in the grid search, reducing the number of degrees of freedom by four or two, respectively. Estimating the number of degrees of freedom is a challenging task. For an appropriate uncertainty analysis, the assumption of band-limited Gaussian noise is required to be justified as reported by Walsh et al. (2013). Thus, taking a fixed value for the degrees of freedom as performed in this study will allow us to compare the reliability of different individual splitting estimates rather than the absolute value of the error bounds.

An example of the basic steps of the RST can be found in Fig. 3 for target stations ES01 and ES35, respectively. Figure 4 present the examples of the obtained splitting parameters at target stations ES19 and ES33, where null or no measurements are reported by Sol et al. (2007). Null splitting may be observed for three reasons: (i) if the incoming polarization direction below an anisotropic layer is parallel to the fast or slow axis; (ii) if the region is isotropic in nature due

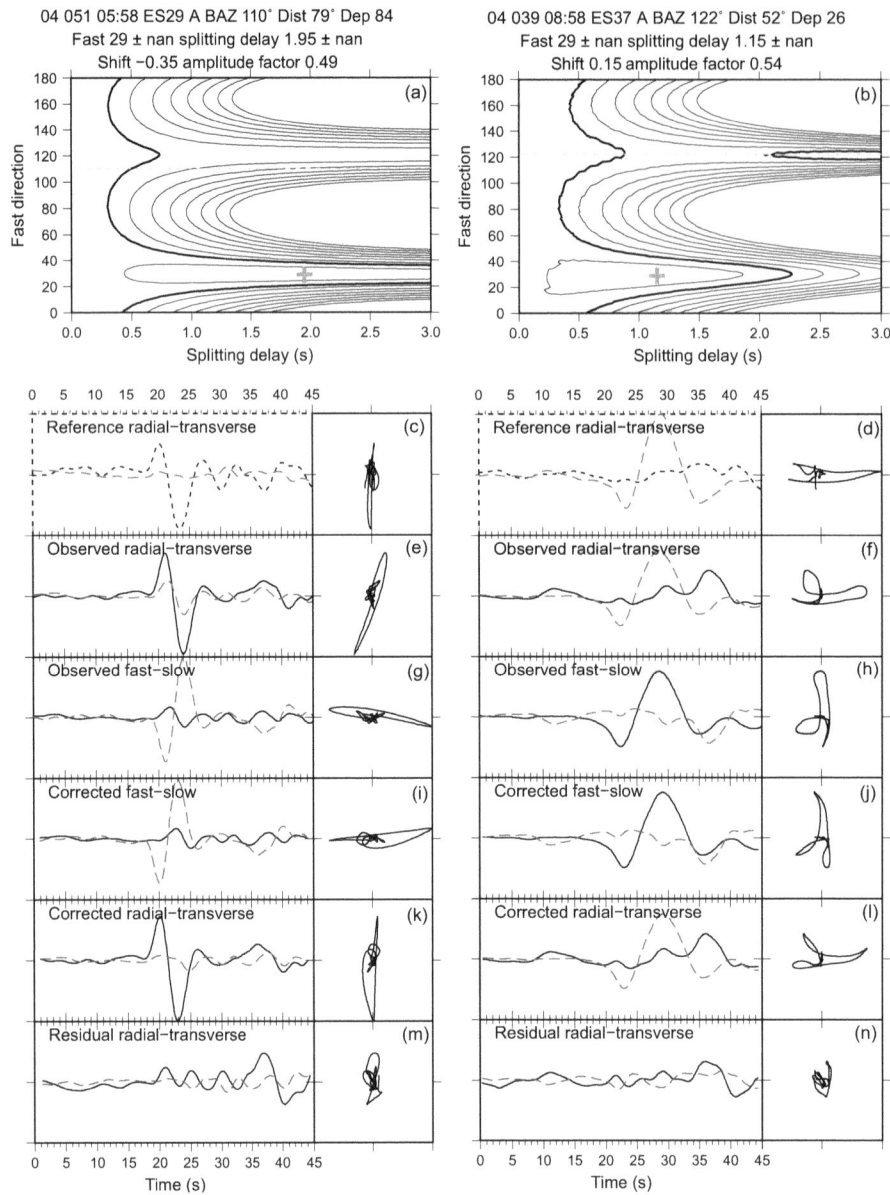

04 051 05:58 ES29 A BAZ 110° Dist 79° Dep 84
Fast 29 ± nan splitting delay 1.95 ± nan
Shift −0.35 amplitude factor 0.49

04 039 08:58 ES37 A BAZ 122° Dist 52° Dep 26
Fast 29 ± nan splitting delay 1.15 ± nan
Shift 0.15 amplitude factor 0.54

Figure 5. An example of null-anisotropy measurement based on the RST at stations ES29 and ES37. Panels on the left side represent the splitting measurement recorded at station pair ES27 (reference station)–ES29 (target station), and those on the right side represent the splitting measurement recorded at station pair ES05 (reference station)–ES37 (target station). The explanation for each panel is the same as in Fig. 3.

to complex anisotropy (e.g. Saltzer et al., 2000; Wüstefeld and Bokelmann, 2007); (iii) if the region itself is isotropic in nature. Following Eken and Tilmann (2014) and Singh et al. (2016), we use the F test with the null-split rejection criteria to be able to avoid the contamination of the null measurement with the good splitting measurements. In this process, we calculate the theoretical residual energy under the assumption of null measurements and compare this with the observed residual energy at the minimum. Figure 5 shows two examples of null-splitting measurements at target stations ES29 and ES37, respectively. To ensure the stability of the results, we

perform a stepwise quality assessment criterion before calculating the average splitting parameters at each station. To achieve this aim, we considered only those waveform pairs that have (i) a normalized residual energy (ΔE) smaller than 0.5; (ii) an amplitude correction factor parameter (a) in between 0.4 and 0.6; and (iii) a 95 % confidence level for null-splitting rejection. We rejected the waveform pairs that have a ϕ error greater than 25° and a delay time error greater than half of the delay time itself. After these quality assessments, we are left with only 3231 waveform pairs. At this stage of the processing, we perform another visual inspection to en-

Table 1. List of the earthquakes used in this study.

Event date	Event time	Latitude (°)	Longitude (°)	Depth (km)	Magnitude	Location site
2003/10/04	14:49:02.7	−07.05	+125.41	532.7	5.5	Banda Sea
2003/10/17	17:19:53.6	−05.08	+102.46	35.1	5.6	Southern Sumatra, Indonesia
2003/11/09	19:23:28.6	+01.56	+127.36	133.9	5.8	Halmahera, Indonesia
2004/02/08	08:58:51.8	−03.66	+135.34	25.7	5.7	Irian Jaya Region, Indonesia
2004/02/20	05:58:45.2	−11.61	+166.45	84.0	5.6	Santa Cruz Islands
2004/03/17	05:21:00.8	+34.59	+023.33	24.5	5.9	Crete, Greece
2004/03/26	15:20:06.6	+41.86	+144.21	22.4	5.7	Hokkaidō, Japan region
2004/04/09	15:23:35.0	−13.17	+167.20	228.4	5.8	Vanuatu
2004/05/28	12:38:44.3	+36.25	+051.62	17.0	6.2	Northern and central Iran
2004/06/22	09:04:43.9	−10.90	+166.26	152.8	5.8	Santa Cruz Islands
2004/06/30	23:37:25.5	+00.80	+124.73	90.8	6.0	Minahassa Peninsula, Sulawesi
2004/07/08	10:30:49.2	+47.20	+151.30	128.5	5.9	Kuril Islands
2004/07/25	14:35:19.1	−02.43	+103.98	582.1	6.8	Southern Sumatra, Indonesia
2003/07/27	06:25:32.0	+47.15	+139.25	470.3	6.3	Primor'ye, Russia
2004/08/02	02:36:54.9	−05.47	+102.62	40.5	5.5	Southern Sumatra, Indonesia
2004/08/07	14:18:35.2	−06.24	+095.67	20.7	5.8	Southwest of Sumatra, Indonesia
2004/08/28	17:00:58.2	−08.69	+157.25	10.0	5.5	Solomon Islands
2003/08/31	23:08:00.3	+43.39	+132.27	481.1	5.5	Primor'ye, Russia
2003/09/11	21:58:25.5	−08.20	+156.16	10.0	5.5	Solomon Islands
2004/09/15	19:10:50.6	+14.22	+120.41	115.4	6.0	Luzon, Philippines
2003/10/11	00:08:49.1	+41.92	+144.36	33.0	5.9	Hokkaidō, Japan region
2003/10/11	01:11:31.2	+43.97	+148.21	51.2	6.2	East of Kuril Islands
2003/10/17	10:19:06.8	−05.47	+154.15	133.0	6.2	Solomon Islands
2003/10/22	11:45:30.8	−06.06	+147.73	53.5	6.2	Eastern New Guinea region
2003/11/12	08:26:43.7	+33.17	+137.07	384.9	6.1	Near S. coast of Honshū

hance the quality of our estimates, yielding only 501 very high-quality waveform pairs. These final waveforms show clear splitting and are free of any distortions due to signal processing. The remaining 501 waveform pairs are extracted from only 25 teleseismic events (Fig. 6) and are used to calculate the average splitting parameter at each station. The list of these 25 teleseismic events is provided in Table 1. We apply the Von Mises approach (Cochran et al., 2003) to calculate the circular mean at each target stations for ϕ and an arithmetic mean is used for δt.

3.1 Results

We present here 501 splitting measurements observed for 42 seismic stations of the XE network between the years 2003 and 2004. The angular average of individual direct-S-derived splitting parameters (ϕ_s and δt_s) at each station is given in Table 2. Station-averaged splitting parameters usually reflect significant anisotropy with large delay times (>1 s, Figs. 7 and 8) compared to those that could be considered negligible based on previously determined SKS-derived anisotropy parameters (Sol et al., 2007). For example, at station ES31 direct S waves provide a relatively large time delay (1.23 s) although SKS splitting analysis performed by Sol et al. (2007) earlier resulted in a much smaller time delay time of about

0.3 s. Across the network, we observe considerable variation in direct S-wave-derived delay times ranging from 0.64 to 1.68 s. In general, we observe the SW–NE to W–E trend in ϕ_s before the edge margin of the southeastern Tibetan region. A consistent change in variation of ϕ_s is observed further east where orientations take a sharp change from a SSW–ENE or W–E to a NW–SE direction (Fig. 7). We find significant splitting (≥ 0.64 s) at seismic stations ES19, ES20, ES22, ES32, ES33, ES34, ES42, and ES45, where previously null or negligible splitting was observed in SKS waves (Sol et al., 2007). This could stem from multi-layered anisotropic orientations or insufficient amount of SKS-derived splitting measurements. This observation is inconsistent with an isotropic nature of the Indian lithosphere and indicates a complex 3-D nature or more complex deformation pattern of the EHS. The scatter plot in Fig. 8 exhibits a comparison between estimated splitting parameters (ϕ_s and δt_s) from the analysis of direct S waves and previous SKS splitting measurements (Sol et al., 2007). The overall trend of the obtained splitting parameters (ϕ_s) is consistent with the previous SKS measurement, whereas we observe larger time delays for the S waves as compared to SKS phase.

We combine our splitting measurements with existing geodetic measurements (GPS velocity vectors and APM velocity vectors). We plot the GPS velocities by using pub-

Table 2. Obtained average splitting parameters (ϕ_S and δt_S) estimated from direct S wave splitting measurement.

Station	Latitude (°)	Longitude (°)	ϕ_S (°)	δt_S (s)	Number of events at a station	Contributing reference stations
ES01	31.26	92.09	73.9	1.5	31	2, 3, 4, 5, 7, 8, 9, 10, 11, 12, 36, 38, 39, 40, 41
ES02	31.00	92.54	75.7	1.5	26	1, 3, 4, 5, 7, 8, 9, 10, 11, 12, 13, 38, 40, 41
ES03	30.75	92.86	81.7	1.2	16	2, 4, 5, 7, 8, 10, 11, 12, 38, 39, 43
ES04	30.65	93.25	91.5	1.2	11	1, 3, 5, 7, 8, 10, 11, 12
ES05	31.68	92.40	71.2	1.1	14	1, 2, 3, 4, 7, 8, 10, 11, 12, 13
ES07	31.48	93.70	93.8	1.0	15	1, 2, 3, 4, 5, 8, 9, 10, 11, 12, 14, 38
ES08	31.28	93.84	106.3	0.9	12	1, 2, 4, 5, 7, 10, 11, 12, 13
ES09	31.91	93.06	81.5	1.1	15	1, 3, 5, 8, 10, 11, 12, 13, 14
ES10	31.84	93.79	103.1	1.0	19	1, 2, 3, 4, 5, 7, 8, 9, 11, 12, 14, 38
ES11	31.91	94.14	96.2	1.2	25	1, 2, 3, 4, 5, 7, 8, 9, 10, 12, 14, 36, 38
ES12	31.59	94.71	101.3	1.2	28	1, 2, 3, 4, 5, 7, 8, 9, 10, 11, 13, 14, 15, 23, 25, 38
ES13	31.54	95.28	88.5	1.4	15	2, 4, 9, 11, 12, 13, 14, 15, 26, 30, 36, 38
ES14	31.25	95.90	102.9	1.2	24	3, 4, 7, 9, 10, 11, 12, 13, 15, 17, 23, 25, 31
ES15	31.19	96.50	104.2	1.0	09	11, 12, 14, 16, 25, 36, 38
ES16	31.18	97.02	117.0	1.4	01	13
ES17	31.27	97.55	135.4	1.0	11	18, 23, 25, 31
ES18	31.30	97.96	122.0	0.8	02	17
ES19	30.81	95.71	106.6	1.5	09	3, 4, 7, 8, 11, 12, 38
ES20	30.73	96.10	108.0	0.9	01	11
ES22	30.81	96.70	110.0	0.9	01	14
ES23	30.69	97.26	96.2	1.0	16	12, 15, 25, 31
ES24	30.50	97.14	106.0	1.1	03	13, 15
ES25	30.12	97.30	121.1	0.9	10	18, 23, 31
ES26	29.96	97.51	142.2	1.0	03	14, 30, 31
ES27	29.64	97.90	130.0	1.1	01	26
ES29	30.01	96.69	84.0	1.1	16	14, 15, 23, 25, 31
ES30	29.32	97.19	110.8	1.1	07	25, 26, 27, 31
ES31	29.51	96.76	82.4	1.2	16	14, 17, 25, 26, 30, 31
ES32	29.76	96.10	103.0	1.4	02	13, 30
ES33	29.77	95.70	67.1	0.6	10	3, 7, 11, 12, 14, 15, 30, 38
ES34	29.91	95.47	84.8	1.2	20	3, 4, 7, 8, 10, 11, 12, 14, 15, 16, 36, 38
ES35	29.96	94.78	111.9	1.1	09	1, 4, 10, 14, 15, 16, 31, 36, 38
ES36	29.81	93.91	86.2	1.1	13	1, 4, 5, 7, 8, 11, 13, 14, 15, 38, 40, 41
ES37	29.90	93.51	80.5	1.4	07	1, 10, 13, 36, 39, 40, 41
ES38	30.02	92.97	72.5	1.2	18	3, 4, 5, 7, 8, 10, 11, 12, 36, 39, 40, 41
ES39	29.87	92.62	75.8	1.6	17	1, 2, 3, 4, 8, 9, 10, 11, 36, 38, 40, 41
ES40	29.71	92.15	78.0	1.4	14	1, 2, 4, 5, 7, 8, 10, 36, 38, 39, 41, 43
ES41	29.19	91.76	70.4	1.7	14	2, 3, 4, 5, 36, 38, 40
ES42	28.90	91.94	108.6	1.6	06	1, 37, 39
ES43	29.04	92.23	75.4	1.2	05	8, 36, 38, 39
ES45	29.12	93.78	65.8	1.3	05	7, 8, 38
ES46	29.25	94.26	89.2	1.1	04	10, 36

lished data of Chen et al. (2000), Zhang et al. (2004), Shen et al. (2005), and Sol et al. (2007). The APM velocities are calculated via a web-based plate motion calculator (https://www.unavco.org/software/geodetic-utilities/plate-motion-calculator/plate-motion-calculator.html) that is based on an integrated global plate motion model (GSRMv1.2) originally developed by Kreemer et al. (2003). Figure 9 shows the correlative analysis of splitting parameters by using direct S phases, APM directions, and GPS mea-

surements in our target region. It suggests that the observed anisotropy is not only due to lithospheric deformation or due to asthenospheric dynamics at the base of the lithosphere but that it is a combined effect of both.

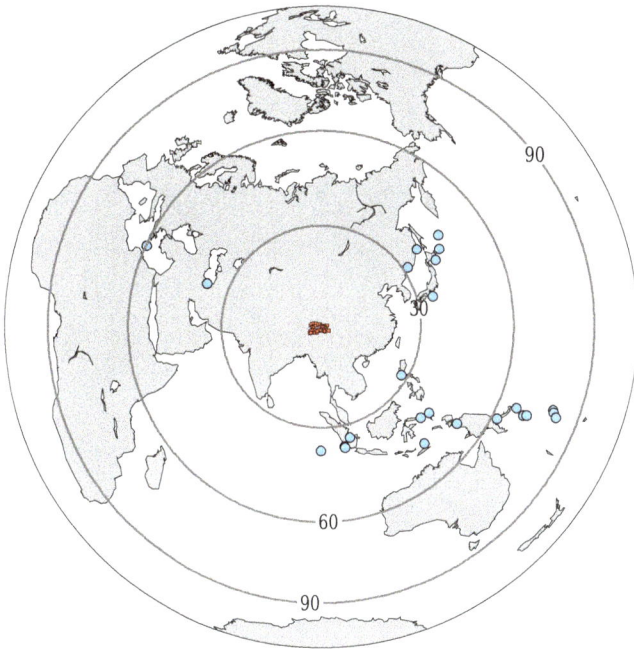

Figure 6. Epicentral distribution of teleseismic earthquakes used in the study (30–90°). Rectangles indicate seismic stations in this study.

3.2 Discussion

3.3 Origin of anisotropy in the southeastern Tibetan region

Our resultant splitting measurements vary over a range of δt_s suggesting the presence of a significant deformation in the region. The fast polarization directions are rather consistent and match well with the surface geology, similar to those observed from the SKS phases in Sol et al. (2007). The fast directions closely follow the strike of the major sutures like BNSZ and ITSZ and surface strain fields as observed through GPS and are under the influence of bending at the EHS (Fig. 7). FPDs that are parallel to the surface geologic features such as faults (e.g. Savage, 1999; Flesch et al., 2005) are indicative of vertically coherent deformation of the crust and upper mantle. This has previously been invoked to explain the anisotropic character in eastern and northeastern Tibet (Holt et al., 2000; León Soto et al., 2012; Eken et al., 2013; Eken and Tilmann, 2014). In the absence of any compelling evidence for crust–mantle coupling, we argue in favour of a large-scale deformation of the crust and upper mantle under similar boundary conditions as a plausible option to explain the observed anisotropy (Flesch et al., 2005; Sol et al., 2007; Holt et al., 2000).

The observed large time delays (> 1 s), in this study, reflect a highly anisotropic region with similar deformation patterns at depths. The presence of a more complex anisotropic structure (e.g. double layer) with different orientations in the

fast axis at various depths may result in smaller delay times (Saltzer et al., 2000). In the western Himalayan region, Vinnik et al. (2007) observed different fast-velocity directions for seismic azimuthal anisotropy that vary from N 60° E at depths between 80 and 160 km to N 150° E at depths between 160 and 220 km depth by using the joint inversion of SKS particle motions and P receiver functions. This provides an argument to explain the null or negligible anisotropy as reported from the same region using SKS phases (Sandvol et al., 1994). Smaller time delays in the Nepal Himalayas and the Sikkim Himalayas are attributed to the combined effect of shear at the base of the lithosphere due to APM-related strain of the Indian Plate and ductile flow along the collision front due to compression, with possibly completely different orientations (Singh et al., 2007). Sol et al. (2007) reported null measurements at a few stations, possibly due to the lack of clear splitting measurements of SKS phases. The transition between deformation types at the boundaries of the Indian and Eurasian lithospheric plates was considered to be the main reason for observed null or negligible anisotropy further west beneath the southern Tibetan Plateau (Chen et al., 2010; Chen and Ozalaybey, 1998; Barruol and Hoffmann, 1999; Zhao et al., 2014). The lack of anisotropy beneath southern Tibet was mainly explained by an isotropic nature of the Indian tectonic plate or a lack in the ability of SKS phases to sample the anisotropy due to a sub-vertical mantle shear strain field created by downwelling Indian lithosphere (Singh et al., 2007; Sandvol et al., 1997). However, the hypothesis of an isotropic nature of the Indian lithosphere was contradicted in various studies (Singh et al., 2006, 2007; Kumar and Singh, 2008), and significant anisotropy is reported beneath Tibet in the region of null measurement (Gao and Liu, 2009; Singh et al., 2016).

Sub-vertical shear strain or complex flow arises due to the subducting Indian slab and may result in null or negligible anisotropy (Sandvol et al., 1997; Fu et al., 2008). Recent tomographic studies (Griot et al., 1998; Huang et al., 2003; Zhou and Murphy, 2005; Yao et al., 2008; Priestley et al., 2006; Singh et al., 2014; Pandey et al., 2014) suggest that in the western Tibetan side, where the N–S extension is less, the Indian lithosphere is supposed to extend as far as the Jinsa River suture zone (JRSZ; Zhao et al., 2010), while in the eastern Tibet side, the Indian lithosphere extends up to the ITSZ (Li et al., 2008; Zhao et al., 2010). A combined study using seismic anisotropy and Bouguer gravity anomalies place the Indian mantle front up at 33° N in central Tibet (Chen et al., 2010). In this segment of the Himalaya–Tibet collision zone, the northern limit of the Indian lithospheric mantle does not seem to extend beyond the ITSZ (Li et al., 2008). The lack of anisotropy reported using SKS/SKKS phases (Sol et al., 2007) at a few seismic stations might be due to insufficient measurements rather than the effects of the downwelling Indian lithosphere as suggested in southern Tibet (Sandvol et al., 1997). By adding a considerably large amount of measurements from direct S waves, we ob-

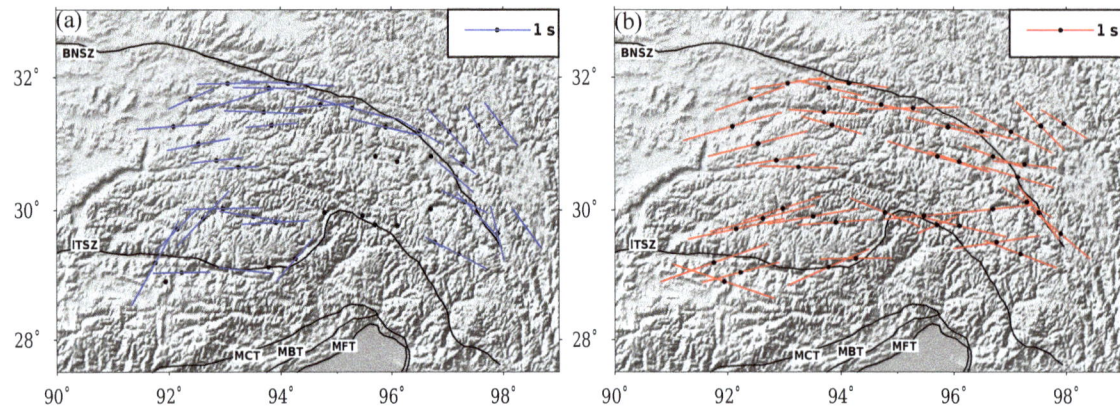

Figure 7. The tectonic and topographic map of the southeastern Tibetan region, which represents the station average splitting parameters: **(a)** the previous SKS-derived splitting measurements (solid blue bar) performed by Sol et al. (2007) and **(b)** the direct *S*-wave-derived splitting measurements (solid red bar, this study). The length of the solid bars in each panel indicates the strength of anisotropy and is scaled by station-averaged splitting time delays. Azimuth of the solid bars indicates the fast polarization direction (FPD). Black circles show location of the seismic stations used in this study. (MFT: Main Frontal Thrust; MBT: Main Boundary Thrust; MCT: Main Central Thrust; ITSZ: Indus–Tsangpo suture zone; BNSZ: Bangong–Nujiang suture zone).

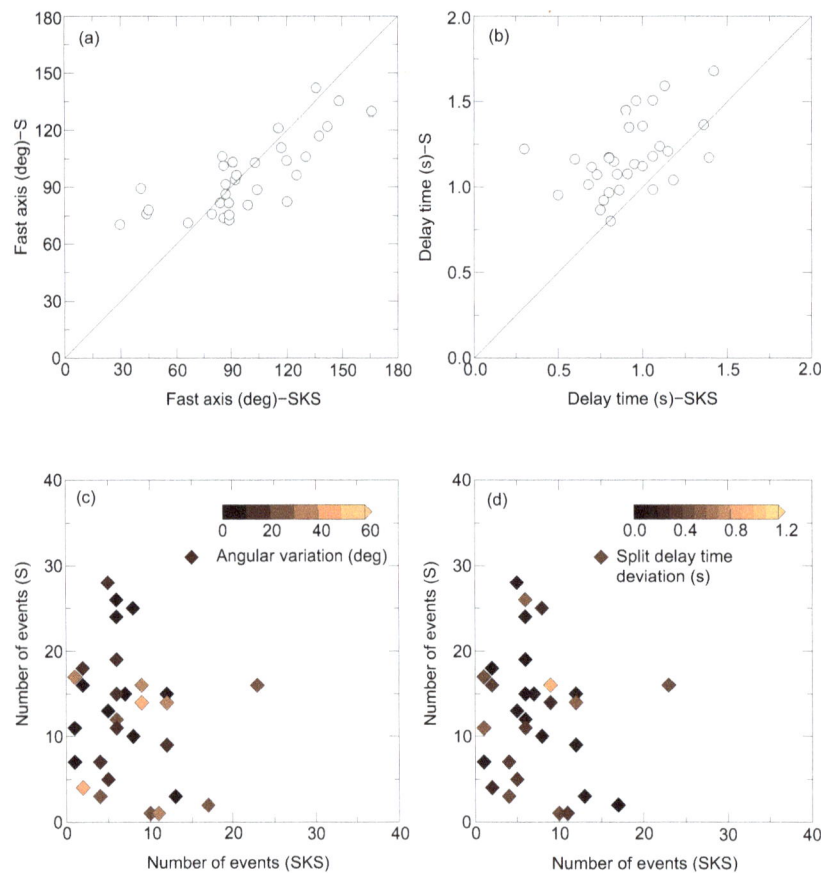

Figure 8. Comparison of SKS and *S*-wave-derived station-averaged splitting parameters in the study region. Panels **(a)** and **(b)**: scatter plots that compare SKS- and *S*-derived FPDs and split time delays (TDs), respectively. **(c)** Scatter plot of the number of individual SKS splitting measurement and the number of events used in direct *S* splitting measurement (this study). Note that each station here is colour-coded by its absolute deviation value that is obtained by subtracting *S*- and SKS-derived FPDs. **(d)** The same plot for the misfit between SKS- and *S*-derived station-averaged split TDs. Average SKS splitting parameters used in this figure are taken from Sol et al. (2007).

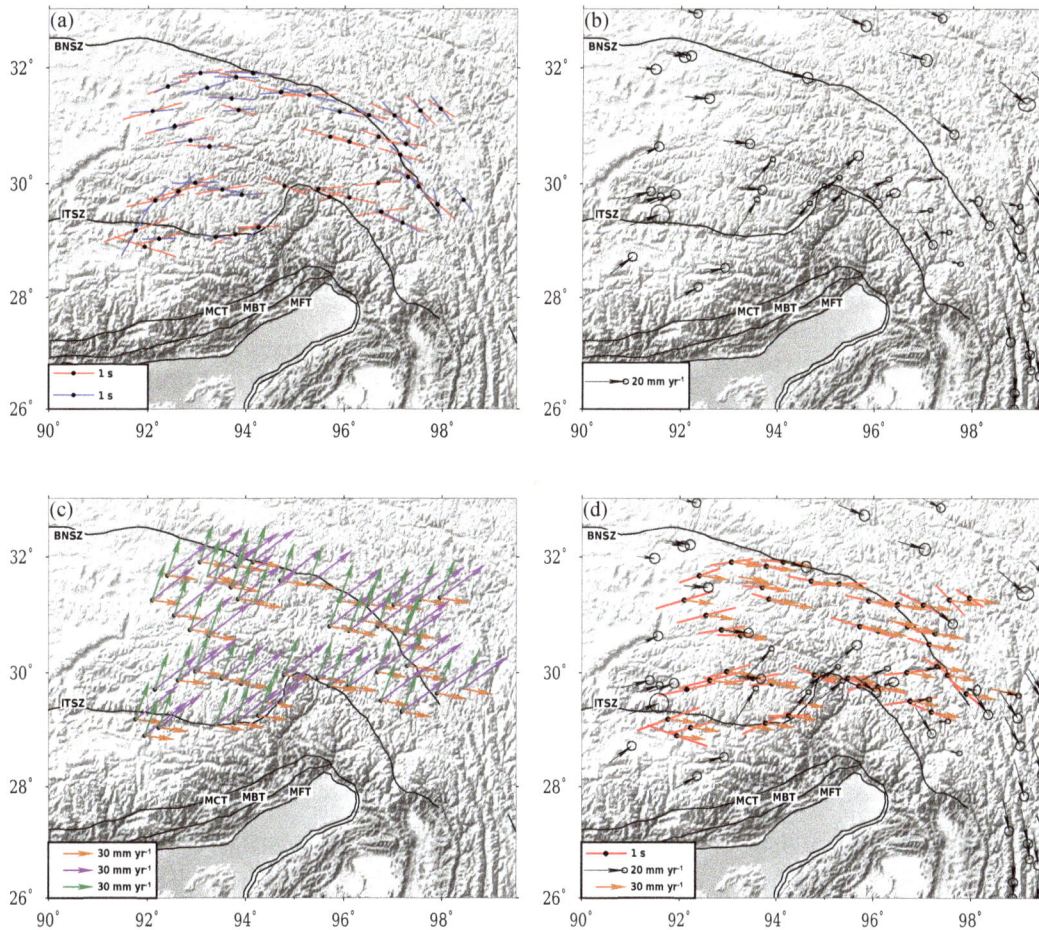

Figure 9. Lateral variations of anisotropic, geodetic, and absolute plate motion data shown over topographic and tectonic features of the study area. **(a)** Map view comparison between the splitting measurement inferred from direct S waves (this study shown by red bars) and SKS splitting measurements (Sol et al., 2007, in blue bars). **(b)** The Global Positioning System (GPS) velocity (mm yr^{-1}) vectors (black arrows) around SE Tibetan region calculated with respect to the South China reference frame and stable Eurasia. GPS data is compiled from several studies including Chen et al. (2000), Zhang et al. (2004), Shen et al. (2005), and Sol et al. (2007). **(c)** Absolute plate motion (APM) velocities calculated through https://www.unavco.org/software/geodetic-utilities/plate-motion-calculator/platemotion-calculator.html by using the GSRM v1.2 (2004) model of Kreemer et al. (2003). Note that arrows in brown, green, and purple represent the APM velocities of the Eurasian Plate in no net rotation frame, of the Indian Plate with respect to the Eurasian Plate, and the motion of the Indian Plate in no net rotation frame, respectively. **(d)** Map view comparisons of the station (black circle) average direct S-wave-derived splitting parameters with GPS velocity (black arrow) and APM (brown arrow) vectors. Abbreviations on the maps: MFT: Main Frontal Thrust; MBT: Main Boundary Thrust; MCT: Main Central Thrust; ITSZ: Indus–Tsangpo suture zone; BNSZ: Bangong–Nujiang suture zone; JRSZ: Jinsa River suture zone.

serve significant anisotropy for the same stations and fast axis deformation, which can be explained by eastward flow in a lithospheric crush zone formed due to the collision of the Indian and Asian tectonic plates as suggested for southern Tibet (Zhao et al., 2014).

The crust beneath Tibet is thick ($\sim$ 80 km; e.g. Singh et al., 2015) and crustal anisotropic effects should be accounted for in the splitting measurements obtained using direct S and SKS/SKKS phases. In the Himalayan region, highly anisotropic crust ($\sim$ 20 %) has been reported using an inversion of receiver functions (Schulte-Pelkum et al., 2005; Singh et al., 2010), while a similar approach at a few seismic

stations covering Tibet suggests approximately 4–14 % seismic anisotropy within the Tibetan crust (Sherrington et al., 2004; Ozacar and Zandt, 2004). Ozacar and Zandt (2004) have accounted for splitting of < 0.5 s over SKS split times due to the observed anisotropy of > 10 % in the crust. Splitting times of 0.2–0.3 s are observed within the eastern Tibet crust using splitting of the Moho-converted Ps phases (receiver functions, Chen et al., 2013). Tomographic (Huang et al., 2002, 2009; Hung et al., 2010; Yao et al., 2008, 2010; Li et al., 2009), magnetotelluric (Bai et al., 2010), and gravity (Jordan and Watts, 2005) studies of the SE Tibetan region suggest ductile flow in the deeper region of the crust. Relatively low seismic velocities are resolved for shear waves in tomographic studies at these crustal depths, indicating localized flow of the crustal material along a network of strike–slip faults in the region (Yao et al., 2010). These types of flow may produce splitting orientations similar to lower lithospheric scales with coherent deformation. A coupled crust and mantle increases the SKS delay times by 0.2–0.5 s due to the effects of crust. The anisotropic orientations observed beneath most parts of Tibet (Sherrington et al., 2004; Ozacar and Zandt, 2004; Chen et al., 2013) within the crust are completely different from the SKS or direct S waves, implying that the types of deformation in the crust and upper mantle could be different; this does not indicate a coherent deformation pattern, at least in some parts of Tibet. A possible explanation for such decoupling could be that the crustal anisotropic parameters are influenced by either current deformation or fossilized fabric with different boundary conditions at mid-crustal and lower crustal levels.

3.4   Comparison between direct S- and SKS-derived splitting parameters

The scatter plot in Fig. 8a, b provides a comparison between estimated splitting parameters (ϕ_S and δt_S) from the analysis of direct S waves and previous SKS splitting measurements (Sol et al., 2007). The obtained splitting parameters (ϕ_S) from the direct S wave measurements show that they are consistent with previous SKS-based measurements. Overall, the consistency between splitting parameters inferred from SKS and direct S waves is most likely because both are derived from the same type of large-scale anisotropic structures. Large differences between SKS- and S-derived δt_S that appear as a move-out in the scatter plot occur for three reasons: (i) longer S wave ray paths as compared to SKS ray paths sampling the same type of large-scale anisotropy, (ii) an increase in the number of the events (Fig. 8c, d) sampling different azimuths that contribute to the direct S wave measurements as compared to the SKS, and (iii) the S-derived type of anisotropy might show strong variation in splitting with incident angle, i.e. an event coming from different azimuths. The precision of our results is evident from small deviations in Fig. 8c, d, and this is most likely due to the involvement of a relatively increased number of observations from both S and SKS phases in splitting measurements. Figure 8c shows that the absolute deviation for ϕ_S is no larger than 25° except at 7 out of 42 stations (namely ES23, ES27, ES31, ES39, ES40, ES41, and ES46), where we observe large deviations of up to 48°. An extreme example of maximum deviation for δt_S is at station ES31. The deviation for ϕ_S is also relatively large (>30°). Because station ES31 does not suffer from a lack of data, (8 observations for SKS and 14 for S), we infer that the mismatch may be a result of the use of an incorrect reference anisotropy when correcting for receiver-side anisotropy. Overall, deviations for δt_S are smaller than 0.56 s. In general, we observe relatively more events for the direct S waves compared to individual SKS phases except at stations ES16, ES18, ES23, ES26, and ES27. For these five stations, we detect deviations for ϕ_S and δt_S of up to 36° and 0.5 s, respectively. In summary, our comparative analysis of splitting parameters shows a good accordance between SKS- and direct-S-derived splitting parameters as previously observed in the Himalaya–Tibet collision zone (e.g. McNamara et al., 1994; Singh et al., 2016; Huang et al., 2011; Eken et al., 2013) and in the Indian shield (Saikia et al., 2010).

3.5   Deformation pattern revealed from the comparison of the GPS, APM, and splitting measurements

Previous studies on seismic anisotropy (McNamara et al., 1994; Sol et al., 2007; Huang et al., 2000, 2007; Wang et al., 2007, 2008; Chen et al., 2013; Zhao et al., 2014; Guilbert et al., 1996; Bai et al., 2009) that compared splitting parameters with APM, GPS, and structural and topographical features provide crucial information concerning the dynamic deformation pattern and possible linkage of the strength of coupling between the crust and lithospheric mantle of the southeastern or eastern Tibetan region. We observe a sharp transition in the spatial distribution of ϕ_S from nearly W–E in the western part of the study region to nearly NW–SE or NNW–SSE near the southeastern Tibetan margin (Fig. 9). The structural and topographical features, such as major suture zones and mountain belts, tend to rotate around the EHS from nearly E–W or ENE–WSW to N–S or NE–SW (Hallet and Molnar, 2001; Booth et al., 2004). The observed ϕ_S and GPS velocity vectors follow a similar trend (Fig. 9a, b). The APM directions are consistent with the present ongoing asthenospheric flow (Vinnik et al., 1992, 1995; Vinnik and Montagner, 1996). By using different plots of the plate motion (APM of Eurasian and Indian plates referenced to the no-net-rotation (NNR) frame or the relative plate motion of the Indian Plate referenced to the Eurasian Plate), we want to examine the contribution of APM to explaining the observed anisotropic variation and to check which plate motion best explains the observed ϕ_S of the splitting measurements. But the observed ϕ_S are not consistent with plate motion. The discrepancy between the ϕ_S and APM may indicate that the obtained splitting, and hence the anisotropic behaviour of the study area, is not only due to asthenospheric dynamics but is a combined effect of the lithospheric deformation and asthenospheric dynamics. The lateral variation of obtained splitting measurements, when taken together with GPS velocity vectors, geological features, and the APM directions, depicts the movement of lithospheric or crustal material of the western and central plateau relative to the Eurasian Plate towards the eastern Tibetan side and clockwise rotation around the EHS. This supports the presence of a deep crustal flow and movement of material from the central and western portion towards the eastern Tibetan side as has been suggested previously by Royden et al. (1997, 2008).

The present-day GPS measurements do not necessarily reveal the deformation of the whole crust but could be associated with deformation of the shallow crust (Chen et al., 2013). Seismic imaging of crustal anisotropy based on receiver function studies (e.g. Sherrington et al., 2004; Chen et al., 2013) supports this argument. The orientation of the GPS velocity vectors and the ϕ_s of the direct S waves only match when the orientation of the different layers of anisotropy within the crust and mantle tend to be similar. Griot et al. (1998), Holt et al. (2000), Fouch et al. (2000), and Sol et al. (2007) discuss the coupling and decoupling of the crust and mantle by making comparisons among ϕ_s, GPS, APM, and surficial features. Sol et al. (2007) report a good coherency between anisotropic and geodetic measurements for the entire southeastern Tibetan region, and on that basis, they discuss the coupling of the crustal and mantle material as similarly observed in the northeast Tibetan Plateau (e.g. Eken et al., 2013; Eken and Tilmann, 2014). The seismic anisotropy directions that were previously obtained from the inversion of receiver functions, however, do not suggest vertically coherent deformation of the crust (e.g. Ozacar and Zandt, 2004; Sherrington et al., 2004). Sherrington et al. (2004) report 4–14 % seismic anisotropy with variable orientations at different depths. They attribute varying patterns of anisotropic directions to both fossilized fabric and more recent deformation. The different orientations at mid and lower crustal levels do not necessarily support a coherent deformation of the crust and upper mantle.

On the basis of driving forces, two kinematic models have been proposed to explain the coupling–decoupling of the crust and lithospheric mantle. The first one is a simple asthenospheric model (Richardson, 1992), proposed to explain the decoupling of the crust and mantle by the intrusion of a mechanically weak layer, such as the asthenosphere, into the crust. Whenever a mechanically weak layer is present in between the crust and mantle, the force acting on the crustal region cannot be transmitted into the mantle. As a result, the crust is decoupled from the mantle due to different driving forces on them. In such models, the velocity difference between the top and bottom of the mechanically weak layer gives rise to mantle deformation, and that difference is parallel to the fast polarization direction. The second model, proposed by Lithgow-Bertelloni and Richards (1998), is the vertically coherent model, and it explains the coupling of the materials within the crust and lithospheric mantle on the basis of the transmission of the buoyancy forces from the crust to the mantle. This model requires a rigid lower part of the crust. In contrast, low shear wave velocity anomalies resolved in various tomographic studies recently (Huang et al., 2002, 2009; Hung et al., 2010; Yao et al., 2008; Li et al., 2009) have indicated a weak layer in the deeper region of the crust beneath the SE Tibetan region. It is noteworthy to mention that we avoid making any comment on the possible linkage between the deformation and coupling of the crust and underlying lithospheric mantle by only using splitting parameters inferred from direct S waves and geodetic measurements, and further study is required.

4   Conclusions

Our splitting observations using direct S waves add new constraints in understanding the deformation pattern and its causes in the southeastern Tibetan region near Namcha Barwa. We list the main concluding remarks from the present study as follows:

1. The observed splitting analysis suggests a highly deformed crust and lithospheric mantle.

2. Significant anisotropy at stations where null or negligible anisotropy is reported based on previous SKS splitting measurements is inconsistent with the hypothesis of an isotropic lithospheric mantle.

3. Our study also provides clear evidence for the development of anisotropy in this region with active geodynamic implications for several tectonic events, i.e. the multistage subduction of the Indian Plate below the Eurasian Plate and the movement of the western central Tibetan lithospheric material towards the southeastern and eastern Tibetan side.

4. The observed splitting delays (0.67–1.68 s) suggest the possible existence of a multi-layered anisotropy structure in the crust and upper mantle. Further understanding of this requires 3-D geodynamic modelling and inversion of multi-frequency datasets to resolve more complex depth-dependent anisotropic structures (e.g. multi-layer anisotropy and dipping axis of symmetry).

Competing interests. The authors declare that they have no conflict of interest.

Acknowledgements. The IRIS data management centre and Project team of Namcha Barwa Seismic Network (PASSCAL) are gratefully acknowledged for making the seismic data available. This work has been performed under the ISIRD project (ASI) of IIT Kharagpur. Tuna Eken thanks the Alexander von Humboldt (AvH) Foundation for the equipment subsidy. We benefited from valuable comments by Frederik Tilmann during the preparation of the revised version of the paper. We thank topical editor Charlotte Krawczyk, reviewer Rob Porritt, and one anonymous reviewer for their valuable comments and suggestions, which helped us to improve the paper substantially.

Edited by: C. Krawczyk

References

Argand, E.: La tectonique de l'Asie, Int. geol, Congr. Rep, 171–322, 1924.

Bai, D., Unsworth, M. J., Meju, M. A., Ma, X., Teng, J., Kong, X., Sun, Y., Sun, J., Wang, L., Jiang, C., Zhao, C., Pengfei, X., and Liu, M.: Crustal deformation of the eastern Tibetan plateau revealed by magnetotelluric imaging, Nat. Geosci., 3, 358–362, doi:10.1038/ngeo830, 2010.

Bai, L., Iidaka, T., Kawakatsu, H., Morita, Y., and Dzung, N.: Seismic anisotropy and shear-wave splitting in lower-crustal and upper-mantle rocks from the Ivrea Zone – experimental and calculated data, Phys. Earth Planet. In., 176, 33–43, doi:10.1016/j.pepi.2009.03.008, 2009.

Barruol, G. and Hoffmann, R.: Upper mantle anisotropy beneath the Geoscope stations, J. Geophys. Res., 104, 10757–10773, doi:10.1029/1999JB900033, 1999.

Becker, T. W. and Faccenna, C.: Mantle conveyor beneath the Tethyan collisional belt, Earth Planet. Sc. Lett., 310, 453–461, doi:10.1016/j.epsl.2011.08.021, 2011.

Booth, A. L., Zeitler, P. K., Kidd, W. S., Wooden, J., Liu, Y., Idleman, B., Hren, M., and Chamberlain, C. P.: U-Pb zircon constraints on the tectonic evolution of southeastern Tibet, Namche Barwa Area, Am. J. Sci., 304, 889–929, doi:10.2475/ajs.304.10.889, 2004.

Chen, W.-P. and Ozalaybey, S.: Correlation between seismic anisotropy and Bouguer gravity anomalies in Tibet and its implications for lithospheric structures, Geophys. J. Int., 135, 93–101, doi:10.1046/j.1365-246X.1998.00611.x, 1998.

Chen, W.-P., Martin, M., Tseng, T.-L., Nowack, R. L., Hung, S.-H., and Huang, B.-S.: Shear-wave birefringence and current configuration of converging lithosphere under Tibet, Earth Planet. Sc. Lett., 295, 297–304, doi:10.1016/j.epsl.2010.04.017, 2010.

Chen, Y., Zhang, Z., Sun, C., and Badal, J.: Crustal anisotropy from Moho converted Ps wave splitting analysis and geodynamic implications beneath the eastern margin of Tibet and surrounding regions, Geophys. Res. Lett., 24, 946–957, doi:10.1016/j.gr.2012.04.003, 2013.

Chen, Z., Burchfiel, B., Liu, Y., King, R., Royden, L., Tang, W., Wang, E., Zhao, J., and Zhang, X.: Global Positioning System measurements from eastern Tibet and their implications for India/Eurasia intercontinental deformation, J. Geophys. Res., 105, 16215–16227, doi:10.1029/2000JB900092, 2000.

Cochran, E. S., Vidale, J. E., and Li, Y.-G.: Near-fault anisotropy following the Hector Mine earthquake, J. Geophys. Res., 108, 2436, doi:10.1029/2002JB002352, 2003.

Confal, J. M., Eken, T., Tilmann, F., Yolsal-Çevikbilen, S., Çubuk-Sabuncu, Y., Saygin, E., and Taymaz, T.: Investigation of mantle kinematics beneath the Hellenic-subduction zone with teleseismic direct shear waves, Phys. Earth Planet. In., 261, 141–151, doi:10.1016/j.pepi.2016.10.012, 2016.

Eken, T. and Tilmann, F.: The Use of Direct Shear Waves in Quantifying Seismic Anisotropy: Exploiting Regional Arrays, B. Seismol. Soc. Am., 104, 2644–2661, doi:10.1785/0120140020, 2014.

Eken, T., Tilmann, F., Mechie, J., Zhao, W., Kind, R., Su, H., Xue, G., and Karplus, M.: Seismic Anisotropy from SKS Splitting beneath Northeastern Tibet, B. Seismol. Soc. Am., 103, 3362–3371, doi:10.1785/0120130054, 2013.

Flesch, L. M., Haines, A. J., and Holt, W. E.: Dynamics of the India-Eurasia collision zone, J. Geophys. Res., 106, 16435–16460, doi:10.1029/2001JB000208, 2001.

Flesch, L. M., Holt, W. E., Silver, P. G., Stephenson, M., Wang, C.-Y., and Chan, W. W.: Constraining the extent of crust – mantle coupling in central Asia using GPS, geologic, and shear wave splitting data, Earth Planet. Sc. Lett., 238, 248–268, doi:10.1016/j.epsl.2005.06.023, 2005.

Fouch, M. J., Fischer, K. M., Parmentier, E. M., Wysession, M. E., and Clarke, T. J.: Shear wave splitting, continental keels, and patterns of mantle flow, J. Geophys. Res., 105, 6255–6275, doi:10.1029/1999JB900372, 2000.

Fu, Y. V., Chen, Y. J., Li, A., Zhou, S., Liang, X., Ye, G., Jin, G., Jiang, M., and Ning, J.: Indian mantle corner flow at southern Tibet revealed by shear wave splitting measurements, Geophys. Res. Lett., 35, L02308, doi:10.1029/2007GL031753, 2008.

Gao, S. S. and Liu, K. H.: Significant seismic anisotropy beneath the southern Lhasa Terrane, Tibetan Plateau, Geochem. Geophy. Geosy., 10, Q02008, doi:10.1029/2008GC002227, 2009.

Garzanti, E. and Van Haver, T.: The Indus clastics: forearc basin sedimentation in the Ladakh Himalaya (India), Sediment. Geol., 59, 237–249, doi:10.1016/0037-0738(88)90078-4, 1988.

Griot, D.-A., Montagner, J.-P., and Tapponnier, P.: Phase velocity structure from Rayleigh and Love waves in Tibet and its neighboring regions, J. Geophys. Res., 103, 21215–21232, doi:10.1029/98JB00953, 1998.

Guilbert, J., Poupinet, G., and Mei, J.: A study of azimuthal P residuals and shear-wave splitting across the Kunlun range (Northern Tibetan plateau), Phys. Earth Planet. In., 95, 167–174, doi:10.1016/0031-9201(95)03120-0, 1996.

Hallet, B. and Molnar, P.: Distorted drainage basins as markers of crustal strain east of the Himalaya, J. Geophys. Res., 106, 13697–13709, doi:10.1029/2000JB900335, 2001.

Henderson, A. L., Najman, Y., Parrish, R., Mark, D. F., and Foster, G. L.: Constraints to the timing of India–Eurasia collision, a re-evaluation of evidence from the Indus Basin sedimentary rocks of the Indus–Tsangpo Suture Zone, Ladakh, India, Earth-Sci. Rev., 106, 265–292, doi:10.1016/j.earscirev.2011.02.006, 2011.

Herquel, G., Wittlinger, G., and Guilbert, J.: Anisotropy and crustal thickness of northern-Tibet. New constraints for tectonic modelling, Geophys. Res. Lett., 22, 1925–1928, doi:10.1029/95GL01789, 1995.

Hirn, A., Jiang, M., Sapin, M., Diaz, J., Nercessian, A., and Lu, Q.: Seismic anisotropy as an indicator of mantle flow beneath the Himalayas and Tibet, Nature, 375, 571–574, doi:10.1038/375571a0, 1995.

Holt, W., Chamot-Rooke, N., Le Pichon, X., Haines, A., Shen-Tu, B., and Ren, J.: Velocity field in Asia inferred from Quaternary fault slip rates and Global Positioning System observations, J.

Geophys. Res., 105, 19185–19209, doi:10.1029/2000JB900045, 2000.

Holt, W. E., Li, M., and Haines, A.: Earthquake strain rates and instantaneous relative motions within central and eastern Asia, Geophys. J. Int., 122, 569–593, doi:10.1111/j.1365-246X.1995.tb07014.x, 1995.

Houseman, G. and England, P.: Finite strain calculations of continental deformation: 1. Method and general results for convergent zones, J. Geophys. Res., 91, 3651–3663, doi:10.1029/JB091iB03p03651, 1986.

Houseman, G. and England, P.: Crustal thickening versus lateral expulsion in the Indian-Asian continental collision, J. Geophys. Res., 98, 12233–12249, doi:10.1029/93JB00443, 1993.

Houseman, G. and England, P.: A lithospheric-thickening model for the Indo-Asian collision, World and Regional Geology, 1, 1–17, 1996.

Huang, J., Zhao, D., and Zheng, S.: Lithospheric structure and its relationship to seismic and volcanic activity in southwest China, J. Geophys. Res., 107, ESE 13-1–ESE 13-14, doi:10.1029/2000JB000137, 2002.

Huang, W.-C., Ni, J. F., Tilmann, F., Nelson, D., Guo, J., Zhao, W., Mechie, J., Kind, R., Saul, J., Rapine, R., and Hearn, T. M.: Seismic polarization anisotropy beneath the central Tibetan Plateau, J. Geophys. Res., 105, 27979–27989, doi:10.1029/2000JB900339, 2000.

Huang, Z., Wang, L., Xu, M., Liu, J., Mi, N., and Liu, S.: Shear wave splitting across the Ailao Shan-Red River fault zone, SW China, Geophys. Res. Lett., 34, L20301, doi:10.1029/2007GL031236, 2007.

Huang, Z., Li, H., Zheng, Y., and Peng, Y.: The lithosphere of North China Craton from surface wave tomography, Earth Planet. Sc. Lett., 288, 164–173, doi:10.1016/j.epsl.2009.09.019, 2009.

Huang, Z., Wang, L., Zhao, D., Mi, N., and Xu, M.: Seismic anisotropy and mantle dynamics beneath China, Earth Planet. Sc. Lett., 306, 105–117, doi:10.1016/j.epsl.2011.03.038, 2011.

Huang, Z.-M., Zhang, Y.-Z., Kotaki, M., and Ramakrishna, S.: A review on polymer nanofibers by electrospinning and their applications in nanocomposites, Compos. Sci. Technol., 63, 2223–2253, doi:10.1016/S0266-3538(03)00178-7, 2003.

Hung, S.-H., Chen, W.-P., Chiao, L.-Y., and Tseng, T.-L.: First multi-scale, finite-frequency tomography illuminates 3-D anatomy of the Tibetan plateau, Geophys. Res. Lett., 37, L06304, doi:10.1029/2009GL041875, 2010.

Jagoutz, O., Royden, L., Holt, A. F., and Becker, T. W.: Anomalously fast convergence of India and Eurasia caused by double subduction, Nat. Geosci., 8, 475–478, doi:10.1038/ngeo2418, 2015.

Jordan, T. and Watts, A.: Gravity anomalies, flexure and the elastic thickness structure of the India–Eurasia collisional system, Earth Planet. Sc. Lett., 236, 732–750, doi:10.1016/j.epsl.2005.05.036, 2005.

Karato, S.-I.: Grain-size distribution and rheology of the upper mantle, Tectonophys, 104, 155–176, doi:10.1016/0040-1951(84)90108-2, 1984.

Karato, S.-I.: Seismic anisotropy due to lattice preferred orientation of minerals: Kinematic or dynamic?, Geoph. Monog. Series, 39, 455–471, doi:10.1029/GM039p0455, 1987.

Karato, S.-I. and Wu, P.: Rheology of the Upper Mantle: A Synthesis, Science, 260, 771–778, doi:10.1126/science.260.5109.771, 1993.

Karplus, M., Zhao, W., Klemperer, S., Wu, Z., Mechie, J., Shi, D., Brown, L., and Chen, C.: Injection of Tibetan crust beneath the south Qaidam Basin: Evidence from INDEPTH IV wide-angle seismic data, J. Geophys. Res., 116, B07301, doi:10.1029/2010JB007911, 2011.

Kennett, B. L. N. and Engdahl, E.: Traveltimes for global earthquake location and phase identification, Geophys. J. Int., 105, 429–465, doi:10.1111/j.1365-246X.1991.tb06724.x, 1991.

Kreemer, C., Holt, W. E., and Haines, A. J.: An integrated global model of present-day plate motions and plate boundary deformation, Geophys. J. Int., 154, 8–34, doi:10.1046/j.1365-246X.2003.01917.x, 2003.

Kumar, M. R. and Singh, A.: Evidence for plate motion related strain in the Indian shield from shear wave splitting measurements, J. Geophys. Res., 113, B08306, doi:10.1029/2007JB005128, 2008.

Kumar, P., Yuan, X., Kind, R., and Ni, J.: Imaging the colliding Indian and Asian lithospheric plates beneath Tibet, J. Geophys. Res., 111, B06308, doi:10.1029/2005JB003930, 2006.

Lavé, J., Avouac, J., Lacassin, R., Tapponnier, P., and Montagner, J.: Seismic anisotropy beneath Tibet: evidence for eastward extrusion of the Tibetan lithosphere?, Earth Planet. Sc. Lett., 140, 83–96, doi:10.1016/0012-821X(96)00045-3, 1996.

León Soto, G., Sandvol, E., Ni, J. F., Flesch, L., Hearn, T. M., Tilmann, F., Chen, J., and Brown, L. D.: Significant and vertically coherent seismic anisotropy beneath eastern Tibet, J. Geophys. Res., 117, B05308, doi:10.1029/2011JB008919, 2012.

Lev, E., Long, M. D., and van der Hilst, R. D.: Seismic anisotropy in Eastern Tibet from shear wave splitting reveals changes in lithospheric deformation, Earth Planet. Sc. Lett., 251, 293–304, doi:10.1016/j.epsl.2006.09.018, 2006.

Li, C., Van der Hilst, R. D., Meltzer, A. S., and Engdahl, E. R.: Subduction of the Indian lithosphere beneath the Tibetan Plateau and Burma, Earth Planet. Sc. Lett., 274, 157–168, doi:10.1016/j.epsl.2008.07.016, 2008.

Li, Y. H., Wu, Q. J., and Tian, X. B.: Crustal structure in the Yunnan region determined by modeling receiver functions, Chinese J. Geophys., 52, 67–80, 2009.

Lithgow-Bertelloni, C. and Richards, M. A.: The dynamics of Cenozoic and Mesozoic plate motions, Rev. Geophys., 36, 27–78, doi:10.1029/97RG02282, 1998.

Long, M. D. and Silver, P. G.: Shear wave splitting and mantle anisotropy: measurements, interpretations, and new directions, Surv. Geophys., 30, 407–461, doi:10.1007/s10712-009-9075-1, 2009.

Mainprice, D., Barruol, G., and Ismail, W. B.: The seismic anisotropy of the Earth's mantle: from single crystal to polycrystal, Earth's Deep Interior: Mineral physics and tomography from the atomic to the global scale, edited by: Karato, S.-I., Forte, A., Liebermann, R., Masters, G., and Stixrude, L., American Geophysical Union, Washington, D. C., 237–264, doi:10.1029/GM117p0237, 2000.

McKenzie, D. and Priestley, K.: The influence of lithospheric thickness variations on continental evolution, Lithos, 102, 1–11, doi:10.1016/j.lithos.2007.05.005, 2008.

McNamara, D. E., Owens, T. J., Silver, P. G., and Wu, F. T.: Shear wave anisotropy beneath the Tibetan Plateau, J. Geophys. Res., 99, 13655–13665, doi:10.1029/93JB03406, 1994.

Molnar, P. and Tapponnier, P.: Cenozoic tectonics of Asia: effects of a continental collision, Science, 189, 419–426, doi:10.1126/science.189.4201.419, 1975.

Nelson, K. D., Zhao, W., Brown, L., Kuo, J., Che, J., Liu, X., Klemperer, S., Makovsky, Y., Meissner, R., Mechie, J., Kind, R., Wenzel, F., Nabelek, N. J., Leshou, C., Tan, H., Wei, W., Jones, A. G., Booker, J., Unsworth, M., Kidd, W. S. F., Hauck, M., Alsdorf, D., Ross, A., Cogan, M., Wu, C., Sandvol, E., and Edwards, M.: Partially molten middle crust beneath southern Tibet: synthesis of project INDEPTH results, Science, 274, 1684–1688, doi:10.1126/science.274.5293.1684, 1996.

Nicolas, A. and Christensen, N. I.: Formation of Anisotropy in Upper Mantle Peridotites-A Review, Composition, structure and dynamics of the lithosphere-asthenosphere system, edited by: Fuchs, K. and Froidevaux, C., American Geophysical Union, Washington, D. C., 111–123, doi:10.1029/GD016p0111, 1987.

Ozacar, A. A. and Zandt, G.: Crustal seismic anisotropy in central Tibet: Implications for deformational style and flow in the crust, Geophys. Res. Lett., 31, L23601, doi:10.1029/2004GL021096, 2004.

Pandey, S., Yuan, X., Debayle, E., Priestley, K., Kind, R., Tilmann, F., and Li, X.: A 3-D shear-wave velocity model of the upper mantle beneath China and the surrounding areas, Tectonophys, 633, 193–210, doi:10.1016/j.tecto.2014.07.011, 2014.

Plomerová, J., Babuška, V., Kozlovskaya, E., Vecsey, L., and Hyvönen, L.: Seismic anisotropy – a key to resolve fabrics of mantle lithosphere of Fennoscandia, Tectonophys, 462, 125–136, doi:10.1016/j.tecto.2008.03.018, 2008.

Priestley, K., Debayle, E., McKenzie, D., and Pilidou, S.: Upper mantle structure of eastern Asia from multimode surface waveform tomography, J. Geophys. Res., 111, B10304, doi:10.1029/2005JB004082, 2006.

Replumaz, A. and Tapponnier, P.: Reconstruction of the deformed collision zone between India and Asia by backward motion of lithospheric blocks, J. Geophys. Res., 108, 2285, doi:10.1029/2001JB000661, 2003.

Richardson, R.: Ridge forces, absolute plate motions, and the intraplate stress field, J. Geophys. Res., 97, 11739–11748, doi:10.1029/91JB00475, 1992.

Rowley, D. B. and Currie, B. S.: Palaeo-altimetry of the late Eocene to Miocene Lunpola basin, central Tibet, Nature, 439, 677–681, doi:10.1038/nature04506, 2006.

Royden, L. H., Burchfiel, B. C., King, R. W., Wang, E., Chen, Z., Shen, F., and Liu, Y.: Surface deformation and lower crustal flow in eastern Tibet, Science, 276, 788–790, doi:10.1126/science.276.5313.788, 1997.

Royden, L. H., Burchfiel, B. C., and van der Hilst, R. D.: The geological evolution of the Tibetan Plateau, Science, 321, 1054–1058, doi:10.1126/science.1155371, 2008.

Saikia, D., Ravi Kumar, M., Singh, A., Mohan, G., and Dattatrayam, R. S.: Seismic anisotropy beneath the Indian continent from splitting of direct S waves, J. Geophys. Res., 115, B12315, doi:10.1029/2009JB007009, 2010.

Saltzer, R. L., Gaherty, J. B., and Jordan, T. H.: How are vertical shear wave splitting measurements affected by variations in the orientation of azimuthal anisotropy with depth?, Geophys. J. Int., 141, 374–390, doi:10.1046/j.1365-246X.2000.00088.x, 2000.

Sandvol, E., Ni, J., Kind, R., and Zhao, W.: Seismic anisotropy beneath the southern Himalayas-Tibet collision zone, J. Geophys. Res., 102, 17813–17823, doi:10.1029/97JB01424, 1997.

Sandvol, E. A., Ni, J. F., Hearn, T. M., and Roecker, S.: Seismic azimuthal anisotropy beneath the Pakistan Himalayas, Geophys. Res. Lett., 21, 1635–1638, doi:10.1029/94GL01386, 1994.

Sato, H., Fehler, M. C., and Maeda, T.: Seismic wave propagation and scattering in the heterogeneous earth, Springer, 496, doi:10.1007/978-3-540-89623-4, 2012.

Savage, M. K.: Seismic anisotropy and mantle deformation: What have we learned from shear wave splitting?, Rev. Geophys., 37, 65–106, doi:10.1029/98RG02075, 1999.

Schulte-Pelkum, V., Monsalve, G., Sheehan, A., Pandey, M. R., Sapkota, S., Bilham, R., and Wu, F.: Imaging the Indian subcontinent beneath the Himalaya, Nature, 435, 1222–1225, doi:10.1038/nature03678, 2005.

Shen, F., Royden, L. H., and Burchfiel, B. C.: Large-scale crustal deformation of the Tibetan Plateau, J. Geophys. Res., 106, 6793–6816, doi:10.1029/2000JB900389, 2001.

Shen, Z.-K., Lü, J., Wang, M., and Bürgmann, R.: Contemporary crustal deformation around the southeast borderland of the Tibetan Plateau, J. Geophys. Res., 110, B11409, doi:10.1029/2004JB003421, 2005.

Sherrington, H. F., Zandt, G., and Frederiksen, A.: Crustal fabric in the Tibetan Plateau based on waveform inversions for seismic anisotropy parameters, J. Geophys. Res., 109, B02312, doi:10.1029/2002JB002345, 2004.

Silver, P. G. and Chan, W. W.: Shear wave splitting and subcontinental mantle deformation, J. Geophys. Res., 96, 16429–16454, doi:10.1029/91JB00899, 1991.

Silver, P. G. and Savage, M. K.: The interpretation of shear-wave splitting parameters in the presence of two anisotropic layers, Geophys. J. Int., 119, 949–963, doi:10.1111/j.1365-246X.1994.tb04027.x, 1994.

Singh, A., Kumar, M. R., Raju, P. S., and Ramesh, D. S.: Shear wave anisotropy of the northeast Indian lithosphere, Geophys. Res. Lett., 33, doi:10.1029/2006GL026106, 2006.

Singh, A., Kumar, M. R., and Raju, P. S.: Mantle deformation in Sikkim and adjoining Himalaya: Evidences for a complex flow pattern, Phys. Earth Planet. In., 164, 232–241, doi:10.1016/0031-9201(93)90156-4, 2007.

Singh, A., Kumar, M. R., and Raju, P. S.: Seismic structure of the underthrusting Indian crust in Sikkim Himalaya, Tectonics, 29, doi:10.1029/2010TC002722, 2010.

Singh, A., Mercier, J.-P., Ravi Kumar, M., Srinagesh, D., and Chadha, R. K.: Continental scale body wave tomography of India: Evidence for attrition and preservation of lithospheric roots, Geochem. Geophy. Geosy., 15, 658–675, doi:10.1002/2013GC005056, 2014.

Singh, A., Singh, C., and Kennett, B.: A review of crust and upper mantle structure beneath the Indian subcontinent, Tectonophys, 644, 1–21, doi:10.1016/j.tecto.2015.01.007, 2015.

Singh, A., Eken, T., Mohanty, D. D., Saikia, D., Singh, C., and Kumar, M. R.: Significant seismic anisotropy beneath southern Tibet inferred from splitting of direct S waves, Phys. Earth Planet. In., 250, 1–11, doi:10.1016/j.pepi.2015.11.001, 2016.

Sol, S., Meltzer, A., Bürgmann, R., Van der Hilst, R., King, R., Chen, Z., Koons, P., Lev, E., Liu, Y., Zeitler, P., Zhang, X., Zhang, J., and Zurek, B.: Geodynamics of the southeastern Tibetan Plateau from seismic anisotropy and geodesy, Geology, 35, 563–566, doi:10.1130/G23408A.1, 2007.

Tapponnier, P., Peltzer, G., Le Dain, A., Armijo, R., and Cobbold, P.: Propagating extrusion tectonics in Asia: New insights from simple experiments with plasticine, Geology, 10, 611–616, doi:10.1130/0091-7613(1982)10<611:PETIAN>2.0.CO;2, 1982.

Tapponnier, P., Zhiqin, X., Roger, F., Meyer, B., Arnaud, N., Wittlinger, G., and Jingsui, Y.: Oblique stepwise rise and growth of the Tibet Plateau, Science, 294, 1671–1677, doi:10.1126/science.105978, 2001.

Tseng, T.-L., Chen, W.-P., and Nowack, R. L.: Northward thinning of Tibetan crust revealed by virtual seismic profiles, Geophys. Res. Lett., 36, L24304, doi:10.1029/2009GL040457, l24304, 2009.

Van Hinsbergen, D. J., Lippert, P. C., Dupont-Nivet, G., McQuarrie, N., Doubrovine, P. V., Spakman, W., and Torsvik, T. H.: Greater India Basin hypothesis and a two-stage Cenozoic collision between India and Asia, P. Natl. Acad. Sci. USA, 109, 7659–7664, doi:10.1073/pnas.1117262109, 2012.

Vinnik, L. and Montagner, J.-P.: Shear wave splitting in the mantle Ps phases, Geophys. Res. Lett., 23, 2449–2452, doi:10.1029/96GL02263, 1996.

Vinnik, L., Green, R., and Nicolaysen, L.: Recent deformations of the deep continental root beneath southern Africa, Nature, 375, 50–52, doi:10.1038/375050a0, 1995.

Vinnik, L., Singh, A., Kiselev, S., and Kumar, M. R.: Upper mantle beneath foothills of the western Himalaya: subducted lithospheric slab or a keel of the Indian shield?, Geophys. J. Int., 171, 1162–1171, doi:10.1111/j.1365-246X.2007.03577.x, 2007.

Vinnik, L. P., Makeyeva, L. I., Milev, A., and Usenko, A. Y.: Global patterns of azimuthal anisotropy and deformations in the continental mantle, Geophys. J. Int., 111, 433–447, doi:10.1111/j.1365-246X.1992.tb02102.x, 1992.

Walsh, B. M., Sibeck, D. G., Nishimura, Y., and Angelopoulos, V.: Statistical analysis of the plasmaspheric plume at the magnetopause, J. Geophys. Res., 118, 4844–4851, doi:10.1002/jgra.50458, 2013.

Wang, C., Chang, L., Lü, Z., Qin, J., Su, W., Silver, P., and Flesch, L.: Seismic anisotropy of upper mantle in eastern Tibetan Plateau and related crust-mantle coupling pattern, Sci. China Ser. D, 50, 1150–1160, doi:10.1007/s11430-007-0053-5, 2007.

Wang, C.-Y., Flesch, L. M., Silver, P. G., Chang, L.-J., and Chan, W. W.: Evidence for mechanically coupled lithosphere in central Asia and resulting implications, Geology, 36, 363–366, doi:10.1130/G24450A.1, 2008.

Wüstefeld, A. and Bokelmann, G.: Null detection in shear-wave splitting measurements, B. Seismol. Soc. Am., 97, 1204–1211, doi:10.1785/0120060190, 2007.

Wüstefeld, A., Bokelmann, G., Zaroli, C., and Barruol, G.: SplitLab: A shear-wave splitting environment in Matlab, Comput. Geosci., 34, 515–528, doi:10.1016/j.cageo.2007.08.002, 2008.

Yao, H., Beghein, C., and Van Der Hilst, R. D.: Surface wave array tomography in SE Tibet from ambient seismic noise and two-station analysis-II, Crustal and upper-mantle structure, Geophys. J. Int., 173, 205–219, doi:10.1111/j.1365-246X.2007.03696.x, 2008.

Yao, H., Van Der Hilst, R. D., and Montagner, J.-P.: Heterogeneity and anisotropy of the lithosphere of SE Tibet from surface wave array tomography, J. Geophys. Res., 115, B12307, doi:10.1029/2009JB007142, 2010.

Yin, A. and Harrison, T. M.: Geologic Evolution of the Himalayan-Tibetan Orogen, Annu. Rev. Earth Planet Sci., 28, 211–280, doi:10.1146/annurev.earth.28.1.211, 2000.

Zhang, P.-Z., Shen, Z., Wang, M., Gan, W., Bürgmann, R., Molnar, P., Wang, Q., Niu, Z., Sun, J., Wu, J., Hanrong, S., and Xinzhao, Y.: Continuous deformation of the Tibetan Plateau from global positioning system data, Geology, 32, 809–812, doi:10.1130/G20554.1, 2004.

Zhao, J., Yuan, X., Liu, H., Kumar, P., Pei, S., Kind, R., Zhang, Z., Teng, J., Ding, L., Gao, X., Xu, Q., and Wang, W.: The boundary between the Indian and Asian tectonic plates below Tibet, P. Natl. Acad. Sci. USA, 107, 11229–11233, doi:10.1073/pnas.1001921107, 2010.

Zhao, J., Murodov, D., Huang, Y., Sun, Y., Pei, S., Liu, H., Zhang, H., Fu, Y., Wang, W., Cheng, H., and Tang, W.: Upper mantle deformation beneath central-southern Tibet revealed by shear wave splitting measurements, Tectonophys, 627, 135–140, doi:10.1016/j.tecto.2013.11.003, 2014.

Zhou, H. and Murphy, M. A.: Tomographic evidence for wholesale underthrusting of India beneath the entire Tibetan plateau, J. Asian Earth Sci., 25, 445–457, doi:10.1016/j.jseaes.2004.04.007, 2005.

Uncertainty assessment in 3-D geological models of increasing complexity

Daniel Schweizer, Philipp Blum, and Christoph Butscher

Karlsruhe Institute of Technology (KIT), Institute for Applied Geosciences (AGW), Kaiserstr. 12, 76131 Karlsruhe, Germany

Correspondence to: Daniel Schweizer (daniel.schweizer@kit.edu)

Abstract. The quality of a 3-D geological model strongly depends on the type of integrated geological data, their interpretation and associated uncertainties. In order to improve an existing geological model and effectively plan further site investigation, it is of paramount importance to identify existing uncertainties within the model space. Information entropy, a voxel-based measure, provides a method for assessing structural uncertainties, comparing multiple model interpretations and tracking changes across consecutively built models. The aim of this study is to evaluate the effect of data integration (i.e., update of an existing model through successive addition of different types of geological data) on model uncertainty, model geometry and overall structural understanding. Several geological 3-D models of increasing complexity, incorporating different input data categories, were built for the study site Staufen (Germany). We applied the concept of information entropy in order to visualize and quantify changes in uncertainty between these models. Furthermore, we propose two measures, the Jaccard and the city-block distance, to directly compare dissimilarities between the models. The study shows that different types of geological data have disparate effects on model uncertainty and model geometry. The presented approach using both information entropy and distance measures can be a major help in the optimization of 3-D geological models.

1 Introduction

Three-dimensional (3-D) geological models have gained importance in structural understanding of the subsurface and are increasingly used as a basis for scientific investigation (e.g., Butscher and Huggenberger, 2007; Caumon et al., 2009; Bis-tacchi et al., 2013; Liu et al., 2014), natural resource exploration (e.g., Jeannin et al., 2013; Collon et al., 2015; Hassen et al., 2016), decision making (e.g., Campbell et al., 2010; Panteleit et al., 2013; Hou et al., 2016) and engineering applications (Hack et al., 2006; Kessler et al., 2008). Overall, 3-D geological models are usually preferable over 2-D solutions because our object of study is intrinsically three-dimensional in space and, therefore, they offer a higher degree of data consistency and superior data visualization. Moreover, they enable the integration of many different types of geological data such as geological maps, cross sections, outcrops, boreholes and data from geophysical (e.g., Boncio et al., 2004) and remote-sensing methods (e.g., Schamper et al., 2014). Nevertheless, input data are often sparse, heterogeneously distributed or poorly constrained. In addition, uncertainties from many sources such as measurement error, bias and imprecisions, randomness, and lack of knowledge are inherent to all types of geological data (Mann, 1993; Bárdossy and Fodor, 2001; Culshaw, 2005). Furthermore, assumptions and simplifications are made during data collection, and subjective interpretation is part of the modeling process (Bond, 2015). Hence, model quality strongly depends on the type of integrated geological data and its associated uncertainties.

In order to assess the quality and reliability of a 3-D geological model as objectively as possible, it is essential to address underlying uncertainties. Numerous methods have recently been proposed that enable estimates, quantification and visualization of uncertainty (Tacher et al., 2006; Wellmann et al., 2010; Lindsay et al., 2012, 2013, 2014; Lark et al., 2013; Park et al., 2013; Kinkeldey et al., 2015). A promising approach is based on the concept of information entropy (Shannon, 1948). Wellmann and Regenauer-Lieb (2012) applied this concept to 3-D geological models. In their study, they evaluated uncertainty as a property of each discrete point of the model domain by quantifying the amount of missing information with regard to the position of a geological unit (Wellmann and Regenauer-Lieb, 2012). They consecutively added new information to a 3-D model and compared uncertainties between the resulting models at discrete locations and as an average value for the total model domain using information entropy as a quantitative indicator. Through their approach, they addressed two important questions: (1) how is model quality related to the available geological information and its associated uncertainties, and (2) how is model quality improved through the incorporation of new information?

Wellmann and Regenauer-Lieb (2012) illustrated their approach using synthetic 3-D geological models, showing how additional geological information affects model uncertainty. The present study goes a step further. It applies the concept of information entropy as well as model dissimilarity to a real case, namely the city of Staufen, Germany, at the eastern margin of the Upper Rhine Graben. In contrast to the previous study, the present study evaluates the effects of the consecutive addition of data from different data categories to an existing model on model uncertainty and overall model geometry. We hypothesize that disparate effects of different data types on model uncertainty exist and that the quantification of these effects provides a trade-off between costs (i.e., data acquisition) and benefits (i.e., reduced uncertainty and therefore higher model quality). Thus, several 3-D geological models of the study site were consecutively built with increasing complexity; each of them based on an increasing amount of (real) categorized data. An approach was developed that uses information entropy and model dissimilarity for the quantitative assessment of uncertainty in the consecutive models. Results indicate that the approach is applicable for complex and real geological settings. The approach has large potential as a tool to support both model improvement through successive data integration and cost–benefit analyses of geological site investigations.

2 Study site

The city of Staufen suffers from dramatic ground heave that resulted in serious damage to many houses (southwest Germany, Fig. 1). Ground heave with uplift rates exceeding $10\,\text{mm month}^{-1}$ started in 2007 after seven wells were drilled to install borehole heat exchangers (BHEs) for heating the local city hall (LGRB, 2010). After more and more houses in the historic city center showed large cracks, an exploration program was initiated by the state geological survey (LGRB – Landesamt für Geologie, Rohstoffe und Bergbau) in order to investigate the case. Results showed that the geothermal wells hydraulically connected anhydrite-bearing clay rocks with a deeper aquifer, and resulting water inflow into the anhydritic clay rock triggered the transformation of the mineral anhydrite into gypsum (Ruch and Wirsing, 2013). This chemical reaction is accompanied by a volume increase that leads to rock swelling, a phenomenon typically encountered in tunneling in such rock (e.g., Einstein, 1996; Anagnostou et al., 2010; Butscher et al., 2011b, 2015; Alonso, 2011), but recently also observed after geothermal drilling (Butscher et al., 2011a; Grimm et al., 2014). The abovementioned exploration program was aimed not only at finding the cause of the ground heave but also at better constraining the complex local geological setting. The hitherto existing geological data were not sufficient to explain the observed ground heave, locate the geological units that are relevant for rock swelling, and plan countermeasures.

Staufen is located west of the Black Forest at the eastern margin of the Upper Rhine Graben (URG). It is part of the "Vorbergzone" (Genser, 1958), a transition zone between the eastern main border fault (EMBF) of the graben and the graben itself. This zone is characterized by staggered fault blocks that were trapped at the graben margin during opening and subsidence of the graben. The strata of this transition zone are often steeply inclined or even vertical (Schöttle, 2005) and are typically displaced by west-dipping faults with a large normal displacement. The fault system, kinematically linked to the EMBF, has a releasing bend geometry and today experiences sinistral oblique movement (Behrmann et al., 2003). The major geological units at the site comprise Triassic and Jurassic sedimentary rocks, which are covered by Quaternary sediments of an alluvial plain in the south (Sawatzki and Eichhorn, 1999) (Fig. 1).

Three geological units play an important role for the swelling problem at the site: the Triassic Gipskeuper ("Gypsum Keuper") formation, which contains the swelling zone, and the underlying Lettenkeuper formation and Upper Muschelkalk formation, which are aquifers providing groundwater that accesses the swelling zone via pathways along the BHE. The Gipskeuper formation consists of marlstone and mudstone and contains the calcium-sulfate minerals anhydrite ($CaSO_4$) and gypsum ($CaSO_4 + H_2O$). The thickness of this formation varies between 50 and 165 m, with an average thickness of 100–110 m (LGRB, 2010), depending on the degree of leaching of the sulfate minerals close to the ground surface. It is underlain by the Lettenkeuper formation (5–10 m thickness), consisting of dolomitic limestone, standstone and mudstone, and the Upper Muschelkalk formation ($\approx 60\,\text{m}$ thickness) dominantly consisting of limestone and dolomitic limestone.

3 Methods

3.1 Input data

Input data for the 3-D geological modeling include all available geological data that indicate (1) boundaries between geological units, (2) the presence of geological units and

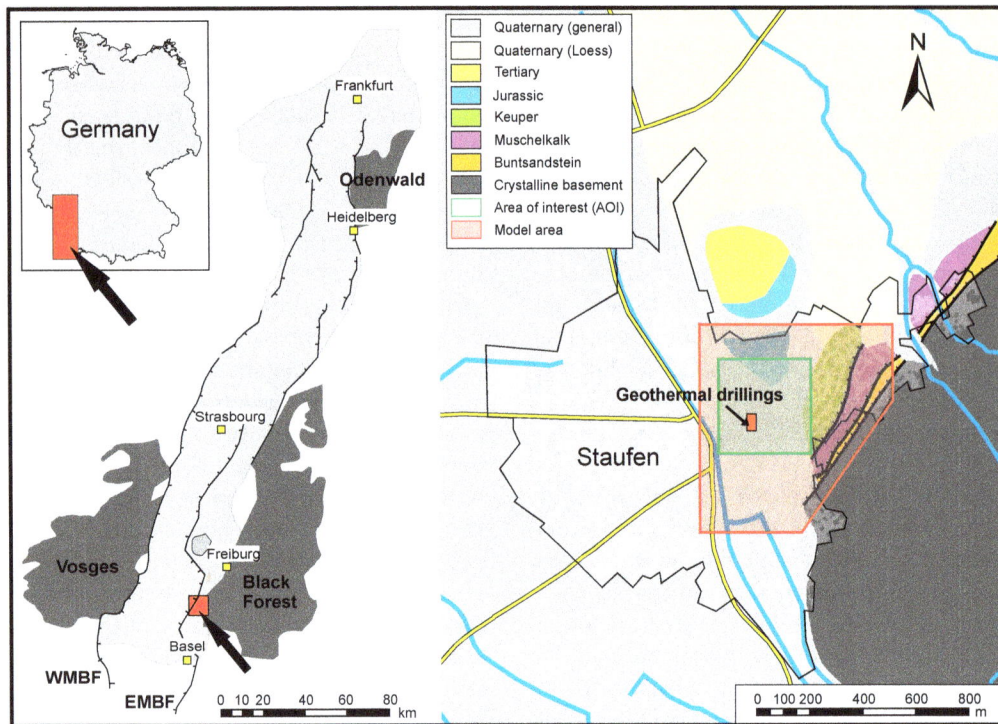

Figure 1. Study site and location of the model area and area of interest (AOI).

faults at a certain positions, and (3) orientation (dip and azimuth) of the strata. These data were classified into four categories (Fig. 2): (1) non-site-specific, (2) site-specific, (3) direct problem-specific data and (4) indirect problem-specific data.

The non-site-specific data category comprises geological data that are generally available from published maps (Sawatzki and Eichhorn, 1999), the literature (Genser, 1958; Groschopf et al., 1981; Schreiner, 1991) and the database of the state geological survey, LGRB. Furthermore, a digital terrain model (DTM) of 1 m grid size is included in the non-site-specific data. Outcrop and borehole data are mostly scarce and irregularly distributed in space. The site-specific data comprise drill logs of the geothermal drillings, which provided a pathway for uprising groundwater that finally triggered the swelling. Problem-specific data comprise all data collected during the exploration program that was conducted after heave at the ground surface caused damage to the local infrastructure (LGRB, 2010, 2012). This exploration program was initiated because geological knowledge of the site was insufficient for an adequate understanding of the swelling process in the subsurface and for planning and implementing suitable countermeasures. The problem-specific data were further divided into direct data from drill cores of the three exploration boreholes (Fig. 2; EKB 1+2 and BB 3), which add very accurate point information, and indirect data from a seismic campaign (Fig. 2; Profile 1–5), which add rather "fuzzy" 2-D information that has to be interpreted.

3.2 3-D geological modeling

The 3-D geological models were constructed using the geomodeling software SKUA/GoCAD® 15.5 by Paradigm. They cover an area of about 0.44 km² and have a vertical extent of 665 m. A smaller area of interest (AOI, 300 m × 300 m, 250 m vertical extent) was defined within the model domain, including the drilled wells and the area where heave at the ground surface was observed and the problem-specific data were collected.

The strata of the models cover 10 distinct geological units including Quaternary sediments, Triassic and Jurassic bedrock, and crystalline basement at the lower model boundary (Fig. 3). The Triassic strata are further divided (from top to bottom) into four formations of Keuper (Steinmergelkeuper, Schilfsandstein, Gipskeuper and Lettenkeuper), two formations of Muschelkalk (Upper Muschelkalk, Middle to Lower Muschelkalk) and the Buntsandstein formation. Figure 3 provides an overview over the modeled geological units and average thicknesses used in the initial models.

Four initial models were consecutively built, according to the four previously described data categories. Model 1 was constructed based only on non-site-specific data (maps, literature, etc.); Model 2 additionally considered site-specific data (drill logs of the seven geothermal drillings); Model 3 also included "direct" problem-specific data (exploration boreholes); and finally, Model 4 included "indirect" problem-specific data (seismic campaign). Through this approach,

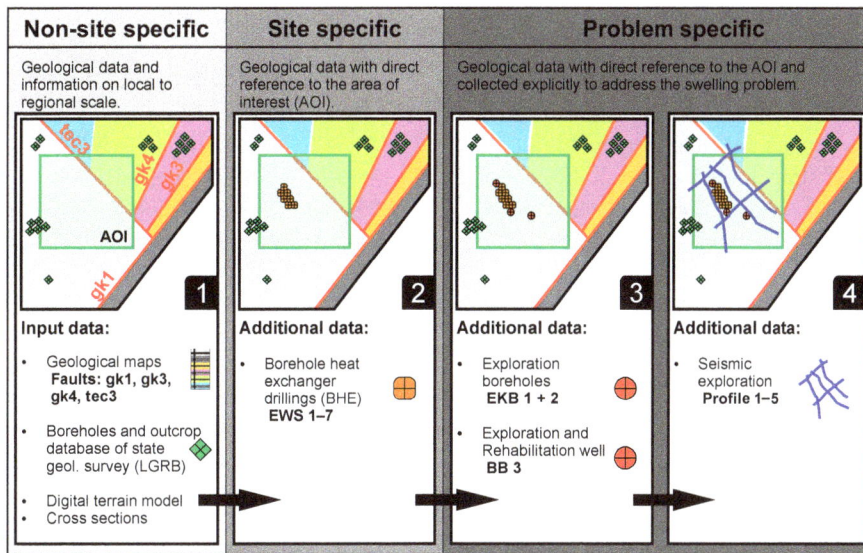

Figure 2. Data categories and geological input data used to build four initial 3-D geological models. The green square indicates the area of interest (AOI), where data were extracted for further analysis. For geological formation color code, see Fig. 1.

Figure 3. Stratigraphic overview of the study area and modeled geological units with average thicknesses.

data density and structural model complexity increase from Models 1 to 4 and the models required successively higher efforts in data acquisition in the field.

First, an explicit modeling approach (Caumon et al., 2009) was used to create representative boundary surfaces for the geological units and faults of the initial model because the available input data were, in terms of spatial coverage, not sufficient to directly use an implicit approach. Discrete smooth interpolation (DSI) provided by GoCAD® was used as the interpolation method (Mallet, 1992), which resulted in

Delaunay-triangulated surfaces for both horizons and faults. Subsequently, based on the explicitly constructed surfaces, a volumetric 3-D model was built by implicit geological modeling, implemented in the software SKUA®. The implicit modeling approach uses a potential field interpolation considering the orientation of strata (Frank et al., 2007), and is based on the U-V-t concept (Mallet, 2004), where horizons represent geochronological surfaces.

3.3 Uncertainty assessment

3.3.1 General approach

Our approach for assessing uncertainties in the 3-D geological models consists of four distinct steps (Fig. 4):

i. Building the initial 3-D geological models of increasing data density and structural complexity (see above).

ii. The definition of fault and horizon uncertainties. Horizon uncertainties were specified in SKUA® by a maximum displacement parameter or by alternative surface interpretations, resulting in a symmetric envelope of possible surface locations around the initial surface. To constrain the shape of generated horizons, SKUA® uses a variogram that spatially correlates perturbations applied to the initial surfaces (Paradigm, 2015). Fault uncertainties were defined by a maximum displacement parameter and a Gaussian probability distribution around the initial fault surface (Caumon et al., 2007; Tertois and Mallet, 2007).

iii. The creation of 30 model realizations for each initial model based on the surface variations defined above, applying the Structure Uncertainty workflow of SKUA®.

iv. The extraction of the geological information from all model realizations for analysis, comparison and visualization. For this purpose, the AOI was divided into a regular 3-D grid of 5 m cell size, resulting in 180 000 grid cells. The membership of a grid cell to a geological unit was defined as a discrete property of each grid cell and extracted for all 30 model realizations. Based on these data, we calculated the probability of each geological unit being present in a grid cell in order to derive the information entropy at the level of (1) a single grid cell, (2) a subset representing the area of extent of a geological unit and (3) the overall AOI. Furthermore, the fuzzy set entropy was calculated to determine the ambiguousness of the targeted geological units Gipskeuper (km1), Lettenkeuper (ku) and Upper Muschelkalk (mo) within the AOI. Calculations were conducted using the statistics package R (R Core Team, 2016). The underlying concepts and equations used to calculate probabilities and entropies are described in the following section.

3.3.2 Information entropy

The concept of information entropy (or Shannon entropy) was first introduced by Shannon (1948) and is well known in probability theory (Klir, 2005). It quantifies the amount of missing information and hence, the uncertainty at a discrete location x, based on a probability function P of a finite data set. When applied to geological modeling, information entropy expresses the "degree of membership" of a grid cell to a specific geological unit. In other words, information entropy quantitatively describes how unambiguously the available information predicts that unit U is present at location x. Information entropy was recently applied to 3-D geological modeling by Wellmann et al. (2010) and Wellmann and Regenauer-Lieb (2012) in order to quantify and visualize uncertainties introduced by the imprecision and inaccuracy of geological input data. A detailed description of the method can be found in the cited references and is briefly summarized here.

By subdividing the model domain M into a regular grid, a discrete property can be assigned to any cell at location x in the model domain. In a geological context, the membership of a grid cell to a geological unit U can be defined as such a property by an indicator function:

$$\mathbf{I}_U(x) = \begin{cases} 1 & \text{if } x \in U \\ 0 & \text{otherwise,} \end{cases} \tag{1}$$

Applied to all n realizations k of the model space M, the indicator function yields a set of n indicator fields $\mathbf{I}$ with each of them defining the membership of a geological unit as a property of a grid cell. Considering the combined information of all indicator fields, it follows that membership is no longer unequivocally defined at a location x and hence has to be expressed by a probability function P_U:

$$P_x(U) = \sum_{k \in n} \frac{\mathbf{I}_{U_k}(x)}{n}. \tag{2}$$

From the probabilities of occurrence $P_x(U)$ the uncertainty (or amount of missing information) associated with a discrete point (grid cell) can be obtained by calculating the information entropy H_x (Shannon, 1948) for a set of all possible geological units $\mathcal{U}$:

$$H_x = -\sum_{U \in \mathcal{U}} P_x(U) \times \log P_x(U). \tag{3}$$

In a next step, information entropy H_M can be calculated as an average value of H_x over the entire model space:

$$H_M = \frac{1}{|M|} \times \sum_{x \in M} H_x, \tag{4}$$

where $|M|$ is the number of elements within M, $H_M = 0$ denotes that the location of all geological units is precisely known (no uncertainty) and H_M is maximal for equally distributed probabilities of the geological units ($P_{U1} = P_{U2} = $

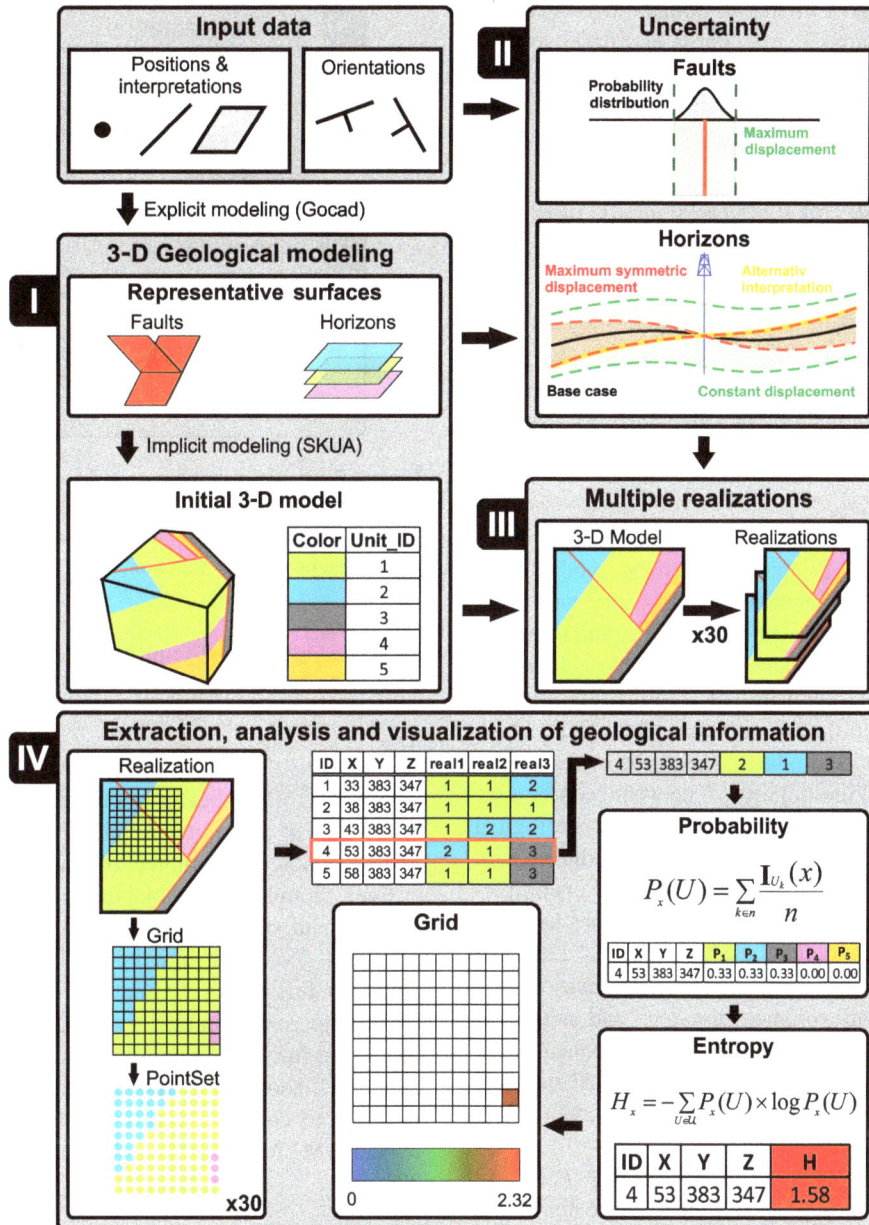

Figure 4. Uncertainty assessment workflow with four distinct steps. This workflow is applied to four initial models that are based on the different data sets illustrated in Fig. 2.

$P_{U3} = \ldots$), which means that a clear distinction between geological units within the model space is not possible. Similarly, average information entropy can also be applied to only a subset of the model space ($S \subseteq M$):

$$H_S = \frac{1}{|S|} \times \sum_{x \in S} H_x. \tag{5}$$

H_S can be used to evaluate the contribution of a specific sub-domain to overall uncertainty. In the case of a drilling campaign, for example, the sub-domain can comprise a targeted depth or a geological formation of specific interest. In

this study, we used the probability function $P_x(U)$ with H_S conditioned by $P_x(U) > 0$ to define subsets within the model space. Thus, each subset represents the probability space of a geological formation of interest, namely the Lettenkeuper (S_{ku}), Gipskeuper (S_{km1}) and Upper Muschelkalk (S_{mo}) formation.

Wellmann and Regenauer-Lieb (2012) also adapted fuzzy set theory (Zadeh, 1965) in order to assess how well-defined a single geological unit is within a model domain. A fuzzy set of n model realizations introduces a certain degree of indefiniteness to a discrete property (e.g., membership of a geological unit), resulting in imprecise boundaries which can be referred to as fuzziness. The fuzziness of a fuzzy set (De Luca and Termini, 1972) in the context of a geological 3-D model can be quantified by the fuzzy set entropy H_U (Leung et al., 1992; Yager, 1995):

$$H_U = -\frac{1}{N} \times \sum_{x=1}^{N} \left[P_x(U) \log P_x(U) + (1 - P_x(U)) \right.$$
$$\left. \log (1 - P_x(U)) \right], \tag{6}$$

where the probability function $P_x(U)$ with an interval $[0,1]$ represents the degree of membership of a grid cell to a fuzzy set. H_U equals 0 when $P_x(U)$ is either 0 or 1 everywhere within the set; and H_U equals 1 when all cells of the set have an equal probability of $P_x(U) = 0.5$.

3.4 Model dissimilarity

The stepwise addition of input data to the models (see Sect. 3.1) not only affects uncertainties associated with a geological unit but also the geometry of the units and therefore their position, size and orientation in space. New data may significantly change the geometry of a geological unit but only marginally change the overall uncertainty. Thus, both model uncertainty and dissimilarity should be evaluated. In order to quantify the dissimilarity d between consecutive models in terms of the probability of a specific geological unit occurring in a given voxel, two measures, the Jaccard and the city-block distance (Fig. 5), are proposed to complement information entropy. However, dissimilarities between models, and therefore, uncertainties, have recently also been addressed very effectively using geo-diversity metrics such as formation depth and volume, curvature and neighborhood relationships together with principal component analysis (Lindsay et al., 2013) and through topological analysis, which quantifies geological relationships in a model (Thiele et al., 2016a, b).

The set of locations for which the probability $P_x(U)$ of belonging to a particular geological unit U is greater than a threshold value t can be defined by

$$Q_M^t = \{x\}_{P_x(U)>t}. \tag{7}$$

A threshold value of $t = 0$ was applied in order to capture and consider the same sample space as in H_U. This definition is highly sensitive to outcomes of small probability and might, in some cases, be more robust using a threshold value greater than 0 (e.g., $t > 0.05$). The Jaccard similarity measure (Webb and Copsey, 2003) is then defined as the size of the intersection divided by the size of the union (overlap) of two sample sets ($M1$, $M2$), which in our case represent the similarity in position of a geological unit U between two models:

$$s_{JAC} = \frac{|Q_{M1}^t \cap Q_{M2}^t|}{|Q_{M1}^t \cup Q_{M2}^t|}. \tag{8}$$

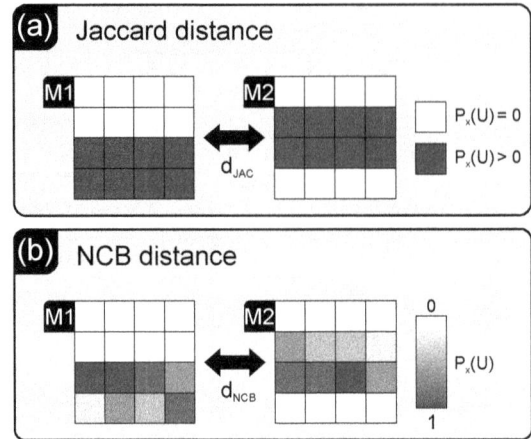

Figure 5. Distance measures used to calculate dissimilarities between models ($M1$, $M2$). **(a)** Jaccard distance (d_{JAC}) using a true/false binary function and **(b)** normalized city-block distance based on a probability function.

Accordingly, the dissimilarity between models can be expressed by the Jaccard distance:

$$d_{JAC} = 1 - s_{JAC}, \tag{9}$$

where $d_{JAC} = 1$ indicates maximum dissimilarity (no match in position of a geological unit U between two models) and $d_{JAC} = 0$ indicates complete overlap.

Even though the use of binary dissimilarities is straightforward and suitable to quantify absolute changes in position of a geological unit between models, it does not account for fuzziness (see Sect. 3.3.2). Hence, the dissimilarity may be overestimated by the Jaccard distance. In order to include fuzziness, the normalized city-block distance was employed, adopting the probability function $P_x(U)$ as a dimension to compare dissimilarities between the two sample sets (M1,M2) (Webb and Copsey, 2003; Paul and Maji, 2014):

$$d_{NCB} = \frac{1}{N} \times \sum_{x=1}^{N} |P_x^{M1}(U) - P_x^{M2}(U)|, \tag{10}$$

where N is the size of $M1 \cup M2$ (i.e, number of grid cells present within the union). The distance is greatest for $d_{NCB} = 1$.

4 Results and discussion

4.1 Initial 3-D models

The four consecutively constructed initial models show a stepwise increase in structural complexity (Fig. 6). Model 1 was based on non-site-specific geological data, and horizon orientations were only constrained by regionally available, isolated outcrop data, which made a general extrapolation of

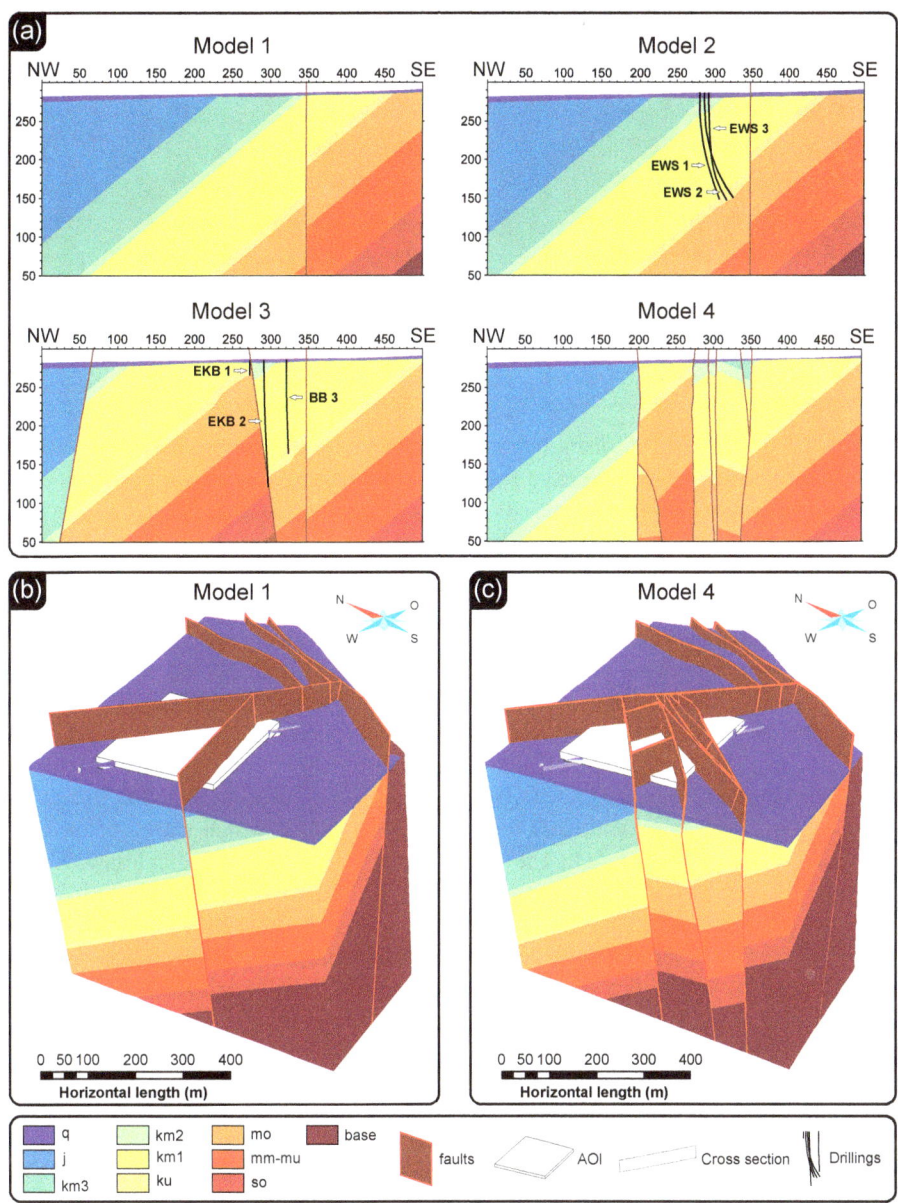

Figure 6. (a) Cross section through the AOI of all four initial geological models with projected borehole tracks (black lines) and 3-D representations of **(b)** Model 1 and **(c)** Model 4.

structures difficult, especially into depth (Jessell et al., 2010). Dip and strike were assumed uniform (40 and 35°) for all horizons across the model domain (see Fig. 6). Information from geological maps and outcrop data revealed a normal fault within the AOI, which was assumed to be ENE–WSW striking with a moderate displacement of about 50 m.

In Model 2, horizon positions of the Schilfsandsteinkeuper (km2), Gipskeuper (km1) and Lettenkeuper (ku) were locally constrained by site-specific information provided by drill logs of the geothermal wells, slightly impacting fault displacement and thickness of the formations. However, changes in model geometry were minor, as no further infor-

mation on horizon orientations was available and no additional faults could be located. By adding the direct problem-specific data from the exploration wells to Model 3, a horst–graben structure was identified that entailed a considerable displacement at two normal faults between and to the northwest of the wells with a displacement of 120 and 70 m, respectively. Furthermore, the drill logs included orientation measurements of the strata, resulting in a shift in position and inclination of layers, compared to the previous models. Thus, large parts of the model domain within the AOI changed from Model 2 to Model 3, and, as a consequence, dissimilarities between these models are particularly high (see Sect. 4.4).

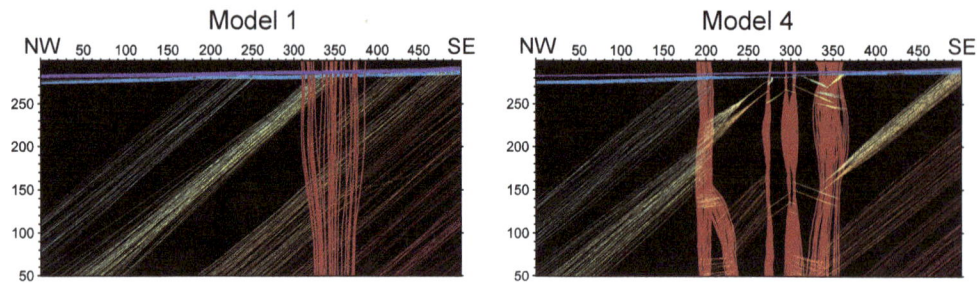

Figure 7. Cross section through Models 1 and 4. The multiple lines show 30 model realizations with shifted faults and horizons (for the location of the cross sections, see Fig. 6). The horizontal lines indicate the land surface (purple) and the base of the Quaternary (blue).

Finally, Model 4, which included data from a seismic campaign, has the highest degree of structural complexity. The information provided by seismic sections revealed uncertainties, which were present previously but not captured by the simpler Models 1 to 3. Ultimately, seismic data force the interpreter to add complexity down to a certain scale. However, seismic surveys are inherently ambiguous and allow alternative interpretations, especially concerning the orientation and number of faults as well as the type of fault contact to a fault network (e.g., branching) (Røe et al., 2014; Cherpeau and Caumon, 2015; Julio et al., 2015). In our case, seismic sections and interpretations were adopted from LGRB (2010). The indirect problem-specific data from the seismic 2-D survey located several additional faults within the AOI, and in some cases caused a shift in position of faults compared to Model 3. The AOI was strongly fragmented by the added faults, and the orientation of layers is no longer uniform but varies strongly between fault blocks. In summary, the stepwise integration of data according to the four data categories improved our general knowledge of subsurface structures at the study site (Fig. 2). In addition, the effect of data integration from different exploration stages on modeled subsurface geometry could be evaluated and visualized.

4.2 Multiple model realizations

The multiple (30) model realizations created by the Structural Uncertainty workflow of SKUA® are illustrated in Fig. 7 using 2-D cross sections of Models 1 and 4 as examples. A total number of 30 realizations and a cell size of 5 m was chosen as a compromise between model detail, lowest practical limit for statistical viability and data handling. For the same reason we did not base our number of realizations on an estimate of convergence. Instead we used the estimate of 30 realizations for a stable fluctuation in fuzzy entropy in a model developed by Wellmann et al. (2010) as a guideline value to our model. Perturbations in horizon location are based on (1) alternative surface interpretations, which reflect a maximum deviation in dip and azimuth ($\pm 5°$) from the initial surface and (2) constant displacement values, which were assigned in order to account for uncertainties in formation thickness and boundary location. For a more detailed expla-

nation of our choice of parameters, assigned probability distributions and specific input modes of the Structural Uncertainty workflow, please refer to the Supplement (Tables S1 and S2). In Model 1, the non-site-specific data set includes minimal constraints, resulting in faults and horizons of the realizations that are widely dispersed but parallel. In contrast, the faults and horizons of the Model 4 realizations are more narrowly dispersed where problem-specific data were available within the AOI. The workflow handles equal uncertainties consistently across models by producing a similar pattern of horizontal displacement in Models 1 and 4. This can be seen in particular for structures located close to the NW boundary, which were not further constrained by consecutively added geological data. However, it is also apparent from the mostly uniform orientation of the surfaces in the 30 realizations of each model that perturbation measures implemented in the Structural Uncertainty workflow did not allow for large variations in dip and azimuth of horizons or faults. Therefore, uncertainty may be systematically underestimated especially at greater depths.

4.3 Uncertainty assessment

4.3.1 Distribution of information entropy

Information entropy, quantified at the level of individual grid cells, can be visualized in 3-D to identify areas of uncertainty and evaluate changes in geometry resulting from successive data integration. Figure 8a shows the distribution of information entropy for Models 1 and 4. It can also be seen that the approach is suitable for locating areas with high degrees of uncertainty, indicated by dark red colors (hot spots) in this figure. Furthermore, Fig. 8b highlights where additional constraints from the data helped to optimize the model by reducing uncertainties ($\Delta H_x < O$) and whether further constraints are needed in locations of specific interest.

The overall distribution of uncertainty was clearly affected by additional geological information from site- and problem-specific input data (Model 4). This effect is highlighted by the changes in entropy between the models (Fig. 8b). Additional constraints on horizon and fault boundaries caused a shift in position and orientation of geological units, followed by a

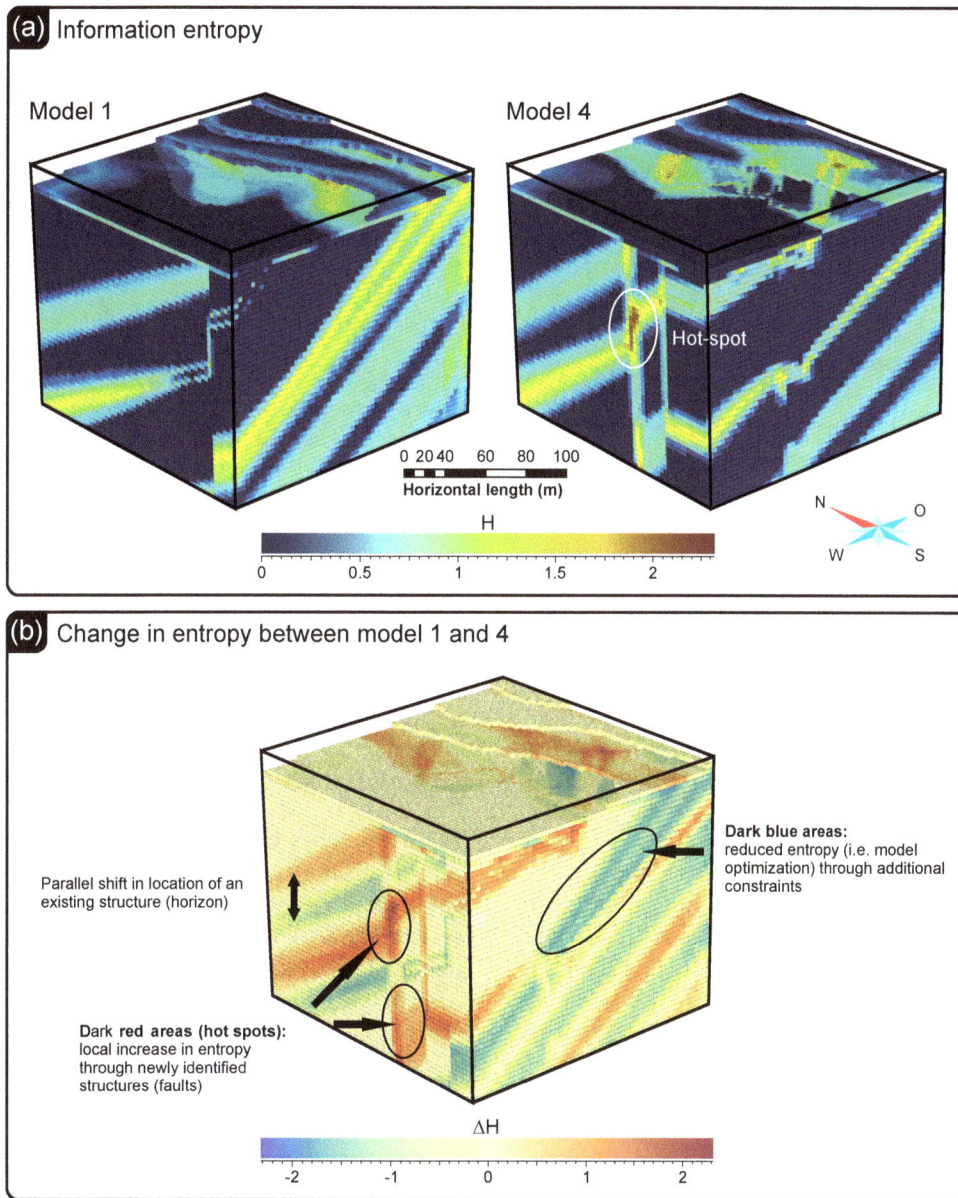

Figure 8. 3-D view of the AOI with a discretization of 5 m for **(a)** average information entropy H_M of Models 1 and 4 and **(b)** change in entropy ΔH_x between both models.

large redistribution of uncertainties, indicated by the changes in entropy. It can be seen that new hot spots of uncertainty were introduced in proximity to the faults identified by the exploration boreholes and the seismic data incorporated into Model 4 (see Fig. 6). However, these new areas of uncertainty can be considered an optimization of the model because large parts of the preceding Model 1 did not reflect the complex local geology. Model 1 (wrongly) predicted low uncertainties for areas where information on unidentified but existing structures (i.e., faults) was missing. This illustrates that epistemic uncertainties at the study site are likely substantial. Even Model 4 will inevitably still underrepresent the

true structural complexity at this site, especially in areas of low data density. In a risk-assessment and decision-making process, this can be problematic because low uncertainty areas might be in fact no-information areas. In such a case, the respective model area would actually be highly uncertain. However, ambiguities in data interpretation (e.g., seismic sections) can lead to incorrectly identified structures and uncertainty in any case, even in areas of high data density. Nevertheless, the approach allows one to assess and visualize uncertainties related to structures that have been identified during site investigation. To lessen the limitations posed by non-sampled locations, Yamamoto et al. (2014) proposed a post-processing method for uncertainty reduction, using multiple indicator functions and interpolation variance in addition to information entropy. Based on information theory, Wellmann (2013) further proposed joint entropy, conditional entropy and mutual information as measures to evaluate correlations and reductions of uncertainty in a spatial context. However, uncertainty from a lack of evidence of a geological structure (e.g., fault), known as imprecise knowledge (Mann, 1993), still depends on the density and completeness of available input data.

4.3.2 Average information entropy

The calculated average information entropy H_T of the consecutive models steadily decreases with higher data specificity (i.e., non-site to problem-specific, see Fig. 2) from Models 1–4 (Fig. 9). Mean values of H_M ranged from 0.56 (Model 1) to 0.39 (Model 4), where $H_M = 0$ would denote no structural uncertainty. The decrease from Models 1 to 4 is approximately linear, indicating that all four categories of geological data had a similar impact on overall model uncertainty, even though the added information resulted in quite different model geometries and, as discussed above, in some cases in a local increase in entropy (see Fig. 8b). A similar but more pronounced trend was observed for the average entropy H_S of the subsets S_{km1}, S_{ku} and S_{mo}, which represent the domain of the three geological units that are of particular importance to the swelling problem. However, entropy, i.e., the amount of uncertainty, is considerably higher within the domain of these geological units than for the overall model space, especially for the subsets S_{ku} and S_{mo}, identifying them as areas of a particularly high degree of uncertainty. Note that these units are the aquifers that have been hydraulically connected to the swellable rocks via the geothermal drillings. Nevertheless, all entropy values are comparably moderate, considering that a maximum of (only) five different geological units was found in any one grid cell across all four models, yielding a possible maximum entropy of $H_M = 2.32$ for an equal probability distribution ($P_1 = P_2 = P_3 = P_4 = P_5$). For comparison: if all 10 geological units would be equally probable, the maximum entropy would be 3.32. Furthermore, median values and interquartile range dropped from 0.51 (0–0.99) in Model 1 to 0 (0–0.84) in Model 4. This helps to illustrate that the amount of grid cells with $H_x = 0$ (indicating no inherent uncertainty), increased notably by 34.8 % from 40.6 (Model 1) to 54.8 % (Model 4) and that the remaining entropies in Model 4 are limited to a considerably smaller number of cells within the model domain.

Overall, comparing the pre- to post-site-investigation situations (Models 1–4), site and problem-specific investigations were all equally successful in adding information to the model and reducing uncertainties in the area of the targeted horizons. While the benefits from the different data are equal, the costs in data acquisition (i.e., work, money and time re-

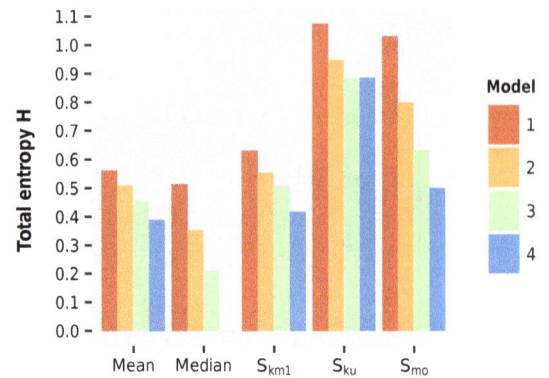

Figure 9. Average entropy H_M calculated for the different models (mean and median) and for subsets of the model space of each model (S_{km1}, S_{ku}, S_{mo}).

quired) may vary considerably, depending on the exploration method (e.g., drillings and seismic survey). An economic evaluation was not within the scope of this study. Nevertheless, the approach presented could improve cost and benefit analyses by quantifying the gain in information through different exploration stages.

4.3.3 Fuzzy set entropy

The fuzzy set entropy was calculated to indicate how well-defined a geological unit is within the model space. Applied to the swelling problem of our case study, a high degree of uncertainty remains with regard to the position of the relevant geological units (km1, ku, mo) after full data integration. We obtained fuzzy set entropy values (H_U) ranging between 0.329–0.504 (Fig. 10). The fuzziness of these geological units only slightly changed from Models 1 to 4, indicating that higher data specificity did not translate into more clearly defined geological units within the model domain. This can be partially attributed to the complex geological setting of the study site. In the process of data integration, additional boundaries between geological units are created at newly introduced faults, increasing the overall fuzziness of a unit.

In the case of the Lettenkeuper formation (unit ku), boundaries are even slightly less well-defined in Model 4 compared to Model 1. This is likely related to the low thickness of the formation (5–10 m, Fig. 3) relative to the mesh size (5 m). A finer grid could reduce this effect; however, computation time would increase significantly. Wellmann and Regenauer-Lieb (2012) propose using unit fuzziness to determine an optimal representative cell size and reduce the impact of spatial discretization on information entropy. As previously discussed in Sect. 4.2, our workflow does not explicitly consider uncertainties through dip and strike variations by a value indicated for this purpose but through perturbations based on alternative surface interpretations, which in our case likely underestimates the fuzziness of the targeted geological units at greater depths. Thus, overall fuzziness,

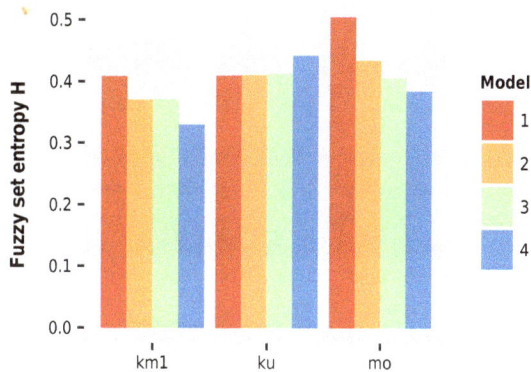

Figure 10. Fuzzy set entropy H_U of the targeted geological units km1, ku and mo of the different models.

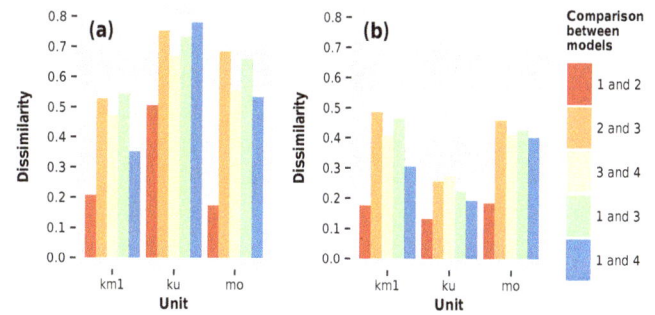

Figure 11. Dissimilarities between the different models expressed by (**a**) Jaccard distance and (**b**) city-block distance.

particularly in Model 1, may be significantly higher than calculated.

4.4 Models dissimilarity

A gain in structural information through newly acquired data usually not only impacts model uncertainty but is also associated with a change in model geometry. The calculated distances between models can identify the data category with the strongest impact on model geometry and make it possible to determine whether model geometry and uncertainty are related. Figure 11 shows the calculated Jaccard and city-block distances between the models with respect to the targeted geological units km1, ku and mo.

Calculated distances between models are rather high, with values of up to 0.78; indicating a pronounced shift in position of the geological units after data were added. The addition of both direct and indirect problem-specific data to Model 3 had a strong impact on model geometry, which can be seen by comparing the calculated distances between Models 2, 3 and 4 for both Jaccard and city block (Fig. 11). In contrast, site-specific data had a much lower effect, with less than a 20 % (0.2) change in unit position, except for ku of the Jaccard distance (see distance between Models 1 and 2).

Overall, the city-block distance, which considers the fuzziness of geological boundaries, shows a similar trend to the Jaccard distance; however, changes are much less pronounced, especially for unit ku. According to the low city-block distance, absolute changes in probability $P_x(U)$ for each grid cell are small, whereas high Jaccard distances indicate a large number of grid cells being affected through newly added data. Thus, the Jaccard distance likely overestimated the actual dissimilarity between models. Comparing unit ku of both distances; the disparity between values hints at a large number of low-degree changes in membership of the grid cells ($\Delta P_x(U) \ll 1$). These predominantly low-degree changes are likely related to the abovementioned high degree of unit boundary fuzziness and the resulting, ill-defined, geological unit ku being shifted within the model

domain. However, a direct comparison of fuzzy set entropy to the corresponding city-block distance yields no quantifiable relationship between model geometry and structural uncertainty.

Nonetheless, both distance measures allow the quantification and assessment of different aspects of dissimilarities and therefore, changes in geometry across models. Nevertheless, the city-block distance is preferable when sets of multiple realizations are compared because it factors in the probability of the occurrence of a geological unit at a discrete location. In recent years, various distance measures have already been applied in other contexts to create dissimilarity distance matrices and compare model realizations in history matching and uncertainty analysis, particularly in reservoir modeling (Suzuki et al., 2008; Scheidt and Caers, 2009a, b; Park et al., 2013). These include the Hausdorff distance which, similar to our approach, directly compares the geometry of different structural model realizations but also more sophisticated measures that calculate distances in realizations based on flow model responses from a transfer function.

5 Summary and conclusions

Prior work has demonstrated the effectiveness of information entropy in assessing model uncertainties and providing valuable insight into the geological information used to constrain a 3-D model. Wellmann and Regenauer-Lieb (2012), for example, evaluated how additional information reduces uncertainty and helps to constrain and optimize a geological model using the measure of information entropy. Their approach focused on a hypothetical scenario of newly added borehole data and cross-section information to a synthetic model. In the present study, information entropy and, in addition, model dissimilarity was used to assess the impact of newly acquired data on model uncertainties using actual site-investigation data in the complex geological setting of a real case.

We presented a new workflow and methods to describe the effect of data integration on model quality, overall structural understanding of the subsurface and model geometry. Our results provide a better understanding of how model quality can be assessed in terms of uncertainties in a data acquisition process of an exploration campaign, showing that information entropy and model dissimilarity are powerful tools to visualize and quantify uncertainties, even in complex geological settings. The main conclusions of this study are as follows:

1. Average and fuzzy set entropy can be used to evaluate uncertainties in 3-D geological modeling and, therefore, support model improvement during a consecutive data integration process. We suggest that the approach could be used to also perform a cost–benefit analysis of exploration campaigns.

2. The study confirms that 3-D visualization of information entropy can reveal hot spots and changes in the distribution of uncertainty through newly added data in real cases. The method provides insight into how additional data reduce uncertainties in some areas and how newly identified geological structures may create hot spots of uncertainty in others. Furthermore, the method stresses that parsimonious models can locally underestimate uncertainty, which is only revealed after new data are available and being considered.

3. Dissimilarities in model geometry across different sets of model realizations can effectively be quantified and evaluated by a single value using the city-block distance. A combination of the concepts of information entropy and model dissimilarity improves uncertainty assessment in 3-D geological modeling.

However, some limitations of the presented approach are noteworthy. Although it was designed to assess uncertainties in the position and thickness of horizons, uncertainty in orientation could only be included indirectly through perturbations based on alternative surface interpretations but not by explicit dip and azimuth parameter values indicated for this purpose. This may result in a systematic underestimation of uncertainties at greater depths of the model domain. Furthermore, our study site (Vorbergzone) is a highly fragmented geological entity, and epistemic uncertainties due to missing information about unidentified but existing geological structures are likely substantial.

Future work should therefore aim to include "fault block uncertainties" more effectively into the workflow, for example by including multiple fault network interpretations (Holden et al., 2003; Cherpeau et al., 2010; Cherpeau and Caumon, 2015) or by considering fault zones that produce a given displacement by a variable number of faults. Finally, all data of the investigated site were collected prior to our analysis; therefore, additional data were not explicitly collected in order to reduce detected uncertainties within the consecutive models. Applying this approach during an ongoing site investigation could improve the targeted exploration and allow a well-founded cost–benefit analysis through uncertainty hot-spot detection.

Author contributions. Daniel Schweizer, Christoph Butscher and Philipp Blum designed the study and developed the methodology. Daniel Schweizer performed the 3-D geological modeling, implemented the approach for uncertainty assessment and analyzed the results. Daniel Schweizer prepared the paper with contributions from all coauthors.

Competing interests. The authors declare that they have no conflict of interest.

Acknowledgements. The financial support for Daniel Schweizer from the German Research Foundation (DFG) under grant number BU 2993/2-1 is gratefully acknowledged. We acknowledge support by the Deutsche Forschungsgemeinschaft and the Open Access Publishing Fund of the Karlsruhe Institute of Technology. Furthermore, we thank the Geological Survey of Baden-Württemberg (LGRB), especially Gunther Wirsing and Clemens Ruch (LGRB), for data provision and the Paradigm support team for their technical support. Finally, we would like to thank Jan Behrmann (GEOMAR) and Thomas Bohlen (KIT) for discussions of the tectonic setting and seismic interpretations, respectively.

Edited by: G. Peron-Pinvidic

References

Alonso, E.: Crystal growth and geotechnics, Paper presented at the Arrigo Croce Lecture, 15 December 2011, Rome, Italy, 46 pp., available at: http://www.associazionegeotecnica.it/sites/default/files/rig/rig_2012_4_013alonso.pdf (last access: 7 December 2015), 2011.

Anagnostou, G., Pimentel, E., and Serafeimidis, K.: Swelling of sulphatic claystones – some fundamental questions and their practical relevance, Geomech. Tunn., 3, 567–572, doi:10.1002/geot.201000033, 2010.

Bárdossy, G. and Fodor, J.: Traditional and New Ways to Handle Uncertainty in Geology, Nat. Resour. Res., 10, 179–187, doi:10.1023/A:1012513107364, 2001.

Behrmann, J. H., Hermann, O., Horstmann, M., Tanner, D. C., and Bertrand, G.: Anatomy and kinematics of oblique continental rifting revealed: A three-dimensional case study of the southeast Upper Rhine graben (Germany), Am. Assoc. Petr. Geol. B., 87, 1105–1121, doi:10.1306/02180300153, 2003.

Bistacchi, A., Massironi, M., Superchi, L., Zorzi, L., Francese, R., Giorgi, M., Chistolini, F., and Genevois, R.: A 3-D Geological Model of the 1963 Vajont Landslide, Ital. J. Eng. Geol. Environ., 2013, 531–539, doi:10.4408/IJEGE.2013-06.B-51, 2013.

Boncio, P., Lavecchia, G., and Pace, B.: Defining a model of 3-D seismogenic sources for Seismic Hazard Assessment applications: The case of central Apennines (Italy), J. Seismol., 8, 407–425, doi:10.1023/B:JOSE.0000038449.78801.05, 2004.

Bond, C. E.: Uncertainty in structural interpretation: Lessons to be learnt, J. Struct. Geol., 74, 185–200, doi:10.1016/j.jsg.2015.03.003, 2015.

Butscher, C. and Huggenberger, P.: Implications for karst hydrology from 3-D geological modeling using the aquifer base gradient approach, J. Hydrol., 342, 184–198, doi:10.1016/j.jhydrol.2007.05.025, 2007.

Butscher, C., Huggenberger, P., Auckenthaler, A., and Bänninger, D.: Risikoorientierte Bewilligung von Erdwärmesonden, Grundwasser, 16, 13–24, doi:10.1007/s00767-010-0154-5, 2011a.

Butscher, C., Huggenberger, P., Zechner, E., and Einstein, H. H.: Relation between hydrogeological setting and swelling potential of clay-sulfate rocks in tunneling, Eng. Geol., 122, 204–214, doi:10.1016/j.enggeo.2011.05.009, 2011b.

Butscher, C., Mutschler, T., and Blum, P.: Swelling of Clay-Sulfate Rocks: A Review of Processes and Controls, Rock Mech. Rock Eng., 49, 1533–1549, doi:10.1007/s00603-015-0827-6, 2015.

Campbell, S. D. G., Merritt, J. E., Dochartaigh, B. E. O., Mansour, M., Hughes, A. G., Fordyce, F. M., Entwisle, D. C., Monaghan, A. A., and Loughlin, S. C.: 3-D geological models and their hydrogeological applications: supporting urban development a case study in Glasgow-Clyde, UK, Z. Dtsch. Ges. Geowiss., 161, 251–262, 2010.

Caumon, G., Tertois, A.-L., and Zhang, L.: Elements for Stochastic Structural Perturbation of Stratigraphic models, in: Proceedings of Petroleum Geostatistics, European Association of Geoscientists & Engineers, doi:10.3997/2214-4609.201403041, 2007.

Caumon, G., Collon-Drouaillet, P., Le Carlier de Veslud, C., Viseur, S., and Sausse, J.: Surface-Based 3-D Modeling of Geological Structures, Math. Geosci., 41, 927–945, doi:10.1007/s11004-009-9244-2, 2009.

Cherpeau, N. and Caumon, G.: Stochastic structural modelling in sparse data situations, Petrol. Geosci., 21, 233–247, doi:10.1144/petgeo2013-030, 2015.

Cherpeau, N., Caumon, G., and Lévy, B.: Stochastic simulations of fault networks in 3-D structural modeling, C. R. Geosci., 342, 687–694, doi:10.1016/j.crte.2010.04.008, 2010.

Collon, P., Steckiewicz-Laurent, W., Pellerin, J., Laurent, G., Caumon, G., Reichart, G., and Vaute, L.: 3-D geomodelling combining implicit surfaces and Voronoi-based remeshing: A case study in the Lorraine Coal Basin (France), Comput. Geosci., 77, 29–43, doi:10.1016/j.cageo.2015.01.009, 2015.

Culshaw, M.: From concept towards reality: developing the attributed 3-D geological model of the shallow subsurface, Q. J. Eng. Geol. Hydroge., 38, 231–284, doi:10.1144/1470-9236/04-072, 2005.

De Luca, A. and Termini, S.: A definition of a nonprobabilistic entropy in the setting of fuzzy sets theory, Inf. Control, 20, 301–312, doi:10.1016/S0019-9958(72)90199-4, 1972.

Einstein, H.: Tunnelling in difficult ground–swelling behaviour and identification of swelling rocks, Rock Mech. Rock Eng., 29, 113–124, doi:10.1007/BF01032649, 1996.

Frank, T., Tertois, A. L., and Mallet, J. L.: 3-D-reconstruction of complex geological interfaces from irregularly distributed and noisy point data, Comput. Geosci., 33, 932–943, doi:10.1016/j.cageo.2006.11.014, 2007.

Genser, H.: Geologie der Vorbergzone am südwestlichen Schwarzwaldrand zwischen Staufen und Badenweiler, PhD thesis, Naturwiss.-Math. Fakultät, Freiburg i. B., Germany, 119 pp., 1958.

Grimm, M., Stober, I., Kohl, T., and Blum, P.: Schadensfallanalyse von Erdwärmesondenbohrungen in Baden-Württemberg, Grundwasser, 19, 275–286, doi:10.1007/s00767-014-0269-1, 2014.

Groschopf, R., Guntram, K., Leiber, J., Maus, H., Ohmert, W., Schreiner, A., and Wimmenauer, W. (Eds.): Erläuterung zur Geologischen Karte von Freiburg im Breisgau und Umgebung 1 : 25 000, 2. edn., Geologisches Landesamt Baden-Württemberg, Stuttgart, Germany, 1981.

Hack, R., Orlic, B., Ozmutlu, S., Zhu, S., and Rengers, N.: Three and more dimensional modelling in geo-engineering, B. Eng. Geol. Environ., 65, 143–153, doi:10.1007/s10064-005-0021-2, 2006.

Hassen, I., Gibson, H., Hamzaoui-Azaza, F., Negro, F., Rachid, K., and Bouhlila, R.: 3-D geological modeling of the Kasserine Aquifer System, Central Tunisia: New insights into aquifer-geometry and interconnections for a better assessment of groundwater resources, J. Hydrol., 539, 223–236, doi:10.1016/j.jhydrol.2016.05.034, 2016.

Holden, L., Mostad, P., Nielsen, B. F., Gjerde, J., Townsend, C., and Ottesen, S.: Stochastic structural modeling, Math. Geol., 35, 899–914, doi:10.1023/B:MATG.0000011584.51162.69, 2003.

Hou, W., Yang, L., Deng, D., Ye, J., Clarke, K., Yang, Z., Zhuang, W., Liu, J., and Huang, J.: Assessing quality of urban underground spaces by coupling 3-D geological models: The case study of Foshan city, South China, Comput. Geosci., 89, 1–11, doi:10.1016/j.cageo.2015.07.016, 2016.

Jeannin, P. Y., Eichenberger, U., Sinreich, M., Vouillamoz, J., Malard, A., and Weber, E.: KARSYS: A pragmatic approach to karst hydrogeological system conceptualisation. Assessment of groundwater reserves and resources in Switzerland, Environ. Earth Sci., 69, 999–1013, doi:10.1007/s12665-012-1983-6, 2013.

Jessell, M. W., Ailleres, L., and de Kemp, E. A.: Towards an integrated inversion of geoscientific data: What price of geology?, Tectonophysics, 490, 294–306, doi:10.1016/j.tecto.2010.05.020, 2010.

Julio, C., Caumon, G., and Ford, M.: Sampling the uncertainty associated with segmented normal fault interpretation using a stochastic downscaling method, Tectonophysics, 639, 56–67, doi:10.1016/j.tecto.2014.11.013, 2015.

Kessler, H., Turner, A. K., Culshaw, M., and Royse, K.: Unlocking the potential of digital 3-D geological subsurface models for geotechnical engineers, in: Eur. econference Int. Assoc. Eng. Geol., Asociacion Espanola de Geologia Aplicada a la Ingenieria, 15–20 September 2008, Madrid, Spain, 15–20, available at: http://nora.nerc.ac.uk/3817/ (last access: 18 April 2016), 2008.

Kinkeldey, C., MacEachren, A. M., Riveiro, M., and Schiewe, J.: Evaluating the effect of visually represented geodata uncertainty on decision-making: systematic review, lessons learned, and recommendations, Cartogr. Geogr. Inf. Sci., 44, 1–21, doi:10.1080/15230406.2015.1089792, 2015.

Klir, G. J.: Uncertainty and Information: Foundations of Generalized Information Theory, John Wiley & Sons, Inc., Hoboken, New Jersey, USA, doi:10.1002/0471755575.ch3, 2005.

Lark, R. M., Mathers, S. J., Thorpe, S., Arkley, S. L. B., Morgan, D. J., and Lawrence, D. J. D.: A statistical assessment of the uncertainty in a 3-D geological framework model, P. Geol. Assoc., 124, 946–958, doi:10.1016/j.pgeola.2013.01.005, 2013.

Leung, Y., Goodchild, M. F., Lin, C. C., Leung, Y., Goodchild, M. F., and Lin, C. C.: Visualization of fuzzy scenes and probability fields, Comput. Sci. Stat., 24, 416–422, 1992.

LGRB: Geologische Untersuchungen von Baugrundhebungen im Bereich des Erdwärmesondenfeldes beim Rathaus in der historischen Altstadt von Staufen i. Br., Tech. rep., Landesamt für Geologie, Rohstoffe und Bergbau (LGRB), available at: http://www.lgrb-bw.de/geothermie/staufen (last access: 5 July 2016), 2010.

LGRB: Zweiter Sachstandsbericht zu den seit dem 01.03.2010 erfolgten Untersuchungen im Bereich des Erdwärmesondenfeldes beim Rathaus in der historischen Altstadt von Staufen i. Br., Tech. rep., Landesamt für Geologie, Rohstoffe und Bergbau (LGRB), available at: http://www.lgrb-bw.de/geothermie/staufen (last access: 5 July 2016), 2012.

Lindsay, M. D., Aillères, L., Jessell, M. W., de Kemp, E. A., and Betts, P. G.: Locating and quantifying geological uncertainty in three-dimensional models: Analysis of the Gippsland Basin, southeastern Australia, Tectonophysics, 546–547, 10–27, doi:10.1016/j.tecto.2012.04.007, 2012.

Lindsay, M. D., Jessell, M. W., Ailleres, L., Perrouty, S., de Kemp, E., and Betts, P. G.: Geodiversity: Exploration of 3-D geological model space, Tectonophysics, 594, 27–37, doi:10.1016/j.tecto.2013.03.013, 2013.

Lindsay, M. D., Perrouty, S., Jessell, M., and Ailleres, L.: Inversion and Geodiversity: Searching Model Space for the Answers, Math. Geosci., 46, 971–1010, doi:10.1007/s11004-014-9538-x, 2014.

Liu, J., Tang, H., Zhang, J., and Shi, T.: Glass landslide: the 3-D visualization makes study of landslide transparent and virtualized, Environ. Earth Sci., 72, 3847–3856, doi:10.1007/s12665-014-3183-z, 2014.

Mallet, J. L.: Discrete Smooth Interpolation in geometric modelling, Comput. Des., 24, 178–191., 1992.

Mallet, J.-L.: Space – Time Mathematical Framework for Sedimentary Geology, Math. Geol., 36, 1–32, doi:10.1023/B:MATG.0000016228.75495.7c, 2004.

Mann, J. C.: Uncertainty in Geology, in: Comput. Geol. – 25 Years Prog., edited by: Davis, J. C. and Herzfeld, U. C., p. 298, Oxford University Press, Inc., New York, USA, 1993.

Panteleit, B. R., Jensen, S., Seiter, K., Budde, H., and McDiarmid, J.: A regional geological and groundwater flow model of Bremen (Germany): an example management tool for resource administration, Z. Dtsch. Ges. Geowiss., 164, 569–580, doi:10.1127/1860-1804/2013/0035, 2013.

Paradigm: SKUA-GOCAD™ – Paradigm® 15.5 User Guide, available at: http://www.pdgm.com/products/skua-gocad/ (last access: 11 April 2017), 2015.

Park, H., Scheidt, C., Fenwick, D., Boucher, A., and Caers, J.: History matching and uncertainty quantification of facies models with multiple geological interpretations, Comput. Geosci., 17, 609–621, doi:10.1007/s10596-013-9343-5, 2013.

Paul, S. and Maji, P.: City block distance for identification of co-expressed microRNAs, Mol. BioSyst., 10, 1509–1523, doi:10.1007/978-3-319-03756-1_35, 2014.

R Core Team: R: A Language and Environment for Statistical Computing, R Foundation for Statistical Computing, Vienna, Austria, available at: https://www.R-project.org/ (last access: 12 April 2017), 2016.

Røe, P., Georgsen, F., and Abrahamsen, P.: An Uncertainty Model for Fault Shape and Location, Math. Geosci., 46, 957–969, doi:10.1007/s11004-014-9536-z, 2014.

Ruch, C. and Wirsing, G.: Erkundung und Sanierungsstrategien im Erdwärmesonden-Schadensfall Staufen i. Br., Geotechnik, 36, 147–159, doi:10.1002/gete.201300005, 2013.

Sawatzki, G. and Eichhorn, F. (Eds.): Vorl. Geol. Karte Baden-Württemberg, 1 : 25 000, Bl. 8112 Staufen im Breisgau, Landesamtes für Geologie, Rohstoffe und Bergbau Baden-Wüttemberg (LGRB), 2. preliminary revised edn., Freiburg i. Br., Germany, 1999.

Schamper, C., Jørgensen, F., Auken, E., and Effersø, F.: Case History Assessment of near-surface mapping capabilities by airborne transient electromagnetic data – An extensive comparison to conventional borehole data, Geophysics, 79, B187–B199, doi:10.1190/Geo2013-0256.1, 2014.

Scheidt, C. and Caers, J.: Representing spatial uncertainty using distances and kernels, Math. Geosci., 41, 397–419, doi:10.1007/s11004-008-9186-0, 2009a.

Scheidt, C. and Caers, J.: Uncertainty Quantification in Reservoir Performance Using Distances and Kernel Methods–Application to a West Africa Deepwater Turbidite Reservoir, SPE J., 14, 680–692, doi:10.2118/118740-PA, 2009b.

Schöttle, M. (Ed.): Geotope im Regierungsbezirk Freiburg, Landesanstalt für Umweltschutz Baden-Württemberg, Karlsruhe, Germany, 2005.

Schreiner, A.: Geologie und Landschaft, in: Markgräflerland – Entwicklung und Nutzung einer Landschaft, edited by: Hoppe, A., 81, 7–24, 6 Abb., Berichte der Naturforschenden Gesellschaft, Freiburg i. Br., Germany, 1991.

Shannon, C. E.: A mathematical theory of communication, Bell Syst. Tech. J., 27, 379–423, doi:10.1145/584091.584093, 1948.

Suzuki, S., Caumon, G., and Caers, J.: Dynamic data integration for structural modeling: Model screening approach using a distance-based model parameterization, Comput. Geosci., 12, 105–119, doi:10.1007/s10596-007-9063-9, 2008.

Tacher, L., Pomian-Srzednicki, I., and Parriaux, A.: Geological uncertainties associated with 3-D subsurface models, Comput. Geosci., 32, 212–221, doi:10.1016/j.cageo.2005.06.010, 2006.

Tertois, A.-L. and Mallet, J.-L.: Editing Faults within tetrahedral volume models in real time, in: Structurally Complex Reservoirs, edited by: Jolley, S. J., Barr, D., Walsh, J. J., and Knipe, R. J., Geol. Society Spec. Publ., 292, 89–101, doi:10.1144/sp292.5, 2007.

Thiele, S. T., Jessell, M. W., Lindsay, M., Ogarko, V., Wellmann, J. F., and Pakyuz-Charrier, E.: The topology of geology 1: Topological analysis, J. Struct. Geol., 91, 27–38, doi:10.1016/j.jsg.2016.08.009, 2016a.

Thiele, S. T., Jessell, M. W., Lindsay, M., Wellmann, J. F., and Pakyuz-Charrier, E.: The topology of geology 2: Topological uncertainty, J. Struct. Geol., 91, 74–87, doi:10.1016/j.jsg.2016.08.010, 2016b.

Webb, A. R. and Copsey, K. D.: Measures of dissimilarity, in: Stat. Pattern Recognit., chap. A1, 419–429, second edn., John Wiley & Sons, Ltd, Chichester, UK, 2003.

Wellmann, J. F.: Information theory for correlation analysis and estimation of uncertainty reduction in maps and models, Entropy, 15, 1464–1485, doi:10.3390/e15041464, 2013.

Wellmann, J. F. and Regenauer-Lieb, K.: Uncertainties have a meaning: Information entropy as a quality measure for 3-D geological models, Tectonophysics, 526–529, 207–216, doi:10.1016/j.tecto.2011.05.001, 2012.

Wellmann, J. F., Horowitz, F. G., Schill, E., and Regenauer-Lieb, K.: Towards incorporating uncertainty of structural data in 3-D geological inversion, Tectonophysics, 490, 141–151, doi:10.1016/j.tecto.2010.04.022,, 2010.

Yager, R. R.: Measures of entropy and fuzziness related to aggregation operators, Inform. Sciences, 82, 147–166, doi:10.1016/0020-0255(94)00030-F, 1995.

Yamamoto, J. K., Koike, K., Kikuda, A. T., Campanha, G. A. D. C., and Endlen, A.: Post-processing for uncertainty reduction in computed 3-D geological models, Tectonophysics, 633, 232–245, doi:10.1016/j.tecto.2014.07.013, 2014.

Zadeh, L.: Fuzzy sets, Inf. Control, 8, 338–353, doi:10.1016/S0019-9958(65)90241-X, 1965.

5

Correcting for static shift of magnetotelluric data with airborne electromagnetic measurements: a case study from Rathlin Basin, Northern Ireland

Robert Delhaye[1,2]**, Volker Rath**[1]**, Alan G. Jones**[1,3]**, Mark R. Muller**[1]**, and Derek Reay**[4]

[1]Geophysics Section, School of Cosmic Physics, Dublin Institute for Advanced Studies (DIAS), 5 Merrion Square, Dublin 2, Ireland
[2]National University of Ireland, Galway, University Road, Galway, Ireland
[3]Complete MT Solutions, Ottawa, Canada
[4]Geological Survey of Northern Ireland (GSNI), Belfast, UK

Correspondence to: Robert Delhaye (rdelhaye@cp.dias.ie)

Abstract. Galvanic distortions of magnetotelluric (MT) data, such as the static-shift effect, are a known problem that can lead to incorrect estimation of resistivities and erroneous modelling of geometries with resulting misinterpretation of subsurface electrical resistivity structure. A wide variety of approaches have been proposed to account for these galvanic distortions, some depending on the target area, with varying degrees of success. The natural laboratory for our study is a hydraulically permeable volume of conductive sediment at depth, the internal resistivity structure of which can be used to estimate reservoir viability for geothermal purposes; however, static-shift correction is required in order to ensure robust and precise modelling accuracy.

We present here a possible method to employ frequency–domain electromagnetic data in order to correct static-shift effects, illustrated by a case study from Northern Ireland. In our survey area, airborne frequency domain electromagnetic (FDEM) data are regionally available with high spatial density. The spatial distributions of the derived static-shift corrections are analysed and applied to the uncorrected MT data prior to inversion. Two comparative inversion models are derived, one with and one without static-shift corrections, with instructive results. As expected from the one-dimensional analogy of static-shift correction, at shallow model depths, where the structure is controlled by a single local MT site, the correction of static-shift effects leads to vertical scaling of resistivity–thickness products in the model, with the corrected model showing improved correlation to existing borehole wireline resistivity data. In turn, as these vertical scalings are effectively independent of adjacent sites, lateral resistivity distributions are also affected, with up to half a decade of resistivity variation between the models estimated at depths down to 2000 m. Simple estimation of differences in bulk porosity, derived using Archie's Law, between the two models reinforces our conclusion that the suborder of magnitude resistivity contrasts induced by the correction of static shifts correspond to similar contrasts in estimated porosities, and hence, for purposes of reservoir investigation or similar cases requiring accurate absolute resistivity estimates, galvanic distortion correction, especially static-shift correction, is essential.

1 Introduction

The electrical resistivity of a volume of rock is highly sensitive to the presence of laterally and vertically varying amounts of electrically conductive fluids connected via pore spaces or fluid conduits. Due to these potentially strong resistivity contrasts between competent host rock and fluid penetrated rock, electromagnetic (EM) methods, and in particular magnetotellurics (MT), have been used with considerable success to image conductive volumes at depth (Chave and Jones, 2012; Simpson and Bahr, 2005).

As with all EM methods, MT data are highly sensitive to rock fluid content and distribution (i.e. porosity and hydraulic permeability) and can be related to other properties relevant to fluid movement. This has made the method particularly interesting for the exploration of geothermal resources. Indeed, geothermal research was the first commercial application of MT in the late 1950s, though the interpretation of the corresponding conductive structures is not always straightforward (Muñoz, 2014).

The MT data set studied here was acquired in the context of a multidisciplinary geothermal research program (IRETHERM), the overarching aim of which is to identify and evaluate low-enthalpy geothermal resources within Ireland. One such resource (Goodman et al., 2004) is the thick, porous, and permeable succession of Permian and Triassic sandstones found within several concealed sedimentary basins in Northern Ireland, with the Rathlin Basin in particular having significantly elevated estimated geothermal gradients in comparison to the remainder of Ireland (Reay and Kelly, 2010).

The island of Ireland was formed during the Caledonian orogeny by the complex accretion of several continental and island arc fragments during the closure of the Iapetus Ocean between the Early Ordovician (485 Ma) and late Silurian (423 Ma), resulting in seven identifiable terranes that comprise the present-day basement across both Ireland and Great Britain (Mitchell, 2004; Hepworth and Sanders, 2009). Our survey area (Fig. 1) lies within the Central Highlands Terrane of Laurentia, the basement of which comprises mainly mid- to late-Neoproterozoic (1000–635 Ma) metasedimentary rocks classified as the Dalradian supergroup, a metasedimentary and igneous rock succession that was deposited on the eastern margin of Laurentia between late Neoproterozoic (≈ 800 Ma) and early Cambrian (≈ 510 Ma) times. Specifically, the basement across the test area is assumed to consist of the latest-Proterozoic (Ediacaran, 635–541 Ma) Argyll Group of psammites and semipelites.

Regional shear and stress during the subsequent late-Paleozoic (350–250 Ma) Variscan orogeny reactivated the Caledonian (490–390 Ma) Tow Valley Fault (TVF), and the ensuing normal and dextral strike-slip faulting resulted in the formation of a rift basin later filled by a succession of sediments to form the Rathlin Basin. Although drilling in the adjacent Magilligan Basin encountered Carboniferous formations at 1347 m total depth, the most basal formations confirmed within the Rathlin Basin are the Permian Enler Group (EG) sandstones and Early-Triassic Sherwood Sandstone Group sandstones (SSG). Both formations are hydrocarbon reservoirs in the Irish Sea to the east (Naylor and Shannon, 2011). The Sherwood Sandstone Group is overlain by the Late-Triassic Mercia Mudstone Group (MMG), which is itself generally overlain by late Jurassic Lower Lias Group (LLG) mudstones. However, in many places significant dolerite and basalt sills (up to approx. 100 m in combined thickness) have been encountered, with poorly known

spatial extent. The final and youngest successions in the basin are chalks of the Cretaceous Ulster White Limestone Formation (UWLF), with the Antrim Lava Group (ALG) concealing the basin entirely.

To date, two deep boreholes have been completed onshore in the Rathlin Basin, namely the Port More 1 (PM1) and Ballinlea 1 (B1) boreholes, drilled in 1967 and 2008 respectively; however, only data from the former are available as information from the latter is not yet in the public domain. The PM1 borehole was drilled to a total depth of 1897 m and terminated in the EG sandstones, with wireline log data acquired in two separate sections due to technical difficulties (Wilson and Manning, 1978). The upper portion of normal resistivity data covers the uppermost 250 m of the hole, including the Antrim Lava Group, Ulster White Limestone Formation, and the upper portion of the Lower Lias Group. These data provide relatively consistent resistivity estimates of ≈ 80 and ≈ 5 Ωm for the UWLF and LLG sedimentary formations respectively, whereas estimates for the ALG vary from ≈ 5–80 Ωm as it comprises a succession of tuffs and basalts. The lower portion of resistivity data spans the depth interval of ≈ 1050–1450 m, covering the boundary between the MMG and SSG at 1320 m, and provides resistivity estimates of ≈ 4 and ≈ 6 Ωm for the respective groups. The estimates within the SSG also show a higher variance, which may be due to the presence of conglomerates and breccias in the upper portion of the group. The stratigraphic column encountered in the PM1 borehole is displayed in Fig. 2, alongside a plot of the borehole resistivity data.

Modelling of regional gravity and magnetic data has been undertaken, with results presented in Mitchell (2004) and Gibson (2004). The density model of Mitchell (2004) (Fig. 3) shows a relatively homogeneous structure along the basin, particularly of the Permo-Triassic section, with a maximum depth to Dalradian basement of approximately 3 km modelled for the Rathlin Basin (located at 33 km distance along profile). This modelling adopts a density of 2.35 Mg m^{-3} for the assumed basal Carboniferous rocks; this value comes from borehole samples in the adjacent Foyle Basin, and Mitchell (2004) advises that this value may be insufficiently dense to represent Rathlin Basin conditions. If Carboniferous sediments in the Rathlin Basin are of a higher density, a greater thickness of overlying lighter sediments (i.e. the target Permian and Triassic sandstones) would be required to be consistent with the observed gravity anomaly. Magnetic and gravity modelling by Gibson (2004) suggests that the Tow Valley Fault zone consists of a series of major fault segments with varying dip of 20–50° to the north-west.

Core samples of the EG and SSG sandstones successions gathered from the Port More 1 borehole show promising reservoir properties, with fractional porosities and hydraulic permeabilities ranging from 0.10 to 0.22 and 1–1000 mD respectively (Mitchell, 2004). Equilibrated temperatures taken from both the PM1 and B1 boreholes have previously been used to estimate geothermal gradients. A temperature of

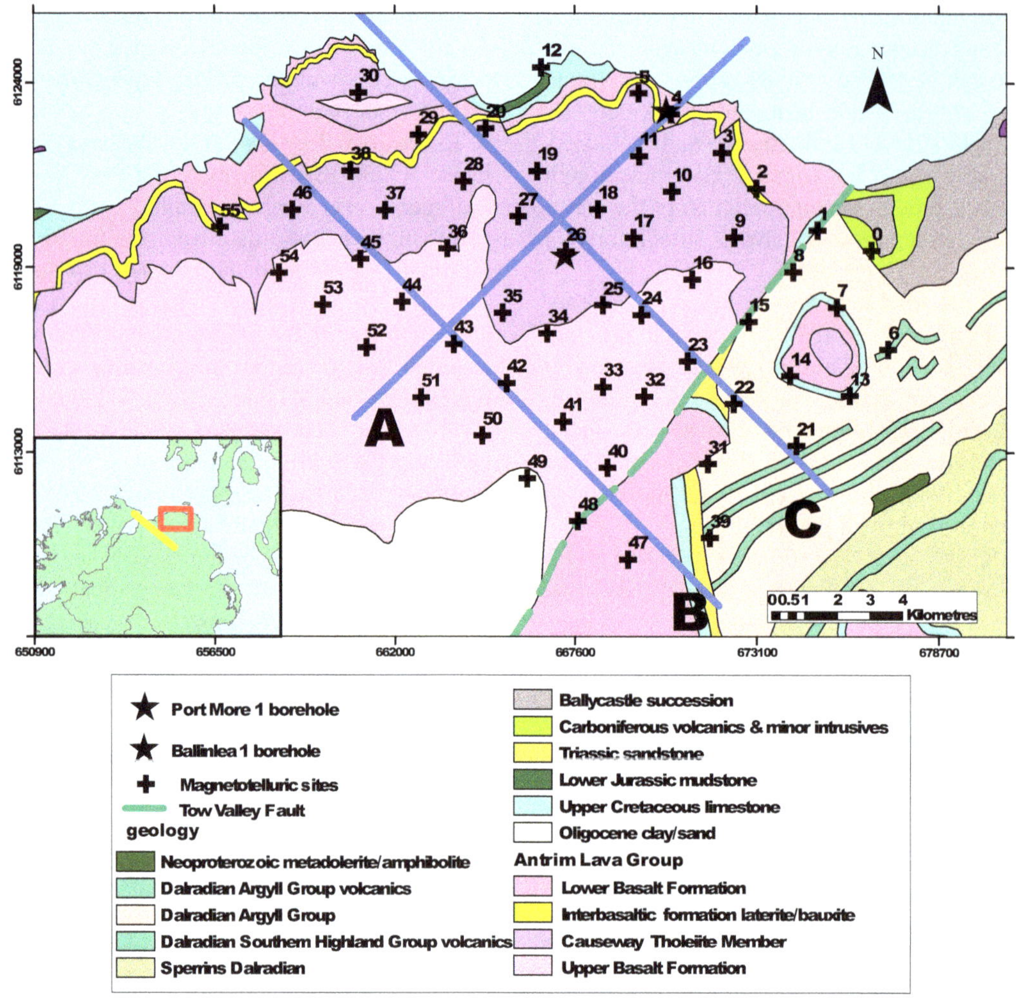

Figure 1. Geological map of Rathlin Basin area, overlain with magnetotelluric acquisition sites numbered and denoted by black crosses. Note the large areal extent of the Antrim Lava Group basalts, with minimal surface expression of the underlying sedimentary basin. Profiles A, B, and C denote the locations corresponding to Figs. 12–14. Inset map shows the northern half of Ireland, with the location of the magnetotelluric survey area shown by the red rectangle. The yellow line indicates the location of the density profile shown in Fig. 3.

35.4 °C was observed at 582 m depth in the PM1 borehole, whereas a temperature of 99 °C was observed at 2650 m in B1 (Reay and Kelly, 2010). Assuming a surface temperature of 10 °C, simple linear estimation gives calculated geothermal gradients of 43.6 (PM1) and 33.6 K km^{-1} (B1), both of which are elevated above the typical estimates of ≈ 20–30 K km^{-1} measured elsewhere across Ireland (Goodman et al., 2004). In conjunction with the promising reservoir properties and basin depth expected from gravity modelling, it has been proposed previously (Goodman et al., 2004; Reay and Kelly, 2010; Pasquali et al., 2010) that the Rathlin Basin may be favourable for geothermal exploitation. As the reservoir potential depends strongly on the intra-basin structure, variations in modelled resistivity may be taken as an excellent proxy for images of the presence of fluids, their distribution, and interconnection within the basin.

The imaging of sub-basalt structures poses difficulties to other commonly employed geophysical methods, particularly seismics (e.g. Martini et al., 2005; Bean and Martini, 2010) due to the negative acoustic impedance contrast at the base of the basalt, and previous reflection experiments struggled to clearly image the sediments through the overlying ALG (Naylor and Shannon, 2011). As the MT method has been successfully applied in sub-basalt investigations both onshore and offshore (Hautot et al., 2007; Jegen et al., 2009; Colombo et al., 2011; Heincke et al., 2014), the method was proposed to study the three-dimensional electrical resistivity distribution of the sedimentary fill of the onshore portion of the Rathlin Basin.

Due to the expected elevated hydraulic properties and saline pore fluids (both factors that increase conductivity) of the proposed hydrothermal aquifer within the basin, it was expected that MT data could be carefully modelled to im-

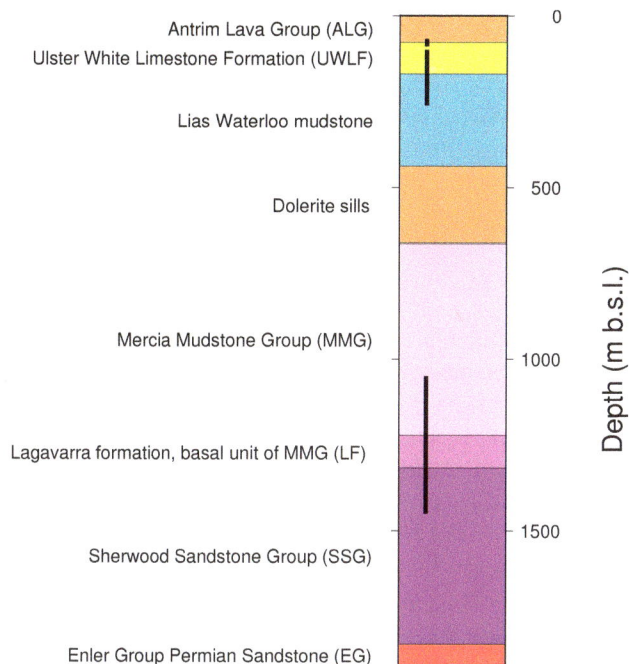

Figure 2. Stratigraphy encountered in the Port More 1 borehole (left), and measured normal resistivity data (right). Stratigraphy encountered includes Paleogene Antrim Lava Group of basalts and tuffs, Cretaceous Ulster White Limestone Formation, Jurassic Lower Lias sandstones, Late-to-Mid-Triassic Mercia Mudstone Group, Early-Triassic Sherwood Sandstone Group, and late-Permian Enler Group sandstones. Due to technical difficulties the resistivity data were acquired in three sections.

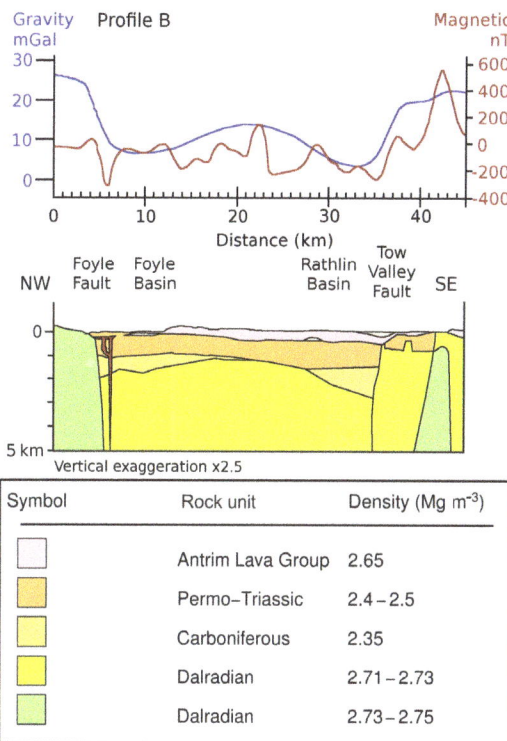

Figure 3. Density model from Mitchell (2004), showing the broad geological structure predicted prior to this study. Density profile locality is shown in Fig. 1 (inset). Although the profile crosses the Rathlin Basin south-west of the MT survey area, a similar lithological sequence is expected elsewhere in the basin, including the survey area.

age the properties and distribution of the aquifer formations. The increase in resistivity observed in wireline data from the MMG to the underlying target sediments implies that, depending upon the thickness of units beneath the MMG, MT may not be able to accurately estimate the units' resistivities, as MT is primarily sensitive to a layer's conductance (i.e. the ratio of a layer's thickness to resistivity), and thinner or less conductive layers may be shielded by overlying conductors (Jones, 1992). However, due to the proven thicknesses and similar lithologies of the SSG and EG sandstones, it is still expected that the SSG and EG sediment fill will cause a sufficiently high resistivity contrast against the resistive metasedimentary country bedrock and provide a viable target for MT. The geometry of this interpreted aquifer structure is expected to be compatible with the gravity model presented in Fig. 3.

The MT method samples the impedance transfer functions that relate the electric and magnetic field components of EM plane waves that propagate into the Earth. As these EM waves attenuate with dependency on the Earth's lateral and vertical resistivity distribution, the observed MT responses can be employed for estimating the underlying 3-D resistivity distribution (Chave and Jones, 2012). The electric and magnetic field components of these EM source waves are each acquired in (preferentially) orthogonal horizontal directions,

allowing the definition of four magnetotelluric transfer functions at the measuring location. These four elements carry information on the value and dimensionality of the subsurface resistivity structure at a range of periods. Many decompositions and analyses have been employed to expose this information (see, e.g., Chave and Jones, 2012, for an overview), with the aim of improving estimation or justifying 1-D (i.e. resistivity varying with depth only) or 2-D (i.e. resistivity varying with depth and one horizontal dimension only) modelling.

Though sensitive to conductive structures at depth, MT data are prone to distortion, primarily of the electric field, due to the presence of galvanic charges on the boundaries of shallow conductivity structures that are unresolvable at the frequency range of the recorded data. One simple form of this galvanic distortion is often easily identified by vertical offsets of the logarithmic apparent resistivity curves and is referred to as the static-shift effect (Jones, 1988), following a similar effect in seismology named "statics". These galvanic signatures are related to inescapable issues in observation of the electric field, wherein point-wise electric field observations, assumed during modelling and inversion, are replaced during MT surveys with voltage difference measurements along finite length dipoles (Poll et al., 1989; Pellerin and Hohmann, 1990), and by issues related to insufficient gridding resolution to describe the lateral variability of the near surface that affect both the electric and magnetic fields (Chave and Smith, 1994; Chave and Jones, 1997). Although the former of these may be handled by appropriate post-processing when modelling and inverting the field measurements, any near-surface inhomogeneity not parameterised in the modelling or inversion process, even at the electrode-scale size, will contribute to the galvanic signatures by distorting the local (primarily) electric fields. For land-based MT data, the magnetic effects of galvanic distortion only occur for a short frequency range of the order of half a decade at most (Chave and Smith, 1994; Chave and Jones, 1997).

Various methods have been proposed to quantify and correct for these static-shift effects, including continuous sampling and filtering of the electric channels (Torres-Verdin and Bostick, 1992), spatial filtering based on mapping of MT data (Berdichevsky, 1989), modelling of parametric homogeneous layers at depth (Jones, 1988), estimation of distortion-related parameters as unknowns during inversion (Sasaki and Meju, 2006; Miensopust, 2010; Avdeeva et al., 2015; De Groot-Hedlin, 1991; DeGroot-Hedlin, 1995), and finally the use of complementary EM geophysical methods (Sternberg et al., 1988; Pellerin and Hohmann, 1990; Miensopust et al., 2014). These methods can be broadly divided into methods that use intrinsic information from MT data and those that use extrinsic information from other geoscientific data. Whereas both families of methods can account for static shifts between MT modes at a single site and improve inter-station shifts, intrinsic information may not yield a correct resistivity in the case of both modes being distorted - as stated in Sternberg et al. (1988), "there is no reason to expect that either of the two MT polarisations will provide the correct resistivity". In contrast, extrinsic methods using purely magnetic measurements (i.e. with no instruments using the subsurface as a component of their circuitry) by definition are unaffected by the electric effects of galvanic distortion and thereby provide a more correct resistivity estimate, albeit generally for much shallower depths than magnetotelluric measurements. The use of extrinsic information may, in some cases, be limited by the requirement that both the extrinsic information and MT data must illuminate some common depth volume within the Earth with a common current system in order to allow reconciliation of the resistivity structure. For example, if the near-surface is three-dimensional (3-D), then the current system from regionally induced currents observed in MT is very different from the current system from locally induced currents from a small EM transmitter–receiver array, and any resistivities estimated by MT would correspond to a different volume than that sampled by the smaller array.

Fortunately, in the case of the Rathlin Basin, broad-scale extrinsic resistivity information is available. As part of the regional Tellus ground and airborne geoscience mapping programme across both Northern Ireland and the Republic of

Ireland, airborne frequency domain electromagnetic (FDEM) data were gathered over the target area (Beamish, 2013) in 2005 and 2006. Sternberg et al. (1988) and Pellerin and Hohmann (1990) describe how time domain electromagnetic (TDEM) data can be used to estimate and correct static-shift distortion of MT data, assuming that the MT data are one-dimensional (1-D) or two-dimensional (2-D) and in the correct geoelectric strike coordinate system, and we propose adapting their method to use the Tellus frequency domain airborne electromagnetic measurement (AEM) data for this purpose. Due to the high density of measurements often available, AEM data allow estimation of near-surface conductivity distributions at resolutions exceeding those of MT, making them an excellent choice for the correction of static-shift effects. The use of airborne EM data for static-shift correction has precedence, as outlined with airborne TDEM data by Crowe et al. (2013).

As three-dimensional inversion of MT data is becoming a more common practice, the effects of static-shift correction on resulting resistivity distributions must be considered. This article focuses on the implementation of this correction scheme for MT data by comparing models found by independent 3-D MT inversion of the observed MT data and the static-shift-corrected MT data. In addition to comparing the model results, the statistical and spatial distribution of calculated static-shift corrections are examined and compared to previous works to verify their validity. Both the absolute resistivity and resistivity gradients are used to evaluate the differences between the two models, with the implications of the differences discussed in the context of the possible geothermal aquifer. Note that the approach presented here is primarily methodological in nature; the geological interpretation is brief and will be investigated at length in a future publication.

2 Electromagnetic methods

EM methods include a wide variety of techniques that observe electromagnetic induction in the Earth and are most commonly used to image the subsurface distribution of electrical resistivity ρ (Ωm) or its inverse, electrical conductivity σ (Sm^{-1}), and, at high frequencies, electrical permittivity ε (Fm^{-1}). Subsurface materials are rarely homogeneous and can generally be described as a mixture of materials with strongly differing properties, as detailed in Nover (2005) and Chave and Jones (2012). The bulk resistivity of a rock is commonly determined by very few of its constituents, with electrically the most important candidates at crustal depths being metallic conductors (sulfides, graphite, iron oxides), clays, and conductive fluids in pore spaces (both saline fluids and partial melts). Note that we use pore space in a general sense for primary and secondary porosity, including pores, fractures, and conduits that may dominate the rock type under consideration. For low-enthalpy geothermal exploration in the upper crust, the most relevant property is the influence of electrolytic conduction by fluids in porous rocks. The relationship between the observed effective resistivity ρ, the pore fluid resistivity ρ_i, and the formation porosity ϕ is classically described for sandstones by Archie's Law (Archie, 1947), with generalisations discussed in Glover (2010). As Archie's Law assumes a clean sandstone matrix with well-established relationships between porosity and hydraulic permeability, its application may not always be appropriate, particularly if clay minerals are present (Mavko et al., 2009; Guéguen and Palciauskas, 1994; Zinszner and Pellerin, 2007).

In a broader sense, electrical conductivity is a proxy measure of hydraulic permeability rather than porosity, as the interconnection of conducting pathways facilitates electric current flow. Due to this strong dependence on the geometry of flow paths on the scale of interest, the relationship between permeability and porosity is highly nonlinear (see, amongst others, Raffensperger, 1996; Pape et al., 1999, 2000; Luijendijk and Gleeson, 2015). These dependencies have a close relationship with the type and geologic history of the rock considered (see, e.g., Bernabe et al., 2003), and the often complex development of geological units can lead to heterogeneities and preferential flow pathways at all scale lengths (Bjørlykke, 2010).

2.1 Magnetotelluric method

The MT method uses impedance transfer functions relating the electric and magnetic field components of vertically propagating EM source field plane waves to image the lateral and vertical resistivity distribution within the Earth. MT signal waves are generated by two sources, namely atmospheric electricity (generating signals of frequency $> 10\,\mathrm{Hz}$) and interactions of the Earth's magnetosphere with solar wind (generating signals of frequency $< 10\,\mathrm{Hz}$). Recent detailed reviews of MT methods, the underlying assumptions, and their application include Simpson and Bahr (2005), Berdichevsky and Dmitriev (2008), and Chave and Jones (2012).

Resistivity information at a range of depths is inferred by considering planar EM waves in the Earth at a range of frequencies, as their attenuation at a given frequency is a function of the resistivity ρ and magnetic permeability μ (Hm^{-1}) of the subsurface material, where the latter is generally assumed not to vary from that of free space, μ_0. In a uniform half-space (i.e. a space with no lateral or vertical resistivity variation) of resistivity ρ, the scale length δ in metres describing this attenuation,

$$\delta = \sqrt{\frac{2\rho}{\omega\mu}} \approx 503\sqrt{\frac{\rho}{f}}, \tag{1}$$

is termed the electromagnetic skin depth and describes the characteristic length over which the amplitude of an EM wave of frequency $\omega = 2\pi f$ decays by a factor of e^{-1}. This quantity is commonly used as a simple measure for the depth

of investigation and radius of influence, although one must beware of its overuse in situations that depart from a uniform half-space, especially in the case of a multidimensional Earth (Jones, 2006).

The resistivity of a select volume of the Earth, as sampled by an EM wave of frequency ω, is determined from complex transfer functions that relate the amplitudes of the horizontal electric E_i (Vm^{-1}) and magnetic H_j (Am^{-1}) field components that constitute the wave, defined as the complex MT impedance tensor $\mathbf{Z}$ (Ω). This is generally formulated in the frequency domain, where the transfer function can be defined by the ratio of the fields:

$$Z_{ij}(\omega) = \frac{E_i(\omega)}{H_j(\omega)}. \tag{2}$$

By considering both orthogonal and parallel pairs of fields, an impedance tensor $\mathbf{Z}$ with four components ($Z_{xx}, Z_{xy}, Z_{yx}, Z_{yy}$) can be defined:

$$\mathbf{Z} = \begin{pmatrix} Z_{xx} & Z_{xy} \\ Z_{yx} & Z_{yy} \end{pmatrix}. \tag{3}$$

These impedances can be restated in more familiar magnitudes and units as an apparent resistivity ρ_a (i.e. equivalent half-space resistivity for a wave of that specific frequency for the orthogonal pairs Z_{xy} and Z_{yx}) and phase lead of the electric field over the magnetic field ϕ (which is $\pi/4$ for a uniform half-space for the orthogonal pairs), defined by

$$\rho_{\mathrm{a},ij} = \frac{1}{2\pi\mu}|Z_{ij}|^2, \tag{4}$$

$$\phi_{ij} = \arg\left(Z_{ij}\right) = \tan^{-1}\left(\frac{\Im(Z_{ij})}{\Re(Z_{ij})}\right). \tag{5}$$

In MT surveying, the electrical field E is measured as a voltage difference over an appropriate distance rather than at a point. For the sampling of the magnetic field components, the electromagnetic properties of the volume sampled within a sensor are known and homogeneous (being the internal properties of the sensor itself), and the averaged field sampled through this sensor volume accurately represents a point magnetic field value (to within a length scale of that of the sensor, typically 1.5 m for coil sensors used in broadband MT). However, the volumes sampled for the electric field components are of the order of 100 m (the typical bipole length in broadband MT (BBMT) surveys to acquire data in the frequency range of 100 to 0.01 Hz), with resistivity variations present within that length scale that may bias observations of the electric field. This distortion of primarily (but not exclusively) the electric field is one form of galvanic distortion and often manifests as frequency-independent multiplicative vertical offsets of MT apparent resistivity data for 1-D or 2-D cases (i.e. where impedance tensor diagonal components Z_{xx} and Z_{yy} are 0), when data are plotted as apparent resistivities on a log–log scale, hence the nomenclature of "static shift". As explained in, e.g., Jones (2011,

2012), whereas real multipliers applied to an impedance tensor with 1-D or 2-D form affect only the magnitudes of the impedances, this is not the case if the impedance tensor has a 3-D form with non-zero diagonal elements. Instead, the applied distortions cause mixing between the diagonal and off-diagonal components, affecting both the magnitudes and phases of the impedances.

2.2 Frequency domain AEM method

The basic theory for AEM can be found in Ward and Hohmann (1988). The FDEM method, as implemented for the Tellus AEM surveys, uses a pair of small coils as the transmitter–receiver (Tx-Rx) pair. The transmitter can be treated as a magnetic dipole that induces eddy currents in the subsurface at discrete frequencies, allowing the resistivity structure to be characterised by comparing the primary and secondary magnetic fields (H_p and H_s respectively). In the Tellus surveys Tx and Rx are oriented in a vertical, coplanar configuration, i.e. the coil axes are horizontal (Leväniemi et al., 2009), with magnetic dipole moments parallel to the flight direction.

AEM data take the form of ratios of the secondary magnetic fields (i.e. formed by current systems in the ground) to the primary magnetic fields (i.e. emitted by the transmitter coil), stated in parts-per-million. Both inphase (i.e. no phase change) and quadrature (i.e. 90° phase change) fields are considered; generally, the quadrature data are sensitive to the overall ground resistivity, whereas the inphase data are more sensitive to strong conductors. Unlike the magnetotelluric signal, the induced current systems and ensuing secondary magnetic fields of the AEM method are very much in the near-field region. As a result, the volume of Earth interrogated by the AEM signal cannot be easily reduced to a measure of skin-depth-type attenuation.

Although multidimensional modelling and inversion methods are available (see Auken et al., 2014, for a recent review), AEM data are commonly treated as representative of a one-dimensional (1-D) Earth, with spatial smoothing or other constraints along 2-D flight lines possible to improve spatial continuity. One-dimensional modelling is usually performed based on analytical solutions for the layered Earth case, which are well known for most Tx-Rx configurations and can be found in many publications (Keller and Frischknecht, 1966; Ward and Hohmann, 1988; Kaufman et al., 2014). The particular analytical solution in a layered half-space for vertical, coplanar configuration of the transmitter and receiver coils as used here (i.e. parallel, horizontal magnetic dipoles) is found in Minsley (2011). With the forward solution known, an inversion algorithm can be applied to determine a suitable resistivity model. The inversion of AEM data for this work uses the standard damped least-squares technique of Jupp and Vozoff (1975) as implemented in Airbeo (Raiche et al., 1985).

3 Rathlin Basin survey

MT data were collected at 56 sites across part of the onshore Rathlin Basin (site locations shown in Fig. 1) in May and June 2012. Seven parallel profiles were aligned perpendicular to the bounding Tow Valley Fault to the southeast (thick blue line in Fig. 1), with profile and site separations each of 2 km in order to obtain a near-regular array of site locations, facilitating three-dimensional modelling and inversion. Both BBMT (i.e. from ≈ 300 to 0.001 Hz) and audio-magnetotelluric (AMT) (i.e. from $\approx 10\,000$ to 10 Hz) data were acquired at each site. Data were recorded with Phoenix Geophysics MTU-5A receivers, with either MTC-50 (for BBMT) or AMTC-30 (for AMT) induction coils used to sample the horizontal magnetic field components (H_x and H_y respectively). Vertical magnetic field components (H_z) were measured at almost all sites using either the appropriate induction coil or an AL-100 airloop, as deemed appropriate given the local ground conditions. The horizontal electric field components E_x and E_y were sampled by non-polarising lead–lead chloride (Pb-PbCl) PE4 Phoenix Geophysics electrodes arranged at each site in a cross configuration with electrode separations of typically 80 m. BBMT measurements were taken over a period of three nights at each site followed by an overnight measurement of AMT data.

Robust estimates of the frequency domain MT transfer functions were determined from the MT time series using commercial processing software from Phoenix Geophysics that implements the technique described in Jones and Jödicke (1984) and Jones et al. (1989). This approach is based on cascade decimation (Wight and Bostick, 1980) and uses a least trimmed squares algorithm (Rousseeuw, 1984; Rousseeuw and Leroy, 2003) to achieve the robustness of the estimate. Whilst a dedicated distant remote reference site was not used, the principle of remote referencing, as described in Gamble (1979) and Gamble et al. (1979), was applied by using the horizontal magnetic field components of each simultaneously acquired site as reference fields and selecting the best available site as reference (typically five sites were recorded simultaneously). Finally, the AMT and BBMT data at each site were merged into a single response, spanning from 10 000 to 0.001 Hz at most sites.

Although data were acquired at an array of sites with the intent of 3-D inversion, as the data were expected to be predominantly 2-D in nature due to the expected strong lateral contrast across the bounding Tow Valley Fault, multi-site, multi-frequency Groom and Bailey distortion analysis was applied to the data on a cross-basinal profile basis using the "strike" analysis tool (McNeice and Jones, 2001). The results of the analysis are presented in Fig. 4 for four depth bands. Analysis of the data with respect to depth in "strike" shows that the data are predominantly 1-D or 2-D to depths of ≈ 2 km, with increasing rms misfits beyond these depths indicative of either a change in strike direction (i.e. still a 2-D structure, but with a different geoelectric strike direction) or

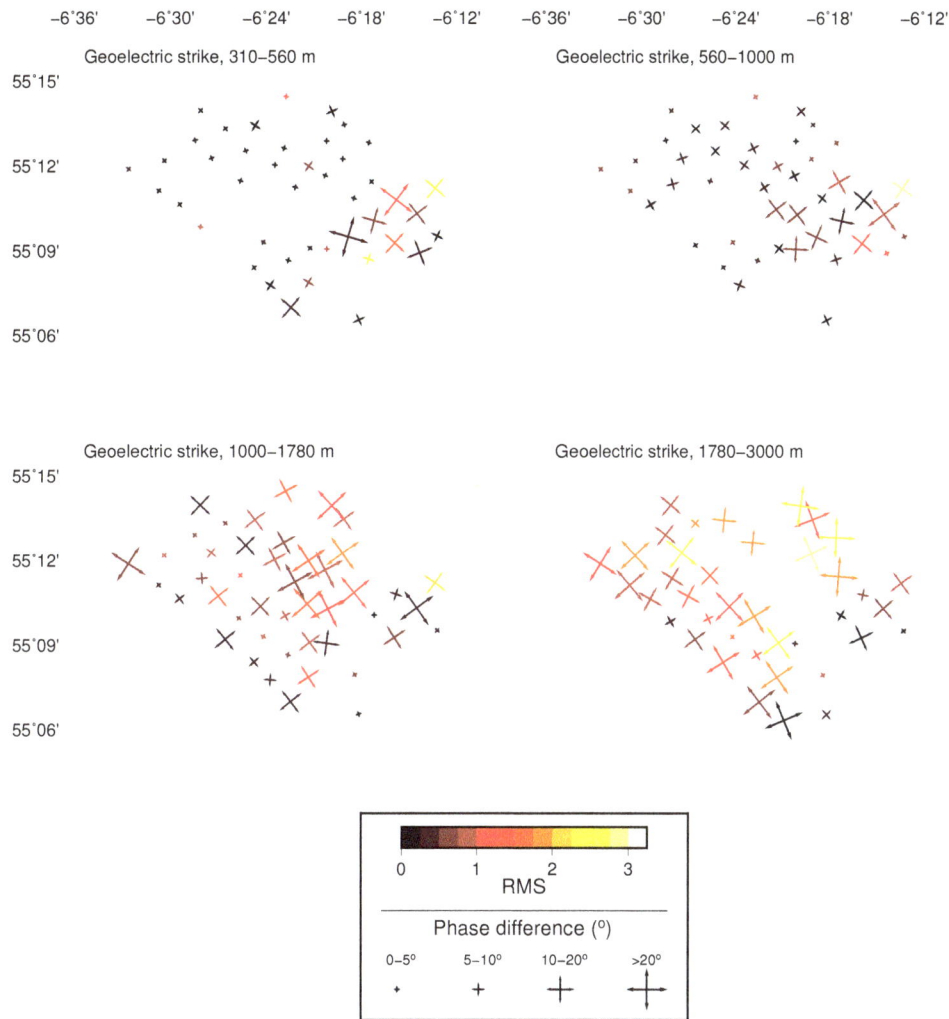

Figure 4. Visualisation of dimensionality of MT data, decomposed using the "strike" analysis program McNeice and Jones (2001) over a range of depth bands (in metres below sea level). The orthogonal vectors at each MT site location indicate the azimuth of best-fitting geoelectric strike direction, coloured by the rms misfit between the observed MT response and the Groom–Bailey model response for the best-fit strike direction. The orthogonal pair of vectors is required as geoelectric strike estimates have a 90° ambiguity. The size of orthogonal vectors is classified by the phase difference, with larger vectors corresponding to larger phase differences. Small phase differences are associated with 1-D resistivity structure, whereas larger phase differences occur with a 2- and 3-D structure. Larger rms misfits indicate that the decomposition to a 2-D structure is potentially invalid and can be caused by either significant noise contamination and distortion of the data or a 3-D structure.

a fully 3-D structure. An estimate of the regional geoelectric strike azimuth was computed by arithmetically averaging the strike estimates from the deepest depth band (1780–3000 m) in the south-west half of the model, as these estimates represent portions of the model less affected by the coastal effect. The mean strike azimuth of the south-western part of the model is ≈ 43° E, with a standard deviation of 10°. The mean geoelectric strike direction is coherent with the strike of the major structural feature, the Tow Valley Fault; however, it should be noted that as geoelectric strike directions inherently possess an ambiguity of ±90°, the mean azimuth could also be interpreted as 47° W (i.e. a bearing of 313°). As the site profile azimuths were aligned perpendicular (≈ 55° W)

to the mapped strike of the TVF in the area (≈ 35° E), both the inversion mesh and input data were rotated to a bearing of 315° (i.e. midway between the mean geoelectric strike direction and dip direction of the TVF). Note that the data were rotated without decomposition, to retain information that does not conform to the 2-D assumption in Groom–Bailey decomposition. For a similar reason, the correction of static shifts was performed before rotation. Note that inversion of non-rotated data and meshes is presented in the Supplement of this article.

The MT responses were inverted for three-dimensional models using the ModEM 3-D MT inversion program (Kelbert et al., 2014; Egbert and Kelbert, 2012), with all four impedance tensor elements and vertical transfer functions as input data. MT data were downsampled to a subset of 28 frequencies, spanning from 1000 to 0.001 Hz (displayed in Fig. 10), with an increased sampling of frequencies in the range most sensitive to the target sediment depths (1– 0.01 Hz). In order to avoid leverage bias in the search for optimum solutions that minimise rms misfit, poor-quality data (typically located near cultural noise sources) were manually identified and removed from the input data. The mesh used for inversion was $59 \times 68 \times 82$ cells in size (X, Y, Z) with the region of interest (the portion covered by MT sites) populated by cells of lateral extent 400 m by 400 m, with layer thicknesses logarithmically increasing beyond the depths required to accurately model bathymetry. Bathymetry was modelled by spanning the first 50 m depth interval with layers of 5 m, increasing to 25 m for the more distant (generally greater than 5000 m from coastline), deeper bathymetry to a total depth of 300 m. Below the bathymetry, layers increased in thickness at a rate of increase of 1.01, increasing to a rate of 1.5 for depths beyond the volume of interest (i.e. beyond 4 km depth) to a total depth exceeding 1500 km (i.e. at least 10 skin depths, given the initial half-space resistivity of 30 Ωm and lowest frequency of 0.001 Hz). The highly efficient but approximate coast-effect forward modelling approach of Booker (described in Burd et al., 2014) was not used as some of our sites are located very close to the coast (well within one skin depth for moderate frequencies), requiring accurate modelling that could not be guaranteed with the approximate approach.

Inversion algorithms determine an appropriate model by iteratively adjusting a resistivity model, computing its forward MT responses, and comparing these responses to the observed data. Whereas the model steps vary depending on the precise algorithm implemented, there are several key parameters that influence an algorithm's behaviour, such as the data errors, type and degree of regularisation, and initial starting and prior models. In particular, the distance between data and model responses, i.e. the sum of the squared residuals is used to measure the distance between the calculated and observed responses. To meet the assumptions of least-squares theory, these residuals must be standardised, i.e. scaled by the variance of the measurements in order to make them normally distributed over $\mathcal{N}(0, 1)$. In order to facilitate convergence of the inversion process, an error floor is commonly applied to input data for MT inversions, wherein the error provided for inversion is defined as the greater of either the observed error or some function of the magnitude of the datum. Separate error floors were used for off-diagonal and diagonal impedance tensor components in this study of 5 and 20 % of the mean magnitudes of (Z_{xy}, Z_{yx}) and (Z_{xx}, Z_{yy}) respectively. As the diagonal components observed were typically an order of magnitude smaller than the off-diagonal components due to generally 2-D resistivity structures, the signal-to-noise ratio was significantly poorer, and applying a

greater error floor here reduces the leveraging of the modelling process by noise-contaminated data.

The regularisation of an inversion describes the weighting between minimising the data residuals and some penalty function, commonly a roughness penalty that enforces smoothness in order to stabilise the resulting model. ModEM allows for the specification of separate regularisation parameters for the x, y, and z directions, with higher values placing greater weight on the penalty function. Several values of these parameters were tested; however, varying regularisation in the z direction had a negligible effect on the model. Laterally isotropic values of 0.15, 0.3, and 0.45 were tested for the x and y directions, and whereas the overall model misfit did not vary significantly, lateral resistivity structure was strongly affected, with a regularisation of 0.15 leading to discrete features with extreme resistivities (i.e. either very high or very low resistivity) beneath MT stations and poor continuity between sites. Resistivity structures in the models obtained with lateral regularisations of 0.3 and 0.45 correlated very well, with a slightly compressed range of resistivities present when regularisation was set to 0.45. In order to reduce over-smoothing of structural boundaries, final inversions were performed with lateral and vertical regularisation parameters of 0.3.

The starting model for each inversion was a 30 Ωm half-space, with seawater to depths defined by coastal bathymetry fixed (i.e. invariant) at 0.3 Ωm (marine sediments were not included). Final resistivity models were obtained by two consecutive inversion runs: a first model was determined by inversion from a uniform half-space starting model; then a second starting model was constructed by logarithmically averaging the resistivities in this first model with those of the starting half-space. The inversion of the averaged starting model was found to improve model fits significantly and result in resistivity distributions of greater range and contrast across interpreted structural boundaries. This work flow was found to preserve broad structural outlines in order to guide the inversion algorithm whilst suppressing finer features (typically associated with local minima in model space).

Frequency domain airborne electromagnetic data in our MT survey area were acquired as part of the regional Tellus survey described in Beamish (2013). The AEM data were obtained in 2005 and 2006 across Northern Ireland by the AEM-05 system described by Levâniemi et al. (2009), giving observations of both inphase and quadrature data at four frequencies (24 510, 11 962, 3005, and 912 Hz). The entirety of Northern Ireland (barring high-flight regions above urban areas and steep topography) is covered by the Tellus AEM data set, which comprises parallel flight lines spaced 200 m apart on a bearing of 345° at a nominal altitude of 56 m, with a spatial sampling rate of one sample approximately every 15 m along the flight lines. The pre-processing work flow of the AEM data is detailed at length in Beamish et al. (2006) and Levâniemi et al. (2009). The inversion of the FDEM data was performed using the one-dimensional Airbeo code from Amira International (Raiche, 1999) that implements the approach of Jupp and Vozoff (1975).

3.1　Correction of static shifts

The method proposed here for the correction of static-shift effects follows the approach of Pellerin and Hohmann (1990), adapted to using airborne FDEM data as extrinsic information. Earlier approaches took advantage of the downwards-propagating transient signal of the TDEM method to directly calculate an empirical, quasi-MT 1-D response (Sternberg et al., 1988), and later work by Pellerin and Hohmann (1990) developed this approach by explicitly modelling the TDEM data to obtain a 1-D resistivity model. The MT forward problem can be solved for this resistivity model and factors δ_{E_x}, δ_{E_y} that correct static shift between the calculated responses and observed data determined by taking the ratio of apparent resistivities. The resulting set of static-shift-corrected MT data may then be modelled as desired. We propose that the approach of Pellerin and Hohmann (1990) can be equally applied with FDEM data, subject to the same constraints, namely that there should be an overlap between the minimum depth of penetration of the MT sounding and the maximum depth of penetration of the extrinsic (to MT) information and that the dimensionality of the two methods should agree. The first constraint mandates the use of high-frequency AMT data rather than BBMT data to ensure overlapping volumes of sensitivity. A flowchart describing the steps taken in correcting the Rathlin Basin MT data with the Tellus AEM data is presented in Fig. 5.

The approach used here is predicated upon certain key assumptions about the near-surface geology and the induced galvanic distortion, and these assumptions clearly show the limits of the approach. Firstly, as mentioned we assume that the near-surface geology is 1-D in structure, i.e. we can treat the off-diagonal impedance tensor elements Z_{xy} and Z_{yx} independently without rotating or otherwise preparing the data. Secondly, we assume that the galvanic distortion affects only the electric field; as the total electric field is represented by E_x and E_y, we require only two corrective factors δ_{E_x} and δ_{E_y} to be applied to the two impedance tensor element pairs corresponding to E_x and E_y, namely (Z_{xx}, Z_{xy}) and (Z_{yx}, Z_{yy}). Finally, whereas the cause of static-shift-type distortion of the MT data is the estimation of the electric field, the airborne FDEM data are observations of the magnetic field (directly proportional to the electric field). Hence, the FDEM data are unaffected by static-shift-type distortion, and any resistivity estimates computed from these data are closer to the true values. For clarity about the effect of the static-shift corrective factors, δ_{E_x} and δ_{E_y} are often presented here on a logarithmic scale; as the corrective factors transform to additive (or subtractive) changes of resistivity on a logarithm scale, their value is intuitively related to whether a correction to more resistive or more con-

ductive true data is required. For example, a (multiplicative) static-shift corrective factor of $\delta_E = 3.16$, indicating that the true resistivity $\rho_T = \delta_E \rho_{\text{obs}} = 3.16 \rho_{\text{obs}}$ corresponds in logarithmic scale to an additive static-shift correction, i.e. $\log_{10}(\rho_T) = \log_{10}(\rho_{\text{obs}}) + 0.5$.

The first step of the implemented procedure is the inversion of the uncorrected MT data to obtain a baseline resistivity model for comparison M_0. The remaining steps describe how static-shift-corrected MT data $\mathbf{Z}_c$ were obtained by solving the MT forward problem for a model of the AEM data, and in turn inverted to obtain a corrected resistivity model M_c.

- Step 1: 3-D MT inversion of observed MT data $\mathbf{Z}_o$ with the ModEM code to obtain model M_0. Both the inversion mesh and input data were rotated to a bearing of 315° for inversion.

- Step 2: Modelling of each four-frequency AEM sounding within the survey area as a single-layer structure (i.e. half-space) with Airbeo (Raiche, 1999), resulting in an apparent resistivity value at each location best reproducing the observed AEM data.

- Step 3: Interpolation of AEM half-space models by inverse-distance-weighted (IDW) averaging of log–resistivity values to populate the uppermost 200 m of an MT forward modelling mesh with cells of 170×170 m. Below the uppermost 200 m, the model reverts to a 100 Ωm half-space.

- Step 4: 3-D solution of MT forward problem for the resistivity model found in Step 3 with ModEM, resulting in a set of high-frequency synthetic MT responses at six frequencies from 10 000 to 1000 Hz for each MT sounding location. The frequencies chosen coincide with those of the downsampled MT data.

- Step 5: Multiplicative static-shift corrective factors δ_{E_x} and δ_{E_y} found by taking the ratio of the apparent resistivities of the up to six high-frequency MT responses found in Step 4 to those of the observed data (i.e. $\delta_{E_x} = \rho_{\text{a},xy}(\text{synthetic})/\rho_{\text{a},xy}(\text{observed})$) over the 10 000–1000 Hz band. Due to either noise contamination or non-1-D behaviour (i.e. violating our assumptions), not all data in the compared frequency band were used; typically, comparison was made using three to four of the six data points that parallel the synthetic responses. The corrective factors are applied to the entire bandwidth of the unrotated observed data to obtain $\rho_{\text{a}}(\text{corrected})$, with δ_{E_x} applied to all elements dependent on E_x, i.e. $\rho_{\text{a},xx}$ and $\rho_{\text{a},xy}$. δ_{E_y} is treated analogously.

- Step 6: The static-shift-corrected data are used as input for 3-D MT inversion with ModEM to obtain an improved resistivity model M_c. Both the inversion mesh

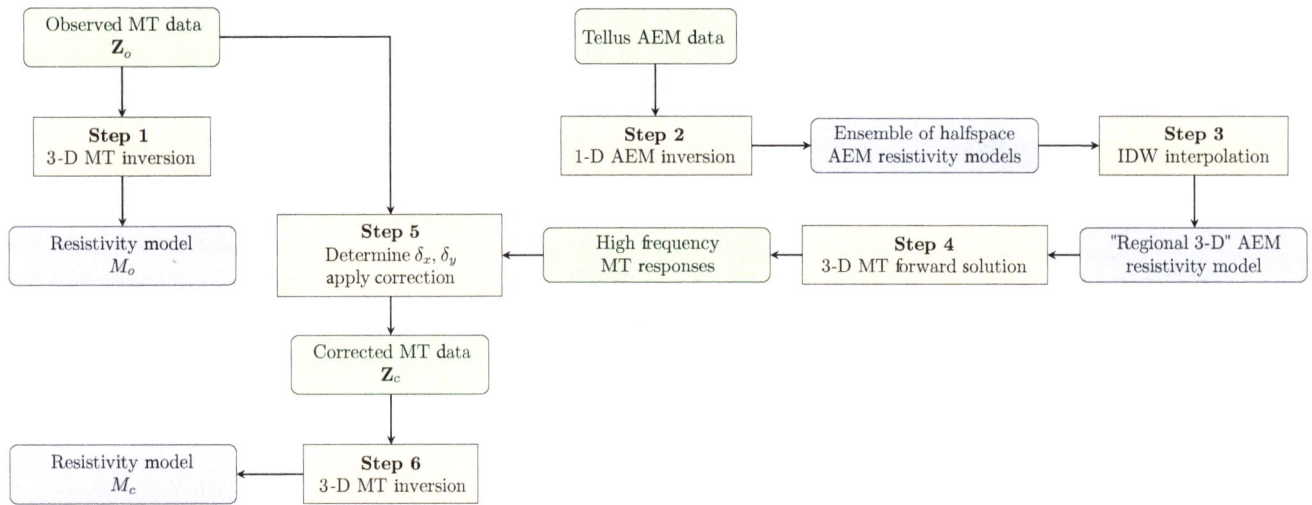

Figure 5. Flowchart illustrating the steps required to (a) find an original, uncorrected resistivity model M_o from MT data (Step 1 only) and (b) find an improved, static-shift-corrected resistivity model M_c from MT data, with coincident airborne FDEM data (data from the Tellus project used for this study). For clarification, the MT data used in Step 5 are in their original coordinate system, whereas the MT data in Steps 1 and 6 are rotated to a bearing of 315° for inversion. IDW in Step 3 refers to inverse distance weighting.

and input data were rotated to a bearing of 315° for inversion.

Whereas multi-layered models that better reproduce the AEM data can also be determined using Airbeo, they were not used in favour of the half-space apparent resistivities for two principal reasons. Firstly, the interpolation of the apparent resistivities to a 3-D MT mesh can be directly computed, whereas multi-layered models require more advanced approaches to reconcile variation in layer thicknesses unless these are explicitly set in the 1-D inversion to facilitate interpolation. Secondly, the depth of sensitivity of the lowest frequency of the AEM data ($912\,\mathrm{Hz} \approx 60\,\mathrm{m}$, from forward modelling) has a moderately narrow overlap with the skin depth of the highest MT frequencies ($10\,800\,\mathrm{Hz}$, skin depth $\approx 50\,\mathrm{m}$ for a typical near-surface resistivity of $\approx 100\,\Omega\mathrm{m}$). Above this overlapping volume, the MT data remain sensitive to but poorly resolve resistivity contrasts, and for the purposes of our work, the added complexity in multi-layered AEM modelling does not significantly improve our results. A map of the half-space models found from the AEM data in Step 2 is shown in Fig. 6. As this work is reliant upon the depth of sensitivity of the FDEM data, we assume that the causes of static shift encountered are locally one- or two-dimensional anomalies that perturb estimates of the electric field estimates E_i, causing frequency-independent static shifts of the MT impedance data. Additionally, as each electric field component is used to compute two forms of data (i.e. $\{Z_{ii}, Z_{ij}\} \propto E_i$), the same correction will be applied to data sharing a common electric field component. Groom and Bailey distortion analysis of the high-frequency MT data showed that the regional near-surface resistivity structure was 1-D or 2-D at most, validating this approach to correct-

ing the static-shift effects. We reiterate that for regions with regionally 3-D structure, this approach would be invalid.

The statistical and spatial distributions of the static-shift multiplicative corrective factors δ_{E_x} and δ_{E_y} were examined for spatially coherent correlation and features that may be indicative of known regional-scale geological structures. The spatial distribution and magnitudes of the corrective factors across the survey area are shown in Fig. 7. Whereas some spatial correlation between static shifts and geology is apparent, static-shift variation mostly does not coincide with mapped surface geological boundaries, which invalidates the use of regional geological units as predictors of static shift. The histograms in Fig. 8 show the distributions of the logarithmic transforms of δ_{E_x} and δ_{E_y}, with statistical quantities shown in Table 1 (the quantities of both $\log\delta_{En}$ and δ_{En} are tabulated, as are those of the mean static-shift correction at each site $\delta_{\overline{E}}$). The distributions of both δ_{E_x} and δ_{E_y} appear close to log-normal, with longer tails towards conductive corrections that likely indicate natural bias introduced by the sampling, such as the geographic location of the MT sites. The bivariate distribution of $\log\delta_{E_x}$ and $\log\delta_{E_y}$ from each site is shown in Fig. 9, where the strong 45° (i.e. $\delta_{E_x} \approx \delta_{E_y}$) clustering indicates that the two electric field components at most sites are similarly affected by static-shift-type effects, with no evident regional preference for static shifts to one polarisation over the other.

From examination of the spatial and statistical distributions of δ_{E_x} and δ_{E_y}, the galvanic distortions present in these data show no consistent anisotropic behaviour and weak spatial correlation with surface geology. As the accepted theory is that galvanic distortion is typically caused by irresolvable near-surface resistivity inhomogeneities below the level of

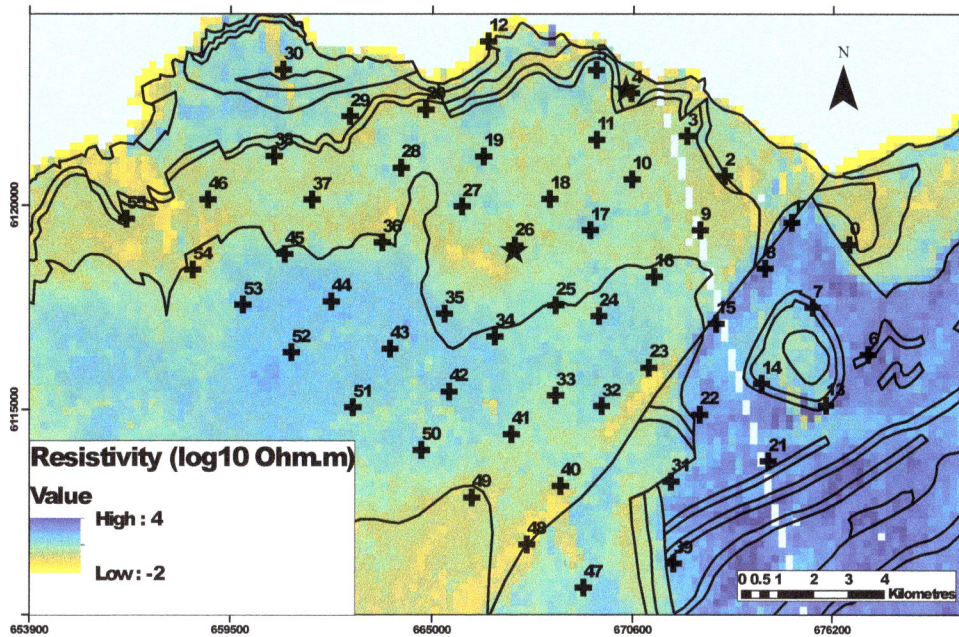

Figure 6. Map of airborne FDEM half-space resistivity models, overlain with geological boundaries (solid black lines, as in Fig. 1) and MT site locations (black cross symbols). The model exhibits good correlation with the mapped boundaries, as well as variation within each geological unit. Note that data from one flight line are missing, resulting in a pale dashed line in the figure. These absences were interpolated across in the synthetic model.

Table 1. Statistical measures of the $\log_{10}$ of the corrective factors δ_{E_x} and δ_{E_y} (i.e. additive in logarithmic domain), as displayed in Fig. 8. In addition to the means, medians, and standard deviations of $\log_{10}\delta_E$, the means and medians of δ_E (i.e. multiplicative) are also shown. Note that $\delta_{\overline{E}}$ is the site-wise mean static shift – i.e. the mean of δ_{E_x} and δ_{E_y} on a site-by-site basis.

	Mean $(\log_{10})$	Median $(\log_{10})$	σ $(\log_{10})$	Mean	Median
δ_{E_x}	-0.035	0.011	0.235	0.923	1.03
δ_{E_y}	-0.035	-0.020	0.236	0.923	0.955
$\delta_{\overline{E}} = \frac{\delta_{E_x}+\delta_{E_y}}{2}$	-0.035	-0.012	0.225	0.923	0.974

resolution of the AMT data, the weak spatial correlation with surficial geology is unlikely to be random and instead reflects the variation in heterogeneity of formations

4 Model evaluation and discussion

Initial comparison of the two models, one with (M_c) and one without (M_o) static-shift correction, shows greater resistivity contrasts and sharper delineation between resistive and conductive volumes in the corrected model M_c, and for these reasons M_c is preferred for interpretation and evaluation of structures in the Rathlin Basin. Due to the static-shift correction, M_c also more likely better represents the correct resistivity distribution of the real Earth. Figures 11–14 present

the resistivity distribution ρ of horizontal and vertical slices taken through the preferred model M_c (profile locations are marked in Fig. 1). Measures of model comparison are also shown on these figures and are discussed below (note that no images of the resistivity distribution of M_o are presented, as the resistivity differences with respect to M_c are subtle and difficult to perceive given the dynamic range of resistivities in the visualisation). The two models generally show good correlation of geoelectric structural geometries but differ in the absolute resistivity estimates and the exact extent of these structures.

The resistivity structure of M_c is generally quite simple, with the most prominent feature being the extensive conductor, of a resistivity less than $10\,\Omega m$, that lies between 900 and 2000 m depth and extends from the Tow Valley Fault in the south-east for a distance of up to 10 km towards the north-west.

We propose that this conductor primarily represents the conductive MMG, with a clearly delineated upper boundary. The interpretation of the lower boundary against the SSG (and equally, the boundary between the SSG and the EG sandstones) is hindered by both the smoothing effects of the regularised inversion approach used and the fact that inductive EM responses are intrinsically sensitive to the tops of conductive units (and their integrated conductivity) rather than to the bottoms of conductive units (i.e. the tops of resistive units). The TVF, which forms the south-eastern boundary of the basin, is clearly defined, although the angles of dip

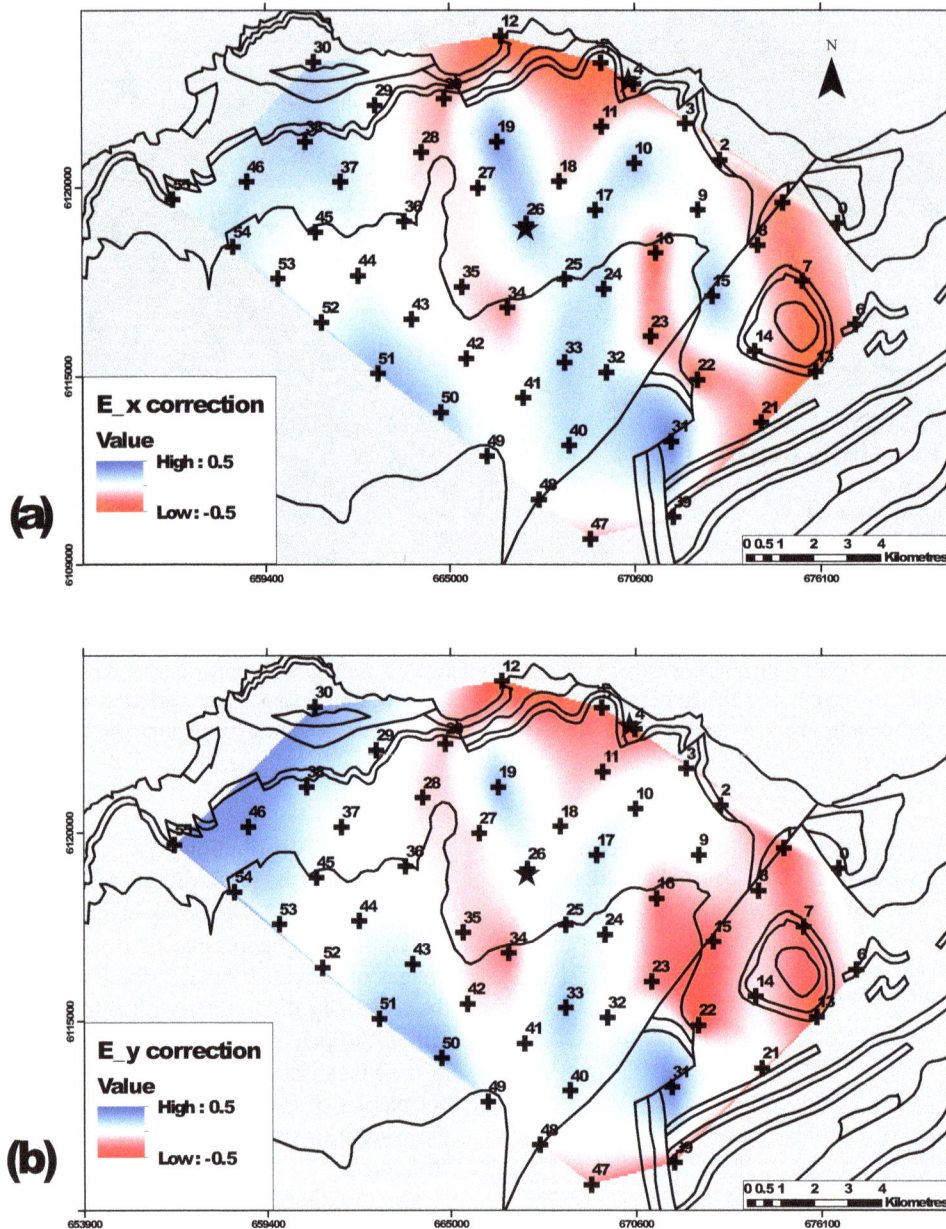

Figure 7. Spatial distribution of static correction factors δ_{E_x} (**a**) and δ_{E_y} (**b**) in decades (i.e. as additive factors, where $+0.5$ corresponds to a multiplicative factor of $10^{0.5}$, ≈ 3). Blue shading indicates resistive corrections to the data, whereas red indicates conductive corrections. Note that for the majority of sites, δ_{E_x} and δ_{E_y} are similar in magnitude and polarity. The magnitudes of static corrections are generally less than 0.5 decades in size, corresponding to apparent resistivity estimates being shifted by factors of less than 3.

modelled are slightly shallower than those modelled in Gibson (2004), again likely due to smoothing within the MT inversion process. The Dalradian metasedimentary horst is represented by the highly resistive volume in the south-east of the model and is visible at all depths and on both cross-basin profiles with a relatively homogeneous structure. All three profiles show the shallowest 500 m as moderately resistive, correlating well with the expected extent and thickness of the ALG basalts and UWLF. Minor conductors of 10–50 Ωm

occurring between depths of 300 and 600 m are interpreted as the LLG, although these depths are slightly greater than the depths observed in the PM1 borehole. The uppermost resistivities within the model correlate well with the mean of the resistivity measurements from the PM1 borehole for the ALG and underlying sediments, though clearly without the same vertical resolution.

In order to demonstrate the lateral heterogeneity of the basin and intra-basinal structures, Fig. 15 shows the resistiv-

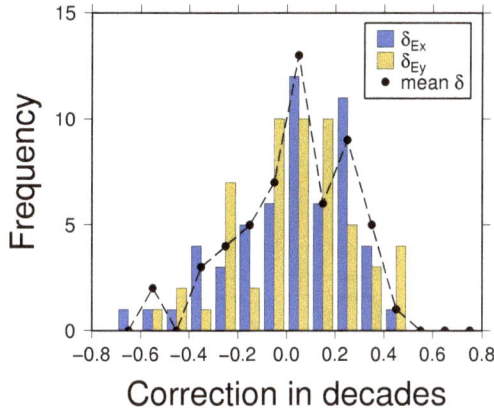

Figure 8. Distribution of calculated static correction factors δ_{E_x} (blue) and δ_{E_y} (yellow) in decades (i.e. as additive factors, where +0.5 corresponds to a multiplicative factor of $10^{0.5}$, ≈ 3). Statistical quantities related to these distributions are tabulated in Table 1.

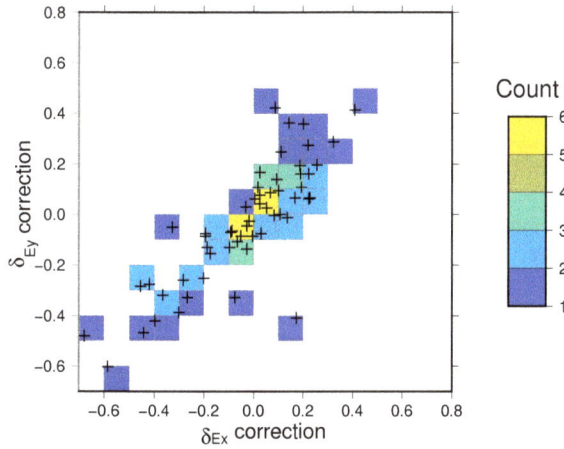

Figure 9. Bivariate distribution of calculated static-shift corrections δ_{E_x} and δ_{E_y} in decades (i.e. as additive factors, where +0.5 corresponds to a multiplicative factor of $10^{0.5}$, ≈ 3), paired by site. The strong clustering and 45° trend (i.e. $\delta_{E_x} \approx \delta_{E_y}$) indicate that the orthogonal static shifts are generally similar sized across the survey area, with no observable regional trend of greater static shift of one polarisation.

ity distribution of the basin portion of M_c as a set of layer-by-layer histograms, each histogram normalised by the number of counts for the mode of that layer. As the model becomes much smoother below depths of approximately 2500 m, several tests were carried out to examine the sensitivity of the model responses to resistivity contrasts beyond these depths. The insertion of synthetic resistive bodies showed that the model responses are sensitive to them but unable to resolve resistivity contrasts below the conductive basin. It remains possible that older sediments with higher resistivity exist below this depth; however, the resistivity contrasts present

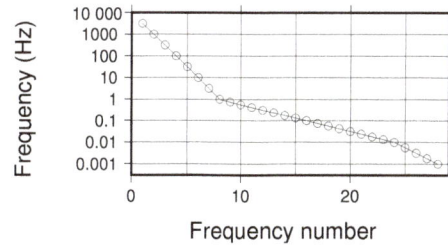

Figure 10. Chart of frequency indices and values used in 3-D inversion process. Frequency spacings per decade were 2 (for 3000–1 Hz), 8 (for 1–0.01 Hz), and 4 (for 0.01–0.001 Hz).

within the MT data are insufficient to clarify their existence or extent.

Two diagnostic measures were used to assess the changes between M_c and the original model M_o, caused by static-shift correction. The first diagnostic measure is the logarithmic resistivity difference Δ between the two models,

$$\Delta(M_c, M_o) = \log_{10} \frac{\rho_{M_c}}{\rho_{M_o}}, \tag{6}$$

and the second is the normalised cross-gradient (NCG) of the two models, being the cross-product of each model's gradient vectors at each cell,

$$\text{NCG}(M_c, M_o) = \frac{|\nabla M_c \times \nabla M_o|}{|\nabla M_c||\nabla M_o|}. \tag{7}$$

The normalised cross-gradient was introduced in Gallardo and Meju (2003) in the context of joint inversion based on structural similarity and has been used by, e.g., Schnaidt and Heinson (2015) and Rosenkjaer et al. (2015) to highlight structural similarities and differences between resistivity models.

As the logarithmic resistivity difference Δ between M_c and M_o highlights discrepancies in the absolute resistivity values of the two models, it is more useful for observing contrasts in formation resistivities (i.e. locations where one model has a relative minimum or maximum) rather than comparing structural boundary locations. The figures presented show that the conductive features in M_c tend to be between 0.25 and 0.50 of a decade more conductive than their equivalent volumes in M_o, whereas the resistive features show similar tendencies towards greater resistivity in M_c. The largest vertical variations in Δ (i.e. from a local maximum to a local minimum or vice versa) are typically seen at the upper boundary of the MMG, making the interpretation of this boundary far less subjective in model M_c. The resistivity difference Δ also shows greater definition of the thin conductors interpreted as the LLG, which are difficult to delineate on the plots of resistivity alone due to weaker resistivity contrasts. The metasedimentary horst shows a range of both conductive and resistive contrasts; however, as this region shows greater geoelectrical heterogeneity and greater static-shift effects in the observed data, the cause of these differences is

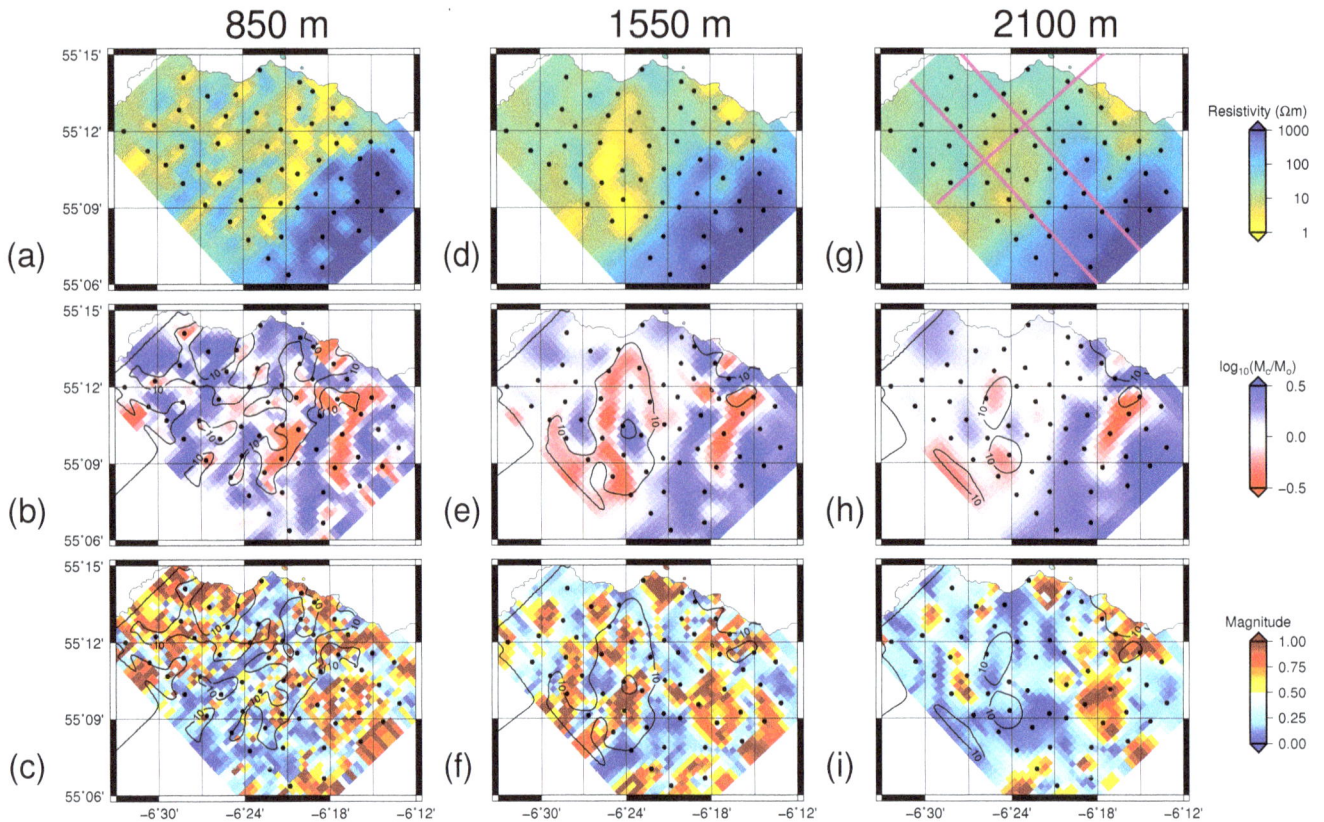

Figure 11. Top row shows resistivity slices through the static-shift-corrected model M_c taken at 850 (**a**), 1550 (**d**), and 2100 m (**g**) below sea level. Middle row shows the resistivity difference Δ between the models in decades ($\Delta(M_c, M_o) = \log_{10}(M_c/M_o)$) for the same depths, where red shows M_c more conductive than M_o and blue more resistive. Bottom row shows the magnitude of the normalised cross-gradient (the cross-product of the gradient vectors of each model ∇M, as defined in Eq. 7), as a diagnostic of structural similarity between the models, with 0 (blue) showing parallel gradient vectors (i.e. very similar structure) and 1 (red) showing orthogonal gradient vectors and structural disagreement. Note that the difference and cross-gradient plots are overlain by the 10 Ωm contour from the corresponding resistivity slice. Magenta lines in panel (**g**) indicate the location of Profiles A, B, and C.

variations in the uncorrected data used to model M_o rather than the relatively uniform horst seen in M_c. The spatial distribution of Δ is insightful, particularly the distribution of points where $\Delta = 0$ (i.e. points where M_o and M_c have the same resistivity) which generally do not coincide with the resistivity minima or maxima of M_c nor with the identifiable resistivity boundaries. Instead, these points in places indicate regions in which the structural geometries differ between models, such as thicker or thinner conductive bodies (e.g. at ≈ 800 m depth, 10 000 m along Profile A, Fig. 12b, where the $\Delta = 0$ line occurs in the middle of a conductor). These differences in geometry are most noticeable as horizontal features in the uppermost 1000 m, where the models are constrained locally by single sites and the static-shift correction affects the model in a generally 1-D manner. The lateral distribution of Δ at these shallow depths is a result of the pseudo-1-D local scaling of the model differing from site to site; as such, we do not expect the distribution of Δ to replicate those of the static-shift corrections themselves (except

for the trivial case of surficial and immediately subsurficial layers).

The NCG is largest where the gradient vectors (i.e. resistivity changes) of the two models are orthogonal, such as differences in structures and locations or similar structure but differing magnitude of resistivity change (i.e. difference in the curvature of the model). Both situations of elevated NCG are evident in the models, especially on the zero-Δ lines (i.e. where the resistivity models intersect and Δ reverses polarity, requiring significantly different gradient vectors), and at resistivity minima and maxima of M_c (and assumingly M_o, although not shown), indicating a difference in curvature due to slight offsets of critical resistivity points or large resistivity differences in co-located critical points. We note again that the significant variation in NCG values computed in the Dalradian horst likely reflects the heterogeneity of this block. Whereas the NCG was originally proposed as a means of determining structural similarity between models, our use of the NCG in this study suggests that it may also be used to

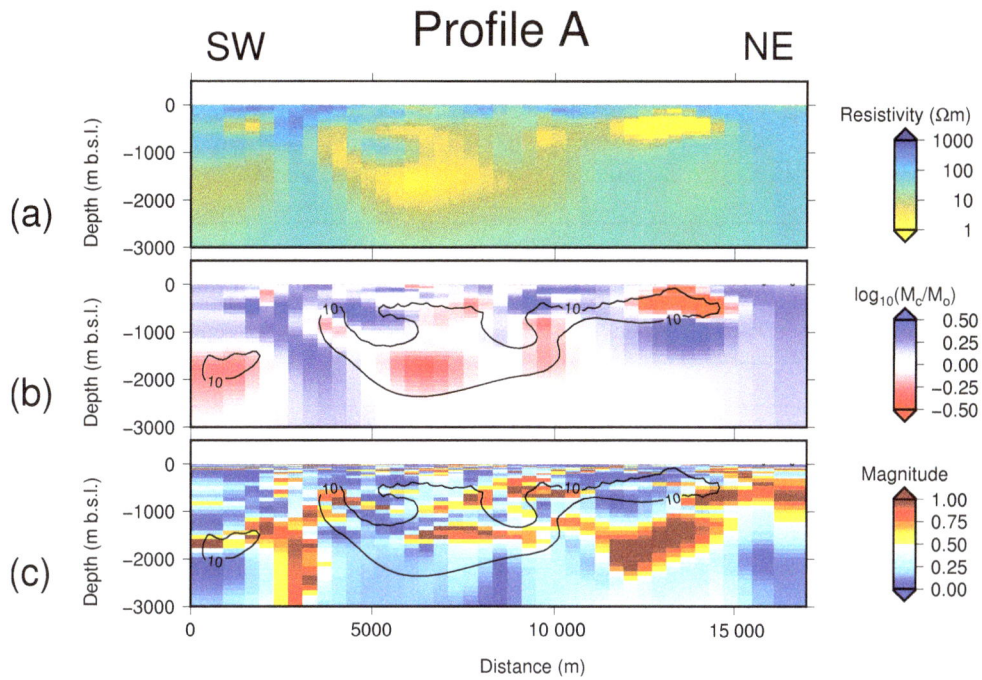

Figure 12. Profile A taken along the axis of the concealed basin through the static-shift-corrected resistivity model M_c (location shown in Figs. 1 and 11g). The resistivity is shown in panel **(a)**, the resistivity difference $\Delta(M_c, M_o)$ is shown in panel **(b)**, and the cross-gradient of M_c and M_o is shown in panel **(c)**. Contours on the difference and cross-gradient plots show the $10\,\Omega$m contour. For presentation a vertical exaggeration of 1.5 is used.

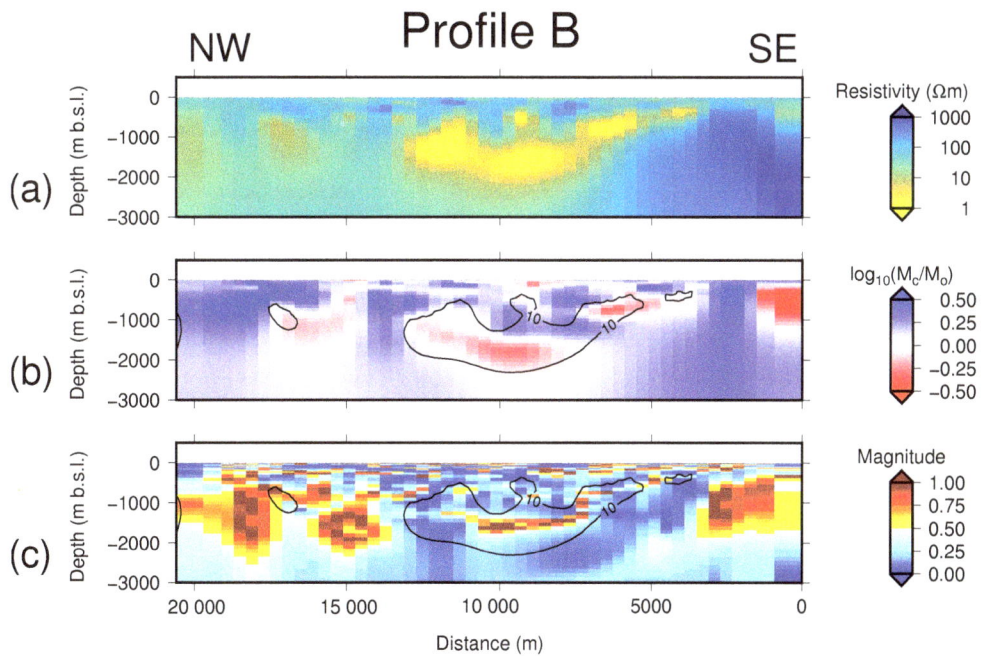

Figure 13. Profile B taken across the static-shift-corrected resistivity model M_c (location shown in Figs. 1 and 11g). The resistivity is shown in panel **(a)**, the resistivity difference $\Delta(M_c, M_o)$ is shown in panel **(b)**, and the cross-gradient of M_c and M_o is shown in panel **(c)**. Contours on the difference and cross-gradient plots show the $10\,\Omega$m contour. For presentation a vertical exaggeration of 1.5 is used.

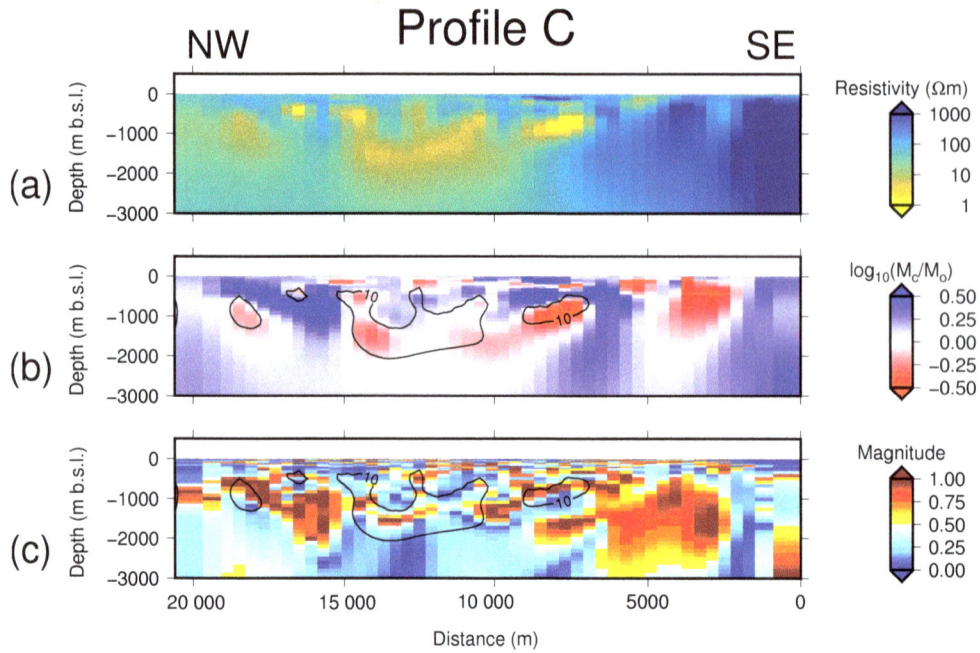

Figure 14. Profile C taken across the static-shift-corrected resistivity model M_c (location shown in Figs. 1 and 11g). The resistivity is shown in panel **(a)**, the resistivity difference $\Delta(M_c, M_o)$ is shown in panel **(b)**, and the cross-gradient of M_c and M_o is shown in panel **(c)**. Contours on the difference and cross-gradient plots show the 10 Ωm contour. For presentation a vertical exaggeration of 1.5 is used.

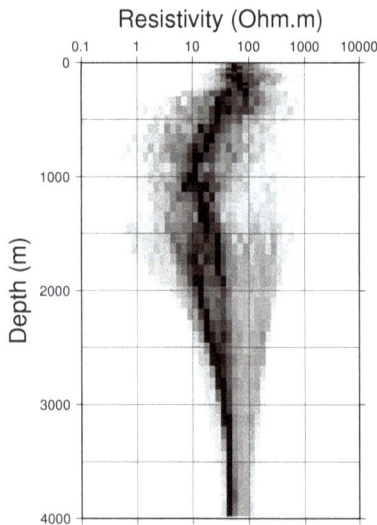

Figure 15. Representative histogram model of the portion of the static-shift-corrected model M_c covering the Rathlin Basin only. Each layer represents the distribution of resistivities present in that layer of the model, normalised by the number of counts for the mode of that layer (black). Whereas a wide spread of resistivities is present from 300 to 2500 m depth, the column of modal resistivities can be viewed as a "typical" 1-D structure of the basin, with a distinct minimum of ≈ 10 Ωm evident at 1100 m depth, the depth interval of the MMG observed in the Port More 1 borehole.

assess the coincidence of similar structure with different resistivities.

The closeness of MT model responses to the data is commonly judged by the normalised root mean square error (nRMS), defined as

$$\text{nRMS} = \sqrt{N^{-1}\sum_{i=1}^{N}\widetilde{r}_i^2,} \tag{8}$$

where $\widetilde{r}$ is the residual between calculated and observed data, normalised by the square root of its variance. In the case of a sufficient number of data N with approximately Gaussian and independent misfits, nRMS error should approach 1 for a model fitting to within 1 standard error; however, the application of an error floor to the observed misfits results in artificially lowered nRMS errors. Additionally, due to the global averaging of residuals, an nRMS error is uninformative about which portion of data are poorly fit.

Regardless, the nRMS error remains a useful metric for comparing the relative goodness of fit of a succession of model responses to the same data set. The two models presented here reproduce the observed and corrected MT responses to similar degrees with overall nRMS of 1.77 and 1.96 for M_c, the corrected model (66 iterations total), and M_o, the original model (52 iterations total), respectively. Figure 16 shows the distribution of misfit between the data and the model responses of M_c for the four impedance tensor elements at each frequency and site, allowing a better insight into which portions of the data are worse fit than can

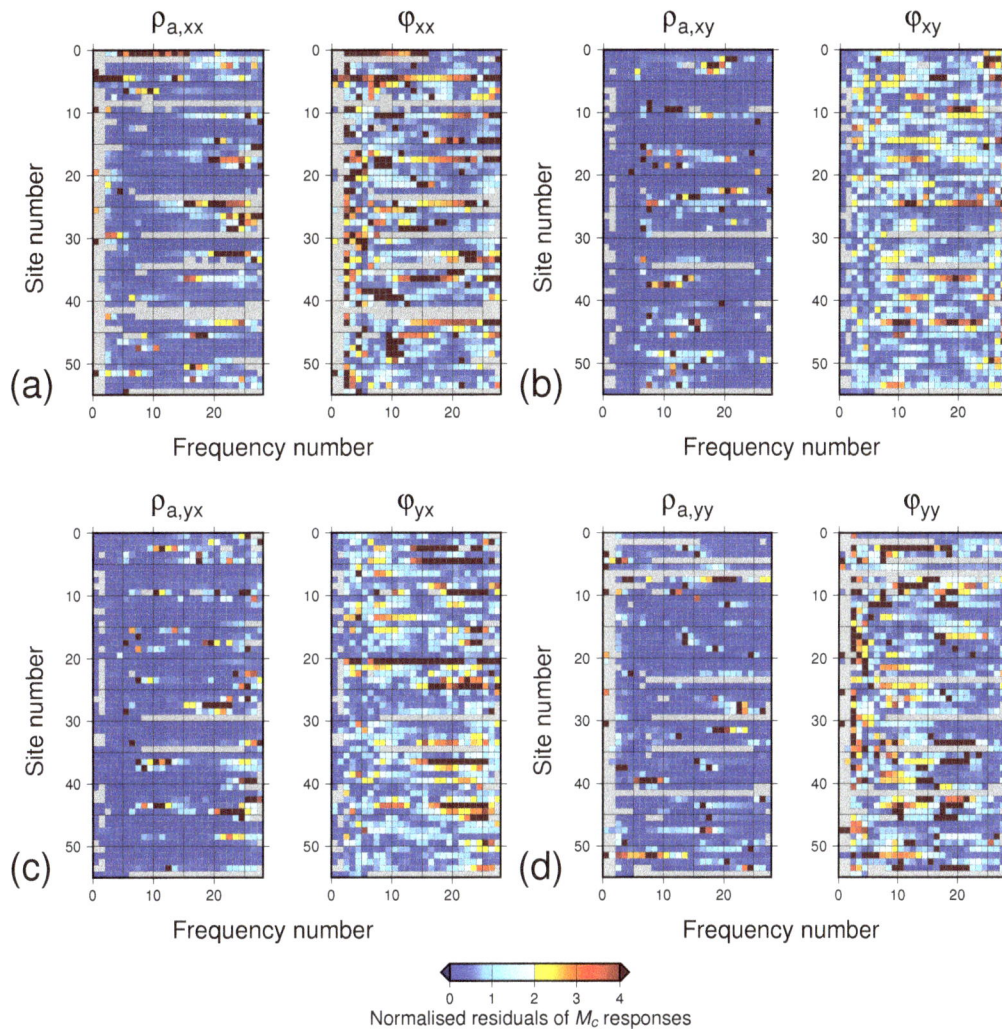

Figure 16. Visualisation of the normalised residuals of the magnetotelluric responses of M_c compared to observed data at each site (numbered as per Fig. 1) and inversion frequency number (as shown in Fig. 10) for each element of the impedance tensor (**a–d**). The normalised residual is determined by taking the absolute value of the difference between model response and datum and normalised by the error used in inversion (i.e. the greater of the observed errors or the applied error floor). Normalised residuals of less than 1 indicate a residual smaller than the error. Note that misfits of apparent resistivity use the logarithmic difference to determine the residual. Values in grey represent the lower-quality data that were masked and not included in the inversion.

be drawn from a single nRMS value for the entire model. The measure of misfit displayed is the magnitude of the residual (the absolute value of the difference between the model response and observed datum), normalised by the error used for inversion (the larger of the experimental errors or the applied error floor), similar to the overall nRMS defined in Eq. (8), with values of 1 or less indicating a residual smaller than the error (i.e. fit to within 1σ). In general, the off-diagonal components are well fit, with a handful of sites showing poorer fits with respect to phase components. As the model responses are noise-free and have internally consistent magnitude and phase behaviours, worse misfits to impedance phases in comparison to magnitudes imply that the affected portions of the data themselves may not entirely

satisfy the assumption of being in the far-field region of the signal, likely due to the proximity of a noise source. Interference from a local noise source contaminates both the magnitude and the phase of the data, and inverting such data can lead to a spurious model structure. Data that showed significant noise contamination (identified by phases trending towards 0°) were removed prior to inversion. The misfits of the diagonal impedance components show similar behaviour in that the magnitude is better fit than the phases; however, the differences in fit between magnitudes and phases are greater for the diagonals. Some increase in misfit is expected at the high-frequency limits due to the relocation of MT sites to the centre of their respective cells for computation; as the cells are 400 m wide within the survey area, the model is effectively 1-D for forward computation of MT responses at such high frequencies, and diagonal impedances are 0 for such resistivity structure. As the observed diagonal MT data can be non-zero due to resistivity variations at scale lengths below those used for modelling purposes, the fit of these data can be improved by the use of a finer mesh, with a corresponding increase in computation time. At longer periods the misfits of diagonal phases are increased for the same reasons as the off-diagonal phases, namely noise contamination in the data; however, due to the generally smaller magnitudes of the diagonal MT data the signal-to-noise ratio can be considerably worse. It is possible that the error floor used for the diagonal components is perhaps too stringent, and lower misfits would be obtained with a higher error floor; however, as the observed data imply a predominantly 2-D structure (as shown in Fig. 4), such a change is unlikely to drastically change the final model.

Although examining the misfits of a single model's responses to the data set used in its determination in such a granular fashion is useful in determining which data components are poorly fit and possibly why, taking a similar approach in order to compare two models is not valid in this case. Due to the application of individual static-shift corrections to each site's MT data, the gradients of the respective data spaces of the models are significantly altered. As most inversion algorithms (the nonlinear conjugate gradient method implemented in ModEM included) rely upon gradients within the data space to determine the direction of line searches as part of their optimisation, the static-shift corrections applied all but guarantee that the two inversions presented in this work are the products of different paths through their data spaces. Hence, although the overall mean nRMS estimates are similar, we cannot categorically state that any differences in misfit at the granular, individual datum level are not simply due to the different gradient progressions.

The key test of the two models lies in comparing the measured resistivity values from the PM1 borehole to the vertical resistivity columns from M_c and M_o corresponding to the location of the borehole (Fig. 17). The input data for the two inversions differ solely in the application of static-shift corrections, assumed to only affect the magnitude of the data; considered as a purely 1-D problem, such a shift results in the rescaling of both layer resistivities and thickness. The nRMS misfits of the two models' responses for this location are 0.97 (M_c) and 1.05 (M_o). The effect in 3-D is more complicated, as the variation occurs in both lateral and vertical directions; however, some geometrical correlation and similar-shaped structures can still be expected in each of the two models. Note that we argue this based upon the perturbation of the data magnitudes rather than the similar nRMS – as mentioned, such an assumption based on misfit is unreasonable. Additionally, for the uppermost extents of the model (i.e. at depths within the inductive volume of only a single MT site), it can be seen that the application of static-shift corrections results in similar effects to a 1-D case, i.e. a

model from data that have been statically shifted upwards to higher apparent resistivities will have increased resistivity–thickness products (thicker, more resistive layers). With this behaviour in mind, the two models M_c and M_o have very similar geometries at shallow depths (the uppermost 200 m), with the resistivities of M_c significantly closer to the mean of the borehole measured resistivities. Due to the regularisation of the inversion process, neither model can adequately reproduce the highly variable near-surface variations measured through the ALG in the borehole. Comparison of the structures at deeper depths shows that whereas M_c has conductive layers at 300 and 1300 m depth, with resistivities of ≈ 3 and 15 Ωm, interpreted as the middle of the LLG sediments and the combined conductances of the MMG and SSG respectively, the equivalent conductors (assuming a scaled geometry) in M_o occur at 450 and 1500 m, with resistivities of ≈ 20 and ≈ 8 Ωm. Similarly, the models show a resistor of ≈ 60 and ≈ 30 Ωm at 600 and 750 m depth in M_c and M_o respectively.

The resistivity columns from the two models at the PM1 borehole site can also be compared on the basis of integrated conductances, i.e. the ratio of a layer's thickness and resistivity. In such a comparison, the LLG sediments are represented as conductances of 31 (M_c) and 13 (M_o) S, the interpreted dolerite sill sequence as conductances of 34 (M_c) and 23 (M_o) S, and the MMG or SSG sediments as conductances of 103 (M_c) and 100 (M_o) S. The greatest contrast in conductance is observed in the LLG, wherein M_c shows a conductance almost half an order of magnitude greater than M_o. The resistivity measurements in the PM1 borehole do not span the entire LLG interval; however, given the measured values the real conductance of the interval is likely greater still than the 31 S recovered in M_c. Hence, although M_c does not fully recover such a conductance, the elevated conductance in comparison to M_o, coupled with the near-identical lower conductances, indicates that M_c is closer to the real structure at this location.

Given the knowledge of the lithology and the measured borehole resistivities, we conclude that M_c is a categorically superior resistivity model in that, of the two models, it more accurately and correctly resolves the central depths and resistivity values of the lithological units encountered. Critically, the limited borehole resistivity data in the MMG and SSG suggest that resistivity increases slightly with depth between the two units, and whereas the resistivity column of M_o more closely matches the absolute resistivity values, the column of M_c better approximates the trend of the observed borehole resistivities. Without further external information to verify the models with, the smaller differences between the M_o model resistivity column and the borehole resistivity measurements of the target sediments cannot lead to the acceptance of the entire model, as the shallower structure is poorly recovered. Conversely, M_c, the static-shift-corrected model, correlates well with the borehole lithological boundaries from the PM1 borehole and shows a trend in resistivity

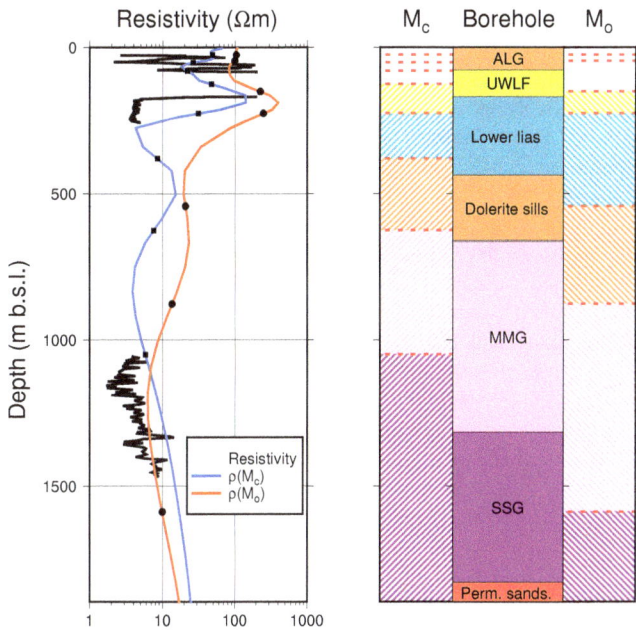

Figure 17. Left-hand panel shows the resistivity columns from both M_c (blue) and M_o (orange) adjacent to the Port More 1 borehole, plotted with the observed normal resistivity data. Due to the regularisation in the inversion process, neither model can reproduce the high variability of resistivities within the ALG observed in the uppermost 100 m of normal resistivity data. Note that the resistivity column of M_c approaches the observed low resistivities of the LLG and reproduces the trend observed in the MMG and SSG resistivities, albeit with resistivities approximately half a decade greater. In contrast, the resistivity column of M_o correlates poorly with the observed normal resistivity data, with overestimated resistivities and layer thicknesses. Right-hand panel shows interpreted structure of each resistivity column (patterned, on either side), with the observed borehole lithology in the centre column. The interpreted M_c structure shows significantly improved correlation with the observed lithology; in particular, the upper boundary of the MMG against the dolerite sills is recovered much closer to the true depth in M_c than in M_o. The static-shift correction factors δ_{E_x} and δ_{E_y} applied at this site were 0.40 and 0.38 (-0.40 and -0.42 in decades) respectively.

with depth that matches the trend evident in the wireline resistivity measurements.

Changes in resistivity at depth from static-shift correction have strong implications for the interpretation of the reservoir potential of the area. Assuming Archie's Law holds and depending on the cementation exponent m (formation dependent, but typically ≈ 2 for reservoir formations), a relative change in resistivity f_ρ between two models ρ_c and ρ_o (i.e. $\rho_c = f_\rho \rho_o$) corresponds to a change in estimated porosity of $\phi_c = f_\phi \phi_o$, where $f_\phi \propto \sqrt[m]{f_\rho}$. The relationship between porosity and hydraulic permeability is known to be complex and highly nonlinear; it would be highly speculative to quantify how the resistivity perturbations from the correction of static-shift effects would affect any attempt to estimate hydraulic permeability. At this stage, given the paucity

of borehole information and unfavourable geoelectric setting (in which the conductive MMG directly overlies the moderately less conductive SSG and EG reservoir targets), the modelled resistivity distribution cannot be used to extend the knowledge of the hydraulic properties of the reservoir targets beyond what is reported from the boreholes within the Rathlin Basin with any useful level of confidence.

The quasi-1-D correction of static-shift effects applied here affects the resulting three-dimensional models to significantly greater depths than expected, with differences between the corrected and uncorrected resistivity models of up to half an order of magnitude present at depths of up to 2000 m (see, for example, the constant depth slice at 2100 m depth in Fig. 11) that do not correlate with the spatial distribution of applied static-shift corrections (Fig. 7). Whereas the structural geometries are very similar between the two models (especially in the shallow regions constrained by single MT sites), interpretations of the target sediment locations in the models differ in significant ways, with both depth to sediments and absolute resistivity affected. As such, it is clear that when seeking accurate resistivity and depth estimates in three-dimensional modelling of MT data, galvanic distortion must be accounted for as its effects are subtle but pervasive.

5 Conclusions

An approach for the correction of static-shift-type galvanic distortion in MT data utilising airborne FDEM data has been tested that follows the use of TDEM data in previous methods. The new approach was tested on an MT data set from Northern Ireland, using a publicly available regional data set of airborne frequency domain electromagnetic data to create a set of corrected MT data. Three-dimensional inversion of each magnetotelluric data set recovers structures with similar geometries; however, structures in the near-surface show scaling of resistivity–thickness products proportional to the static-shift correction applied. When compared to geophysical borehole logs it is clear that the model from static-shift-corrected data reproduces the observed resistivity with significantly greater fidelity. Significant suborders of magnitude variations in resistivity are caused in the model by the correction of the static-shift components of galvanic distortion, not only in the near-surface but extending down to the target sediment depths ($\approx$1500 m). As the test area is a prospective geothermal exploitation site (a sedimentary aquifer with elevated temperatures), electrical resistivity could be used to infer the heterogeneous distribution of hydraulic properties within the reservoir; however, the suborder of magnitude perturbations caused by inadequate consideration of galvanic distortion would lead to a gross misestimation of these physical properties.

The model determined by the inversion of static-shift-corrected data was found to better recover the resistivity structure observed in a nearby borehole in comparison to the model from observed data. Based on these observations, we conclude that airborne FDEM data provide sufficiently accurate resistivity estimates to allow the correction of static-shift effects in MT data. Note that this approach as discussed here is valid only for locales where the near-surface resistivity distribution is approximately one-dimensional. Given the often regional acquisition and open availability of such AEM data, it is hoped that the approach demonstrated here could be further tested with other MT surveys. Pending further case studies, FDEM could in future be considered as another alternative method to evaluate and correct static-shift-type distortion. Additionally, whereas our approach assumes one-dimensional, single-layer models for the AEM data in deriving the static-shift corrections, future advances could investigate what effect more advanced AEM modelling (i.e. multiple layers or, where applicable due to AEM acquisition specifications, full 3-D modelling) would have on the computed forward MT responses and associated static-shift corrections.

Code availability. Aside from the processing codes used to convert the measured MT time series data to robust impedance data (implemented in propriety programs from Phoenix Geophysics), codes used in this article are available for academic and non-commercial purposes. The Airbeo program for AEM modelling is part of the P223 software suite of Amira International, assembled by CSIRO, and available from Amira International's website. The "strike" program for distortion analysis of MT data can be obtained by contacting A. G. Jones, the co-author of McNeice and Jones (2001). The ModEM code for three-dimensional inversion of MT data is available for non-commercial use from its authors' website, hosted by Oregon State University http://blogs.oregonstate.edu/modem3dmt/.

Author contributions. R. Delhaye, A. G. Jones, and M. R. Muller designed the MT acquisition plan, with contributions from D. Reay in terms of prior geological and geophysical knowledge. R. Delhaye and M. R. Muller carried out the acquisition, with the assistance of the acknowledged IRETHERM MT Team. MT data processing was done by R. Delhaye, under instruction from A. G. Jones and M. R. Muller. The workflow for AEM modelling and MT static-shift correction was planned and executed by R. Delhaye and V. Rath. MT inversions were performed by R. Delhaye, with instructive direction from A. G. Jones and V. Rath. The paper was prepared by R. Delhaye and V. Rath, with insightful comments and review from A. G. Jones and M. R. Muller.

Competing interests. The authors declare that they have no conflict of interest.

Acknowledgements. We would like to acknowledge Science Foundation of Ireland for the financial support for the IRETHERM project (10/IN.1/I3022) to A. G. Jones and particularly the student support to R. Delhaye. MT data acquisition was only possible with the assistance of the IRETHERM MT Team (S. Blake, T. Farrell, C. Hogg, J. Vozar, C. Yeomans). G. Egbert, A. Kelbert, and N. Meqbel are very gratefully thanked for making their ModEM code available to the community, especially N. Meqbel for installing it on our clusters and those of the Irish Centre for High-End Computing (ICHEC). The Geological Survey of Ireland and the Geological Survey of Northern Ireland are thanked for providing access to Tellus project data and other complementary information. M. Dessisa of the GSNI is especially thanked for his assistance with the latter. Amira International is thanked for providing open access to the P223 modelling suite. We also acknowledge the work of the Geological Survey of Ireland-funded project GSI-sc-04, "Spatially constrained Bayesian inversion of frequency and time domain airborne electromagnetic data from the Tellus projects" in advancing the available tools and utilities for the handling of Tellus project airborne data. ICHEC is thanked for providing the computational capability required for us to perform our inversions. Lastly, we gratefully thank A. Junge and an anonymous reviewer for their comments and suggestions for improving this paper.

Edited by: C. Krawczyk

References

Archie, G. E.: Electrical Resistivity an Aid in Core-Analysis Interpretation, AAPG Bulletin, 31, 350–366, doi:10.1306/3d93395c-16b1-11d7-8645000102c1865d, 1947.

Auken, E., Vest Christiansen, A., Kirkegaard, C., Fiandaca, G., Schamper, C., Behroozmand, A. A., Binley, A., Nielsen, E., Effersø, F., Christensen, N. B., Sørensen, K., Foged, N., and Vignoli, G.: An overview of a highly versatile forward and stable inverse algorithm for airborne, ground-based and borehole electromagnetic and electric data, Explor. Geophys., 46, 223–235, doi:10.1071/EG13097, 2014.

Avdeeva, A., Moorkamp, M., Avdeev, D., Jegen, M., and Miensopust, M.: Three-dimensional inversion of magnetotelluric impedance tensor data and full distortion matrix, Geophys. J. Int., 202, 464–481, doi:10.1093/gji/ggv144, 2015.

Beamish, D.: The bedrock electrical conductivity structure of Northern Ireland, Geophys. J. Int., 194, 683–699, doi:10.1093/gji/ggt073, 2013.

Beamish, D., Cuss, R. J., Lahti, M., Scheib, C., and Tartaras, E.: The Tellus Airborne Geophysical Survey of Northern Ireland: Final Processing Report, Internal Report IR/06/136, British Geological Survey, Nottingham, UK, 2006.

Bean, C. J. and Martini, F.: Sub-basalt seismic imaging using optical-to-acoustic model building and wave equation datuming processing, Mar. Petrol. Geol., 27, 555–562, doi:10.1016/j.marpetgeo.2009.09.007, 2010.

Berdichevsky, M.: Methods used in the U.S.S.R. to reduce near-surface inhomogeneity effects on deep magnetotelluric sounding, Phys. Earth Planet. In., 53, 194–206, doi:10.1016/0031-9201(89)90003-4, 1989.

Berdichevsky, M. N. and Dmitriev, V. I.: Models and Methods of Magnetotellurics, Springer, Berlin-Heidelberg, Germany, 2008.

Bernabe, Y., Mok, U., and Evans, B.: Permeability-porosity Relationships in Rocks Subjected to Various Evolution Processes, Pure Appl. Geophys., 160, 937–960, 2003.

Bjørlykke, K.: Petroleum Geoscience: From Sedimentary Environments to Rock Physics, Springer, Berlin-Heidelberg, Germany, 2010.

Burd, A. I., Booker, J. R., Mackie, R., Favetto, A., and Pomposiello, M. C.: Three-dimensional electrical conductivity in the mantle beneath the Payún Matrú Volcanic Field in the Andean backarc of Argentina near 36.5° S: evidence for decapitation of a mantle plume by resurgent upper mantle shear during slab steepening, Geophys. J. Int., 198, 812–827, doi:10.1093/gji/ggu145, 2014.

Chave, A. D. and Jones, A. G.: Electric and Magnetic Field Galvanic Distortion Decomposition of BC87 Data, J. Geomagn. Geoelectr., 49, 767–789, doi:10.5636/jgg.49.767, 1997.

Chave, A. D. and Jones, A. G.: The Magnetotelluric Method: Theory and Practice, Cambridge University Press, Cambridge, UK, doi:10.1017/cbo9781139020138, 2012.

Chave, A. D. and Smith, J. T.: On electric and magnetic galvanic distortion tensor decompositions, J. Geophys. Res., 99, 4669–4682, doi:10.1029/93jb03368, 1994.

Colombo, D., Keho, T., Janoubi, E., and Soyer, W.: Sub-basalt imaging with broadband magnetotellurics in NW Saudi Arabia, in: Expanded Abstracts of the SEG San Antonio 2011 Annual Meeting, 18–23 September 2011, in San Antonio, Texas, vol. 619–623, 2011.

Crowe, M., Heinson, G., and Dhu, T.: Magnetotellurics and airborne electromagnetics – a combined method for assessing basin structure and exploring for unconformity-related uranium, ASEG Extended Abstracts 2013: 23rd Geophysical Conference, 11–14 August 2013, in Melbourne, Australia, 1–5, doi:10.1071/ASEG2013ab225, 2013.

DeGroot-Hedlin, C.: Inversion for regional 2-D resistivity structure in the presence of galvanic scatterers, Geophys. J. Int., 122, 877–888, doi:10.1111/j.1365-246X.1995.tb06843.x, 1995.

De Groot-Hedlin, C. D.: Removal of static shift in two dimensions by regularized inversion, Geophysics, 56, 2102–2106, doi:10.1190/1.1443022, 1991.

Egbert, G. D. and Kelbert, A.: Computational recipes for electromagnetic inverse problems, Geophys. J. Int., 189, 251–267, doi:10.1111/j.1365-246x.2011.05347.x, 2012.

Gallardo, L. A. and Meju, M. A.: Characterization of heterogeneous near-surface materials by joint 2-D inversion of dc resistivity and seismic data, Geophys. Res. Lett., 30, 1658, doi:10.1029/2003GL017370, 2003.

Gamble, T. D.: Magnetotellurics with a remote magnetic reference, Geophysics, 44, 53–68, doi:10.1190/1.1440923, 1979.

Gamble, T. D., Goubau, W. M., and Clarke, J.: Error analysis for remote reference magnetotellurics, Geophysics, 44, 959–968, doi:10.1190/1.1440988, 1979.

Gibson, P. J.: Geophysical Characteristics of the Tow Valley Fault Zone in North-East Ireland, Irish J. Earth Sci., 22, 1–13, 2004.

Glover, P. W. J.: A generalized Archie's law for n phases, Geophysics, 75, E247–E265, doi:10.1190/1.3509781, 2010.

Goodman, R., Jones, G. L. I., Kelly, J., Slower, E., and O'Neill, N.: Geothermal Energy Resource Map of Ireland, Final Report, Tech. rep., Sustainable Energy Authority of Ireland, Dublin, Ireland, 2004.

Guéguen, Y. and Palciauskas, V.: Introduction to the physics of rocks, Princeton University Press, New Jersey, USA, 1994.

Hautot, S., Single, R. T., Watson, J., Harrop, N., Jerram, D. A., Tarits, P., Whaler, K., and Dawes, G.: 3-D magnetotelluric inversion and model validation with gravity data for the investigation of flood basalts and associated volcanic rifted margins, Geophys. J. Int., 170, 1418–1430, doi:10.1111/j.1365-246X.2007.03453.x, 2007.

Heincke, B., Jegen, M., Moorkamp, M., and Hobbs, R. W.: Joint-inversion of magnetotelluric, gravity and seismic data to image sub-basalt sediments offshore the Faroe-Islands, SEG Technical Program Expanded Abstracts, 2014, 770–775, doi:10.1190/segam2014-1401.1, 2014.

Hepworth, C. and Sanders, I. S. (Eds.): The Geology of Ireland: Second Edition, Dunedin Academic Press Ltd., 2 Edn., 2009.

Jegen, M. D., Hobbs, R. W., Tarits, P., and Chave, A.: Joint inversion of marine magnetotelluric and gravity data incorporating seismic constraints. Preliminary results of sub-basalt imaging off the Faroe Shelf, Earth Planet. Sci. Lett., 282, 47–55, 2009.

Jones, A. and Jödicke, H.: Magnetotelluric transfer function estimation improvement by a coherence-based rejection technique, in: SEG Technical Program Expanded Abstracts 1984, SEG Technical Program Expanded Abstracts, Society of Exploration Geophysicists, 51–55, doi:10.1190/1.1894081, 1984.

Jones, A. G.: Static shift of magnetotelluric data and its removal in a sedimentary basin environment, Geophysics, 53, 967–978, doi:10.1190/1.1442533, 1988.

Jones, A. G.: Electrical properties of the lower continental crust, in: Continental Lower Crust, vol. 23, 81–143, Elsevier Amsterdam, the Netherlands, 1992.

Jones, A. G.: Electromagnetic interrogation of the anisotropic Earth: Looking into the Earth with polarized spectacles, Phys. Earth Planet. In., 158, 281–291, doi:10.1016/j.pepi.2006.03.026, 2006.

Jones, A. G.: Three-dimensional galvanic distortion of three-dimensional regional conductivity structures: Comment on "Three-dimensional joint inversion for magnetotelluric resistivity and static shift distributions in complex media" by Yutaka Sasaki and Max A. Meju, J. Geophys. Res., 116, B12104+, doi:10.1029/2011jb008665, 2011.

Jones, A. G.: Distortion of magnetotelluric data: its identification and removal, Cambridge University Press, doi:10.1017/cbo9781139020138, 2012.

Jones, A. G., Chave, A. D., Egbert, G., Auld, D., and Bahr, K.: A Comparison of Techniques for Magnetotelluric Response Function Estimation, J. Geophys. Res., 94, 14201–14213, doi:10.1029/jb094ib10p14201, 1989.

Jupp, D. L. B. and Vozoff, K.: Stable Iterative Methods for the Inversion of Geophysical Data, Geophys. J. Roy. Astr. S., 42, 957–976, 1975.

Kaufman, Alekseev, D., and Oristaglio, M.: Principles of Electromagnetic Methods in Surface Geophysics, Elsevier, Amsterdam, the Netherlands, 2014.

Kelbert, A., Meqbel, N., Egbert, G. D., and Tandon, K.: ModEM: A modular system for inversion of electromagnetic geophysical data, Comput. Geosci., 66, 40–53, doi:10.1016/j.cageo.2014.01.010, 2014.

Keller, G. V. and Frischknecht, F. C.: Electrical methods in geophysical prospecting, Pergamon Press, Oxford, UK, 1966.

Leväniemi, H., Beamish, D., Hautaniemi, H., Kurimo, M., Suppala, I., Vironmäki, J., Cuss, R. J., Lahti, M., and Tartaras, E.: The JAC airborne EM system: AEM-05, J. Appl. Geophys., 67, 219–233, doi:10.1016/j.jappgeo.2007.10.001, 2009.

Luijendijk, E. and Gleeson, T.: How well can we predict permeability in sedimentary basins? Deriving and evaluating porosity-permeability equations for noncemented sand and clay mixtures, Geofluids, 15, 67–83, doi:10.1111/gfl.12115, 2015.

Martini, F., Hobbs, W. W., Bean, C. J., and Single, R.: A complex 3-D volume for sub-basalt imaging, First Break, 23, 41–51, 2005.

Mavko, G., Mukerji, T., and Dvorkin, J.: The Rock Physics Handbook. Tools for Seismic Modelling of Porous Media, Cambridge University Press, Cambridge, UK, 2009.

McNeice, G. W. and Jones, A. G.: Multisite, multifrequency tensor decomposition of magnetotelluric data, Geophysics, 66, 158–173, doi:10.1190/1.1444891, 2001.

Miensopust, M. P.: Multidimensional Magnetotellurics. A 2-D case study and a 3-D approach to simultaneously invert for resistivity structure and distortion parameters, PhD thesis, Faculty of Science, National University of Ireland, Galway, Ireland, 2010.

Miensopust, M. P., Jones, A. G., Hersir, G. P., and Vilhjálmsson, A. M.: The Eyjafjallajökull volcanic system, Iceland: insights from electromagnetic measurements, Geophys. J. Int., 199, 1187–1204, doi:10.1093/gji/ggu322, 2014.

Minsley, B. J.: A trans-dimensional Bayesian Markov chain Monte Carlo algorithm for model assessment using frequency-domain electromagnetic data, Geophys. J. Int., 187, 252–272, doi:10.1111/j.1365-246x.2011.05165.x, 2011.

Mitchell, W. I.: The Geology of Northern Ireland: Our Natural Foundation, Geological Survey of Northern Ireland, Belfast, UK, 2004.

Muñoz, G.: Exploring for Geothermal Resources with Electromagnetic Methods, Surv. Geophys., 35, 101–122, doi:10.1007/s10712-013-9236-0, 2014.

Naylor, D. and Shannon, P.: Petroleum geology of Ireland, Dunedin Academic Press Ltd, Edinburgh, UK, 2011.

Nover, G.: Electrical Properties of Crustal and Mantle Rocks – A Review of Laboratory Measurements and their Explanation, Surv. Geophys., 26, 593–651, doi:10.1007/s10712-005-1759-6, 2005.

Pape, H., Clauser, C., and Iffland, J.: Permeability prediction based on fractal pore-space geometry, Geophysics, 64, 1447–1460, 1999.

Pape, H., Clauser, C., and Iffland, J.: Variation of Permeability with Porosity in Sandstone Diagenesis Interpreted with a Fractal Pore space Model, Pure Appl. Geophys., 157, 603–619, 2000.

Pasquali, R., O'Neill, N., Reay, D., and Waugh, T.: The geothermal potential of Northern Ireland, in: Proceedings World Geothermal Congress, 25–30 April 2010 in Bali, Indonesia, 2010.

Pellerin, L. D. and Hohmann, G. W.: Transient electromagnetic inversion; a remedy for magnetotelluric static shifts, Geophysics, 55, 1242–1250, doi:10.1190/1.1442940, 1990.

Poll, H. E., Weaver, J. T., and Jones, A. G.: Calculations of voltages for magnetotelluric modeling of a region with near-surface inhomogeneities, Phys. Earth Planet. Inter., 53, 287–297, 1989.

Raffensperger, J. P.: Numerical simulation of basin-scale hydrochemical processes, in: Advances in Porous Media, edited by: Corapcioglu, M. Y., vol. 3, 185–305, Elsevier, New York, USA, 1996.

Raiche, A.: A flow-through Hankel transform technique for rapid, accurate Green's function computation, Radio Sci., 34, 549–555, doi:10.1029/1998rs900037, 1999.

Raiche, A. P., Jupp, D. L. B., Rutter, H., and Vozoff, K.: The joint use of coincident loop transient electromagnetic and Schlumberger sounding to resolve layered structures, Geophysics, 50, 1618–1627, doi:10.1190/1.1441851, 1985.

Reay, D. and Kelly, J.: Deep Geothermal Energy Resource Potential in Northern Ireland, European Geologist, 29, 14–18, 2010.

Rosenkjaer, G. K., Gasperikova, E., Newman, G. A., Arnason, K., and Lindsey, N. J.: Comparison of 3-D MT inversions for geothermal exploration: Case studies for Krafla and Hengill geothermal systems in Iceland, Geothermics, 57, 258–274, 2015.

Rousseeuw, P. J.: Least Median of Squares Regression, J. Am. Stat. Assoc., 79, 871–880, doi:10.1080/01621459.1984.10477105, 1984.

Rousseeuw, P. J. and Leroy, A. M.: Robust Regression and Outlier Detection, Wiley-Interscience, 1 Edn., 2003.

Sasaki, Y. and Meju, M. A.: Three-dimensional joint inversion for magnetotelluric resistivity and static shift distributions in complex media, J. Geophys. Res., 111, B05101, doi:10.1029/2005JB004009, 2006.

Schnaidt, S. and Heinson, G.: Bootstrap resampling as a tool for uncertainty analysis in 2-D magnetotelluric inversion modelling, Geophys. J. Int., 203, 92–106, 2015.

Simpson, F. and Bahr, K.: Practical Magnetotellurics, Cambridge University Press, Cambridge, UK, 2005.

Sternberg, B. K., Washburne, J. C., and Pellerin, L.: Correction for the static shift in magnetotellurics using transient electromagnetic soundings, Geophysics, 53, 1459–1468, doi:10.1190/1.1442426, 1988.

Torres-Verdin, C. and Bostick, F. X.: Principles of spatial surface electric field filtering in magnetotellurics; electromagnetic array profiling (EMAP), Geophysics, 57, 603–622, doi:10.1190/1.1443273, 1992.

Ward, S. H. and Hohmann, G. W.: Electromagnetic Methods in Applied Geophysics, Tulsa, Oklahoma, USA, vol. 1, Theory, 131–312, 1988.

Wight, D. and Bostick, F. X.: Cascade decimation – A technique for real time estimation of power spectra, in: Acoustics, Speech, and Signal Processing, Acoustics, Speech, and Signal Processing, IEEE International Conference on ICASSP '80, 9–11 April 1980, Denver, Colorado, USA, vol. 5, 626–629, IEEE, doi:10.1109/icassp.1980.1170868, 1980.

Wilson, H. E. and Manning, P. I.: Geology of the Causeway Coast, British Geological Survey, Nottingham, UK, 1978.

Zinszner, B. and Pellerin, F. M.: A Geoscientist's Guide to Petrophysics, Editions Technip, Paris, France, 2007.

6

The response of Opalinus Clay when exposed to cyclic relative humidity variations

Katrin M. Wild, Patric Walter, and Florian Amann

Department of Earth Sciences, Geological Institute, ETH Zurich, 8092 Zurich, Switzerland

Correspondence to: Florian Amann (florian.amann@erdw.ethz.ch)

Abstract. Clay shale specimens were exposed to cyclic relative humidity (RH) variations to investigate the response of the material to natural environmental changes. Opalinus Clay, a clay shale chosen as host rock for nuclear waste disposal in Switzerland, was utilized. The specimens were exposed to stepwise relative humidity cycles in which they were alternately allowed to equilibrate at 66 and 93 % relative humidity. Principal strains were monitored throughout the experiments using strain gauges. After each relative humidity cycle, Brazilian tensile strength tests were performed to identify possible changes in tensile strength due to environmental degradation.

Results showed that Opalinus Clay follows a cyclic swelling–shrinkage behaviour with irreversible expansion limited to the direction normal to bedding, suggesting that internal damage is restricted along the bedding planes. The Brazilian tensile strength in direction parallel and normal to bedding as well as the water retention characteristic remained unaffected by the RH variations.

1 Introduction

In Switzerland, Opalinus Clay, a Mesozoic shale formation of about 180 My in age, has been selected as the host rock for the disposal of high-level nuclear waste (BFE, 2011). Opalinus Clay features several beneficial properties such as its low permeability, the high radionuclide retention, and the potential for self-sealing. However, the favourable characteristics of the rock mass may change during tunnel excavation. Excavation is accompanied by stress redistribution and the development of an excavation damage zone (EDZ). The evolution of the EDZ is an important factor for the long-term safety of a nuclear repository as it may significantly influence the permeability of the confining host rock and offer pathways for radionuclide transport. Unloading and/or exposure to atmospheric conditions with a low relative humidity (RH) may lead to suction and, if the air-entry value is exceeded, to desaturation of the rock mass close to the tunnel. These processes can lead to shrinkage and the formation of desiccation cracks (Tsang et al., 2012). During the open-drift stage of a nuclear repository, seasonal atmospheric changes, especially RH variations, may alter the rock mass and influence the long-term crack evolution. Möri et al. (2010) measured crack apertures of Opalinus Clay in the framework of the cyclic deformation (CD) experiment at Mont Terri Underground Rock Laboratory (URL) located in the Jura Mountains, Switzerland. They found that the cracks close during summer (i.e. when the RH is high) and open during winter (i.e. when the RH is low). Crack closure and opening are associated with swelling and shrinkage of Opalinus Clay. Möri et al. (2010) also observed a net closure of the cracks over several seasonal cycles, which indicates an irreversible deformation component that is likely associated with time-dependent processes such as consolidation, creep, or slaking. These irreversible deformation components can contribute to both tunnel convergence and self-sealing of the EDZ. The self-sealing effect is the ability of clay shales to close previously developed cracks and therefore reduce their permeability through hydro-mechanical, hydro-chemical, and/or hydro-biochemical processes (Bernier et al., 2007). Among others, the adsorption of water on clay minerals and related volumetric expansion can be associated with this effect.

Numerous studies have been conducted to show the influence of drying–wetting cycles on clay or clay shale specimens (e.g. Chu and Mou, 1973; Popescu, 1980; Chen and

Ma, 1987; Osipov et al., 1987; Dif and Bluemel, 1991; Day, 1994; Al-Homoud et al., 1995; Basma et al., 1996; Pejon and Zuquette, 2002). In those studies, however, swelling was performed by allowing the specimens to fully soak in water. Few studies exist in which the influence of cycles in RH on the drying–swelling characteristics of clay shales has been investigated (e.g. Grice, 1968; Van Eeckhout, 1976; Olivier, 1979; Pham et al., 2007; Farulla et al., 2010; Cardoso et al., 2011; Yang et al., 2012; Pineda et al., 2014). Grice (1968) noted that specimens of Utica shale that were immersed in water disintegrated completely after oven drying. Specimens that were exposed to RH fluctuations between 60 and 90 % for a period of 9 months, however, showed only minor cracking. Van Eeckhout (1976) equilibrated specimens of Beatrice coalmine shale to various levels of RH to study the mechanisms of reduction in rock strength resulting from variations in RH. During moisture absorption he measured a volumetric expansion in the order of 0.2–1 %. The strains were larger in direction normal to bedding and occurred mostly between 48 and 100 % RH. Subsequent drying of the specimens to the initial level of RH showed that about 0.25 % of the strains were not recoverable. Van Eeckhout (1976) identified these expansion–contraction characteristics and the associated lengthening in internal cracking as a possible cause for the lowering in strength he observed due to humidity fluctuations. Similar observations have been made by Olivier (1979) for specimens of a Lower Triassic mudrock. With the help of water retention curves for several wetting–drying cycles between 10 and 99 % RH, Cardoso et al. (2011) showed that the air-entry value of an Upper Jurassic marl decreases with an increasing number of cycles. The decrease is accompanied by an increase in void ratio indicating a degradation of the material. Pham et al. (2007) subjected specimens of mudstones from Bure to one cycle of RH from 98 to 32 % and back to 98 %. The measurement of strains as well as the ultrasonic velocity showed a hysteresis between drying and wetting curves. Additionally, a non-linearity has been observed as more strain was induced by a change in RH at high levels of RH (i.e. between 76 and 98 %) than for the same change at lower RH. Yang et al. (2012) used digital image correlation techniques to study the deformation behaviour of Callovo–Oxfordian argillaceous rock specimens subjected to axial load (between 0.3 and 8.5 MPa) and RH cycles (between 39 and 85 %). A linear relationship between RH and strain has been observed for RH smaller than 75 %. Furthermore, the strains were reversible during cycles of hydration and dehydration at low axial stress (0.3 and 2 MPa), whereas irreversible strains (i.e. a net shrinkage) have been measured for RH cycles at 8.5 MPa axial load. Pineda et al. (2014) experimentally investigated the influence of RH cycles on the degradation of Lilla claystone in a long-term RH cycling experiment using ultrasonic wave velocity measurements and Brazilian tensile strength tests. The applied RH cycles caused an irreversible increase in the specimens' volumes as swelling always exceeded the amount of shrink-

age. Pineda et al. (2014) found that higher peak-to-peak amplitudes in RH cycles (cycles between 20 and 99 % were compared to cycles between 50 and 99 %) led to larger volumetric swelling. This effect was less pronounced for specimens that were tested under higher confinements. With the help of microstructural analyses, cracking has been identified as the main cause for irreversible swelling for Lilla claystone. Furthermore, the degradation of the material was manifested in a decrease in tensile strength from 2.9 to 0.2 MPa after four cycles and a decrease in dynamic Young's modulus by more than 50 %. For both quantities, the reduction was largest for the first cycle; afterwards a decreasing degradation rate was observed.

All studies mentioned above showed that cyclic variations in RH can have a significant influence on rock mechanical parameters such as tensile strength and can lead to irreversible volume changes which might contribute to the destabilization of underground excavations but also favour processes that are considered to control self-sealing.

This study aims at contributing to the understanding of the influence of RH variations on the mechanical and hydromechanical behaviour of Opalinus Clay. A series of specimens were exposed to RH variations under unstressed conditions and tested for their Brazilian tensile strength. The study focuses on answering the question of whether Opalinus Clay shows a damage evolution when exposed to RH cycles that affects the tensile strength and causes irreversible volumetric expansion that might be relevant for long-term deformations and/or self-sealing.

2   Tested material and experimental procedure

2.1   Material description

For this study, samples from the shaly facies of Opalinus Clay from the Mont Terri URL, Switzerland, were used. The formation is part of the Mont Terri anticline that formed during the folding of the Jura Mountains. The present overburden at the URL lies between 230 and 330 m but is estimated to have reached up to 1350 m in the late Tertiary (Thury and Bossart, 1999; Mazurek et al., 2006). The shaly facies of Opalinus Clay contains a clay content of 50–80 % (Mazurek, 1998; Klinkenberg et al., 2009; Nagra, 2002; Bossart, 2005). The clay minerals can be subdivided into illite (15–25 %), illite–smectite mixed-layer phases (10–15 %), kaolinite (20–30 %), and chlorite (5–15 %). Beside the clay minerals, Opalinus Clay consists of quartz (10–20 %), feldspar (0–5 %), carbonates (5–25 %), pyrite (0–3 %), and organic material (0–1 %). The clay particles are aligned sub-parallel to each other leading to a distinct macroscopic bedding. As a result, the physical properties of Opalinus Clay show a strong transversely isotropic behaviour. The hydraulic conductivity, for example, varies between 10^{-12} m s^{-1} parallel to bedding and 10^{-14} m s^{-1} normal to bedding (Marschall et al., 2004). The

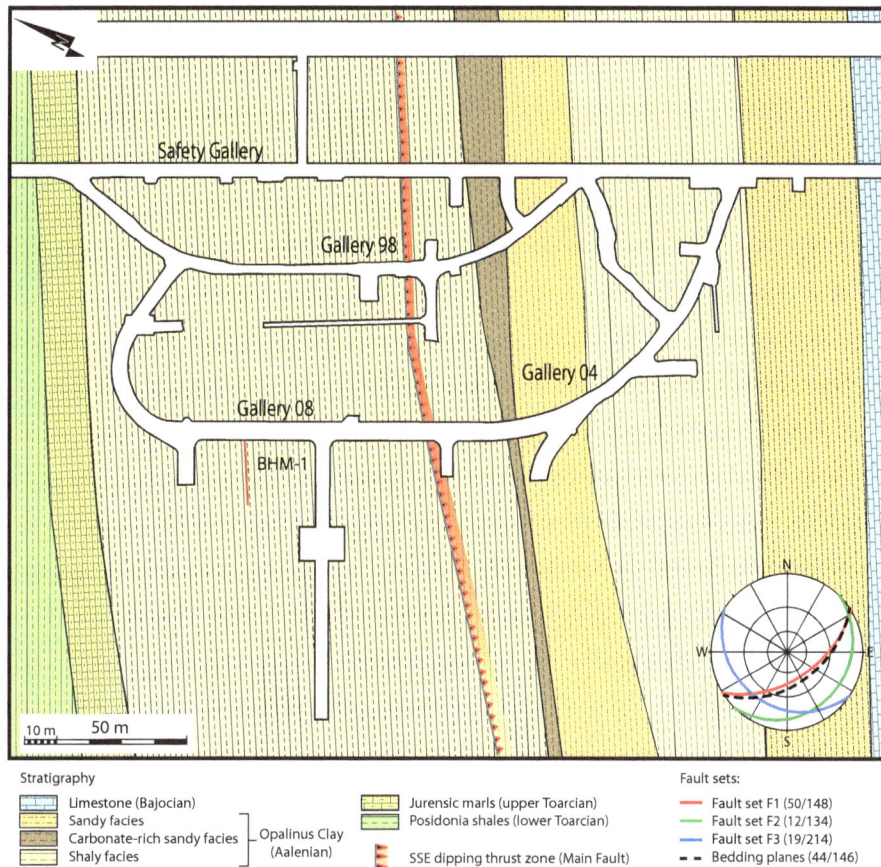

Figure 1. Geological map of the Mont Terri URL (modified from Nussbaum et al., 2011). The specimens used in this study were obtained from the borehole BHM-1 that was drilled parallel to bedding in Gallery 08. The location of the borehole is indicated approximately.

water loss porosity of the shaly facies is in the order of 15–19 % (Bossart, 2005; Amann et al., 2011, 2012; Wild et al., 2015; Wild, 2016). The pore water of the Opalinus Clay at the Mont Terri URL can be classified as Na-Cl water and, with regard to its composition, is still close to that of the seawater where the material was deposited (Thury and Bossart, 1999; Pearson et al., 2003).

2.2 Sampling and specimen preparation

In total, 31 specimens were taken from two 67.5 mm diameter bore cores obtained from a 25 m long borehole (BHM-1) that was drilled in Gallery 08 in the shaly facies of Opalinus Clay at the Mont Terri URL (Fig. 1). A triple-tube core barrel with compressed air cooling was used. The bore axis of BHM-1 was oriented parallel to the bedding. The samples were immediately sealed in vacuum-evacuated aluminium foil after core extraction. Core samples were cut under dry conditions to a diameter-to-length ratio of approximately 2 : 1. Sample sections from 2.5–3.3 m depth (N specimens) and 8.5–9.4 m depth (P specimens) were selected. Both sections were located outside of the EDZ as no EDZ-related fractures were detected at these depths. Additionally,

a cubic specimen (S4) was cut in such a way as to allow for the measurement of the principal strains (i.e. the strains perpendicular, ε_1, and parallel, ε_2, to the bedding plane orientation and the strain in the plane of isotropy, ε_3). The layout of the electric resistive strain gauges (HBM type: K-LY46) that were directly glued onto the specimen's faces is shown in Fig. 2a. Furthermore, one cylindrical specimen (E19) was used to measure the strain perpendicular and parallel to bedding (Fig. 2b). The environmental exposure time of the specimens during the installation of the strain gauges was minimized to about 30 min.

2.3 Experimental layout

The specimens were exposed in desiccators to an alternating sequence of low and high RH levels under unstressed conditions. The RH was controlled by using supersaturated salt solutions. Sodium nitrite ($NaNO_2$, RH = 66 % at 20 °C) and ammonium di-hydrogen phosphate ($NH_4H_2PO_4$, RH = 93 % at 20 °C) were used based on the seasonal variations at the Mont Terri URL between 1997 and 2011 reported by Swisstopo (2014). It was found that the RH follows a cyclic annual variation following a sine curve with maximum humid-

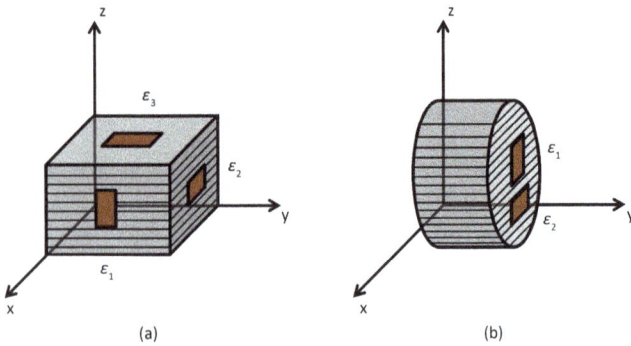

Figure 2. Illustration of strain gauge arrangement for measurements of principal strains: **(a)** on a cubic Opalinus Clay specimen (S4) measuring all principal strains individually; **(b)** on a cylindrical Brazilian test specimen (E19) measuring only two principal strains.

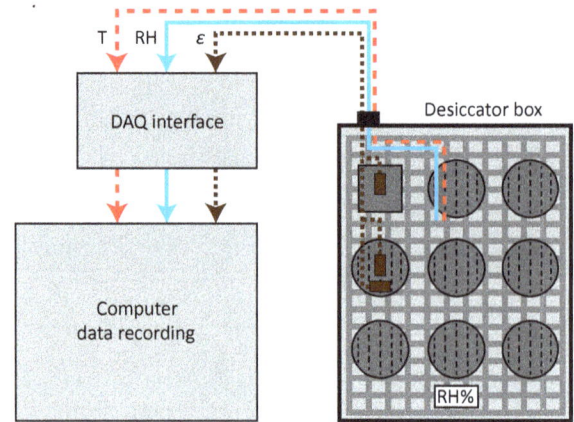

Figure 3. Schematic illustration of the experimental setup of the stepwise cyclic RH laboratory experiment.

ity values of 93.8 % during summer and minimum humidity values of 62.6 % during winter. At each level of RH, the specimens were allowed to equilibrate with the environment. Equilibration was achieved when the weight of the specimens (periodically measured every 1–7 days, accuracy 0.01 g) remained constant. Homogeneity of the ambient conditions within the desiccator boxes was ensured by installing computer fans at the back of the boxes. Brazilian tensile strength tests were conducted after each cycle when the specimens were equilibrated at 66 % RH. In total 4.5 cycles were applied, starting with an equilibration phase at 93 % RH to establish the same initial conditions for all specimens. The experimental setup is schematically shown in Fig. 3. For the monitoring of the RH, a Honeywell HIH-4000-001 sensor (sampling rate of about 12 h, accuracy ±3.5 %) was used. Temperature was monitored by a resistance thermometer (Pt100) (sampling rate of about 12 h, accuracy 0.015 °C). The temperature in the laboratory was kept between 19 and 23 °C throughout the experiment.

2.4 Suction and strain calculations

The water content was determined according to the International Society for Rock Mechanics (ISRM)-suggested methods (ISRM, 1979). From the RH and the temperature that were monitored during the experiment, the total suction can be calculated according to Kelvin's relationship:

$$\Psi = -\frac{RT}{V_{w0}\omega_w} \ln\left(\frac{p}{p_0}\right), \tag{1}$$

where Ψ is the suction in pascal, R is the ideal gas constant (i.e. $8.314\,\mathrm{J\,mol^{-1}\,K^{-1}}$), T the absolute temperature in kelvin, V_{w0} the specific volume of water (i.e. about $0.001\,\mathrm{m^3\,kg^{-1}}$), ω_w the molecular mass of water vapour (i.e. $0.018\,\mathrm{kg\,mol^{-1}}$), p the vapour pressure of water in the system in megapascal, and p_0 the vapour pressure of pure water in megapascal. The ratio p/p_0 equals the RH.

On the cubic specimen (S4), strains were measured in all three principal directions. Thus, the volumetric strain (ε_v) can be calculated by adding all three principal strains (i.e. $\varepsilon_v = \varepsilon_1 + \varepsilon_2 + \varepsilon_3$). On the cylindrical specimen (E19), the strain parallel (ε_2) and perpendicular (ε_1) to the bedding was recorded. Assuming a transversely isotropic material, the strain parallel to bedding (ε_2) equals the strain in the plane of isotropy (ε_3). Hence, the volumetric strain can be calculated from the sum of the strain perpendicular to bedding and twice the strain parallel to bedding (i.e. $\varepsilon_v = \varepsilon_1 + 2\varepsilon_2$). For all strains, expansion is taken as positive.

2.5 Mechanical testing procedure

Brazilian tensile strength tests were conducted at ETH Zurich utilizing a modified 2000 kN servo-hydraulic rock-testing machine (Walter and Bai, Switzerland). The tests were conducted according to the ISRM-suggested methods (ISRM, 1978) immediately after removal from the desiccator. Load was applied parallel (P specimens) or normal to bedding (N specimens) (Fig. 4) using a constant loading rate of $0.08\,\mathrm{kN\,s^{-1}}$ (except for specimens N1–N6, for which a loading rate of $0.05\,\mathrm{kN\,s^{-1}}$ was used). The accuracy is < 1 % of the actual reading.

3 Results

The specimens' dimensions and initial properties are given in Table 1. The initial water contents of the specimens range between 6.95 and 7.34 %, which is comparable with the water content measured on cores right after core extraction (Pearson et al., 2003; Amann et al., 2011; Wild, 2016). The initial saturation was estimated from the initial water content, the bulk dry density, and the porosity of the specimens according to the ISRM-suggested methods (ISRM, 1979). A grain density of $2.73\,\mathrm{g\,cm^{-3}}$ was used to calculate the porosity (Pearson et al., 2003; Bossart, 2005; own data). Values of

Table 1. Properties and test configurations of specimens. Water content, dry density, porosity, and initial saturation were determined according to ISRM (1979). Furthermore, the direction of the applied load with respect to bedding during the Brazilian tensile strength tests is indicated.

Specimen no.	Diameter	Thickness	Initial water content	Bulk dry density	Porosity	Initial saturation	Direction of applied stress
(–)	(mm)	(mm)	(%)	(g cm^{-3})	(%)	(%)	(–)
A1	67.32	35.35	6.95	2.30	15.85	100.8	parallel
A2	67.20	34.78	7.13	2.28	16.47	98.7	parallel
A3	67.66	34.70	7.33	2.26	17.10	97.0	parallel
A4	67.61	32.33	7.23	2.24	17.96	90.1	parallel
A5	67.64	35.76	7.02	2.27	16.69	95.7	parallel
A6	67.37	35.94	6.97	2.28	16.47	96.5	parallel
A7	67.59	34.49	7.04	2.28	16.49	97.3	parallel
A8	67.44	35.01	7.01	2.30	15.74	102.4	parallel
A9	67.39	35.27	7.34	2.26	17.09	97.2	parallel
A10	67.58	35.50	7.21	2.27	16.80	97.4	parallel
A11	67.52	34.77	7.20	2.28	16.33	100.7	parallel
A12	67.63	35.06	7.16	2.27	16.96	95.7	parallel
A14	67.54	35.16	7.26	2.26	17.29	94.8	parallel
A15	67.51	34.89	7.27	2.27	17.02	96.8	parallel
E1	67.06	36.47	7.23	2.30	15.59	106.9	normal
E2	67.12	34.51	7.24	2.28	16.35	101.1	normal
E3	67.08	35.17	7.25	2.30	15.74	105.9	normal
E4	67.1	35.18	7.14	2.30	15.82	103.7	normal
E6	67.14	34.79	7.21	2.30	15.85	104.5	normal
E7	67.02	35.74	7.13	2.29	15.97	102.4	normal
E8	67.15	35.06	7.07	2.29	16.15	100.2	normal
E9	67.21	35.64	7.12	2.28	16.42	98.9	normal
E10	67.24	35.17	7.10	2.28	16.62	97.3	normal
E11	67.47	35.07	7.23	2.26	17.29	94.4	normal
E12	67.43	34.98	7.15	2.27	16.79	96.8	normal
E13	67.56	36.08	7.19	2.24	18.08	88.9	normal
E14	67.50	36.06	7.16	2.27	16.73	97.2	normal
E15	67.20	35.78	7.15	2.28	16.65	97.7	normal
E16	67.08	34.23	7.06	2.28	16.41	98.1	normal
E17	67.28	35.40	7.04	2.28	16.57	96.7	normal
E18	67.25	35.26	7.05	2.27	16.73	95.8	normal

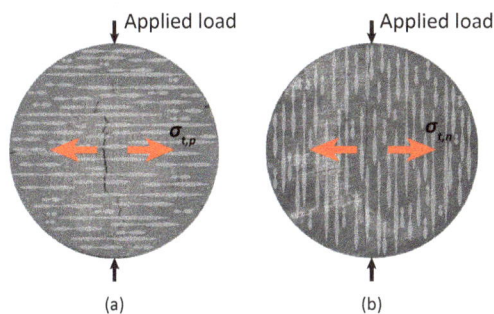

Figure 4. Loading configuration for the Brazilian tests with respect to bedding (bedding orientation is indicated by the light grey pattern). In panel (**a**) the load is applied normal to bedding, allowing the measurement of the Brazilian tensile strength parallel to bedding ($\sigma_{t,p}$). In panel (**b**) the load is applied parallel to bedding, allowing the measurement of the Brazilian tensile strength normal to bedding ($\sigma_{t,n}$).

saturation that exceed 100 % can be related to the uncertainty in the grain density (± 0.03 g cm^{-3}) and the specimen's volume.

Figure 5 shows the results of the RH, temperature, and strain measurements for the N specimens. The specimens were first equilibrated to a RH of 93 % and then subjected to 4.5 cycles with peak–peak amplitudes of between 30 and 36 % (i.e. RH variation between 63 and 94 %; Fig. 5a). The resulting suction applied to the specimens was calculated according to Eq. (1) and is plotted together with the corresponding response of the water content in Fig. 5b. Similar trends with respect to the water content changes were observed for all specimens. A constant water content and a small change in volumetric strain were observed during the first equilibration phase, indicating a high initial saturation degree of the specimens. During cycling, the water content changed by ± 2.2–2.4 %. Except for the first drying period, in which 0.6–0.8 % water content was lost, the water con-

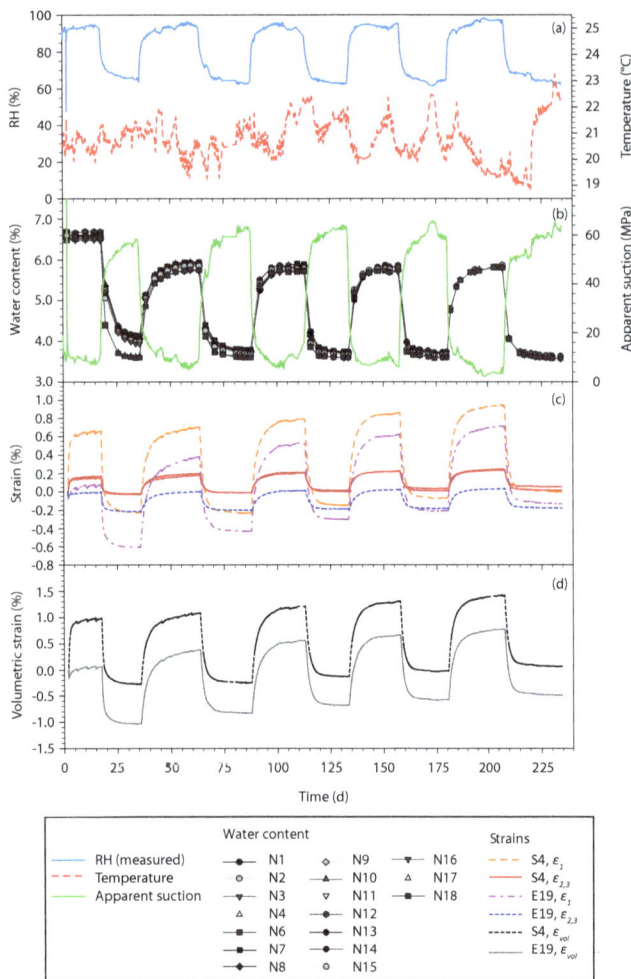

Figure 5. Results of the stepwise cyclic RH experiment for the N specimens, including **(a)** the changes in relative humidity (RH) and temperature, **(b)** the water content and apparent suction calculated from the measured RH, **(c)** the principal strains normal to bedding (ε_1) and parallel to bedding (ε_2 and ε_3) of specimens S4 and E19, and **(d)** the volumetric strains of specimens S4 and E19.

tent was reversible. A comparable response was observed for the series of P specimens (Fig. 6) which were subjected to the same testing procedure, although RH or strains were not measured explicitly. Strain measurements on the specimens E19 and S4 showed an immediate response to changes in RH (Fig. 5c, d). Swelling occurred during wetting, shrinkage during drying phases. The magnitude of swelling exceeded the shrinkage, which accumulated to an irreversible volumetric strain of 0.55–0.75 % at the end of the experiment (Fig. 5d). The individual strain measurements showed that mainly deformations normal to bedding contributed to the overall expansion of the rock specimen. The strain measured parallel to bedding was significantly smaller and approximately reversible.

Although irreversible strain was measured, no significant change in Brazilian tensile strength was observed (Fig. 7).

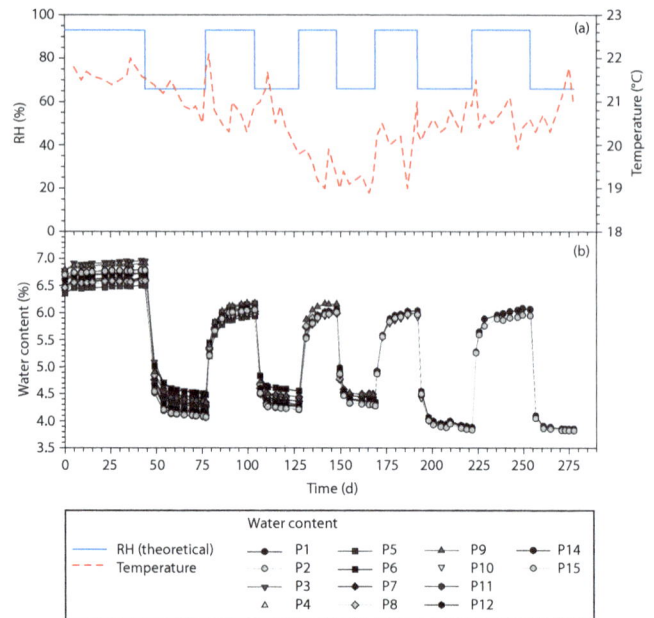

Figure 6. Results of the stepwise cyclic RH experiment for the P specimens, including **(a)** the theoretically applied levels of relative humidity (RH) and the measured temperature (in the laboratory) and **(b)** the water content of the specimens.

The Brazilian tensile strength parallel to bedding remained constant over three to five cycles, while corresponding values for the direction normal to bedding only indicate insignificant decreasing trends that lie within the strength values reported by Wild et al. (2015) for specimens that were equilibrated to 56–87 MPa suction. Thus, a change in Brazilian tensile strength as a response to the RH variations was not measurable or insignificant.

4   Discussion

4.1   Water retention

Figure 8 shows the relationship between suction and water content for specimen E7 during the stepwise cyclic RH experiment. The system is in equilibrium at the highest and lowest suction values (turning points between wetting and drying paths) but not in between. Also shown are the main drying and wetting paths reported by Wild et al. (2015).

The first drying path for the specimen follows the main drying path as it represents the drying of the intact rock starting from initial conditions which were comparable to the study of Wild et al. (2015). Since the specimens were not dried to their residual water content, the following scanning curves lie between the main drying and main wetting paths. Hysteresis can be observed between drying and wetting path caused by non-homogeneous pore size distribution, different contact angles between wetting and drying, or entrapped

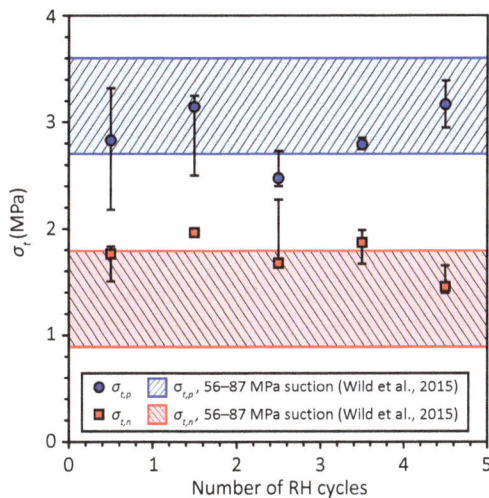

Figure 7. The median value and the associated ranges of the Brazilian tensile strength of specimens from the stepwise cyclic RH experiment with tension parallel ($\sigma_{t,p}$) and normal ($\sigma_{t,n}$) to bedding. Furthermore, the range of Brazilian tensile strength values reported by Wild et al. (2015) for specimens that were equilibrated to 56–87 MPa suction are indicated.

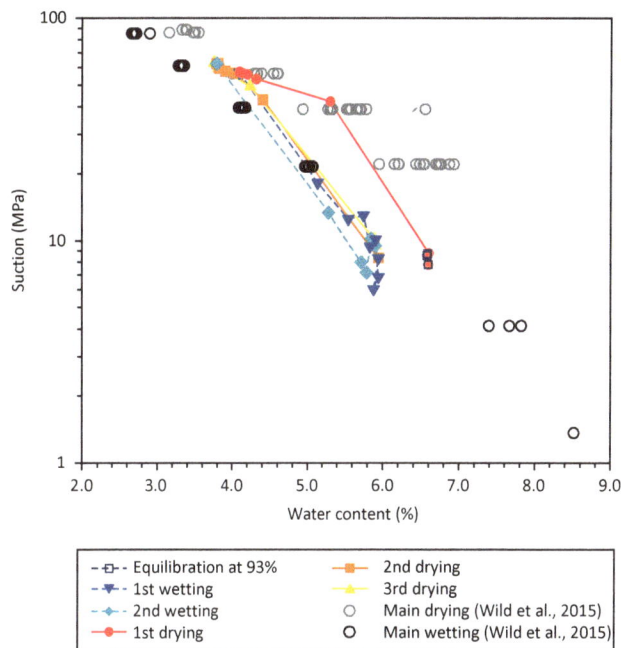

Figure 8. Water retention curve for specimen N7 displaying suction as a function of the water content. Also shown are the water retention curves (main drying and main wetting) reported by Wild et al. (2015) for specimens which were equilibrated to the applied level of RH at any point.

air bubbles during wetting (Birle et al., 2008). Therefore, the initial water content cannot be re-established anymore and a water loss of 0.6–0.8 % occurred. However, the scanning curves of the specimen subjected to the stepwise RH cycles approximately lie within the main drying and wetting paths, indicating that the water retention characteristics are not significantly affected by the variations in RH. This is consistent with findings by Pineda et al. (2014).

4.2 Strain and damage

Strain results of the dynamic and stepwise cyclic RH experiments (Fig. 5c) indicate that the Opalinus Clay follows a cyclic expansion and contraction associated with water absorption and desorption processes. Thereby, the Opalinus Clay shows a strongly transversely isotropic deformation behaviour where the strain in the direction normal to bedding (ε_1) dominates the bulk deformation. These results are consistent with findings by Minardi et al. (2016), who also found an anisotropic response of strain for an Opalinus Clay specimen subjected to one cycle of wetting and drying. This observation can be related to the absorption of water into parallel-orientated clay interlayers (i.e. parallel to the bedding planes) leading to swelling in normal direction. Moreover, according to Houben et al. (2013), the pores of Opalinus Clay are elongated along the bedding.

Irreversible volumetric expansion took place during the stepwise cyclic RH exposure. Many studies on single mineral types (e.g. Na-montmorillonite) have demonstrated that clay minerals show distinct hydration states when exposed to different levels of RH (e.g. Mering, 1946; Mooney et

al., 1952; Gillery, 1959; Emerson, 1962; Van Olphen, 1965; Glaeser and Mering, 1968; Chipera et al., 1997; Ferrage et al., 2005; Likos and Lu, 2006). These hydration stages reflect the intercalation of one to four discrete layers of water molecules between the mineral interfaces and are driven by the hydration of the cations of the clay minerals (Norrish, 1972). During transition between these stages, the interlayer spacing can increase by up to a factor of 2 (Norrish, 1954). For the hydration of Na-montmorillonite, for example, the interlayer spacing increases from 10 to 12.5 Å between 0 and 20 % RH, from 12.5 to 15.5 Å between 50 and 70 % RH, and further to about 19 Å for RH > 98.5 % (Mooney et al., 1952; Gillery, 1959; Emerson, 1962; Glaeser and Mering, 1968). Between 70 and 95 % a two-layer hydration state is present for both Ca- and Na-montmorillonite (Seedsman, 1985). Similar stages for other clay minerals are given by Gillery (1959). They all indicate a relatively stable state between 80 and 90 % RH. Furthermore, sorption and adsorption paths for clay minerals show hysteresis, indicating that crystalline swelling is an irreversible thermodynamic process (Laird et al., 1995).

This might explain the accumulated irreversible volumetric expansion as most of the clay minerals transition between the one- and two-layer hydration state in the RH range covered within the experiments in this study. This is also supported by macroscopically detectable cracking that was observed during the experiment.

Although slight cracking of the specimens was detected and irreversible volumetric strain was observed, no significant influence on the Brazilian tensile strength was observed after three to five cycles for both experiments. It is therefore concluded that the observed degradation caused by the cyclic variations of RH in this study is not sufficient to cause severe damage that influences the strength of the material. The lower degradation potential for Opalinus Clay compared to other clay shales when subjected to RH cycling is in agreement with findings reported by Pineda et al. (2011). Compared to Lilla Claystone (Pineda et al., 2008, 2014), the (tensile) strength and (dynamic) stiffness of Opalinus Clay is significantly less affected by cyclic RH variations.

5   Conclusions

This study demonstrates that cyclic RH variations have the potential to internally damage the Opalinus Clay leading to irreversible volumetric expansion. Internal damage mainly takes place along the bedding, supported by the fact that irreversible strain was almost exclusively observed in the direction normal to the bedding.

The Brazilian tensile strength of Opalinus Clay seems to be unaffected by cyclic RH variations (i.e. a change was not measurable or insignificant). The Brazilian tensile strength parallel to bedding remained constant over three to five cycles while corresponding values for the direction normal to bedding only indicate insignificant decreasing trends. Water retention characteristics of Opalinus Clay were not significantly altered by the observed environmental degradation.

The experimental study demonstrates that RH variations can lead to irreversible volumetric strains and therefore supports the hypothesis that long-term environmental variations might contribute to long-term deformations of underground excavations and favour processes that are considered to control self-sealing in Opalinus Clay.

Author contributions. The experiment was carried out by Patric Walter under the supervision of Florian Amann and Katrin M. Wild. Katrin M. Wild prepared the paper with contributions from all co-authors.

Competing interests. The authors declare that they have no conflict of interest.

Acknowledgements. This study was funded by the Swiss Federal Nuclear Safety Inspectorate ENSI. The authors would like to thank Reto Seifert and Stewart Bishop (ETH Zurich) for their support with the mechanical and electrical challenges during the setup of the experiment. We are also grateful to Matthew Perras and Wilfried Winkler (ETH Zurich) for fruitful discussions. Furthermore, we would like to thank Claudio Madonna (ETH Zurich) for help provided during the laboratory work and for feedback on this paper.

Edited by: M. Oliva

References

Al-Homoud, A. S., Basma, A. A., Husein Malkawi, A. I., and Al Bashabsheh, M. A.: Cyclic Swelling Behavior of Clays, J. Geotech. Eng., 121, 562–565, 1995.

Amann, F., Button, E. A., Evans, K. F., Gischig, V. S., and Blümel, M.: Experimental study of the brittle behavior of clay shale in rapid unconfined compression, Rock Mech. Rock Eng., 44, 415–430, 2011.

Amann, F., Kaiser, P. K., and Button, E. A.: Experimental study of the brittle behavior of clay shale in rapid triaxial compression, Rock Mech. Rock Eng., 45, 21–33, 2012.

Basma, A. A., Al-Homoud, A. S., Husein Malkawi, A. I., and Al-Bashabsheh, M. A.: Swelling-shrinkage behavior of natural expansive clays, Appl. Clay Sci., 11, 211–227, 1996.

Bernier, F., Li, X. L., Bastiaens, W., Ortiz, L., Van Geet, M., Wouters, L., Fireg, B., Blümling, P., Desrues, J., Viaggiani, G., Coll, C., Chanchole, S., De Greef, V., Hamza, R., Malinsky, L., Vervoort, A., Vanbrabant, Y., Debecker, B., Verstraelen, J., Govaerts, A., Wevers, M., Labiouse, V., Escoffier, S., Mathier, J.-F. Gastaldo, L., and Bühler, Ch.: Fracture and Self-healing within the excavation Disturbed Zone in Clays (SELFRAC), Final report, European Commission Report, Luxembourg, EUR 22585, 62 pp., 2007.

BFE (Bundesamt für Energie): Press Release 01.12.2011, Bern, Switzerland, 2011.

Birle, E., Heyer, D., and Vogt, N.: Influence of the initial water content and dry density on the soil-water retention curve and the shrinkage behavior of a compacted clay, Acta Geotech., 3, 191–200, 2008.

Bossart, P.: Characteristics of Opalinus Clay at Mont Terri, available at: http://www.mont-terri.ch/internet/mont-terri/de/home/geology/key_characteristics.html (last access: 11 July 2016), 2005.

Cardoso, R., Della Vecchia, G., Jommi, C., and Romero, E.: Water retention curve for evolving marl under suction cycles, in: Unsaturated Soils, edited by: Alonso, E. and Gens, A., Proceedings of the 5th International Conference on Unsaturated Soils, 6–8 September 2010, Barcelona, Spain, 2, 1451–1457, 2011.

Chen, F. H. and Ma, G. S.: Swelling and Shrinkage Behavior of Expansive Clays, in: Proceedings of the 6th International Conference on Expansive Soils, 1–4 December 1987, New Delhi, India, 127–129, 1987.

Chipera, S. J., Carey, J. W., and Bish, D. L.: Controlled-humidity XRD analyses: Application to the study of smectite expansion/contraction, Adv. X Ray Anal., 39, 713–722, 1997.

Chu, T. Y. and Mou, C. H.: Volume change characteristics of expansive soils determined by con-trolled suction tests, in: Proceedings of the 3rd International Conference on Expansive Soils, 30 July–1 August 1973, Haifa, Israel, 2, 177–185, 1973.

Day, R. W.: Swell-shrink behavior of compacted clay, J. Geotech. Eng., 120, 618–623, 1994.

Dif, A. E. and Bluemel, W. F.: Expansive Soils under Cyclic Drying and Wetting, Geotech. Test. J., 14, 96–102, 1991.

Emerson, W. W.: The swelling of Ca-montmorillonite due to water absorption, 1. Water uptake in the vapour phase, J. Soil Sci., 13, 31–39, 1962.

Farulla, C. A., Ferrari, A., and Romero, E.: Volume change behaviour of a compacted scaly clay during cyclic suction changes, Can. Geotech. J., 47, 688–703, 2010.

Ferrage, E., Lanson, B., Sakharov, B. A., and Drits, V. A.: Investigation of smectite hydration properties by modeling experimental X-ray diffraction patterns: Part I. Montmorillonite hydration properties, Am. Mineral., 90, 1358–1374, 2005.

Gillery, F. H.: Adsorption-desorption characteristics of synthetic montmorillonoids in humid atmospheres, Am. Mineral., 44, 806–818, 1959.

Glaeser, R. and Mering, J.: Homogeneous hydration domains of the smectites, Cr. Acad. Sci. D Nat., 267, 463–466, 1968.

Grice, R. H.: The Effect of Temperature-Humidity on the Disintegration of Nonexpandable Shales, Bull. Asso Eng. Geol., 5, 70–77, 1968.

Houben, M. E., Desbois, G., and Urai, J. L.: Pore morphology and distribution in the Shaly facies of Opalinus Clay (Mont Terri, Switzerland): Insights from representative 2D BIB-SEM investigations on mm to nm scale, Appl. Clay Sci., 71, 82–97, 2013.

ISRM: Suggested Methods for Determining Tensile Strength of Rock Materials, Int. J. Rock Mech. Min., 15, 99–103, 1978.

ISRM: Suggested Methods for Determining Water Content, Porosity, Density, Absorption and Related Properties and Swelling and Slake-Durability Index Properties, Int. J. Rock Mech. Min., 16, 141–156, 1979.

Klinkenberg, M., Kaufhold, S., Dohrmann, R., and Siegesmund, S.: Influence of carbonate micro-fabrics on the failure strength of claystones, Eng. Geol., 107, 42–54, 2009.

Laird, D.A., Shang, C., and Thompson, M.L.: Hysteresis in Crystalline Swelling of Smectites, J. Colloid Interf. Sci., 171, 240–245, 1995.

Likos, W. J. and Lu, N.: Pore-scale analysis of bulk volume change from crystalline interlayer swelling in Na^+- and Ca_2^+-smectite, Clay Clay Miner., 54, 515–528, 2006.

Marschall, P., Croisé, J., Schlickenrieder, L., Boisson, J.-Y., Vogel, P., and Yamamoto, S.: Synthesis of Hydrogeological Investigations at Mont Terri Site (Phase 1 to 5), in: Mont Terri Project – Hydrogeological Synthesis, Osmotic Flow, edited by: Heitzmann, P., Reports of the Federal Office for Water and Geology, Geology Series, 6, 7–92, 2004.

Mazurek, M.: Mineralogical composition of Opalinus Clay at Mont Terri – a laboratory inter-comparison, unpublished Mont Terri Technical Note TN 98-41, 1998.

Mazurek, M., Hurford, A. J., and Leu, W.: Unravelling the multistage burial history of the Swiss Molasse Basin: integration of apatite fission track, vitrinite reflectance and biomarker isomerisation analysis, Basin Res., 18, 27–50, 2006.

Mering, J.: On the hydration of montmorillonite, T. Faraday Soc., 42, 205–219, 1946.

Minardi, A., Crisci, E., Ferrari, A., and Laloui, L.: Anisotropic volumetric behaviour of Opalinus Clay shale upon suction variation, Géotechnique Letters, 6, 144–148, 2016.

Mooney, R. W., Keenan, A. G., and Wood, L. A.: Adsorption of Water Vapor by Montmorillonite. II. Effect of Exchangeable Ions and Lattice Swelling as Measured by X-Ray Diffraction, J. Am. Chem. Soc., 74, 1371–1374, 1952.

Möri, A., Bossart, P., Matray, J.-M., Müller, H., and Frank, E.: Mont Terri project, cyclic deformations in the Opalinus Clay, in: ANDRA, Clays in Natural & Engineered Barriers for Radioactive Waste Confinement, 4th International Meeting, 29 March–1 April 2010, Nantes, France, 81–82, 2010.

Nagra: Projekt Opalinuston. Synthese der geowissenschaftlichen Untersuchungsergebnisse, Nagra Technischer Bericht NTB 02-03, Wettingen, Switzerland, 2002.

Norrish, K.: The Swelling of Montmorillonite, T. Faraday Soc., 18, 120–134, 1954.

Norrish, K.: Forces between clay particles, in: Proceedings of the International Clay Conference, edited by: Serratosa, J. M. and Sanchez, A., 23–30 June 1972, Madrid, Spain, 1, 375–383, 1972.

Nussbaum, C., Bossart, P., Amann, F., and Aubourg, C.: Analysis of tectonic structures and excavation induced fractures in Opalinus Clay, Mont Terri underground rock laboratory (Switzerland), Swiss J. Geosci., 104, 187–210, 2011.

Olivier, H. J.: Some aspects of the influence of mineralogy and moisture redistribution on the weathering behavior of mudrocks, in: Proceedings of the 4th ISRM Congress, 2–8 September 1979, Montreux, Switzerland, 467–474, 1979.

Osipov, V. I., Bik, N. G., and Rumjantseva, N. A.: Cyclic Swelling of Clays, Appl. Clay Sci., 2, 363–374, 1987.

Pearson, F. J., Arcos, D., Bath, A., Boisson, J.-Y., Fernández, A. M., Gäbler, H.-E., Gautschi, A., Griffault, L., Hernán, P., and Waber, H. N.: Mont Terri Project – Geochemistry of Water in the Opalinus Clay Formation at the Mont Terri Rock Laboratory, Reports of the Federal Office for Water and Geology (FOWG), Bern, Switzerland, 5, 321 pp., 2003.

Pejon, O. J. and Zuquette, L. V.: Analysis of cyclic swelling of mudrocks, Eng. Geol., 67, 97–108, 2002.

Pham, Q. T., Vales, F., Malinsky, L., Nguyen Minh, D., and Gharbi, H.: Effects of desaturation-resaturation on mudstone, Phys. Chem. Earth, 32, 646–655, 2007.

Pineda, J. A., Arroyo, M., Romero, E., and Alonso, E. E.: Dynamic tracking of hydraulically induced claystone degradation, in: Deformational Characteristics of Geomaterials, edited by: Burns, S. E., Mayne, P. W., and Santamarina, C. J., Proceedings of the 4th International Symposium on Deformation Characteristics of Geomaterials, 22–24 September 2008, Atlanta, Georgia, USA, 2, 809–816, 2008.

Pineda, J. A., Mitaritonna, G., Romero, E., and Arroyo, M.: Effects of hydraulic cycling on the stiff-ness response of a rigid clay, in: Unsaturated Soils, edited by: Alonso, E. and Gens, A., Proceedings of the 5th International Conference on Unsaturated Soils, 6–8 September 2010, Barcelona, Spain, 2, 1465–1470, 2011.

Pineda, J. A., Alonso, E. E., and Romero, E.: Environmental degradation of claystones, Géotechnique, 64, 64–82, 2014.

Popescu, M.: Behaviour of expansive soils with a crumb structure, in: Proceedings of the 4th International Conference on Expansive Soils, edited by: Snethen, D., 16–18 June 1980, Denver, Colorado, USA, 1, 158–171, 1980.

Seedsman, R.: The behavior of clay shales in water, Can. Geotech. J., 23, 18–22, 1985.

Swisstopo (Bundesamt für Landestopographie): Relative humidity data at Mont Terri rock laboratory. Data acquisition Swisstopo, Mont Terri project, St-Ursanne, Switzerland, 2014.

Thury, M. and Bossart, P.: The Mont Terri rock laboratory, a new international research project in a Mesozoic shale formation, in Switzerland, Eng. Geol., 52, 347–359, 1999.

Tsang, C. F., Barnichon, J. D., Birkholzer, J., Li, X. L., Liu, H. H., and Sillen, X.: Coupled thermo-hydro-mechanical processes in the near field of a high-level radioactive waste repository in clay formations, Int. J. Rock Mech. Min., 49, 31–44, 2012.

Van Eeckhout, E. M.: The Mechanisms of Strength Reduction due to Moisture in Coal Mine Shales, Int. J. Rock Mech. Min., 13, 61–67, 1976.

Van Olphen, H.: Thermodynamics of interlayer adsorption of water in clays. I. Sodium Vermiculite, J. Colloid Sci., 20, 822–837, 1965.

Wild, K. M.: Evaluation of the hydro-mechanical properties and behavior of Opalinus Clay, Dissertation, ETH Zurich, Switzerland, 222 pp., 2016.

Wild, K. M., Wymann, L. P. Zimmer, S., Thoeny, R., and Amann, F.: Water Retention Characteristics and State-Dependent Mechanical and Petro-Physical Properties of a Clay Shale, Rock Mech. Rock Eng., 48, 427–439, 2015.

Yang, D. S., Bornert, M., Chanchole, S., Charbi, H., Valli, P., and Gatmiri, B.: Dependence of elastic properties of argillaceous rocks on moisture content investigated with optical full-field strain measurement techniques, Int. J. Rock Mech. Min., 53, 45–55, 2012.

7

Evaluating of the spatial heterogeneity of soil loss tolerance and its effects on erosion risk in the carbonate areas of southern China

Yue Li[1,2], Xiao Yong Bai[1,2,4], Shi Jie Wang[1,2], Luo Yi Qin[1,2], Yi Chao Tian[1,3], and Guang Jie Luo[1,3]

[1]State Key Laboratory of Environmental Geochemistry, Institute of Geochemistry, Chinese Academy of Sciences, Guiyang, Guizhou, 550002, PR China
[2]Puding Comprehensive Karst Research and Experimental Station, Institute of Geochemistry, CAS and Science and Technology Department of Guizhou Province, Puding, Guizhou, 562100, PR China
[3]Graduate School of Chinese Academy of Sciences, Beijing 100029, PR China
[4]Institute of Mountain Hazards and Environment, Chinese Academy of Sciences,Chengdu, Sichuan, 610041, PR China

Correspondence to: Xiao Yong Bai (baixiaoyong@126.com)

Abstract. Soil loss tolerance (T value) is one of the criteria in determining the necessity of erosion control measures and ecological restoration strategy. However, the validity of this criterion in subtropical karst regions is strongly disputed. In this study, T value is calculated based on soil formation rate by using a digital distribution map of carbonate rock assemblage types. Results indicated a spatial heterogeneity and diversity in soil loss tolerance. Instead of only one criterion, a minimum of three criteria should be considered when investigating the carbonate areas of southern China because the "one region, one T value" concept may not be applicable to this region. T value is proportionate to the amount of argillaceous material, which determines the surface soil thickness of the formations in homogenous carbonate rock areas. Homogenous carbonate rock, carbonate rock intercalated with clastic rock areas and carbonate/clastic rock alternation areas have T values of 20, 50 and 100 t/(km^2 a), and they are extremely, severely and moderately sensitive to soil erosion. Karst rocky desertification (KRD) is defined as extreme soil erosion and reflects the risks of erosion. Thus, the relationship between T value and erosion risk is determined using KRD as a parameter. The existence of KRD land is unrelated to the T value, although this parameter indicates erosion sensitivity. Erosion risk is strongly dependent on the relationship between real soil loss (RL) and T value rather than on either erosion intensity or the T value itself. If $RL \gg T$, then the erosion risk is high despite of a low RL. Conversely, if $T \gg RL$, then the soil is safe although RL is high. Overall, these findings may clarify the heterogeneity of T value and its effect on erosion risk in a karst environment.

1 Introduction

The fragile ecological environment of karst areas is closely related to surface soil (Keesstra et al., 2016; Novara et al., 2016; Comino et al., 2016; Li et al., 2016; Debolini et al., 2013). However, this factor is less associated with the total lack of inherent soil in such areas (Gessesse et al., 2014; Luo et al., 2016; Ligonja and Shrestha, 2013). Soil is continuously distributed through erosion, and rocky desertification landscapes are frequently generated (Molla and Sisheber, 2016; Tian et al., 2016; Bai et al., 2013). Determining soil loss tolerance (T value) is one of the most important criteria in controlling erosion and restoring ecosystems; therefore, this factor must be measured with accuracy. T is expressed in terms of annual soil loss (t km^{-2} a) and reflects the maximum level of soil erosion that can occur while allowing the land to sustain an indefinite, economic level of crop productivity (Wischmeier and Smith, 1965). This value is an important criterion in determining the potential erosion risk of a particular soil and often serves as the ultimate erosion control criterion to preserve long-term soil productivity (Duan et al., 2012). Thus, a scientifically determined T value is among the most significant aspects in planning soil erosion control on agricultural lands and other types of lands (Liu et al., 2003). The concept of this value was first proposed in the United States in 1956, and the top 10 influencing factors were identified for a particular soil. Although the determination of T value is often modified, the soil formation rate remains a typical and necessary factor. Early researchers (Hays et al., 1941; Klingebiel, 1961; Pretorius and Cooks, 1989) generated empirical proofs to compute this value. In the 1980s, Fran Pierce suggested the use of a soil productivity model to calculate the T value and initiated the quantitative study of this factor. Worldwide T values obtained based on the soil productivity method range from 116 to 9300 t/(km^2 a) depending on the location (Pierce et al., 1983, 1984). In India, the default soil loss tolerance limit is set at 11.2 Mg ha^{-1} yr^{-1} for soil conservation activities. Scholars who examined related topics suggested that criteria should be established to determine the T value limits, and that these values should differ for each soil series (Bhattacharyya et al., 2008). William and Smith (1964) proposed a notion model of an estimated T value in relation to the strength of both soil properties and soil formation rates (William and Smith, 1964). Skidmore (1982) improved the concept model and calculated this value using soil thickness instead of soil characteristics (Skidmore, 1982). Both high and low T limits are incorporated in this approach. According to Bazzoffi (2009), the notion of tolerance erosion based on only soil productivity and soil reformation rate is declining, and the off-site effects of soil erosion should be considered. Therefore, he suggested expanding the concept of hydrogeological risk to soil erosion by implementing the notion of T alongside a new concept, namely, the environmental risk of soil erosion. Scholars agree that soil loss should stabilize soil fertility and long-term soil productivity in addition to maintaining the balance between soil loss and soil formation rates (Pierce et al., 1983; Alexander, 1988). Purple soils (entisols) derived from limestone bedrock in China have faster formation rates than other soils. When exposed at the surface, the maximum weathering rate of this soil type is 15 000 Mg km^{-2} yr^{-1} (Nadalromero and Regüés, 2010). Purple soils are ideal for T research conducted over a short period because of their high formation rates. In the carbonate mountain areas of southern China, soil thickness generally ranges from 30 to 50 cm. Once the soil is lost, the underlying basement rock is exposed, and karst rocky desertification (KRD) land appears (Wang et al., 2004). This occurrence, which is caused by soil erosion, is among the most serious eco-environmental problems in this region. Mineralogical and geochemical studies indicate that soil layers are predominantly derived from residues (argillaceous material) that remain after the dissolution of the underlying carbonate rocks and of the thin argillaceous layers interbedded among these rocks (Wang et al., 1999; Pak et al., 2016). Owing to the low concentrations of acid-insoluble components, the volume of carbonate rocks tends to decrease sharply with the formation of weathering crusts. Highly pure carbonate rocks correspond to low acid-insoluble substance content; therefore,

the weathering–pedogenesis of carbonate rocks is the most fundamental and common geological–geochemical process (Drever and Stillings, 1997; Liu et al., 2009). This process is also the main soil formation factor used in subtropical carbonate regions. The severity of soil erosion depends strongly on the soil formation rate, which is also dependent on the local geological setting. Therefore, the T value in carbonate areas can be determined according to this rate. The objectives of this research are as follows: (1) discover the spatial heterogeneity and diversity of soil erosion tolerance in the carbonate areas of southern China and disprove the old "one region, one T value" concept; and (2) propose a new viewpoint stating that in karst regions, a large soil erosion modulus does not correspond to severe soil erosion and clarify the heterogeneity of T values and its effects on the erosion risk in karst environments.

2   Study area

The study area is located across the Yangtze River and Pearl River in south-western China. The approximate coordinates are 22°01′–33°16′ N and 98°36′–116°05′ E. The area covers Guizhou, Yunnan, Guangxi Zhuang Autonomo, Hunan, Hubei, Sichuan, Chongqing Municipalities and Guangdong Province (Fig. 1). Moreover, the study area is one of the tropical moist and subtropical moist regions, and the annual average temperature is 11.0–19.0 °C. Considering the plenty rain, more than 80 % of the area's average annual total precipitation is between 1100 and 1300 mm. The quantity of rain throughout the seasons is uneven; more rainfall occurs in May–October, and the precipitation from June to August accounts for approximately half of the total. Carbonate rock covers an outcropped area of 522 100 km^2 from the Sinian to Triassic. The thick carbonate formation was deposited in the study area. Yunnan, Qianxi–Qiannan and western Guangxi are mainly covered with a thick layer of bare limestone, dolomite and limestone. North-eastern Guizhou, Chongqing, Hubei and Xiangxi trough valley areas are mainly composed of interbedded dolomite and clastic rocks. The middle part of Hunan, central Guilin area, south-eastern Guangxi and northern Guangdong are covered with carbonate rock. The west of Sichuan and Yunnan consist primarily of buried limestone. The south-western karst mountainous areas are characterized by limestone soil, and the distribution of this soil varies considerably. This area also contains inland plateau lands.

3   Materials and methods

3.1   Construction of a carbonate rock assemblage distribution map

A 1 : 500 000 scale digital geological map is constructed to show the distribution of carbonate rock assemblage types

Figure 1. Map showing the location and distribution of carbonate regions in southern China.

in the carbonate areas of southern China. An officially published map is used as a data source.

The method of constructing a carbonate rock assemblage distribution map is identical to our previously used technique (Wang et al., 2004). The amount of argillaceous material in formations is an indicator that distinguishes rock assemblages because it indicates the surface soil thickness. Thus, assemblages can be divided into three types as follows.

1. Homogenous carbonate rock (HC): >90 % carbonate rock, <10 % argillaceous material and no clear clastic interbed. Based on the composition, HC can be categorized into three subtypes: homogenous limestone (HL), homogenous dolomite (HD) and mixed dolomite/limestone (HDL).

2. Carbonate rock intercalated with clastic rock (CI): 70–90 % carbonate rock, 10–30 % argillaceous material and has a clear clastic interbed. Based on composition, CI can be divided into two subtypes, namely, limestone interbedded with clastic rock (LI) and dolomite interbedded with clastic rock (DI).

3. Carbonate/clastic rock alternations (CA): 30–70 carbonate and 70–30 % clastic rocks. Based on composition, CA can be categorized into two subtypes, namely, limestone/clastic rock alternations (LA) and dolomite/clastic rock alternations (DA).

The argillaceous material can be computed based on 5, 20 and 50 % for HC, CI and CA. In addition, carbonate rock can be computed based on 95, 80 and 50 % for HC, CI and CA. This information is shown in Table 1.

3.2 Method of computing soil information rate

The soil information rate of carbonate rocks is related to temperature, precipitation, hydrology, vegetation and other environmental conditions. This rate changes annually, monthly, daily and even hourly on the same day (over daytime and night time). The average soil information rate can reflect the

overall characteristics but does not represent the specific position and special time. The soil information rate ranges from 30.00–89.70 mm ka^{-1} with a mean rate of 55.27 mm ka^{-1} in the carbonate areas of southern China as per a long-term field observation. As per the results of an in-house laboratory investigation, the densities of calcite carbonatite and dolomite carbonatite are 2.75 and 2.86 t m^{-3}. The soil formation rate of other rock types is 200 t/(km^2 a) (Li et al., 2006), and the rates of different rock type assemblages serve as their T values.

Specific T value can be calculated with the following equation:

$$T = v Q \rho C + R (1 - C), \tag{1}$$

where T is the soil loss tolerance (t km^{-2} yr^{-1}), v is the dissolution velocity of carbonate rocks (m^3 km^{-2} yr^{-1}), Q is the content of acid-insoluble components (%), ρ is the carbonate density (t m^{-3}), C is the proportion of carbonate and R is the soil formation rate of other rock types.

3.3 Construction of a KRD land distribution map in Guizhou Province in 2000

Based on the classification scheme shown in Table 2 and in combination with the corresponding 1 : 100 000 scale digital land use maps, the human–computer interactive interpreting method is used to construct a 1 : 100 000 scale digital hydrogeology, relief, soil distribution and karst rock desertification (KRD) land use maps in the year 2000 from the Landsat images.

Guizhou Province has an area of 176 000 km^2 and lies in the centre of the south-eastern Asian karst zone (Fig. 2). Carbonate rock is widespread and accounts for 62 % of the total land area. KRD is a serious problem in this region (Wang et al., 2004). Therefore, the relationship between KRD and T value is determined using Guizhou Province as an example. As per this classification, a 1 : 100 000 scale digital map that shows KRD land distribution overlaps with a T distribution map is developed. The spatial relationship between these two maps is then analysed.

4 Results and discussion

4.1 Spatial distribution of carbonate rock assemblages

As shown in Fig. 2a and Table 3, the total area is 527 196 km^2, of which 109 416, 108 828 and 81 772 km^2 belong to Guizhou, Yunan and Guangxi, respectively. Carbonate is mainly concentrated in Guizhou, eastern Yunan, central and western Guangxi, western Hubei, south-eastern Chongqing, southern Hunan, northern Guangdong and south-western Sichuan. HL covers 134 996 km^2 and is primarily distributed in western, southern and south-western Guizhou, eastern Yunan and western Guangxi. However,

Table 1. Division of rock type assemblage.

	Continuity carbonate rocks assemblage (Homogenous carbonate rock > 90 %)		
Carbonate rocks	homogenous limestone	homogenous dolomite	mixed dolomite/limestone
	carbonate rock intercalated with clastic rock (carbonate rock: 70–90 %)		
	limestone interbedded with clastic rock		dolomite interbedded with clastic rock
	carbonate/clastic rock alternations (carbonate rock: 30–70 %)		
	limestone/clastic rock alternations		dolomite/clastic rock alternations
Clastic rocks	siliceous rock, metamorphic rock, magmatic rock		

Table 2. Classification criterion and characteristic code of KRD types (Hu et al., 2008).

Classification and code of KRD type	Proportion percentage of bare rock (%)	Distribution character of the exposed rock	Colour of the RS image
No KRD (NKRD)	< 20	Star	Scarlet
Potential KRD (PKRD)	20–30	Star, Line	Shocking pink
Present KRD (AKRD)	> 31	Patch	Pink, grey, white

Note: colours of the RS image displayed with Landsat TM bands 4, 3 and 2 are displayed as red, green and blue.

this limestone is slightly scattered in Hunan. HD covers 58 723 km^2 and is exposed in the form of elongated belts in various places; other assemblage types are scarce. HDL covers 63 819 km^2 and is mainly found in Guangxi and Hunan and northern central and southern Guizhou. LI covers 148 577 km^2 and is the most widespread type of carbonate rock. DI covers 22 889 km^2 and is chiefly detected in central Guizhou and south-western Sichuan. LA covers 55 527 km^2 and is mainly detected in southern Guizhou and western Hubei. Finally, DA covers only 42 665 km^2 and is primarily found in south-west Sichuan and eastern Yunan.

4.2 Determination of T value and assessment of soil erosion risk

Figure 2b shows the calculated T values of different carbonate rock assemblages according to Eq. (1). The T values in the HC, HL and HDL areas are 17.22, 17.51 and 17.36 t/(km^2 a), whereas the T values in the LI and DI areas are 46.08 and 46.02 t/(km^2 a). The T values in LA and DA areas are 103.80 and 107.95 t/(km^2 a). These values indicate the spatial heterogeneity in the carbonate areas of southern China; such heterogeneity is closely related to the amount of argillaceous material, which determines the surface soil thickness in the formations. The "one region, one T value" concept cannot fully reflect the essence and the real circumstances in the area. This inadequacy may be attributed to the diverse results obtained by different researchers. An incor-

rect value is typically obtained regardless of the calculated T value, and three criteria should be considered instead of only one criterion.

The T values of the HC, CI and CA areas are 20, 50 and 100 t/(km^2 a). These areas contain the least, lesser and great amounts of argillaceous materials, and thus are extremely, severely and moderately sensitive to soil erosion. Hence, the T values in the carbonate areas of southern China are spatially heterogeneous (Table 4).

In addition, the T values of limestone and dolomite have similar amounts of argillaceous material. However, according to the results of our in-house laboratory investigation (Zhang et al., 2007), the dissolution velocity of calcite is 16 times that of dolomite. These two types of mineral constituent rocks differ by 1.5–2 times both in in-house laboratory and field observations (Cao et al., 2009). In the same season and under similar spring conditions, the carbonate content of the dolomite area in the water exceeds that of the limestone area (Jiang et al., 1997). In terms of lithology, the dolomite voidage is uniform and dense; thus, the specific surface area of water–rock interaction can be increased. As a result, conditions are set for water retention and interaction time extension (Cao et al., 2009). Given its uniformity, dolomite weathering is extremely intense, induces the loosening and easy formation of storage cataclasites and establishes conditions for plant growth. Biological processes further accelerate the dissolution velocity. In addition, dolomite

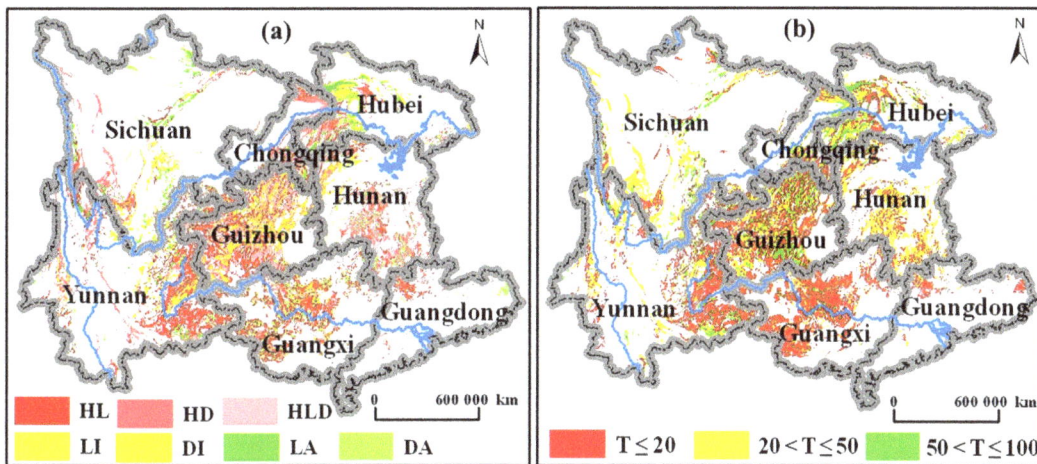

Figure 2. Distribution map of carbonate rock assemblage types (**a**) and T value (**b**) in carbonate areas of southern China. Homogenous limestone is HL, homogenous dolomite is HD, mixed limestone/dolomite is HLD, limestone interbedded with clastic rock is LI, dolomite interbedded with clastic rock is DI, limestone/clastic rock alternations is LA, dolomite/clastic rock alternations is DA, soil loss tolerance is T.

Table 3. Distribution areas of different carbonate rock assemblage types in carbonate areas of southern China (km^2).

	Chongqing	Guangdong	Guangxi	Guizhou	Hubei	Hunan	Sichuan	Yunan	Study area
Total	82 400	179 800	236 300	176 100	185 900	21 1875	485 000	394 000	1 951 375
Carbonate	29 896	10 440	81 772	109 416	53 146	65 780	67 918	108 828	527 196
HL	6722	4603	34 309	30 677	5184	9087	7579	36 835	134 996
HD	2474	0	3131	22 991	10 393	4101	3458	12 175	58 723
HDL	2006	3143	26 162	3690	4694	12 071	7484	4568	63 819
LI	11 114	2694	12 355	19 340	14 641	35 683	26 085	26 666	148 577
DI	58	0	260	7210	2664	3193	7730	1774	22 889
LA	6835	0	5517	25 231	6374	483	1889	9197	55 527
DA	687	0	38	276	9196	1161	13 693	17 613	42 665

releases abundant magnesium ions during the weathering–pedogenesis of carbonate rocks as its main contribution to the formation of clay mineral. By contrast, limestone cannot supply a sufficient amount of such ions. These phenomena accelerate the dissolution velocity of dolomite and supplement the deficiency. This mechanism may explain the similarity in the T values of limestone and dolomite (Feng et al., 2013).

4.3 Effect of T value on karst rocky desertification

As shown in Table 5, the AKRD land measured 18 491, 10 955 and 9456 km^2 in the extremely, severely and moderately sensitive areas. KRD land is concentrated in the extremely sensitive area ($T = 20$) and covers over 47 % of the total area in Guizhou Province. Of the total AKRD land, 28.16 % is severely sensitive ($T = 50$), and 24.31 % is moderately sensitive ($T = 100$).

These findings suggest that a low T value corresponds to a large KRD area. The KRD land area is coherent in relation to the T value criterion. Nonetheless, the relationship between NKRD land and T value is unchanged. Based on the information provided above, the areas of background value in different T value regions ($T = 20, 50, 100$) were 57 375, 26 558 and 25 515 km^2. The distribution area of KRD is strongly affected by the area of geological environment. During the Cretaceous Period of the Mesozoic Era, this area was a shallow marine sedimentary environment and experienced many transgressive and regressive phenomena, which were alternately caused by marine and continental sediments. The formation of a variety of rock types in Karst, Yanshan and Himalayan movements caused the deposition of Karst rock uplift to the surface and the formation of folds, faults and broken steep landform patterns. The development of rock and soil in Karst is slow, thereby resulting in the shallow soil layer, slow growth of vegetation and loss of soil. Therefore, the AKRD land area might not reflect the appearance of this land in different regions, although this area indicates the distribution situation.

Table 6 shows the generation of KRD land relative to the different regions that are sensitive to soil erosion. This oc-

Table 4. Criteria of T value and sensitivity of soil erosion in carbonate areas of southern China.

Carbonate rock assemblages	T value t/(km² a)	Area (km²)	Proportion (%)	Sensitivity of soil erosion
Homogenous carbonate rock	20	257 538	48.85 %	Severe
Carbonate rock intercalated with clastic rock	50	171 466	32.52 %	Moderate
Carbonate/clastic rock alternations	100	98 192	18.63 %	Low

Table 5. Karst rocky desertification area under different sensitivities.

	AKRD (km²)	PKRD (km²)	NKRD (km²)
Moderate sensitivity	9457	7889	8169
Severe sensitivity	10 955	6004	9599
Utmost sensitivity	18 491	17 926	20 957

Note: karst rocky desertification (AKRD), potential karst rocky desertification (PKRD) and no karst rocky desertification (NKRD).

Table 6. KRD area percentage under different sensitivities.

	AKRD (%)	PKRD (%)	NKRD (%)
Moderate sensitivity	37.06	22.61	32.02
Severe sensitivity	41.25	22.61	36.14
Utmost sensitivity	32.23	31.24	36.53

Table 7. Criterion for risk assessment of soil erosion in carbonate areas of southern China.

Types	Range	RL/T value	Erosion risk grade
Low	Above critical	$R > 2$	Utmost safety
		$1.5 < R \leq 2$	Severe safety
		$1 < R \leq 1.5$	Moderate safety
Moderate	Equal	$R = 1$	Critical point
High	Below critical	$0.5 \leq R < 1$	Utmost danger
		$0.2 \leq R < 0.5$	Severe danger
		$R < 0.2$	Moderate danger

currence is maximized at 41.25, 37.06 and 32.23 % in the severely, moderately and extremely sensitive areas. This finding confirms that the occurrence of AKRD land is unrelated to the T value. In other words, this value is not the real factor that determines the KRD appearance in carbonate areas. Thus, the T value cannot reflect the soil erosion risk although it reflects the sensitivity of soil erosion.

Erosion risk depends on the relationship between RL and T value rather than on soil erosion intensity or T value itself. If RL $>> T$, then the risk is high but RL is low. Conversely, if RL $<< T$, then the soil is safe but RL is high (Table 7).

The occurrence of KRD is highest in the severely sensitive area (41.25 %). This result indicates that RL is considerably greater than the T value, and that the situation is extremely dangerous. However, these values do not necessarily imply that RL remains considerably smaller than T value in the moderately and extremely sensitive areas. Conversely, the occurrences of KRD land are 37.06 and 32.23 % in these areas; such values clearly indicate high degrees of soil erosion. Thus, the severely sensitive area is the most hazardous area.

4.4 T value criteria in different regions

T values with reference to the different conditions of their respective regions are subsequently proposed to establish

an accurate T value standard. The United States Department of Agriculture Soil Conservation Bureau established a systematic T value system that ranged between 220 and 1120 t/(km² a) in 1973. This standard is still being used at present. In central Africa, the sand and clay T values are 150 and 180 t/(km² a). In Russia, a T value range of 340–1090 t/(km² a) was reported, whereas a range of 450–1120 t/(km² a) was established in India. In China, T values of 1000, 200 and 500 t/(km² a) are reported for the Loess Plateau, the phaeozem regions of north-eastern China and northern Rocky Mountain and the hilly red soil regions of southern China and south-western Rocky Mountain, respectively. In this work, the T values in the HC, CI and CA areas are 20, 50 and 100 t/(km² a) (Yang et al., 2004)

Some senior scholars and scientists conducted preliminary studies on soil erosion. Duan et al. (2012) modified soil productivity index model to calculate a quantitative T value for different black soil species in the black soil region of north-eastern China. The T values of the 21 black soil species in the study area ranged from 68 to 358 t/(km² a) with an average of 141 t/(km² a). This average T value is 29.5 % less than the current national standard. The T values of the three different soil subgroups in the study area are albic black soil, 106 t/(km² a); typical black soil, 129 t/(km² a); and meadow black soil, 184 t/(km² a) (Duan et al., 2012). Based on the soil nutrient balance and test data, Shui et al. (2003) suggested that soil loss tolerance in Q_2 red-clay-derived red earth should be lower than 300 t/(km² a). Yuan et al. (2005) determined the soil loss tolerance of less than 120 t/(km² a) in the hilly purple rock area in central Hunan (2005). Based on theoretical analysis, field examination and investigation, Chen et al. (2003) reported that the 200 t/(km² a) is the rational soil loss tolerance of sloping field in semi-arid hill–gully area of the Loess Plateau over a long period according to soil formation velocity, top soil nutrient balance, land productivity stability in sloping field, sediment transport tolerance of the Yellow River course and regional economic development (2003).

Some scholars conducted countless studies in karst areas. According to the corroded ratio and content rate of carbonate rocks, Chai (1989) calculated the estimated amount of soil loss tolerance at $68\,t/(km^2\,a)$ in the karst area of Guangxi Autonomous Region. Chen (1993) measured the accumulated and loss amounts of soil nutrient for the top layer soil in forest land and analysed the balance of NPK (nitrogen, phosphorus and potassium) and the rate of soil formation, which approached the amount of allowed soil loss. Under the upper reaches of the Changjiang River climatic conditions, the upper line of allowed soil loss is $50\,t/(km^2\,a)$ for developing soil from limestone, whereas $100\,t/(km^2\,a)$ for developing soil from non-carbonaceous rock. Wei (1996) reported that the T values of the calcareous soil area in the karst area ranged from 0.522 to $1.285\,t/(km^2\,a)$; if the eluviation and normal erosion in soil-forming process were not considered, then the scope of the T value ranges from 3.24 to $8.10\,t/(km^2\,a)$. However, the soil loss tolerance of some parts of the argillaceous limestone, such as the non-pure carbonate rocks, can be increased to 16.2–$40.5\,t/(km^2\,a)$, and the upper line of soil allowed loss is $50\,t/(km^2\,a)$ for the karst area (Wei, 1996). Li (2006) reported that with $49.67\,mm\,ka^{-1}$ as the average weathering dissolving rate of carbonate rocks in Guizhou, the pedogenesis rates of different petrologic assemblages in carbonate area were calculated and used as the values of soil loss tolerance in carbonate areas. The soil loss tolerance in HC area was lower than 6.84, 45.53 in CI areas and $103.46\,t/(km^2\,a)$ in CA areas (2006).

In this study, T value was calculated using digital-distribution map of carbonate rock assemblage types based on the pedosphere system theory. Results indicated spatial heterogeneity and diversity in such values. The T value is proportionate to the amount of argillaceous material, which determines the surface soil thickness in the formations of HC areas. The values are 20 and $50\,t/(km^2\,a)$ in CI areas and $100\,t/(km^2\,a)$ in carbonate/clastic rock alternation areas. Erosion risk is strongly dependent on the relationship between real soil loss (RL) and T value rather than on either erosion intensity or the T value itself. These findings may clarify the heterogeneity of T value and its effect on erosion risk in a karst eco-environment. Hence, innovative technological assessment solutions are not required. In summary, this paper presents a method that provides experience and data for reference on the related research of soil erosion of karst landform areas of international counterparts. However, this study has limitations; it cannot fully consider the dry and wet deposition in atmosphere and the contribution of acid rain to soil forming rate. Such a restriction might affect the accuracy.

5 Conclusions

This study might clarify the heterogeneity of T values and its effects on erosion risk in a karst eco-environment, providing an alternative to inventing innovative technological assessment solutions. Our main findings are listed as follows:

1. T values are spatially heterogeneous, and a minimum of three criteria should be considered instead of only one when investigating the carbonate areas of southern China. Apparently, the "one region, one T value" concept might not be applicable to this region.

2. T value is proportionate to the amount of argillaceous material, which determines the surface soil thickness in the formations. The HC, CI and CA areas have T values of 20, 50 and $100\,t/(km^2\,a)$, and are respectively extremely, severely and moderately sensitive to soil erosion.

3. The generation of KRD land is unrelated to T value, although this value reflects erosion sensitivity. Erosion risk depends strongly on the relationship between RL and T value instead of on erosion intensity or the T value itself. If $RL \gg T$, then the risk is high despite the low RL. On the contrary, if $RL \ll T$, then the soil is safe despite the high RL.

In summary, we first report the following discoveries. The T values are spatially heterogeneous, and a minimum of three criteria should be considered instead of only a single criterion in karst areas. Our findings disprove the old "one region, one T value" concept. Secondly, we proposed a new viewpoint, which states that in karst regions, a large soil erosion modulus does not correspond to severe soil erosion. Although the T value can reflect soil sensitivity, this value cannot indicate soil erosion risk. Thus, a low T value indicates that the local soil is highly sensitive; however, the soil erosion risk is not necessarily high. Therefore, this risk depends strongly on the ratio between RL and T value instead of on erosion intensity or on T value itself.

Given that the determination time of natural erosion and environmental background conditions is poorly understood, the research object, method and consideration factors of soil loss tolerance are different. Therefore, further efforts should focus on defining and specifying the connotation and research methods of natural erosion and soil loss tolerance, as well as comprehensively and systematically studying the natural erosion and soil loss tolerance in different types of soil and water loss.

Competing interests. The authors declare that they have no conflict of interest.

Acknowledgements. This research work was supported jointly by the National Key Research Program of China (No. 2016YFC0502300, 2016YFC0502102, 2013CB956700 & 2014BAB03B02), United Fund of karst science research center (No. U1612441), International cooperation research projects of the national natural science fund committee (No. 41571130074 & 41571130042), Science and Technology Plan of Guizhou Province of China (No. 2012-6015 & 2013-3190 & 201742920512120000), Science and technology cooperation projects (No. 2014-3).

Edited by: M. Oliva

References

Alexander, E. B.: Rates of soil formation: Implications for soil-loss tolerance, Soil Sci., 145 37–45, 1988.

Bai, X., Zhang, X., Long, Y., Liu, X., and Zhang, S.: Use of ^{137}Cs and ^{210}Pb$_{ex}$, measurements on deposits in a karst depression to study the erosional response of a small karst catchment in southwest china to land-use change, Hydrol Process., 27, 822–829, 2013.

Bazzoffi, P.: Soil erosion tolerance and water runoff control: minimum environmental standards, Reg. Environ. Change, 9, 169–179, 2009.

Bhattacharyya, P., Bhatt, K. V., and Mandal, D.: Soil loss tolerance limits for planning of soil conservation measures in Shivalik-Himalayan region of India, Catena, 73, 117–124, 2008.

Cao, J. H., Jiang, Z. C., Yang, D. S., Tong, L. Q., Pei, J. G., Luo, W. Q., and Yang, H.: Soil and water loss, desertification controlled by karst environment in Guizhou Karst region, Soil Water Conserv. China, 1, 20–23, 2009 (in Chinese).

Chai, Z. X.: Soil Erosion in Karst Area of Guangxi Autonomous Region, Mountain Research, 7, 255–260, 1989 (in Chinese).

Chen, L. J.: Study on the Amount of soil Allowed Loss for Forest Land, J. Soil Water Conserv., 7, 8–22, 1993 (in Chinese).

Chen, Q. B., Wang, K. Q., Qi, S., Sun, L. D., and Wu, X. W.: Soil Loss Tolerance of Sloping Field in Semi-arid Hilly-gully Area of Loess Plateau, B. Soil Water Conserv., 23, 1–4, 2003 (in Chinese).

Comino, J. R., Quiquerez, A., Follain, S., Raclot, D., Bissonnais, Y. L., Casalí, J., Giménez, R., Cerdà, A., Keesstra, S. D., Brevik, E. C., Pereira, P., Senciales, J. M., Seeger, M., Ruiz Sinoga, J. D., and Ries, J. B.: Soil erosion in sloping vineyards assessed by using botanical indicators and sediment collectors in the Ruwer-Mosel valley, Agr. Ecosyst. Environ., 233, 158–170, 2016.

Debolini, M., Schoorl, J. M., Temme, A., Galli, M., and Bonari, E.: Changes in Agricultural Land Use Affecting Future Soil Redistribution Patterns: A case study in Southern Tuscany (Italy), Land Degrad Dev., 26, 574–586, 2013.

Drever, J. I. and Stillings, L. L.: The role of organic acids in mineral weathering, Colloids & Surfaces A Physicochemical & Engineering Aspects, 120, 167–181, 1997.

Duan, X. W., Xie, Y., and Liu, B. Y.: Soil loss tolerance in the black soil region of Northeast China, J. Geogr. Sci., 22, 737–751, 2012.

Feng, Z. G., Ma, Q., Li, S. P., Wang, S. J., Huang, W., Liu, J., and Shi, W. G.: Weathering Mechanism of Rock-Soil Interface in Weathering Profile Derived from Carbonate Rocks: Prelimi-

nary Study of Leaching Simulation in Rock Powder Layer, Acta Geol. Sin., 87, 199–132, 2013.

Gessesse, B., Bewket, W., and Bräuning, A.: Model-Based Characterization and Monitoring of Runoff and Soil Erosion in Response to Land Use/Land Cover Changes in the Modjo watershed, Ethiopia, Land Degrad. Dev., 26, 711–724, 2014.

Hays, O. E. and Clark, N.: Cropping system that helps control erosion. USA: University of Wisconsin, Bull No. 452, 1941.

Hu, Y. C., Liu, Y. S., Wu, P. L., and Zou, X. P.: Rocky desertificationin Guangxi karst mounminous area: its tendency, formation causes and rehabilitation, Transactions of the Chinese Society of Agricultural Engineering, 24, 96–101, 2008.

Jiang, Z. C., Xie, Y. Q., Zhang, C., and Weng, J. T.: A projected protection scheme for the karst. Geological landscape in Xishan region, Beijing, Nat Resour., 6, 60–73, 1997 (in Chinese).

Keesstra, S., Pereira, P., Novara, A., Brevik, E. C., Azorin-Molina, C., Parras-Alcántara, L., Jordán, A., and Cerdà, A.: Effects of soil management techniques on soil water erosion in apricot orchards, Sci. Total Environ., 551–552, 357–366, 2016.

Klingebiel, A. A.: Soil factors and soil loss tolerance, in: Soil Loss Prediction, North and South Dakota, Nebraska, and Kansas, USA: Soil Conservation Service, United States Department of Agriculture, 1961.

Lan, L., Zhou, Z. H., and Liu, G. C.: The present situation and conceive of soil loss tolerance study, Adv. Earth Sci., 20, 65–72, 2005 (in Chinese).

Li, Y. B., Wang, S. J., Wei, C. F., and Long, J.: The spatial distribution of soil loss tolerance in carbonate area in Guizhou province, Earth Environ., 34, 36–40, 2006.

Ligonja, P. J. and Shrestha, R. P.: Soil Erosion Assessment in Kondoa Eroded Area in Tanzania using Universal Soil Loss Equation, Geographic Information Systems and Socioeconomic Approach, Land Degrad. Dev., 26, 367–379, 2013.

Liu, G. C., Li, L., Wu, L., Wang, G., Zhou, Z., and Du, S.: Determination of soil loss tolerance of an Entisol in Southwest China, Soil Sci. Soc. Am. J., 73, 412–417, 2009.

Liu, Q. M., Wang, S. J., Piao, H. C., and Ouyang, Z. Y.: The changes in soil organic matter in a forest-cultivation sequence traced by stable carbon isotopes, Aust. J. Soil. Res., 41, 1317–1327, 2003.

Luo, G. J., Wang, S. J., Bai, X. Y., Liu, X. M., and Cheng, A. Y.: Delineating small karst watersheds based on digital elevation model and eco-hydrogeological principles, Solid Earth, 7, 457–468, doi:10.5194/se-7-457-2016, 2016.

Molla, T. and Sisheber, B.: Estimating soil erosion risk and evaluating erosion control measures for soil conservation planning at Koga watershed in the highlands of Ethiopia, Solid Earth, 8, 13–25, doi:10.5194/se-8-13-2017, 2017.

Nadalromero, E. and Regüés, D.: Geomorphological dynamics of subhumid mountain badland areas-weathering, hydrological and suspended sediment transport processes: a case study in the Araguás catchment (Central Pyrenees) and implications for altered hydroclimatic regimes, Prog. Phys. Geogr., 34, 123–150, 2010.

Novara, A., Keesstra, S., Cerdà, A., Pereira, P., and Gristina, L.: Understanding the role of soil erosion on CO$_2$-C loss using ^{13}C isotopic signatures in abandoned mediterranean agricultural land, Sci. Total Environ., 550, 330–336, 2016.

Pak, T., Butler, I. B., Geiger, S., Van Dijke, M. I. J., Jiang, Z., and Surmas, R.: Multiscale pore-network representation of heterogeneous carbonate rocks, Water Resour. Res., 52, 2016.

Pierce, F. J., Dowdy, R. H., and Larson, W. E.: Soil productivity in the Corn belt: An assessment of erosion's long-term effects, J. Soil Water Conserv., 39, 131–136, 1984.

Pierce, F. J., Larson, W. E., and Dowdy, R. H.: Productivity of soils: Assessing long-term changes due to erosion, J. Soil Water Conserv., 38, 39–44, 1983.

Pretorius, J. R. and Cooks, J.: Soil Loss Tolerance Limits: An Environmental Management Tool, Geol. J., 19, 67–75, 1989.

Shui, J. G., Ye, Y. L., Wang, J. H., and Liu, C. C.: Regularity of Erosion and Soil Loss Tolerance in Hilly Red-Earth Region of China, J. Integr. Agr., 36, 179–183, 2003.

Skidmore, E. L.: Soil loss tolerance, in: Determinants of Soil Loss Tolerance, American Society of Agronomy, ASA Special Publication No. 45, 87–94, 1982.

Tian, Y. C., Wang, S. J., Bai, X. Y., Luo, G. J., and Xu, Y.: Trade-offs among ecosystem services in a typical Karst watershed, SW China, Sci. Total Environ., 1297, 566–567, 2016.

Wang, S. J., Ji, H. B., Ouyang, Z. Y., Zhou, D. Q., Zheng, L. P., and Li, T. Y.: Preliminary study on weathering and pedogenesis of carbonate rock, Sci. China Ser. D., 29, 441–449, 1999 (in Chinese).

Wang, S. J., Liu, M., and Zhang, D. F.: Karst rocky desertification in southwestern China: Geomorphology, landuse, impact and rehabilitation, Land Degrad Dev., 15, 115–121, 2004.

Wang, Z., Johnson, D. A., Rong, Y., and Wang, K.: Grazing effects on soil characteristics and vegetation of grassland in northern China, Solid Earth, 7, 55–65, doi:10.5194/se-7-55-2016, 2016.

Wei, Q. P.: Soil Erosion in Karst Region of South China and Its Control, Res. Soil Water Conserv., 4, 72–76, 1996 (in Chinese).

William, L. and Smith, R. M.: A conservation definition of erosion tolerance, Soil Sci., 97, 183–186, 1964.

Wischmeier, W. H. and Smith, D. D.: Predicting Rainfall Erosion Losses from Cropland East of the Rocky Mountains: a guide to conservation planning, 1965.

Yang, C. Q., Cai, Q. G., and Fan, H. M.: Process of Soil Loss Tolerance Research-in the Phaeozem Region of Northeast China, Res. Soil Water Conserv., 11, 66–96, 2004 (in Chinese).

Yuan, Z. K., Zhu, G., Tian, D. L., Yuan, H. B., Zhang, C. M., and Liu, W. R.: Process of Loss of Soil and Water in Red Soil and Purple Soil Areas of Recovering Plants, J cent south forest univ., 25, 1–7, 2005.

Zhang, T. F., Cui, Z. N., Qian, Y. X., Xie, S. Y., and Bao, Z. Y.: Dissolution Kinetic Characteristics of Ordovician Marine Carbonate in Central Tarim Basin, Geol. Sci. Technol. Info., 26, 19–25, 2007.

8

Application of a new model using productivity coupled with hydrothermal factors (PCH) for evaluating net primary productivity of grassland in southern China

Zheng-Guo Sun[1]**, Jie Liu**[2]**, and Hai-Yang Tang**[1]

[1]College of Agro-grassland Science, Nanjing Agricultural University, 1 Weigang, Nanjing, Jiangsu 210095, People's Republic of China
[2]Department of Environmental Science, Hokkaido University, Sapporo 060-0810, Japan

Correspondence to: Zheng-Guo Sun (sunzg@njau.edu.cn)

Abstract. Grassland ecosystems play important roles in the global carbon cycle. The net primary productivity (NPP) of grassland ecosystems has become the hot spot of terrestrial ecosystems. To simulate grassland NPP in southern China, a new model using productivity coupled with hydrothermal factors (PCH) was built and validated based on data recorded from 2003 to 2014. The results show a logarithmic correlation between grassland NPP and mean annual temperature and a linear positive correlation between grassland NPP and mean annual precipitation in southern China, both highly significant relationships. There was a highly significant correlation between simulated and measured NPP ($R^2 = 0.8027$). Both RMSE and relative root mean square error (RRMSE) were relatively low, showing that the simulation results of the model were reliable. The NPP values in the study area had a decreasing trend from east to west and south to north. Mean NPP was $471.62\,\mathrm{g\,C\,m^{-2}}$ from 2003 to 2014. Additionally, the mean annual NPP of southern grassland presented a rising trend, increasing $3.49\,\mathrm{g\,C\,m^{-2}\,yr^{-1}}$ during the past 12 years. These results document performance and use of a new method to estimate the grassland NPP in southern China.

1 Introduction

Grassland is one of the major biological communities in the world. It covers more than 40 % of the total land area on the planet and plays an important role in the global biogeo-chemical cycle and energy transformation process (Chen and Zhang, 2000; Mosier et al., 1991). Meanwhile, grassland also plays a role in water and soil conservation, wind breaking and sand fixation, biodiversity maintenance, and shaping soil from surface to depth, and it shows a close connection with human survival and development (Brevik et al., 2015). The root system of grassland vegetation occurs in soil, and thus the direct link between soil and vegetation can be discovered. Most soil functions have strong ties to vegetation, including biomass production; biodiversity pooling; and storing, filtering, and transforming nutrients, substances, and water (Keesstra et al., 2016). In the soil – grassland vegetation – atmosphere continuum, grassland acts as the center of ecological functions on the ecosystem scale. The impacts of the climate on grasslands are quite complicated. On the one hand, different types of grasslands have their own spatial distributions controlled by temperature and precipitation; on the other hand, a rise in temperature will alter some processes in the ecosystem (such as evapotranspiration, decomposition, and photosynthesis). Therefore, temperature exerts a significant effect on biological community productivity (Douglas and Geoffrey, 1997). Net primary productivity (NPP) is an indicator that measures the production capacity and economically and socially significant products of the plant community under natural conditions (Sun et al., 2013). Changes in NPP directly reflect the response of ecosystems to climatic conditions; therefore, it can be used as a research index in the relationship between ecosystem function and climate change (Zhou et al., 2014). It also has an important theoretical and practical significance for evaluating the environmental quality of terrestrial ecosystems, regulating ecological processes, and estimating the terrestrial carbon sink to master the interannual variation rule of terrestrial NPP (Cao et al., 2013; Richardson et al., 2012; Picard et al., 2005; Zhang et al., 2011; Xu et al., 2012).

Estimation methods, most based on models, to calculate grassland NPP were discussed in previous research (Gill et al., 2002). Models demonstrate advantages over other methods in global, regional, and other large-scale studies, becoming an important tool in macro-ecological research of grasslands. Grassland NPP estimation models have been used by some researchers for dynamic monitoring and forecasting (Raich et al., 1991; Matsushita and Tamura 2002), providing theoretical and technical support for ecological improvement and recovery of grasslands (Christenson et al., 2014). A large number of studies were conducted by domestic and foreign scholars to understand the influence of climate change on ecosystem processes, including grassland productivity and grassland C circulation. Although many researchers have studied the influences on a national or regional scale (Parton et al., 1995; Hall et al., 1995; Braswell et al., 1997; Cao and Woodward, 1998; Fang et al., 2001; Ni, 2002; Mantgem and Stephenson, 2007; Wunder et al., 2013; Gang et al., 2015), there has been little research on relationships between grassland NPP and climate factors in southern China. Grassland resources are abundant in China, with an area of nearly 400 million ha, nearly one-sixth of which is in southern China. As the grassland in northern areas continues to deteriorate and become desert, the ecological system of grassy hills and slopes in southern China is becoming increasingly important. Study of the relationship between NPP and climatic factors, together with their dynamic simulation, will provide insights on the effective management and reasonable utilization of grasslands in southern China and the promotion of global change research. Our objectives were the following: (1) to build a model using productivity coupled with hydrothermal factors (PCH) based on the statistical analysis of the relationship between measured NPP, precipitation, and temperature; (2) to modify the adjustment coefficient and the parameter of the model based on the grassland types and their ecological characteristics; (3) to simulate NPP using the PCH model and analyze its changing trends in spatial and temporal patterns from 2003 to 2014; (4) to verify the accuracy of the PCH model by comparing it with field observation data; and (5) to explore the dominant hydrothermal factor for determining the NPP change in the study area.

2   Materials and methods

2.1   Study area

The grassy hills and slopes of southern China, centered at 110°0′ E, 27°30′ N, were the focus of our research. The site

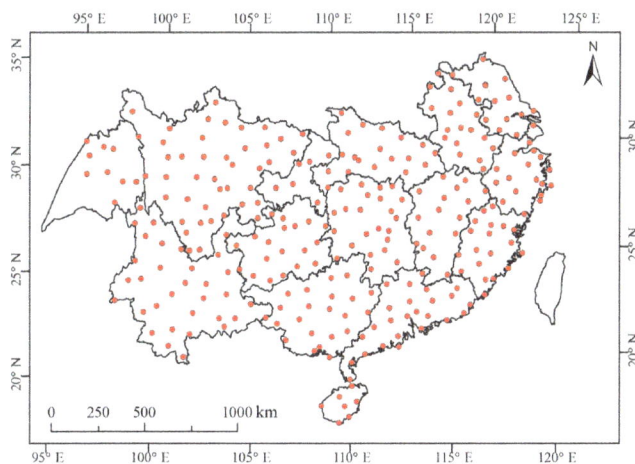

Figure 1. Study area and meteorological stations in southern China (the black boundary lines indicate the provincial boundary; the red dots represent the locations of the meteorological stations).

encompassed 17 provinces and an area of about 60 million ha (Fig. 1). The grasslands of southern China are mainly composed of typical grassland, wetland grassland, lowland meadow, and upland meadow. The southern grasslands are scattered and distributed among areas of forest land and cultivated land and are mostly located on slopes. Most regions of the southern grasslands are managed with grazing and some regions with enclosure and cutting. The climate characteristics in this area include hot and rainy summers and mild and rainy winters, with the frost-free period being more than 300 days per year. The annual mean precipitation is between 800 and 1600 mm and the annual mean temperature is greater than 15 °C. These climate conditions contribute to a suitable environment for grassland.

2.2   Data acquirement and processing

NPP data were acquired in July of 2011, 2012, and 2013, 66 sample plots were investigated in several provinces of the study area. Five quadrats (1 m × 1 m) were set on corners and in the center of each representative sample plot (10 m × 10 m). Aboveground biomass and the latitude and longitude information were recorded in each small quadrat, with an average level calculated after sampling. Every 2.2 g of dry matter was converted into 1 g carbon, leading to the grass NPP in each sample area, represented in the form of carbon (grams of carbon per square meter) (Fang et al., 2001).

Climate data acquired includes temperature and precipitation data from the years 2003 to 2014 from the ground stations of China Meteorological Data Service center (http://data.cma.cn/site/index.html) (Fig. 1). Kriging interpolation from the geographic information system (GIS) interpolation tool was utilized to analyze meteorological data according to the latitude and longitude of each station. Then the image projection transformation converted data into a raster image with a latitude and longitude network and 1000 m resolution. Finally, temperature and precipitation information was extracted according to latitude and longitude corresponding to the investigation points.

The 1980 Chinese grassland resource inventory and MOD12Q1 data acquired in 2004 were used to generate the land cover, land use map and the grassland distribution map (Fig. 2). Open shrubs, woody savannas, savannas, grasslands, and permanent wetlands were included as the grassland of southern China based on the land use and land cover classification project proposed by the International Geosphere–Biosphere Programme (IGBP).

2.3   Model establishment and validation

Based on the statistical analysis of the relationship between measured NPP, precipitation, and temperature, the preliminary structure of the model was developed. Then the nonlinear fitting algorithm was utilized to optimize and determine the parameters of the model.

In order to verify the reliability of the simulation results, both RMSE and relative root mean square errors (RRMSE) were applied to the model for testing and evaluating the simulation effects. RMSE and RRMSE were expressed as

$$RMSE = \sqrt{\frac{1}{n}\sum_{i=1}^{n}(O_i - S_i)^2} \tag{1}$$

$$RRMSE = \frac{\sqrt{\frac{1}{n}\sum_{i=1}^{n}(O_i - S_i)^2}}{O_a}, \tag{2}$$

where O_i was the real value, S_i was the simulated value, O_a was the average of real value, and n was the total number of samples.

3   Results

3.1   Relationship between grassland NPP and temperature

Grassland NPP is a joint result of the regional light, temperature, precipitation, soil, and other natural conditions, which reflects the ability of using natural environmental resources (Gang et al., 2015). Under natural conditions, temperature and precipitation were the two dominate influential factors in grassland NPP in southern China (Sun et al., 2014). The results of the analysis of the relationship between grassland NPP and temperature in southern China showed that (1) between 10 and 20 °C there was a linear positive correlation between temperature and the NPP and (2) a para-curve relationship was found between 20 and 30 °C. Generally, the relationship between temperature and grassland NPP was

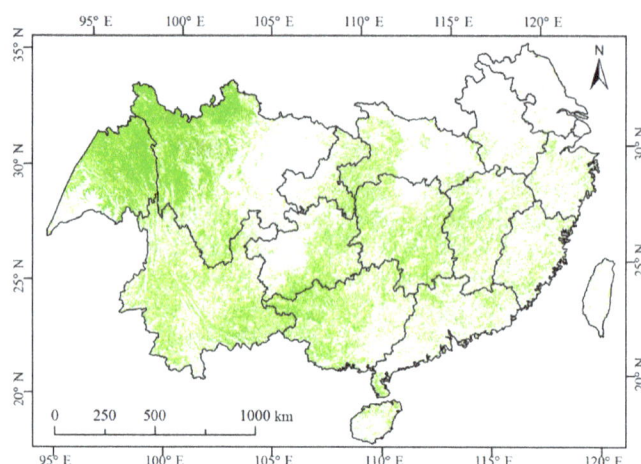

Figure 2. The distribution map of grasslands of southern China (the black boundary lines indicate the provincial boundary, the green zone represents the grassland area, and the colorless region represents non-grassland in the study area).

logarithmic, with correlation coefficient r being 0.4629 and reaching a significant level ($P < 0.01$). As a result, the relationship could be presented as a logarithmic equation.

3.2   Relationship between grassland NPP and precipitation

Precipitation is a key factor in many NPP estimation models (Huston, 2012; Yu et al., 2008). Mean monthly precipitation in the grassland ecological system of southern China presented a large range throughout a year, with minimum precipitation being 40 mm and the maximum being over 200 mm. NPP also showed a regular distribution according to the precipitation, with a typical linear positive correlation. The correlation coefficient r was 0.7836, reaching a very significant level ($P < 0.01$). Therefore, the influences of precipitation on grassland NPP could be expressed as a linear equation.

3.3   Estimation model of grassland NPP

3.3.1   Model establishment

According to the analysis results, a positive relationship existed between grassland NPP and mean annual temperature and annual precipitation in southern China. Thus, it is feasible to express the relationship with logarithmic and linear equations, respectively. However, the results varied greatly when temperature was directly used as the equation factor and any data below 0 °C failed to be processed. Thus, it was necessary to introduce a temperature adjustment coefficient, described here as

$$T_a = \mathrm{Ln}\left(\frac{T}{t_1} + a_1\right), \tag{3}$$

where T_a was the temperature adjustment coefficient, T was the mean annual temperature (°C), t_1 was the model parameter, and a_1 was a constant, set to 2.5 in the paper.

Precipitation showed a similar trend. Growth stopped when moisture was below a certain level. Thus, another adjustment coefficient was introduced and expressed as the following:

$$W_a = \text{Sqrt}\left(\frac{W}{w_1} + a_2\right), \tag{4}$$

where W_a was the adjustment coefficient, W was the mean annual precipitation (mm), w_1 was the model parameter, and a_2 was a constant, set to 0.5 in the paper.

According to the information above, the estimation model of grassland NPP in southern China could be written as the following:

$$\text{NPP} = T_a \times W_a \times \left(T + \frac{W}{6}\right). \tag{5}$$

The PCH model was built to simulate grassland NPP of southern China based on the principle of grassland productivity coupled with hydrothermal factors. In order to improve the applicability in the grassland of southern China and the accuracy of the simulation results, adjustment coefficients related to temperature and precipitation were introduced into the model. Though the PCH model has not been applied to simulate the NPP of different types of vegetation, the establishment and application of the model were based on the specific spatial distribution of grassland and the complicated hydrothermal conditions in southern China. Each model has its advantages and limitations depending on different study targets and scales. The limitation of the model is that fewer influential factors were introduced into the model compared with other ecological models. The future analysis and explication of this will be carried out in the discussion part of this paper. The strength of the PCH model lies in the origin of the model establishment and the focalization and directness of assessing the NPP of grassland in southern China. The novelty of this model is mainly embodied in the process of hydrothermal assimilation in comparison to other models. Understanding controls over NPP will be crucial in developing models of these processes at larger spatial scales. Thus, the PCH model combines the hydrothermal parameter and ecosystem process approach to quantify the carbon flow of grassland in southern China (Gill et al., 2002).

3.3.2   Calculation of model parameters

The acquisition of model parameters was a complicated process and would directly affect the accuracy of the final results. Based on the measured data from 2009 to 2010, by adopting the contraction expansion algorithm of the nonlinear fitting and MATLAB programs (Conway and Wilcox, 1970), those parameters were calculated as $t_1 = 5.8$ and $w_1 = 560.4$.

3.3.3   Model validation

The measured grassland NPP data from 2014 in southern China were used to validate the simulation results. The results indicated that there was a strong and significant correlation between the simulated and measured NPP ($R^2 = 0.802$, $P < 0.01$). The RMSE of the simulation was $58.351\,\text{g C m}^{-2}$, the RRMSE was 0.326, and both were small. All these results indicate that the simulation of precipitation and temperature model for southern grassland NPP was feasible. The trends of the simulated and measured grassland NPP were similar (Fig. 3), which also indicated that the results were reliable.

3.4   Spatiotemporal variations of grassland NPP from the years 2003 to 2014

The spatial distribution map of grassland NPP produced by the estimation model was beneficial in monitoring the grassland resource. This paper built the spatial distribution map of southern grassland NPP using the estimation model of grassland NPP based on climatic conditions (Fig. 4). It showed that the minimum of mean annual NPP of southern grassland was $57.83\,\text{g C m}^{-2}$ and the maximum was $1328.06\,\text{g C m}^{-2}$ in the last 12 years. The NPP of southern grassland had an obvious zonal distribution. The NPP value was lower in northwestern regions and higher in southeastern and southern regions, especially in Jiangxi, Guangdong, and Hainan provinces.

The variation of mean annual NPP and the relevant statistical indices of southern grassland in the last 12 years were shown in Fig. 5. The trend of mean annual NPP presented an increasing tendency of the whole southern grassland from 2003 to 2014. The variation range of the mean annual NPP was from 430.31 to $519.82\,\text{g C m}^{-2}$, and the mean was $471.62\,\text{g C m}^{-2}$. The minimum of the mean annual NPP appeared in 2006, and the maximum value appeared in 2013. The tilt rate of the mean annual NPP of southern grassland in the last 12 years was $3.49\,\text{g C m}^{-2}\,\text{yr}^{-1}$, which indicated that the NPP increased about $3.49\,\text{g C m}^{-2}$ every year ($P < 0.05$).

4   Discussion

The parameters of the model were set to mediate the abnormal values from the model inputs and thus keep the stability of the model results. They were determined and constrained using multi-observation results. Hence, the model parameters are not associated with specific grassland types or the corresponding ecological characteristics. To incorporate remote sensing information into this model, we propose applying a remote sensing dataset as a spatially explicit scalar for the model parameterization, and thus enhancing the prediction ability of the future version of the model.

Research on the relationships between the NPP and climate factors in global or regional ecological systems started in the mid-1800s (Nemani et al., 2003; Zhou et al., 2014). As

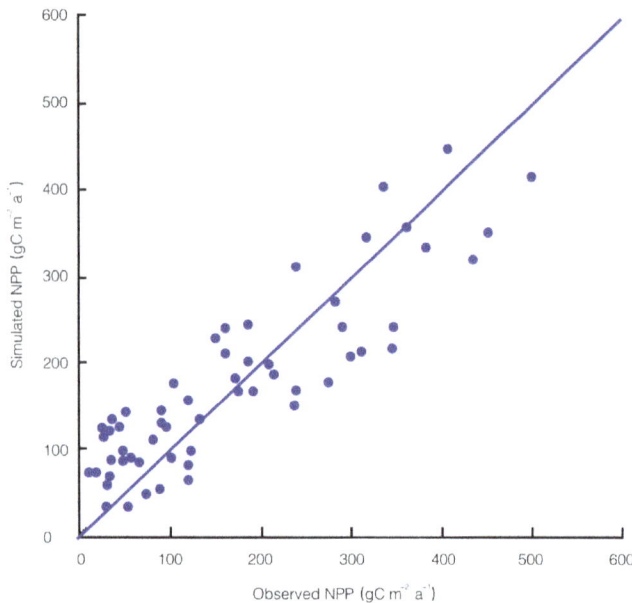

Figure 3. Comparison between simulated and observed grassland NPP (net primary productivity) in southern China.

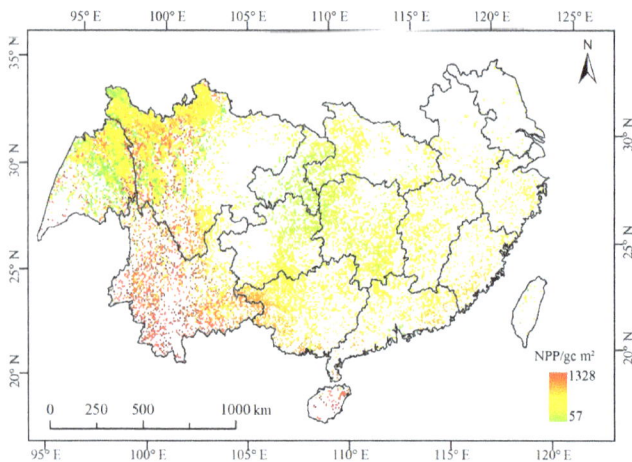

Figure 4. Spatial characteristics of grassland NPP in southern China from 2003 to 2014.

revealed in these studies, the vegetation index showed periodic variations with corresponding climate indices, including temperature and precipitation, during the growth process of most plants. Temporal and spatial variations were quite distinct in grassland NPP, since climatic factors, especially precipitation and temperature, were factors directly linked to periodic variations (Ronnenberg and Wesche, 2010). This study showed that a temperature rise would cause a certain level of rise in the grassland NPP in southern China, especially in the high-temperature zones. However, these results differed from some previous reports (Mcguire et al., 1993). In addition, there was a significant positive correlation between precipitation and NPP. When mean annual precipitation increases,

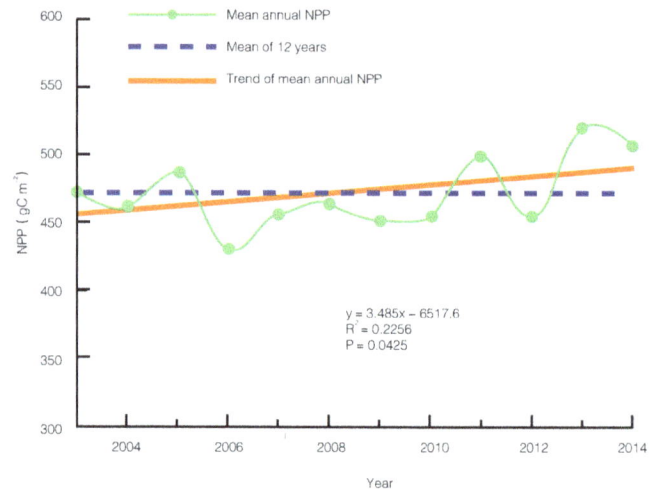

Figure 5. The interannual variation of grassland NPP in southern China from 2003 to 2014.

grassland NPP would also increase significantly. This conclusion is consistent with previous studies (Sala et al., 2000; Knapp and Smith, 2001; Mohamed et al., 2004).

The ultimate goal of those studies regarding the relationship between climate and terrestrial ecosystem NPP is to predict the possible impacts on climate change and to take scientific countermeasures (Pablo et al., 2007), and establishing a model is an efficient means to make these predictions. Through modeling and simulation, one could reveal the quantitative change and trend of NPP caused by climate change. That was why the research of the NPP model attracted a vast amount of attention (Ren et al., 2011). This study establishes an estimation model for the grassland NPP in southern China by using the statistical analysis of the relationship between the southern grassland NPP and precipitation and temperature, combined with biological process. The relationship between simulated and observed values reached a highly significant level. This and the low RMSE validated the reliability of the model. Therefore, it was feasible to estimate the grassland NPP in southern China by using the PCH model described in this paper.

The estimation of grassland NPP is a complex process. It is not only affected by climatic factors such as precipitation and temperature but also by the grassland vegetation's own inner physiological processes, fire severity, slope position and aspect, grazing, human activities, cutting frequency, and grassland ecotypes (Pereira et al., 2016; Shaw et al., 2016; Lu et al., 2015; Lin et al., 2015; Poeplau et al., 2016; Roosendaal et al., 2016).

Grassy hills and slopes in southern China had a wide distribution with various vegetation types; therefore, the NPP distribution was uneven. Although the model estimation worked well, some imperfections exist. Firstly, a classification for grass hills and slopes is needed, without which the NPP estimation fell into a single type (Hu et al., 2016). Secondly, the NPP estimation results were representative of the entire year, while arbitrary NPP estimation for a single month has not been verified yet. Thirdly, as an important ecological parameter, the MODIS normalized difference vegetation index needs to be added into the model (Gong et al., 2015). Then precision of the model could be improved in the process of evaluating the changes of grassland in southern China. Fourth, grassland soil coarseness needs to be taken into account as a result of nutrient cycling and respiration in grassland (Lü et al., 2016). The last issue concerns sensitivity. The study indicated that the simulation results from the PCH model were large in a small fraction of areas with relatively low NPP, while they were small in an area with high NPP. This may be caused by the limited time span and other factors, including the influences from different types of grasslands. Hence, there might be some uncertainty in estimating the lower or higher grassland NPP using the estimation model. Further study is required to solve these problems.

5 Conclusion

In this study, a new model using productivity coupled with hydrothermal factors (PCH) was built to simulate the NPP in southern China's grasslands. The PCH model uses productivity coupled with hydrothermal factors that can be expressed by the transformation of the model parameters, mean annual temperature and mean annual precipitation, which are the two most critical factors affecting the NPP of southern China's grasslands. The results show that there is a logarithmic correlation between grassland NPP and mean annual temperature, and there is a linear positive correlation between grassland NPP and mean annual precipitation in southern China. There was a very significant correlation between simulated and measured NPP ($R^2 = 0.8027$). Meanwhile, both RMSE and RRMSE stayed at a relatively low level, showing that the simulation results of the model were reliable. The NPP values in the study area had a decreasing trend from east to west and south to north. The mean NPP was $471.62\,\mathrm{g\,C\,m^{-2}}$ from 2003 to 2014. Additionally, the mean annual NPP of southern grassland presented a rising trend and the rate of change was $3.49\,\mathrm{g\,C\,m^{-2}\,yr^{-1}}$ in the last 12 years.

Competing interests. The authors declare that they have no conflict of interest.

Acknowledgements. We are grateful to the chief editor and anonymous reviewers for their illuminating comments. We would also like to thank Kenneth A. Albrecht (Department of Agronomy, University of Wisconsin–Madison, WI 53706, USA) for his helpful comments on the draft of this paper. This work was supported by the project of Natural Science Fund of Jiangsu Province (BK20140413) and the Key Project of the Chinese National Programs for Fundamental Research and Development (973 Program, 2010CB950702).

Edited by: A. Jordán

References

Braswell, B. H., Schimel, D. S., Linder, E., and Moore III., B.: The response of global terrestrial ecosystems to interannual temperature variability, Science, 278, 870–872, doi:10.1126/science.278.5339.870, 1997.

Brevik, E. C., Cerdà, A., Mataix-Solera, J., Pereg, L., Quinton, J. N., Six, J., and Van Oost, K.: The interdisciplinary nature of SOIL, SOIL, 1, 117–129, doi:10.5194/soil-1-117-2015, 2015.

Cao, L., Xu, J., Chen, Y., Li, W., Yang, Y., Hong, Y., and Li, Z.: Understanding the dynamic coupling between vegetation cover and climatic factors in a semiarid region-a case study of Inner Mongolia, China, Ecohydrology, 6, 917–926, doi:10.1002/eco.1245, 2013.

Cao, M. K. and Woodward, F. I.: Dynamic responses of terrestrial ecosystem carbon cycling to global climate change, Nature, 393, 249–252, doi:10.1038/30460, 1998.

Chen, Z. X. and Zhang, X. S.: Value of ecosystem services in China, Chinese Sci. Bull., 45, 17–22, doi:10.1007/BF02886190, 2000.

Christenson, L. M., Mitchell, M. J., Groffman, P. M., and Lovett, G. M.: Cascading effects of climate change on forest ecosystems: biogeochemical links between trees and moose in the northeast USA, Ecosystems, 3, 1–16, doi:10.1007/s10021-013-9733-5, 2014.

Conway, G. R. and Wilcox, J. C.: Fitting nonlinear models to biological data by Marquardt's algorithm, Ecology, 3, 503–507, doi:10.2307/1935386, 1970.

Douglas, G. G. and Geoffrey, M. H.: A technique for monitoring ecological disturbance in tall grass prairie using seasonal NDVI trajectories and a discriminate function mixture model, Remote Sens. Environ., 61, 270–278, doi:10.1016/S0034-4257(97)00043-6, 1997.

Fang, J. Y., Chen, A. P., Peng, C. H., Zhao, S. Q., and Ci, L. J.: Changes in forest biomass carbon storage in China between 1949 and 1998, Science, 292, 2320–2322, doi:10.1126/science.1058629, 2001.

Gang, C., Zhou, W., Wang, Z., Chen, Y., Li, J., Chen, J., Qi, J., Odeh, I., and Groisman, P. Y.: Comparative Assessment of Grassland NPP Dynamics in Response to Climate Change in China, North America, Europe and Australia from 1981 to 2010, J. Agron. Crop Sci., 1, 57–68, doi:10.1111/jac.12088, 2015.

Gill, R. A., Kelly, R. H., Parton, W. J., Day, K. A., Jackson, R. B., Morgan, J. A., Scurlock, J. M. O., Tieszen, L. L., Castle, J. V., Ojima, D. S., and Zhang, X. S.: Using simple environmental variables to estimate below-ground productivity in grasslands, Global Ecol. Biogeogr., 1, 79–86, doi:10.1046/j.1466-822X.2001.00267.x, 2002.

Gong, Z., Kawamura, K., Ishikawa, N., Goto, M., Wulan, T., Alateng, D., Yin, T., and Ito, Y.: MODIS normalized difference vegetation index (NDVI) and vegetation phenology dynamics in the Inner Mongolia grassland, Solid Earth, 6, 1185–1194, doi:10.5194/se-6-1185-2015, 2015.

Hall, D. O., Ojima, D. S., Parton, W. J., and Scurlock, J. M. O.: Response of Temperate and Tropical Grasslands to CO_2 and Climate Change, J. Biogeogr., 22, 537–547, doi:10.2307/2845952, 1995.

Hu, G., Liu, H., Yin, Y., and Song, Z.: The Role of Legumes in Plant Community Succession of Degraded Grasslands in Northern

China, Land Degrad. Dev., 27, 366–372, doi:10.1002/ldr.2382, 2016.

Huston, M. A.: Precipitation, soils, NPP, and biodiversity: resurrection of albrecht's curve, Ecol. Monogr., 3, 277–296, doi:10.1890/11-1927.1, 2012.

Keesstra, S. D., Bouma, J., Wallinga, J., Tittonell, P., Smith, P., Cerdà, A., Montanarella, L., Quinton, J. N., Pachepsky, Y., van der Putten, W. H., Bardgett, R. D., Moolenaar, S., Mol, G., Jansen, B., and Fresco, L. O.: The significance of soils and soil science towards realization of the United Nations Sustainable Development Goals, SOIL, 2, 111–128, doi:10.5194/soil-2-111-2016, 2016.

Knapp, A. K. and Smith, M. D.: Variation among biomes in temporal dynamics of aboveground primary production, Science, 291, 481–484, doi:10.1126/science.291.5503.481, 2001.

Lin, L., Li, Y. K., Xu, X. L., Zhang, F. W., Du, Y. G., Liu, S. L., Guo, X. W., and Cao, G. M.: Predicting parameters of degradation succession processes of Tibetan Kobresia grasslands, Solid Earth, 6, 1237–1246, doi:10.5194/se-6-1237-2015, 2015.

Lü, L., Wang, R., Liu, H., Yin, J., Xiao, J., Wang, Z., Zhao, Y., Yu, G., Han, X., and Jiang, Y.: Effect of soil coarseness on soil base cations and available micronutrients in a semi-arid sandy grassland, Solid Earth, 7, 549–556, doi:10.5194/se-7-549-2016, 2016.

Lu, X., Yan, Y., Sun, J., Zhang, X., Chen, Y., Wang, X., and Cheng, G.: Short-term grazing exclusion has no impact on soil properties and nutrients of degraded alpine grassland in Tibet, China, Solid Earth, 6, 1195–1205, doi:10.5194/se-6-1195-2015, 2015.

Mantgem, P. J. V. and Stephenson, N. L.: Apparent climatically induced increase of tree mortality rates in a temperate forest, Ecol. Lett., 10, 909–916, doi:10.1111/j.1461-0248.2007.01080.x, 2007.

Matsushita, B. and Tamura, M.: Integrating remotely sensed data with an ecosystem model to estimate net primary productivity in East Asia, Remote Sens. Environ., 81, 58–66, doi:10.1016/S0034-4257(01)00331-5, 2002.

Mcguire, A. D., Joyce, L. A., Kicklighter, D. W., Melillo, J. M., Esser, G., and Vorosmarty, C. J.: Productivity response of climax temperate forests to elevated temperature and carbon dioxide: a North American comparison between two global models, Climatic Change, 4, 287–310, doi:10.1007/BF01091852, 1993.

Mohamed, M. A., Babiker, I. S., Chen, Z. M., Ikeda, K., Ohta, K., and Kato, K.: The role of climate variability in the inter-annual variation of terrestrial net primary production (NPP), Sci. Total Environ., 332, 123–137, doi:10.1016/j.scitotenv.2004.03.009, 2004.

Mosier, A. R., Schimel, D., Valentine, D., Bronson, K., and Parton, W.: Methane and nitrous oxide fluxes in native, fertilized and cultivated grasslands, Nature, 350, 330–332, doi:10.1038/350330a0, 1991.

Nemani, R. R., Keeling, C. D., Hashimoto, H., Jolly, W. M., Piper, S. C., Tucker, C. J., Myneni, R. B., and Running, S. W.: Climate-driven increases in global terrestrial net primary production from 1982 to 1999, Science, 300, 1560–1563, doi:10.1126/science.1082750, 2003.

Ni, J.: Effects of climate change on carbon storage in boreal forests of china: a local perspective, Climatic Change, 1–2, 61–75, doi:10.1023/A:1020291220673, 2002.

Pablo, M., Thomas, H., David, P. R., Benjamin, S., and Martin, T. S.: Changes in European ecosystem productivity and carbon balance driven by regional climate model output, Glob. Change Biol., 1, 108–122, doi:10.1111/j.1365-2486.2006.01289.x, 2007.

Parton, W. J., Scurlock, J. M. O., Ojima, D. S., Schimel, D. S., and Hall, D. O.: Impact of climate change on grassland production and soil carbon worldwide, Glob. Change Biol., 1, 13–22, doi:10.1111/j.1365-2486.1995.tb00002.x, 1995.

Pereira, P., Cerdà, A., Lopez, A. J., Zavala, L. M., Mataix-Solera, J., Arcenegui, V., Misiune, I., Keesstra, S., and Novara, A.: Short-term vegetation recovery after a grassland fire in Lithuania: the effects of fire severity, slope position and aspect, Land Degrad. Dev., 27, 1523–1534, doi:10.1002/ldr.2498, 2016.

Picard, G., Quegan, S. N., Lomas, M. R., Toan, T. L., and Woodward, F. I.: Bud-burst modelling in Siberia and its impact on quantifying the carbon budget, Glob. Change Biol., 12, 2164–2176, doi:10.1111/j.1365-2486.2005.01055.x, 2005.

Poeplau, C., Marstorp, H., Thored, K., and Kätterer, T.: Effect of grassland cutting frequency on soil carbon storage – a case study on public lawns in three Swedish cities, SOIL, 2, 175–184, doi:10.5194/soil-2-175-2016, 2016.

Raich, J. W., Rastetter, E. B., Melillo, J. M., Kicklighter, D. W., and Steudler, P. B. J.: Potential net primary productivity in South America: application of a global model, Ecol. Appl., 4, 399–429, doi:10.2307/1941899, 1991.

Ren, W., Tian, H., Tao, B., Chappelka, A., Sun, G., Lu, C., Liu, M., Chen, G., and Xu, X.: Impacts of tropospheric ozone and climate change on net primary productivity and net carbon exchange of china's forest ecosystems, Global Ecol. Biogeogr., 3, 391–406, doi:10.1111/j.1466-8238.2010.00606.x, 2011.

Richardson, A. D., Anderson, R. S., Arain, M. A., Barr, A. G., Bohrer, G., and Chen, G.: Terrestrial biosphere models need better representation of vegetation phenology: results from the north American carbon program site synthesis, Glob. Change Biol., 2, 566–584, doi:10.1111/j.1365-2486.2011.02562.x, 2012.

Ronnenberg, K. and Wesche, K.: Effects of fertilization and irrigation on productivity, plant nutrient contents and soil nutrients in Southern Mongolia, Plant Soil, 1–2, 239–251, doi:10.1007/s11104-010-0409-z, 2010.

Roosendaal, D., Stewart, C. E., Denef, K., Follett, R. F., Pruessner, E., Comas, L. H., Varvel, G. E., Saathoff, A., Palmer, N., Sarath, G., Jin, V. L., Schmer, M., and Soundararajan, M.: Switchgrass ecotypes alter microbial contribution to deep-soil C, SOIL, 2, 185–197, doi:10.5194/soil-2-185-2016, 2016.

Sala, O. E., Chapin III., F. S., Armesto, J. J., Berlow, E., Bloomfield, J., Dirzo, R., Huber-Sanwald, E., Huenneke, L. F., Jackson, R. B., Kinzig, A., Leemans, R., Lodge, D. M., Mooney, H. A., Oesterheld, M., Poff, N. L., Syke, M. T., Walker, B. H., Walker, M., and Wall, D. H.: Global biodiversity scenarios for the year 2100, Science, 287, 1770–1774, doi:10.1126/science.287.5459.1770, 2000.

Shaw, E. A., Denef, K., Milano de Tomasel, C., Cotrufo, M. F., and Wall, D. H.: Fire affects root decomposition, soil food web structure, and carbon flow in tallgrass prairie, SOIL, 2, 199–210, doi:10.5194/soil-2-199-2016, 2016.

Sun, Z. G., Long, X. H., Sun, C. M., Zhou, W., Ju, W. M., and Li, J. L.: Evaluation of net primary productivity and its spatial and temporal patterns in Southern China's grasslands, Rangeland J., 3, 331–338, doi:10.1071/RJ12061, 2013.

Sun, Z. G., Sun, C. M., Zhou, W., Ju, W. M., and Li, J. L.: Classification and Net Primary Productivity of the Southern China's Grasslands Ecosystem Based on Improved Comprehensive and Sequential Classification System (CSCS) Approach, J. Integr. Agr., 4, 893–903, doi:10.1016/S2095-3119(13)60415-3, 2014.

Wunder, J., Fowler, A. M., Cook, E. R., Pirie, M., and Mccloskey, S. P. J.: On the influence of tree size on the climate–growth relationship of New Zealand Kauri (Agathis australis): insights from annual, monthly and daily growth patterns, Trees, 4, 937–948, doi:10.1007/s00468-013-0846-4, 2013.

Xu, X., Niu, S. L., Sherry, R. A., Zhou, X. H., Zhou, J. Z., and Luo, Y. Q.: Interannual variability in responses of belowground net primary productivity (NPP) and NPP partitioning to long-term warming and clipping in a tall grass prairie, Glob. Change Biol., 18, 1648–1656, doi:10.1111/j.1365-2486.2012.02651.x, 2012.

Yu, D., Zhu, W., and Pan, Y.: The role of atmospheric circulation system playing in coupling relationship between spring NPP and precipitation in East Asia area, Environ. Monit. Assess., 1–3, 135–143, doi:10.1007/s10661-007-0023-6, 2008.

Zhang, G. G., Kang, Y. M., Han, G. D., and Sakurai, K.: Effect of climate change over the past half century on the distribution, extent and NPP of ecosystems of Inner Mongolia, Glob. Change Biol., 17, 377–389, doi:10.1111/j.1365-2486.2010.02237.x, 2011.

Zhou, W., Gang, C., Zhou, L., Chen, Y., Li, J., Ju, W., and Odeh, I.: Dynamic of grassland vegetation degradation and its quantitative assessment in the northwest China, Acta Oecol., 55, 86–96, doi:10.1016/j.actao.2013.12.006, 2014.

9

Electric resistivity and seismic refraction tomography: a challenging joint underwater survey at Äspö Hard Rock Laboratory

Mathias Ronczka[1], **Kristofer Hellman**[1], **Thomas Günther**[2], **Roger Wisén**[1], and **Torleif Dahlin**[1]

[1]Engineering Geology, Lund University, Lund, Sweden
[2]Leibniz Institute for Applied Geophysics, Hanover, Germany

Correspondence to: Mathias Ronczka (mathias.ronczka@tg.lth.se)

Abstract. Tunnelling below water passages is a challenging task in terms of planning, pre-investigation and construction. Fracture zones in the underlying bedrock lead to low rock quality and thus reduced stability. For natural reasons, they tend to be more frequent at water passages. Ground investigations that provide information on the subsurface are necessary prior to the construction phase, but these can be logistically difficult. Geophysics can help close the gaps between local point information by producing subsurface images. An approach that combines seismic refraction tomography and electrical resistivity tomography has been tested at the Äspö Hard Rock Laboratory (HRL). The aim was to detect fracture zones in a well-known but logistically challenging area from a measuring perspective.

The presented surveys cover a water passage along part of a tunnel that connects surface facilities with an underground test laboratory. The tunnel is approximately 100 m below and 20 m east of the survey line and gives evidence for one major and several minor fracture zones. The geological and general test site conditions, e.g. with strong power line noise from the nearby nuclear power plant, are challenging for geophysical measurements. Co-located positions for seismic and ERT sensors and source positions are used on the 450 m underwater section of the 700 m profile. Because of a large transition zone that appeared in the ERT result and the missing coverage of the seismic data, fracture zones at the southern and northern parts of the underwater passage cannot be detected by separated inversion. Synthetic studies show that significant three-dimensional (3-D) artefacts occur in the ERT model that even exceed the positioning errors of underwater electrodes. The model coverage is closely connected to the resolution and can be used to display the model

uncertainty by introducing thresholds to fade-out regions of medium and low resolution. A structural coupling cooperative inversion approach is able to image the northern fracture zone successfully. In addition, previously unknown sedimentary deposits with a significantly large thickness are detected in the otherwise unusually well-documented geological environment. The results significantly improve the imaging of some geologic features, which would have been undetected or misinterpreted otherwise, and combines the images by means of cluster analysis into a conceptual subsurface model.

1 Introduction

Underground structures have become an increasingly important part of modern infrastructure, and the possibilities to improve construction approaches have attracted much attention. With constantly reduced space for new structures on the surface, underground space is attractive for use in the transportation sector to challenge the increase in traffic in and around cities or as underground storage facilities. Geological uncertainties increase the risk of delays and thus the costs of underground construction. A detailed subsurface model is essential for reducing risks and for a successful project. In order to ensure a smooth construction phase, a critical point is to locate weak zones, especially those that can generate a large inflow of water, causing problems and slowing down the construction progress. Except for southwestern Scania and the islands Gotland and Öland, crystalline bedrock is the dominant material for underground infrastructure construction in Sweden. For these geologic conditions, weakness zones that are important for the underground design are normally indicated by dry, water-bearing or sediment-filled fractures.

Two methods for site investigation in crystalline bedrock are drilling and surface-based or borehole geophysics. Drilling is often the first choice since it provides high resolution and accuracy at any given depth. Nevertheless, drilling is expensive and delivers only point information. Therefore, surface-based geophysical methods have gained more attention, since they provide continuous models that reveal the extreme points and an opportunity for extrapolation into 2-D or 3-D space. The usage of geophysics has increased lately to obtain more continuous and comprehensive subsurface models. Recently, the Swedish transportation authority has provided funding for research in an increasing number of projects with the aim of developing site investigations based on additional geophysical measurements for mapping the structure and quality of the rock mass.

Dahlin et al. (1999) report a case in which electrical resistivity tomography (ERT) has been used successfully to map weak and permeable rock in an onshore railway tunnel project in Sweden. Ha et al. (2010) used different geoelectrical applications to detect weak zones of approx. 40 m × 40 m during underground construction. In the Norwegian R & D project "Tunnels for the citizens", which was funded by the road administration, several publications (Karlsrud et al., 2003; Palmstrøm et al., 2003; Rønning et al., 2013; Wisén et al., 2012; Lindstrøm and Kveen, 2004) report that elaborate site investigations are important in a controlled tunnelling process, but also that further studies are needed. Rønning et al. (2013) assessed ERT, refraction seismics, very low frequency (VLF) electromagnetics and the AMAGER method (aeromagnetic and geomorphological relations). They concluded that these are all able to locate fracture zones and state that ERT is able to give more hints to the fracture width, dip and depth extent compared to the other methods used. They also suggest a quantitative rock quality measure on the basis of resistivity values. Refraction seismics has long been an established method for information on fracture width and seismic p-wave velocity; the latter has an obvious coupling to the hardness of the rock and hence to rock quality (Bergman et al., 2006). During an investigation for a road tunnel in Norway, Ganerød et al. (2006) found that 2-D resistivity and refraction seismics are the most suitable geophysical methods, while ERT gave detailed results at lower costs compared to seismics. Diaz et al. (2014) successfully conducted seismic refraction and ERT surveys and associated resistivity and velocity changes with the main and secondary structures of a major fault zone. The final velocity and resistivity models were also consistent with deformed sedimentary units. Another multidisciplinary geophysical approach for mapping a fault zone is given by Malehmir et al. (2016). Heincke et al. (2010) used seismic and electric tomography to assess the rock quality on a hard rock slope in Norway.

Several methods are generally combined to overcome the limits of the natural resolution and corresponding ambiguity in inversion and interpretation. One example for synthetic and field data is given by Garofalo et al. (2015). Seismic data and ERT were used to reduce model ambiguities to improve the estimation of the geophysical parameters. With the joint inversion approach, data sets from different methods are used to constrain each other. The general assumption is that subsurface structures lead to parameter changes for the different methods, i.e. imaging the same underlying geologic conditions. Smoothness constraints often prevent the correct mapping of sharp interfaces. Different joint inversion approaches exist. The cross-gradient method for ERT and seismic data is explained in Gallardo and Meju (2004). Although a significant improvement in the results compared to separated 2-D inversions was observed, it can only be applied on regular grids, which makes an accurate incorporation of topography difficult. Juhojuntii and Kamm (2015) present a joint inversion algorithm using seismic refraction and ERT data, assuming a fixed number of layers. The derived subsurface models agreed well with independent in situ tests, but the authors also stated that their approach would lead to misinterpretations in environments with smooth subsurface variations.

This paper describes a representative case study for the combination of geoelectric and refraction seismics in typical Scandinavian geologic conditions at a coastal region. The survey was conducted at Äspö Hard Rock laboratory (HRL) and designed to perform a joint inversion on the data. The main objective was the localisation and characterisation of fracture zones under challenging conditions, which are the extreme variation in electrode coupling, possible 3-D effects on ERT data and high acoustic damping due to gasbearing sediments. Dahlin and Wisén (2016) and Günther and Südekum (2007) showed that underwater field surveys are possible and quite promising. Loke and Lane (2004) demonstrated that the water layer has a large effect on apparent resistivities, but subsurface resistivity can be recovered if water resistivity and seabed topography are properly incorporated in the finite-element meshes. The methodical approach of a structurally coupled joint inversion presented in this study shows how results can be improved such that an easier and more unique interpretation of the underground models is possible. In order to increase the reliability of the results, a combined inversion and interpretation was investigated. This was done by joint inversion followed by a cluster analysis as an additional integrated interpretation approach. After describing the site conditions and the numerical background, we show a synthetic study on the 3-D effects and the influence of the seabed topography on ERT data before the analysis and interpretation of the field data is presented.

2 Site description

The Swedish Nuclear Fuel and Waste Management Company (Svensk Kärnbränslehantering AB; SKB) started to design a solution for the deep final disposal of nuclear fuel. Äspö HRL is SKB's underground facility for research that tests concepts for the final disposal of nuclear waste material in hard rock (Rhén et al., 1997). The laboratory has provided a full-scale test environment for different technological solutions. It has now mainly fulfilled its purpose so that the laboratory has also become available for other branches of research. The facility provides a research opportunity in a well-documented and relatively undisturbed geological environment that is representative of many Swedish metropolitan areas.

The Äspö Hard Rock Laboratory is located on the east coast of the Baltic Sea, about 400 km south of Stockholm (see Fig. 1). From 1990 to 1995, the excavation of a 3600 m tunnel that connects the nuclear power plant with the disposal at approximately 450 m of depth was conducted. During the construction phase, a detailed site characterisation was carried out that included geological, hydrogeological and geochemical investigations.

The Äspö bedrock is part of the Transscandinavian Igneous Belt (TIB) that extends from southern Sweden toward the north and northwest. Generally, granitoids and volcanic rocks can be found in the TIB. Four rock types are dominant: the Äspö diorites, Ävrö granite, greenstone and fine-grained granite. Wikberg et al. (1991) found that continuous magma mixing processes supported the development of dikes and mafic inclusions, which form an inhomogeneous rock mass. The crystalline bedrock exhibits porosities of 0.4–0.45 % for the Äspö diorite and 0.23–0.27 % for the fine-grained granite (Stanfors et al., 1999). During the preinvestigation of Äspö HRL, fracture zones were divided into major (width > 5 m) and minor (width < 5 m) categories. The majority of the fractures are oriented northwest–southeast (Berglund et al., 2003). All fracture zones that are important for this field survey are depicted as black lines in Fig. 1.

The filling material of the fractures was extracted from drill cores and analysed. Missing unconsolidated material that might have been additionally filling the fractures was probably washed away and thus not taken into account in these analyses. The crystallised calcite in the fractures was possibly formed by hydrothermal processes and can be used as an indicator for water paths in the rock (Wikberg et al., 1991). This indicated that fractures in the N–S and E–W directions most likely conduct or formerly conducted water. According to Wikberg et al. (1991) all fracture zones are at least partly water bearing. They also gave a judgement of the fracture zones according to Bäckblom et al. (1990). Based on that, the most critical fracture zone along the measured profile is NE-1, which is judged as "certain". EW-3 is also judged "certain", but hydraulically of minor importance. NE-3 and NE-4 are judged as "certain" as well. Both consist of

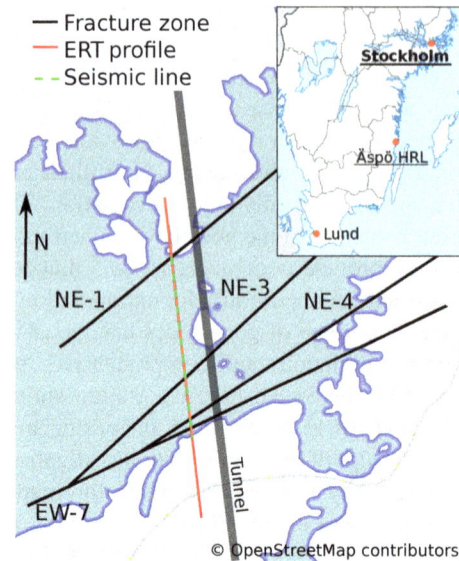

Figure 1. The location, major fracture zones (black lines) after Stanfors et al. (1999), the scheduled ERT profile (solid red line) and the seismic (dashed green) line at Äspö Hard Rock Laboratory.

several subzones that are one to a few metres wide, some of which are open fractures that are hydraulically highly conductive. In general, the fracture zones NE-3, NE-4 and EW-7 are judged to be "probable" in a hydraulic sense (Wikberg et al., 1991). The authors also stated that the Quaternary sediments on top of the bedrock were supposed to be scarce at the Äspö test site. Due to the deep target of the Äspö HRL within the bedrock, no detailed investigation of the Quaternary sediments was carried out. Vidstrand (2003) stated that the unconsolidated overburden should rarely exceed 5 m in thickness and consists mainly of clay, sand and gravel.

2.1 Electrical resistivity tomography

ERT measurements were carried out along a profile in the N–S direction simultaneously with the seismic survey on 20–24 April 2015. The profile lies between Hålö and Äspö (see Fig. 1) to the west of the tunnel line, about 10 m away from a small island. Electrodes were placed onshore and underwater with a 5 m electrode spacing along a 780 m profile. Data were recorded using the multichannel instrument ABEM Terrameter LS (Guideline Geo, Sundbyberg, Sweden). A multiple-gradient array (Dahlin and Zhou, 2006) was employed to ensure fast measuring progress as it can fully exploit the recording channels. The resistivity of the water was measured with a micro Wenner alpha array at different depths with the ABEM Terrameter LS. The collected ERT data were first published in Fennvik (2015). A model based on accurate bathymetry measurements was used to determine the heights of the sensor positions at the seabed. A nearby power plant caused a high noise level in the ERT data. Large variations in the contact impedance between the water and the rock outcrops created a technically difficult measuring situation. Contact resistances, including cable resistance, started from $100\,\Omega$ for electrodes in brackish water and exceeded $100\,k\Omega$ on rock outcrops. An over-amplification of the signals was avoided due to an automated gain control of the instrument. Furthermore, the input channels are galvanically separated; i.e. one channel can have a high gain and the next channel a low gain, avoiding any problems. The full wave form of the transmitted and received signals was recorded in order to recover possibly valuable IP signals from the data. However, the signal-to-noise ratio was sufficiently good for recovering DC resistivity but not IP data. About 6700 data points were gathered during the ERT survey. While the raw data were processed, combinations with uncoupled electrodes were identified and all combinations containing these electrodes were deleted. To account for the variable data quality of the individual data, usually a data error is estimated by a fixed percentage and a voltage error. They can be retrieved by analysing reciprocal measurements (Udphuay et al., 2011), which were not available here. Therefore, we used the default values of 3 % noise and a voltage error of 0.1 mV.

2.2 Seismic refraction tomography

The green dashed line in Fig. 1 marks the profile for the seismic refraction. Hydrophone streamers were laid out with 91 hydrophones in total and a 5 m spacing along a 450 m profile line. For data acquisition, the instruments ABEM Terraloc (Guideline Geo) and Geometrics Stratavisor (San Jose, CA, USA) were used, both with 48 channels and a 5-channel overlap of the two streamers. Hydrophone positions were determined by a differential GNSS, while the topography of the seabed was mapped with a multibeam echo sounder (MBES). For all underwater sensors (electrodes and hydrophones), a very accurate DTM (digital terrain model) from the MBES survey was used for the heights. The positions of the sensors (E/S) are coincident and measured with sufficient accuracy. For the excitation of seismic p-waves, small explosives were placed approximately 0.5 m above the seabed. Shots were performed every 20 m. Data were first processed and published in Lasheras Maas (2015). Due to time constraints, not all planned shots were fired, and hence there are two small gaps in the data coverage in the northern part of the data set. Raw data processing revealed that the seismic signal quality was significantly reduced in the southern part of the profile, which made it difficult to pick first arrivals. However, no additional filters were used during the raw data processing. About 650 first-arrival times were semi-automatically picked and manually checked using the software package Rayfract (http://www.rayfract.com).

3 Numerical modelling and inversion

We used the open-source ERT software package BERT (Boundless Electrical Resistivity Tomography) for ERT inversion (Günther et al., 2006b) using irregular triangle meshes to accurately take into account both the surface and submarine topography (Rücker et al., 2006). Furthermore, we used the underlying framework pyGIMLi (Python Geophysical Inversion and Modelling Library; http://www.pygimli.org) for the refraction tomography and the implementation of the coupled inversion.

3.1 Inversion

Geophysical inversion describes the process of estimating a model with a forward response that fits the observed data. The linearised problem for ERT is given in Eq. (1a) and for seismic in Eq. (1b). Here, the model parameters are either the logarithmic resistivities or the velocity/slowness held in the model vector $\boldsymbol{m}$:

$$\mathbf{J}\Delta\boldsymbol{m} = \boldsymbol{d} - f(\boldsymbol{m}), \tag{1a}$$

$$\mathbf{A}\Delta\boldsymbol{m} = \Delta\boldsymbol{t}. \tag{1b}$$

The Jacobian matrices $\mathbf{J}$ and $\mathbf{A}$ contain the partial derivatives $\partial\rho_{a,i}/\partial m_j$ (ERT) or $\partial t_i/\partial m_j$. Apparent resistivities (ρ_a) are held in $\boldsymbol{d}$ and travel times in $\boldsymbol{t}$. The inversion of ERT and SRT (seismic refraction tomography) was done by a smoothness-constrained minimisation using the cost function

$$\Phi = \Phi_d + \lambda\Phi_m, \tag{2a}$$

$$= \sum_{i=1}^{N}\left(\frac{d_i - f_i(\boldsymbol{m})}{\epsilon_i}\right)^2 + \lambda \parallel \boldsymbol{Cm}\parallel_2^2, \tag{2b}$$

containing an error-weighted data misfit Φ_d and a model roughness Φ_m weighted by the regularisation parameter λ. As the travel time t between source and receiver along a ray path is given by $t = \sum_{i=1}^{n} l_i/v_i$, it is a linear combination of the path length l_i and the slowness $1/v_i$ for a segment i. The difference between the individual data points d_i and the corresponding forward responses $f_i(\boldsymbol{m})$, both as logarithmic apparent resistivities or travel times, is weighted by their individual errors ϵ_i.

The roughness (second term in Eq. 2b) consists of the derivative matrix $\mathbf{C}$ applied to the model $\boldsymbol{m}$ (Günther et al., 2006b), whereas each row in $\mathbf{C}$ is associated with a boundary. Additional model constraints can be incorporated in the object function by extending Φ_m from Eq. (2b) with a weighted model functional $\mathbf{W}_c$ (Rücker, 2011), resulting in

$$\Phi_m = \parallel \mathbf{W}_c\boldsymbol{Cm}\parallel_2^2. \tag{3}$$

The weighting matrix $\mathbf{W}_c$ is diagonal and contains the elements w_i, representing penalty factors for the different model cell boundaries (Günther et al., 2006a). Very small values can

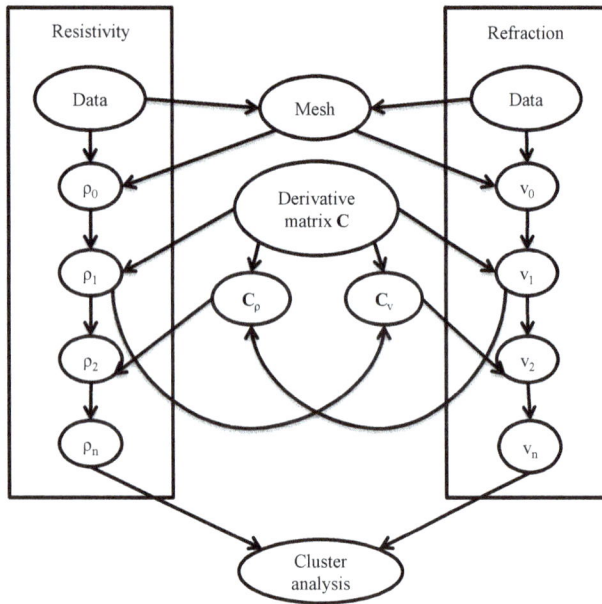

Figure 2. Scheme of the coupled inversion approach, where the roughness **C** of one inversion is influenced by the other (Günther et al., 2006a).

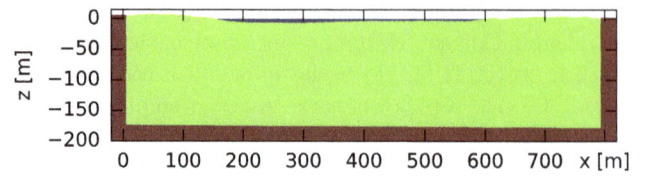

Figure 3. Cropped mesh used for the structurally coupled inversion of ERT and seismic data. Three regions are used: (i) the background (in red; much bigger) to prevent influences of the boundaries, (ii) the parameter domain (green) on which the inversion is done and (iii) the water region (blue), which was fixed.

lead to sharper boundaries. The limited amount and quality of recorded data leads to a non-unique inversion result. Due to the model smoothing needed for mixed determined problems, it is possible that sharp boundaries appear as transition zones that lead to misinterpretations. A structurally coupled joint inversion finds common structures and allows the models to emphasise these and reduce smoothing effects (Gallardo and Meju, 2004). Here, the roughness vector $r = \mathbf{C}\mathbf{W}_m m$ is used to calculate the mutual penalty factors w_i using Eq. (4) after Günther et al. (2010):

$$w_i = \left(\frac{a}{|r_i| + a} + b \right)^c. \tag{4}$$

The parameters a, b and c are used to adjust the coupling strength and the influence of the gradients. Differently from the latter approach, we multiply the w_i of the different methods and calculate one weighting matrix for both methods.

A certain number of separated iterations is done before the coupling starts so that each method can first independently develop structures before their similarity is promoted. A schematic sketch of the structurally coupled joint inversion is shown in Fig. 2.

Forward modelling and inversion are done accurately on unstructured finite-element (FE) meshes that allow for the incorporation of both the surface and the underwater topography. The finite-element mesh used for the joint inversion of seismic and ERT data is shown in Fig. 3.

The shown mesh consists of three regions that present the background (red), the water (blue) and the parameter domain (green) on which the data inversion is conducted. The orig-

inal mesh extension is 1250 m in the x and approximately 420 m in the z direction and is clipped for display reasons. In situ water conductivity measurements showed resistivity values of about $1.4 \, \Omega$m and negligible variation with position or depth. As the seismic velocity of water is constant (about $1400 \, \mathrm{m \, s^{-1}}$), the water region can be assumed as homogeneous and is incorporated as a single region with a fixed resistivity or velocity so that the correct values are used for the forward calculation but are not subject to inversion. The parameter domain is extended to approximately 790 m in the x and 190 m in the z direction. Additionally, an outer background region is needed for accurate forward calculation using approximate boundary conditions (Rücker et al., 2006). Although the seismic line is shorter than the ERT, the shown mesh was used for both data sets in the joint inversion. The parameter domain consists of about 3500 cells, which is the number of model parameters. More details on region-based inversion can be found in Rücker (2011).

3.2 Model appraisal and display

All shown inversion results are faded out using the coverage to point out the contribution of the model parts to the data. The calculation of the coverage is based on the sensitivity, which is the partial derivative $S_{i,j}(\mathbf{m}^n) = \frac{\partial f_i(\mathbf{m}^n)}{\partial m_j}$. Whereas $m = \log \rho$ are the model parameters and $f = \log \rho_a$ is the forward response (both logarithmically transformed) for ERT, the seismic model parameters are $m = 1/v$ (slowness) with the corresponding forward response $f = t$ (travel times). The summation of all sensitivities for each model parameter gives the coverage for the model cell assigned with this parameter. Unlike ERT, a normalised coverage is calculated for seismics, which is either 1 or 0 depending on whether a ray crosses a cell or not. Additionally, resolution radii after Friedel (2003) were calculated for the purpose of a comparison with the coverage. For that the model resolution matrix $\mathbf{R}^M$ is required, which can be calculated by a singular value decomposition (SVD) of the Jacobian matrix of the final model, i.e. the resistivity or velocity/slowness distribution. The radius $r_j = \sqrt{A_{\mathrm{cell}}/(R^M_{j,j}\pi)}$ is calculated for each model cell j and is an equivalent of the cell area A_{cell} to a

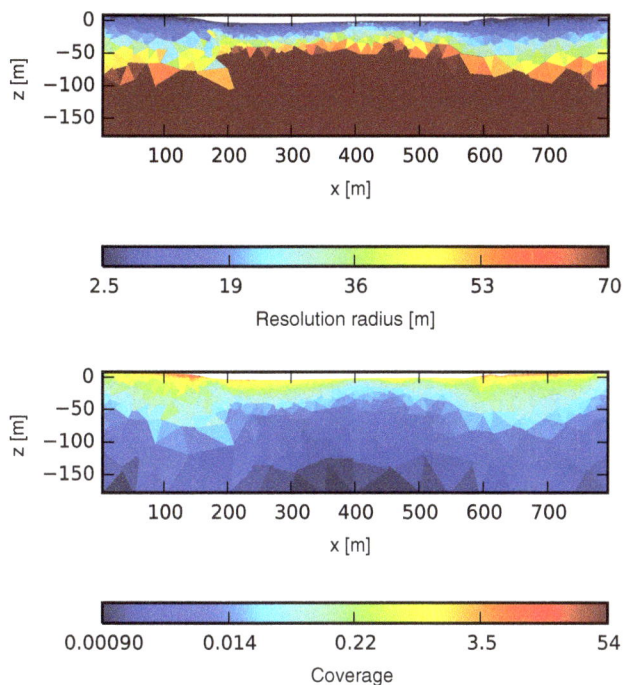

Figure 4. Calculated model resolution radii (**a**) and coverage (**b**) for ERT using the Jacobian matrix of the final model.

perfectly resolved sphere. The coverage and resolution radii distributions are shown in Fig. 4 for ERT.

In general, low-resolution radii correspond with a high coverage and vice versa. The water and the low-resistive sediments between $x = 180$ m and $x = 550$ m lead to a reduced investigation depth with high-resolution radii starting at approx. 50 m of depth. Compared to that, the onshore model parts at $x < 180$ m and $x > 550$ m show a medium reliability up to 70 m of depth with resolution radii < 50 m. Figure 4 clearly shows the existing relationship between the coverage and the model resolution radii. Thus, the coverage can also be used as a resolution measure to display well-, medium- and poorly resolved model regions. In the following, two coverage thresholds were used to define regions of high, medium and low certainty. The low-certain region is completely blanked out and considered as untrustworthy, while the region of high certainty is imaged without shading and can thus be judged as trustworthy.

3.3 Synthetic study on 3-D effects and seabed topography

We follow a strict two-dimensional (2-D) scheme, i.e. assuming constant values perpendicular to the profile. For the given test site, it can be assumed that the seismic refraction data are not or only minimal corrupted by 3-D effects. By picking first arrivals, only signals that took the shortest way or travelled in the fastest medium are taken into account. The small island next to the profile consists of the

same bedrock as that directly in the profile line. Assuming the same velocity, the recorded first arrival is still from the signal travelling in the profile line, because it is the shortest way. The small bay (water body) north of the profile would be a low-velocity anomaly because the velocity in water is lower than in bedrock. Therefore, the bay can be ignored because a refraction only appears for an increasing velocity. Three-dimensional (3-D) effects occur if significant resistivity changes perpendicular to a 2-D profile are present. According to the test site map in Fig. 1, severe 3-D effects can be expected near the small island in the middle of the profile and in the northern part, where the water continues just a few metres next to the profile. The latter is not expected to have a significant effect on the first-arrival times, since these are related to the smallest distance to the layers. It will, however, have an effect on the measured apparent resistivity by all materials present within the measured volume. In order to appraise the expected shapes and magnitudes of 3-D effects, we generated a simplified model based on the Äspö geometry. The underlying model used for generating synthetic data is shown in Fig. 5. The water body is simulated by a cube with an extension of 450 m in the x direction starting at $x = 100$ m, being 10 m in depth and infinite in the y direction. A large cube simulating the bedrock (brown) surrounds the water cube, with an infinite extension in the x, y and z direction. The water (blue) is assigned with a resistivity of 3 Ωm, while the bedrock is assigned with 3000 Ωm. Two anomalies are inserted representing the island in the middle and the small bay at the northern end of the ERT profile. The island (red) is a 10 m thick cube, with an extension of 90 m in the x and 70 m in the y direction, placed between $x = 370$ m and $x = 460$ m with a distance of 10 m to the ERT profile. The small bay at the northern part (green) is incorporated by a rectangular cube with an edge length of 100 m (x, y direction) and 3 m of depth. It starts directly after the water cube at $x = 550$ m with a distance of 5 m to the profile. The ERT line consists of 153 electrodes, starts at $x = 15$ m and $y = 0$ m and is aligned along the x direction. The simulated survey is identical to the field measurements except that the electrodes are assumed to be at the surface and topography is neglected. The ERT profile is marked with red spheres in Fig. 5.

For reference, we additionally calculated data from a 2-D model, where the island is assigned with a water resistivity of 3 Ωm and the bay with 3000 Ωm (bedrock), i.e. with no 3-D effects. Both data sets were corrupted with Gaussian noise with an error level consisting of 3 % plus a voltage error of 100 μV. A smoothness-constrained inversion was performed to estimate resistivity models from the two synthetic data sets. Figure 6a and b show the inversion results from the data set with and without 3-D effects. The ratio between those two is shown in Fig. 6c.

Figure 6a shows the expected smooth resistivity distribution with a horizontal interface between the simulated bedrock and water. When the island and the small bay are included in the underlying model, serious 3-D effects occur.

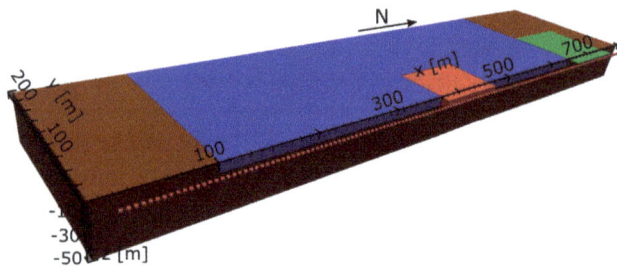

Figure 5. Sketch of the synthetic model used to generate synthetic data. It reflects a simplified version of the Äspö test site conditions. The red spheres mark electrode positions, the blue coloured areas simulate a low-resistive body, like sea water, and the brown parts mark highly resistive bodies, like bedrock.

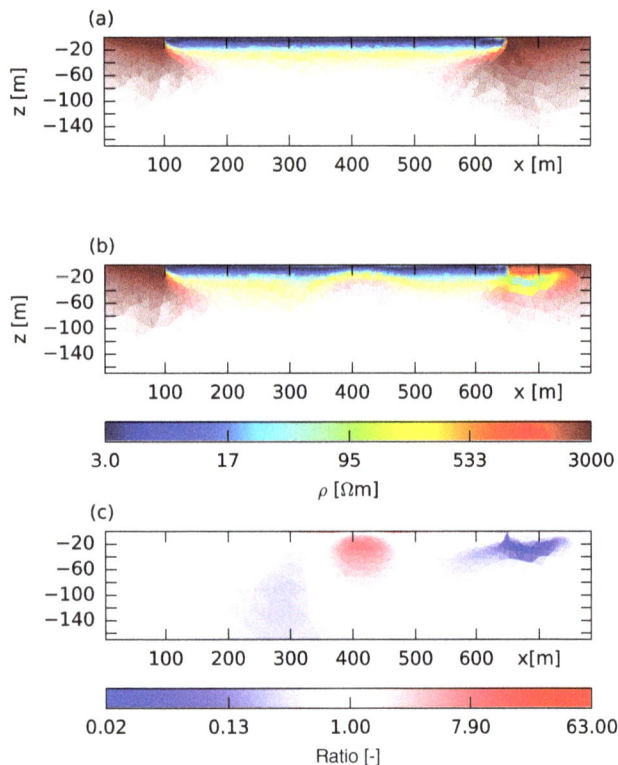

Figure 6. Inversion results of the synthetic case with **(a)** a pure 2-D model, **(b)** the incorporated island and the small bay causing 3-D effects and **(c)** the ratio between **(a)** and **(b)**.

These lead to higher resistivities in the middle of the profile where the island was included with additional low-resistive compensation artefacts next to it. The small water-filled bay at the end of the profile leads to a characteristic low-resistive feature at intermediate depths. Both anomalies, including the possible compensation artefacts, are more visible in the ratio plot given in Fig. 6c.

The second synthetic study investigates the effect of the seabed topography on ERT data. Three different geometries were used for generating synthetic data. Based on the

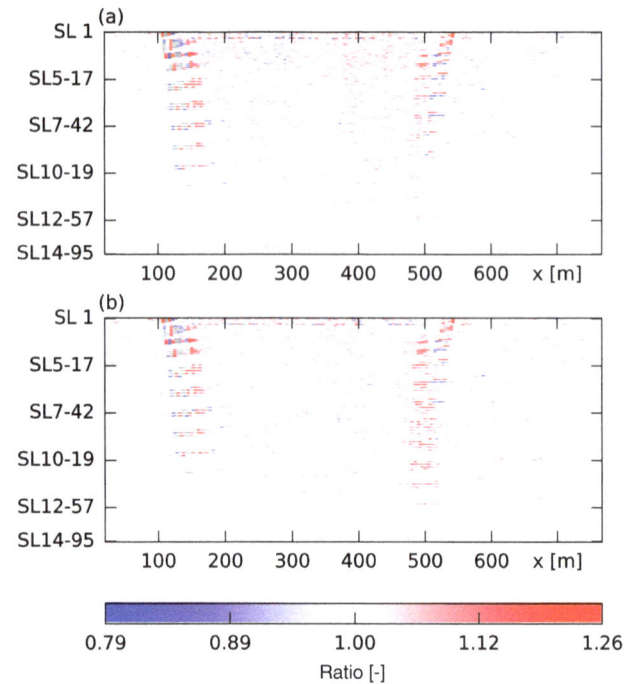

Figure 7. Pseudo-section of the synthetic data for the gradient array. The ratios between case one and the reference model **(a)** and between case two and the reference model **(b)**.

Äspö case, a water-filled valley with a depth of 10 m and a length of 550 m was used. The reference model contains a flat seabed, whereas cases one and two contain a depth variation of ± 0.30 m. For the first case, the depth of the seabed was set to -10.3 m between $x = 230$ m and $x = 300$ m and to -9.7 m between $x = 380$ m and $x = 450$ m. The depth varies, alternating from -10.3 to 9.7 m for 250 m $\leq x \leq 395$ m for the second case. While the data were generated, 2 % Gaussian noise was added. Afterwards, geometric factors for the first and second case were replaced with the reference data in order to simulate a flat seabed. One mesh for the data inversion was used with a flat seabed. The ratios between the two cases and the reference data set are shown as pseudo-sections of the simulated data in Fig. 7

The deviation due to the changed seabed topography is in the range of ± 20 %. Figure 7a shows a clear pattern due to the changed model, which is a lower apparent resistivity for a slight downwards shift of the seabed and an increased apparent resistivity for an upwards shift. Compared to that, an alternately varying seabed leads to a rather random pattern (Fig. 7b). This simple synthetic study confirms that the ERT data gathered at the Äspö test site are contaminated or distorted by 3-D effects that have to be taken into account when interpreting the results.

4   Results

A smoothness-constrained inversion was done with the abort criterion $\chi^2 = \Phi_d/N = 1$; i.e. the data are fitted within their errors. A visual inspection of the data misfit ensured that there was no more unresolved structure. The L_1 norm data (robust) inversion was used to account for remaining outliers in the ERT data set that lead to poor data fits. Nevertheless, the apparent resistivities cover several orders of magnitude (3–47 000 Ωm) and extraordinarily high resistivity variations occur, which is challenging for ERT inversion. The ERT inversion result is shown in Fig. 8a using the coverage (sum of the absolute Jacobian values over all data for each model) for alpha shading. In the middle of the profile, the penetration depth is limited due to the low-resistive water body and the anomalies below.

Outcrops of the bedrock lead to high resistivities of about 35 000 Ωm at the northern and southern ends of the profile. A low-resistive zone appears at $x = 200$–600 m directly below the sea. The depth varies between approximately 80 m at $x = 270$ m and 30 m at $x = 450$–600 m. As such a deep weathering zone seems implausible and the resistivity is too low for usual weathering, we interpret this structure as a deep valley filled with sediments. This has not been documented by previous investigations conducted in the construction phase of the test nuclear waste disposal. The low-resistive zone is extended diagonally downwards towards the north for $x > 600$ m at a depth range of 50–100 m. Although the coverage is low for this part, it is still possible that this feature indicates fractured water-bearing bedrock.

Resistivities of about 500 Ωm at $x = 100$–200 m and a depth of 100 m indicate a larger transition zone that continues below the sediment body. This could possibly lead to an incorrect depth of the sediment-filled valley and thus the bedrock interface. It also prevents any further interpretation regarding possible fracture zones.

The inversion result of the refraction seismics shown in Fig. 8b images the interface to the bedrock more accurately. However, the poor signal quality in the southern part results in a lower coverage and thus larger uncertainty. To display the inversion result, a standardised coverage was calculated, which is either 0 or 1 depending on whether any ray travels through a model cell or not.

According to Fig. 8, the crystalline bedrock appears as a high-velocity zone of about 5600 m s^{-1}, which agrees with the velocity for intact crystalline rock at Äspö HRL given by Wikberg et al. (1991). Brodic et al. (2016) recently showed that the velocity decreases from > 5000 m s^{-1} for intact rock down to approx. 4200–4700 m s^{-1} for fracture zones. Towards the northern part, the velocity of the bedrock decreases down to 5000 m s^{-1}. At the southern part between $x = 200$ m and $x = 300$ m, the result shows a low-velocity zone down to 60 m of depth, which is extended towards the north for shallow parts of the model above 20 m of depth. This finding coincides with the low-resistive part in the ERT result. The

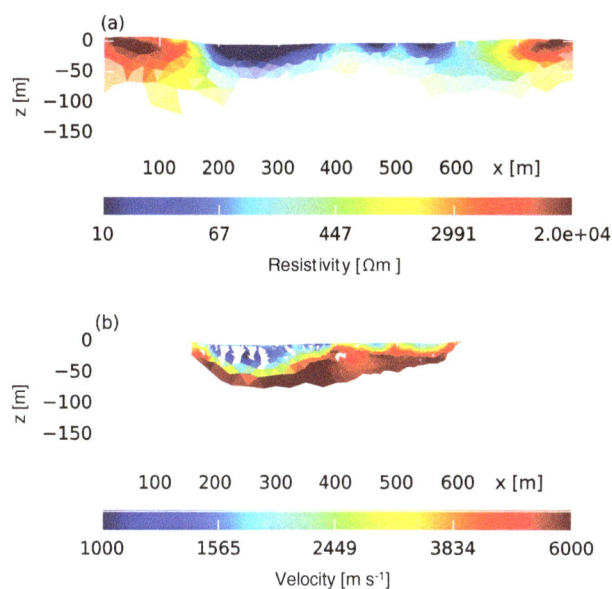

Figure 8. Separated inversion results of the ERT data set (**a**) and the refraction seismic data (**b**). The shading is based on the coverage.

sediments exhibit a minimum velocity of about 1000 m s^{-1}, which is below the velocity of water (1400 m s^{-1}). A reason could be gas contained in the sediments, which reduces the acoustic velocity for frequencies below 1 kHz (Wilkens and Richardson, 1998). This is supported by the presence of gas bubbles rising to the water surface during the blasting. Gas-bearing sediments were also reported by Dahlin et al. (2014) near Stockholm, which has a similar geologic history. It is assumed that the gas-bearing sediments lead to the poor data quality in the southern part by damping the seismic signals. No further low-velocity zones appear at larger depths.

To summarise, a (possibly gas-bearing) sediment body could be identified, which appears as a zone of low resistivities and velocities. Furthermore, the interface towards the bedrock could be found by the joint interpretation of the separated inversion results. However, the bedrock appears with a low resistivity due to the large transition zone. Fracture zones are not visible in the separated inversion results (Fig. 8) because of a low coverage in the refraction model and a large transition zone in the resistivity model.

In order to improve the results and enable further interpretation, a structurally coupled joint inversion of the ERT and seismic data was performed. To ensure that common structures are present in the models, the first four iterations were done separately. A robust data fit, i.e. L_1 norm, was used for ERT data inversion, while the first arrivals were fitted using the L_2 norm (least squares). Both data sets were fitted within their errors, i.e. with $\chi^2 = 1.1$ for ERT and $\chi^2 = 1.3$ for refraction data. In this case, the RMSE (root mean square error) for the first-arrival fit was about 2.4 ms. The result is shown in Fig. 9.

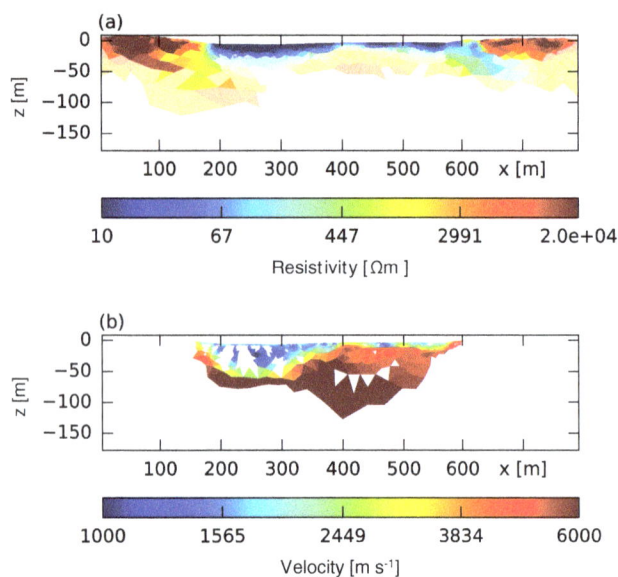

Figure 9. Joint inversion result with resistivity (top panel) and velocity (bottom panel) distribution. The shading is based on the coverage of each model cell.

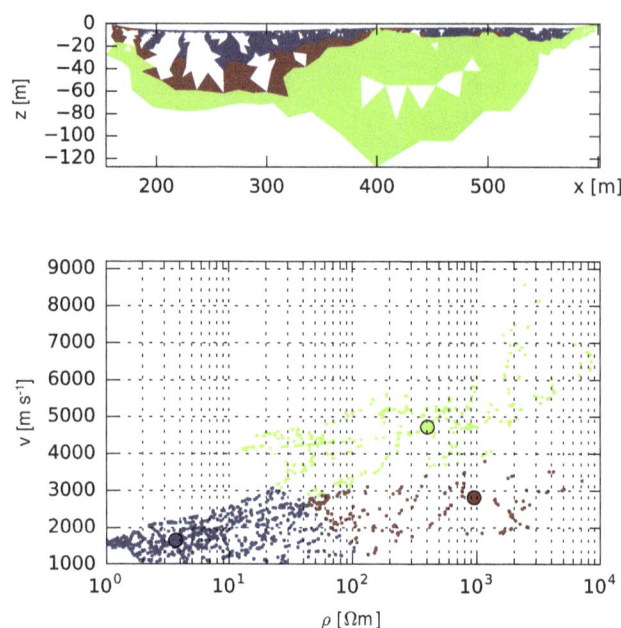

Figure 10. Cluster analysis of the joint inversion result using tree clusters. The upper picture shows the spatial distribution of the clusters and the lower one shows the parameter distribution within each cluster.

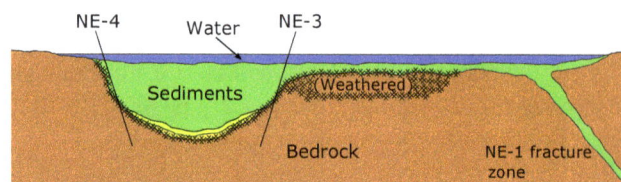

Figure 11. Conceptual model based on the geophysical results and known geologic interpretations of the test site at Äspö. The hash signature at the bedrock interface indicates a higher uncertainty.

Both models show significant changes compared to the separated inversions and allow for further interpretations. Generally, most changes occur in the resistivity model, while the velocity model shows only small improvements. The low-resistive zone, which corresponds to the sediment-filled valley, appears thinner followed by a much smaller transition zone. This reduces the ambiguity in estimating the bedrock interface. The bedrock is also assigned with a higher resistivity, which is more realistic as it agrees with the resistivity of near-surface rock outcrops at the northern and southern ends of the profile.

Additional structural constraints that moved from the velocity to the resistivity model pointed out the diagonal low-resistive zone in the northern part in more detail. This anomaly matches very well with the water-bearing fracture zone NE-1 in the northern part of the profile. The southern fracture zones NE-3, NE-4 and EW-1 cannot be identified directly. Possible explanations could be that these are (i) too small to be detected from the surface or (ii) filled with a material so that no parameter contrast appears.

According to the synthetic study, the low-resistive feature directly at the surface at $x = 610$ m and, in part, the diagonal low-resistive zone at $x = 600$ m are most likely caused by 3-D effects and should not be interpreted any further. Following Günther et al. (2006a), a post-processing of the two inversion models was done using a cluster analysis to obtain a simplified result (Fig. 10). For clustering the resistivity and velocity model, a modified mean-shift algorithm approach was used, which is described in Comaniciu and Meer (2002). The input for this algorithm is a feature space that consists in this case of resistivities and velocities. In order to analyse the feature

space, a window or bandwidth is needed. The bandwidth can be determined by an estimator that uses a selected quantile as input, whereas the quantile is defined between 0 and 1. In general, a low quantile will produce a larger number of clusters than a high quantile. In contrast to cluster number-driven algorithms such as the K-means algorithm (see Joydeep and Alexander, 2009), the input is data and a window to the data. Therefore, the selection of clusters is driven only by data and not by an arbitrary number of clusters.

As data input for the clustering, we only used model parameters included by the coverage of the seismic result (displayed cells in Fig. 9b) because the seismically covered volume is also covered by ERT.

The data-driven cluster algorithm divided the model parameters into three clusters that represent sedimentary deposits, the bedrock and the transition zone between those two. It can most likely be assumed that the interface between the sediments and the bedrock is within the third cluster.

As a final interpretation of the presented ERT and seismic results, a conceptual model was developed (Fig. 11). The primary origin of the deep sedimentary deposits can be explained by glacial erosion. The small valley was formed between the fracture zones NE-3 and NE-4. It might have been easier to erode the bedrock along zones with an already low rock quality. Two possible explanations can be given for the remaining transition zone at the bottom of the sedimentary valley. The first is that the bedrock–sediment interface is (i) fractured or weathered to a certain extent, and the second is (ii) that coarse sediments could have been deposited before fine-grained marine material was sedimented above. The latter possibility is visualised by the dark yellow and orange parts at the bottom of the valley in Fig. 11. As the medium velocities north of the sedimentary valley appear slightly thicker, the most probable explanation could be weathered bedrock. During an earlier investigation, it was found that the NE-1 fracture zone in the northern part of the model is water bearing at its boundaries and dry in its core due to clay deposits. Thus, it appears as a zone of lower resistivities and velocities. Only the NE-1 fracture zone could be identified by this survey, although the fracture zones NE-3, NE-4 and EW-3 are also partly water bearing according to Wikberg et al. (1991). As shown in Fig. 1, NE-4 and EW-7 are close to each other at the profile line, which means that they most likely cannot be imaged separately by ERT measurements. In addition, the low-resistive sediments are a complicating factor that may mask the fracture zones by reducing the model resolution such that it is not sufficient to resolve the fracture zones NE-3, NE-4 and EW-7.

5   Conclusions and outlook

A combination of refraction seismics and ERT data has been tested on an underwater profile crossing a water passage along part of the access tunnel that connects surface facilities with an underground test laboratory at the Äspö Hard Rock Laboratory. The aim was to detect fracture zones in a well-known but logistically challenging area. Co-located sensor positions for ERT and seismics were used on a 450 m underwater section of the 700 m ERT profile.

A synthetic study inspired by the geologic conditions of the Äspö test site showed that significant 3-D effects are expected that contaminate the ERT data and thus influence the obtained inversion result. This was taken into account to prevent the misinterpretation of the final inversion results. The results of the separated inversions showed a previously unknown sediment-filled valley that appeared as a zone with low resistivities and low velocities, even in an unusually well-documented geological environment. The poor coverage of the seismic model in the northern and southern parts of the profile in conjunction with the large transition zone of the ERT result prevent further detailed interpretations. However, the water-bearing fracture zone NE-1 could be iden-

tified by the results of the structurally coupled joint inversion. The evaluation shows that the joint inversion approach combining ERT and seismics has very promising results for three reasons: (i) the decreased extent of the transition zone, (ii) the more reliable interpretation of two independent parameters and (iii) their combination by a clustering approach. Although the refraction seismic does not cover the fraction zone NE-1, the additional constraints by the joint inversion helped to determine the fracture zone with ERT. The southern fracture zones NE-3, NE-4 and EW-1 could not be detected due to the missing parameter contrast and/or the model resolution. The latter is considered to be the main reason, which was shown by the distribution of the model resolution radii and coverage. The reduced investigation depth of ERT is due to the fact that the current preferably flows through low-resistive bodies (water or sediments) and is the major disadvantage of this method.

The comparison of the joint inversion with the separated inversion result shows significant improvements. Therefore, the combination of geoelectric and seismic refraction is recommended as the standard tool for site investigations under geologic conditions similar to those presented.

Competing interests. The authors declare that they have no conflict of interest.

Acknowledgements. Thanks to all who participated in the field survey and made the measurements possible. Per-Ivar Olsson, Erik Fennvik and Nayeli Lasheras Maas were also instrumental in the field data acquisition. Special thanks go to Marcus Wennermark, who planned the survey and was responsible for seismic instrumentation and data collection in the field campaign. We are grateful to SKB (Swedish Nuclear Fuel and Waste Management Company) for logistic support during the field campaign. The funding which made this work possible was provided by Nova FoU, BeFo (Swedish Rock Engineering Research Foundation; refs. 314 and 331), SBUF (the Development Fund of the Swedish Construction Industry; refs. 12718 and 12719) and Formas (the Swedish Research Council for Environment, Agricultural Sciences and Spatial Planning; ref. 2012-1931) as part of the Geoinfra-TRUST framework (http://www.trust-geoinfra.se/).

Edited by: U. Werban

References

Bäckblom, G., Gustafsson, G., Stanfors, R., and Wikberg, P.: A synopsis of predictions before the construction of the Äspö Hard Rock Laboratory and the process of their validation, Tech. rep., SKB, Stockholm, 1990.

Berglund, J., Curtis, P., Eliasson, T., Olsson, T., Starzec, P., and Tullborg, E.: Äspö Hard Rock Laboratory, Update of the geological model 2002, Tech. rep., SKB (Swedish Nuclear Fuel and Waste Management Company), Stockholm, 2003.

Bergman, B., Tryggvason, A., and Juhlin, C.: Seismic tomography studies of cover thickness and near-surface bedrock velocities, Geophysics, 71, U77–U84, https://doi.org/10.1190/1.2345191, 2006.

Brodic, B., Malehmir, A., and Juhlin, C.: Fracture System Characterization Using Wave-mode Conversions and Tunnel-surface Seismics, in: EAGE Near Surface Geophusics, EAGE, Barcelona, Spain, 2016.

Comaniciu, D. and Meer, P.: Mean Shift : A Robust Approach Toward Feature Space Analysis, IEEE T. Pattern Anal. Mach. Intell., 24, 603–619, 2002.

Dahlin, T. and Wisén, R.: Underwater ERT Surveys for Urban Underground Infrastructure Site Investigation in Central Stockholm, in: 17th Nordic Geotechnical Meeting, 25–28 May 2016, Reykjavik, 2016.

Dahlin, T. and Zhou, B.: Multiple-gradient array measurements for multichannel 2D resistivity imaging, Near. Surf. Geophys., 4, 113–123, 2006.

Dahlin, T., Bjelm, L., and Svensson, C.: Use of electrical imaging in site investigations for a railway tunnel through the Hallandså s Horst, Sweden, Q. J. Eng. Geol., 32, 163–173, 1999.

Dahlin, T., Loke, M. H., Siikanen, J., and Höök, M.: Underwater ERT Survey for Site Investigation for a New Line for Stockholm Metro, Near Surface Geoscience, in: 20th European meeting of Environmental and Egineering Geophysics, 14–18 September 2014, Athens, Greece, 2014.

Diaz, D., Maksymowicz, A., Vargas, G., Vera, E., Contreras-Reyes, E., and Rebolledo, S.: Exploring the shallow structure of the San Ramón thrust fault in Santiago, Chile ($\sim 33.5° S$), using active seismic and electric methods, Solid Earth, 5, 837–849, https://doi.org/10.5194/se-5-837-2014, 2014.

Fennvik, E.: Resistivitets- och IP-mätningar vid Äspö Hard Rock Laboratory, bSc thesis 438, Dept. of Geology, Lund University, Lund, 2015.

Friedel, S.: Resolution, stability and efficiency of resistivity tomography estimated from a generalized inverse approach, Geophys. J. Int., 153, 305–316, https://doi.org/10.1046/j.1365-246X.2003.01890.x, 2003.

Gallardo, L. A. and Meju, M. A.: Joint two-dimensional DC Resistivity and Seismic travel time inversion with cross-gradients constraints, J. Geophys. Res., 109, B03311, https://doi.org/10.1029/2003JB002716, 2004.

Ganerød, G., Rønning, J. S., Dalsegg, E., Elvebakk, H., Holmøy, K., Nilsen, B., and Braathen, A.: Comparison of geophysical methods for sub-surface mapping of fault and fracture zones in a section of the Viggja road tunnel, Norway, Bull. Eng. Geol. Environ., 65, 231–243, https://doi.org/10.1007/s10064-006-0041-6, 2006.

Garofalo, F., Sauvin, G., Socco, L. V., and Lecomte, I.: Joint inversion of seismic and electric data applied to 2D media, Geophysics, 80, 93–104, https://doi.org/10.1190/GEO2014-0313.1, 2015.

Günther, T. and Südekum, W.: New Approach for Routine Investigation of the Shallow Sea Floor with Bottom-towed DC Resistivity Measurements, in: 14th European Meeting of Environmental and Egineering Geophysics, 15–17 September 2007, Kraków, 2007.

Günther, T., Bentley, L. R., and Hirsch, M.: A new joint inversion algorithm applied to the interpretation of DC resistivity and refraction data, in: XVI International Conference on Computa-

tional Methods in Water Recources, 19–22 Jun 2006, Copenhagen, Denmark, 2006a.

Günther, T., Rücker, C., and Spitzer, K.: Three-dimensional modeling and inversion of dc resistivity data incorporating topography – Part II: Inversion, Geophys. J. Int., 166, 506–517, https://doi.org/10.1111/j.1365-246X.2006.03011.x, 2006b.

Günther, T., Dlugosch, R., Holland, R., and Yaramanci, U.: Aquifer Characterization using coupled inversion of DC/IP and MRS data on a hydrogeophysical test-site, in: SAGEEP, Symposium on the Application of Geophysics to Engineering and Environmental Problems 2010, 302–307, https://doi.org/10.4133/1.3445447, 2010.

Ha, H. S. K., D. S., and Park, I. J.: Application of electrical resistivity techniques to detect weak and fracture zones during underground construction, Environ. Earth Sci., 60, 723–731, 2010.

Heincke, B., Günther, T., Dahlsegg, E., Rønning, J. S., Ganerød, G., and Elvebakk, H.: Combined three-dimensional electric and seismic tomography study on the Åknes rockslide in western Norway, J. Appl. Geophys., 70, 292–306, https://doi.org/10.1016/j.jappgeo.2009.12.004, 2010.

Joydeep, G. and Alexander, L.: The Top Ten Algorithms in Data Mining, in: chap 2: K-Means by Joydeep and Alexander, edited by: Wu, X. and Kumar, V., Chapman and Hall/CRC, 21–35, https://doi.org/10.1201/9781420089653.ch2, 2009.

Juhojuntii, N. and Kamm, J.: Joint inversion of seismic refraction and resistivity data using layered models – Applications to groundwater investigation, Geophysics, 80, EN43–EN55, 2015.

Karlsrud, K., Erikstad, L., and Snilsberg, P.: Tunnels for the citizens – investigations of and restrictions on water leakage to maintain the environment, Tech. Rep. 103, Public Road Administration, Oslo, 2003.

Lasheras Maas, N.: Site characterisation at the Äspö Hard Rock Laboratory through seismic refraction, Civil Engineering, bSc thesis, Lund University, Lund, 2015.

Lindstrøm, M. and Kveen, A.: Tunnels for the citizens – final report, Tech. Rep. 105, Public Road Administration, Oslo, 2004.

Loke, M. H. and Lane, J. W. J.: Inversion of data from electrical resistivity imaging surveys in water-covered areas, Explor. Geophys., 35, 266–271, 2004.

Malehmir, A., Andersson, M., Mehta, S., Brodic, B., Munier, R., Place, J., Maries, G., Smith, C., Kamm, J., Bastani, M., Mikko, H., and Lund, B.: Post-glacial reactivation of the Bollnäs fault, central Sweden – a multidisciplinary geophysical investigation, Solid Earth, 7, 509–527, https://doi.org/10.5194/se-7-509-2016, 2016.

Palmstrøm, A., Nilsen, B., Pedersen, K. B., and Grundt, L.: Correct extent of site investigations for underground facilities, Tech. rep., Directorate of Public Roads, Oslo, 2003.

Rhén, I., Rhen, I. G., Gustafsson, G., Stanfors, R., and Wikberg, P.: Äspö HRL – Geoscientific evaluation 1997/2, Results from pre-investigation and detailed site characterisation, Summary report, Tech. rep., SKB (Swedish Nuclear Fuel and Waste Management Company), Stockholm, 1997.

Ronczka, M., Hellmann, K., Guenther, T., Dahlin, T., and Wisen, R.: Data and codes for Äspö field case, https://doi.org/10.5281/zenodo.806753, June 2017.

Rønning, J. S., Ganerød, G., Dalsegg, E., and Reiser, F.: Resistivity mapping as a tool for identification and characterisation of weakness zones in crystalline bedrock: definition and testing of an interpretational model, Bull. Eng. Geol. Environ., 73, 1225–1244, 2013.

Rücker, C.: Advanced Electrical Resistivity Modelling and Inversion using Unstructured Discretization, PhD thesis, University of Leipzig, Leipzig, http://nbn-resolving.de/urn:nbn:de:bsz:15-qucosa-69066 (last access: 1 May 2017), 2011.

Rücker, C., Günther, T., and Spitzer, K.: 3-d modeling and inversion of DC resistivity data incorporating topography – Part I: Modeling, Geophys. J. Int., 166, 495–505, https://doi.org/10.1111/j.1365-246X.2006.03010.x, 2006.

Stanfors, R., Rhen, I., Tullborg, E. L., and Wikberg, P.: Overview of geological and hydrogeological conditions of the Äspö hard rock Laboratory site, Appl. Geochem., 14, 819–834, 1999.

Udphuay, S., Günther, T., Everett, M., Warden, R., and Briaud, J.-L.: Three-dimensional resistivity tomography in extreme coastal terrain amidst dense cultural signals: application to cliff stability assessment at the historic D-Day site, Geophys. J. Int., 185, 201–220, 2011.

Vidstrand, P.: Äspö Hard Rock Laboratory, Update of the hydrogeological model 2002, Tech. rep., SKB (Swedish Nuclear Fuel and Waste Management Company), Stockholm, 2003.

Wikberg, P., Gustafsson, G., Rhén, I., and Stanfors, R.: Äspö hard rock laboratory, Evaluation and conceptual modelling based on the pre-investigation 1986–1990, Tech. rep., SKB (Swedish Nuclear Fuel and Waste Management Company), Stockholm, 1991.

Wilkens, R. H. and Richardson, M. D.: The influence of gas bubbles on sediment acoustic properties: in situ, laboratory, and theoretical results from Eckernförde Bay, Baltic sea, Cont. Shelf Res., 18, 1859–1892, https://doi.org/10.1016/S0278-4343(98)00061-2, 1998.

Wisén, R., Mykland, J., Rønning, J. S., Elvebakk, H., and Olsen, F.: Experience from Multidisciplinary Geophysical Survey for Tunneling In Norway – Refraction Seismic, Resistivity Profiling and Borehole Logging, in: 16th Nordic Geotechnical Meeting, 9–12 May 2012, Copenhagen, 2012.

10

Structure of the Suasselkä postglacial fault in northern Finland obtained by analysis of local events and ambient seismic noise

Nikita Afonin[1], **Elena Kozlovskaya**[2,3], **Ilmo Kukkonen**[4], **and DAFNE/FINLAND Working Group**[*]

[1]Federal Centre for Integrated Arctic Research RAS, Arkhangelsk, Russia
[2]Oulu Mining School, POB-3000, 90014, University of Oulu, Finland
[3]Geological Survey of Finland, P.O. Box 96, 02151, Espoo, Finland
[4]Department of Physics, University of Helsinki, P.O. Box 64, 00014, Helsinki, Finland
[*]A full list of authors and their affiliations appears at the end of the paper.

Correspondence to: Nikita Afonin (afoninnikita@inbox.ru)

Abstract. Understanding the inner structure of seismogenic faults and their ability to reactivate is particularly important in investigating the continental intraplate seismicity regime. In our study we address this problem using analysis of local seismic events and ambient seismic noise recorded by the temporary DAFNE array in the northern Fennoscandian Shield. The main purpose of the DAFNE/FINLAND passive seismic array experiment was to characterize the present-day seismicity of the Suasselkä postglacial fault (SPGF), which was proposed as one potential target for the DAFNE (Drilling Active Faults in Northern Europe) project. The DAFNE/FINLAND array comprised an area of about 20 to 100 km and consisted of eight short-period and four broadband three-component autonomous seismic stations installed in the close vicinity of the fault area. The array recorded continuous seismic data during September 2011–May 2013. Recordings of the array have being analysed in order to identify and locate natural earthquakes from the fault area and to discriminate them from the blasts in the Kittilä gold mine. As a result, we found a number of natural seismic events originating from the fault area, which proves that the fault is still seismically active. In order to study the inner structure of the SPGF we use cross-correlation of ambient seismic noise recorded by the array. Analysis of azimuthal distribution of noise sources demonstrated that during the time interval under consideration the distribution of noise sources is close to the uniform one. The continuous data were processed in several steps including single-station data analysis, instrument response removal and time-domain stacking. The data were used to estimate empirical Green's functions between pairs of stations in the frequency band of 0.1–1 Hz and to calculate corresponding surface wave dispersion curves. The S-wave velocity models were obtained as a result of dispersion curve inversion. The results suggest that the area of the SPGF corresponds to a narrow region of low S-wave velocities surrounded by rocks with high S-wave velocities. We interpret this low-velocity region as a non-healed mechanically weak fault damage zone (FDZ) formed due to the last major earthquake that occurred after the last glaciation.

1 Introduction

In studying of mechanisms of large earthquakes investigations of seismogenic fault structure and properties are of particular importance. One group of seismological studies concentrates on mapping the seismic source using recordings of seismic events. This includes mapping of the fault plane using distribution of hypocentres of earthquakes originating from the fault and also calculating orientation and dip of fault planes from seismograms of earthquakes (fault plane solutions, centroid moment tensor solutions). Another group of methods investigates the inner structure of fault zones using structural geology, palaeoseismology, seismic reflection and refraction experiments and geodetic measurements. The studies of the second group show that the inner structure of seismogenic faults is complex (Davis and Reynolds, 1996) and the main slip planes are surrounded by so-called fault

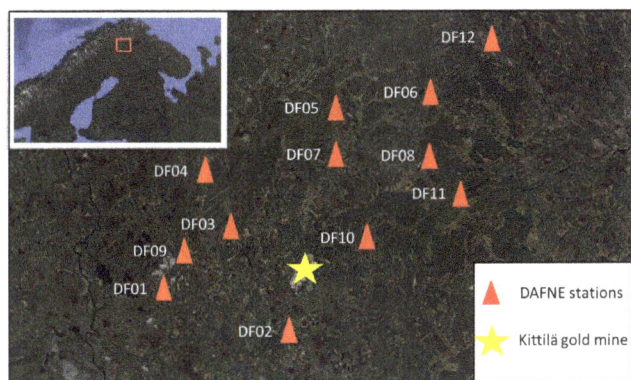

Figure 1. Position of the DAFNE temporary array on the geographical map.

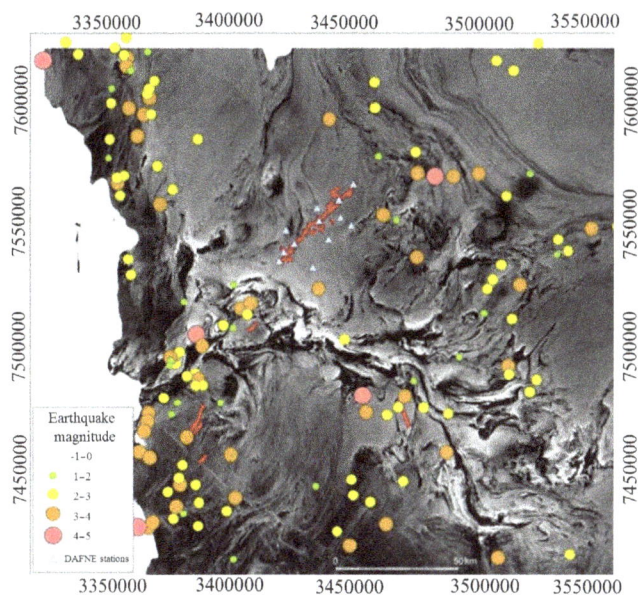

Figure 2. Position of DAFNE temporary array (blue triangles) on the aeromagnetic field map of Finland. Epicentres of local earthquakes detected by regional seismic networks in Fennoscandia prior to DAFNE/FINLAND experiment are shown by circles with size proportional to the magnitude of the event. Postglacial faults in the area are shown by red lines. The coordinate system is the Finnish National Coordinate System (KKJ).

damage zones (FDZs) which arise along the fault as a result of brittle deformation and cracking in rock in response to stress (Chester and Logan, 1986; Shipton and Cowie, 2003). Kim et al. (2004) demonstrated that damage zones show very similar geometries across a wide range of scales and fault types, including strike-slip, normal and thrust faults. These fault damage zones are mechanically weaker than surrounding rock and they can be detected as zones of low seismic velocities. Some recent investigations of FDZs produced by large earthquakes have demonstrated that the width of them can vary from several dozen metres to 1–2 km (Vidale and Li, 2003; Cochran et al., 2009). It is still debatable whether FDZs persist over a full earthquake cycle, which may last hundreds to thousands of years (Cochran et al., 2009) or the FDZs are healing during the years or decades following a main shock (Crone et al., 2003; Vidale and Li, 2003).

The question of longevity of faults is particularly important in investigating the continental intraplate seismicity regime (Stein, 2007). It is known that most such earthquakes can be related to detected fault zones, but continents contain many such features, of which only few are active. As suggested by Stein (2007) and McKenna et al. (2007), long-lived, wide and mechanically weak damage zones concentrate intraplate strain release; hence, hypocentres of future earthquakes would be located along these zones. However, if the damage zones are healed after the major shock, and are not significantly weaker than surrounding, then the intraplate seismicity would be a transient phenomenon that migrates among many fossil weak zones. That is why study of damage zones of the faults proved to be not active during historical time may provide new information about the intraplate seismicity phenomenon.

In our study we address the problem using an example of the Suasselkä postglacial fault located in northern Fennoscandia (Fig. 2). Postglacial (PG) faults there were formed during the last stages of the Weichselian glaciation (ca. 9000–15 000 years BP), when reduced ice load and relaxation of accumulated tectonic stress resulted in rapid up-

lift in Fennoscandia and large-magnitude earthquakes with M_w of 7–8.2 (Wu et al., 1999; Olesen et al., 2004; Kukkonen et al., 2010). The length of the PGFs may vary from 2 to 150 km and the maximum height of the fault scarps from 1 to 12 m, and up to 30 m in the extreme cases (see compilation in Olesen et al., 2004). However, postglacial faulting is not confined to northern Fennoscandia, but has also been reported in northernmost Germany (Brandes et al., 2012) and central parts of Sweden and Finland (Malehmir et al., 2016; Mikko et al., 2015; Palmu et al., 2015; Juhlin and Lund, 2011; Juhlin et al., 2010).

The Suasselkä postglacial fault (SPGF) of total length of 48 km is the longest PGF in Finland with a strike of 35–50° and scarp of 0–3 m. Geomorphological studies revealed multiple and differently oriented surface signatures of the SPGF (Fig. 2). The fault zone has been studied using magnetic and electromagnetic measurements by Paananen (1987) and Kuivamäki et al. (1998), who suggested that the fault dips southeastward. The sense of motion was thrusting from southeast. The origin of the structure is probably much older than postglacial as suggested by the coincidence of the fault scarp with magnetic anomalies of the Palaeoproterozoic bedrock. The southeastern part of the fault is located in pre-existing fracture zone in the southeast, whereas existence of the oldest structure in the northwest remains uncertain. Based on the radiocarbon-dated buried organic materials, the time frame of fault activity ranges from 9730 to 5055

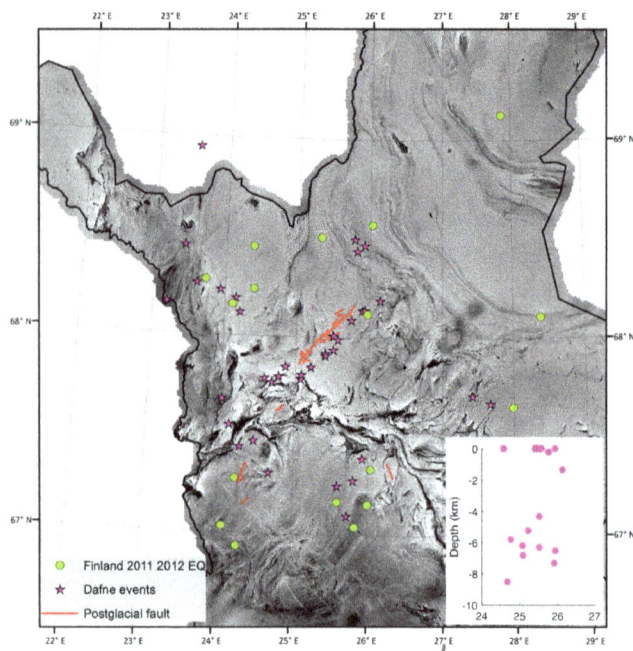

Figure 3. Epicentres of local earthquakes (magenta stars) identified from recordings of the DAFNE/FINLAND temporary during September 2011–May 2012 shown on the aeromagnetic map of Finland. Epicentres of local earthquakes detected by regional seismic networks in Fennoscandia during September 2011–May 2012 are shown with green dots. Red lines indicate postglacial faults in the area. Right corner inset shows depth distribution of the events originating from the SPGF area. Precision of hypocentre coordinate determination is of the order of 2–5 km. The average error of depth determination is about 3–4 km.

cal. years BP (Sutinen et al., 2014). The reprocessed seismic data along several reflection profiles crossing the fault area (Ahmadi et al., 2015) demonstrated the complex structure of the fault area down to a depth of 2–3 km, with two sets of segmented and discontinuous dipping reflectors.

In our study we investigate the inner structure of the SPGF using distribution of hypocentres of local seismic events and analysis of ambient seismic noise, recorded by the temporary DAFNE array. The ambient seismic noise analysis has been used recently by Hillers et al. (2014) in order to investigate the inner structure and properties of Calico Fault Zone in eastern California. Zigone et al. (2015) applied ambient seismic noise tomography in order to study the Southern California Plate Boundary Region.

2   Data

The main objective of the DAFNE/FINLAND seismic passive experiment was to answer two major questions: (a) is the Suasselkä postglacial fault (SPGF) still seismically active, and (b) if it is active, what is the geometry of its seismogenic zone and the depth to it? The project was initiated by several

organizations in Finland (Geological Survey of Finland, Sodankylä Geophysical Observatory of the University of Oulu and Institute of Seismology of the University of Helsinki). The main purpose of the field experiment was to install an array of autonomous seismic stations in the target area of the SPGF in order to collect continuous seismic data for the period of September 2011–May 2013.

Selection of the DAFNE sites was done in August 2011. Four sites with the permanent electric power supply were found in the vicinity of the target area (SPGF). The other eight sites were selected taking into consideration surface position of the fault and the possible dip of the fault to the southeast. In the first part of September 2011 the sites were prepared for installation and autonomous seismic recording instruments were installed to the sites during the second part of September 2011. Coordinates of all stations of the DAFNE array and description of the instrumentation, used in the experiment, are given in Table 1. The instruments were provided by the Institute of Seismology of the University of Helsinki and by the Sodankylä Geophysical Observatory. Location of stations of the DAFNE array is shown in Fig. 1.

During the data acquisition period, all the stations were regularly (once per 2 months on average) served by two staff members of the Sodankylä Geophysical Observatory. The regular service included changing of batteries and media and basic data quality control. Raw continuous data files from temporary stations of the DAFNE array were copied to the data server of the SGO in their original formats. The main steps for data pre-processing included applying time corrections, conversion to the miniSEED format and merging the data files into miniSEED files of 24 h length using IRIS PASSCAL software for processing and visualization of seismological data (http://www.iris.edu/manuals/SEEDManual_V2.4.pdf). Analysis of continuity of the data and estimation of noise level at the sites was performed after each field trip. Generally, the noise level was higher during daytime, and seismograms, recorded during the daytime, were seriously contaminated by signals from production blasts in numerous mines in northern Sweden, Finland and Russia. Major problems were detected at stations DF10 and DF07, where the high noise level was due to technical problems with sensors and cables.

3   Detection and location of seismic events

At the beginning of the data analysis, we compiled a preliminary list of local seismic events using monthly seismic bulletins, published by the Institute of Seismology of the University of Helsinki (FENCAT: http://www.seismo.helsinki.fi/english/bulletins/index.html). The criteria for events selection were coordinates of epicentres (from 67.5 to 69° N and from 23 to 28° E) and the time frame corresponding to the DAFNE data acquisition period. The epicentres of these events are shown in Fig. 3. As can be seen, regional perma-

Table 1. Information about seismic stations of the DAFNE/FINLAND temporary array.

Name	Location	Coordinates			Operation started					Group	Sensor type	Logger	Sampling rate (sps)
		Lat.	Long.	Height (m)	Day	Month	Year	Hour	Min.				
DF01	Rautuskylä	67.8758	25.0514	241	13	9	2011	7	17	Oulu	Trillium compact	PR6-24	100
DF02	Lehto	67.8566	25.3862	227	23	9	2011	13	50	Oulu	Trillium compact	Reftek 130	100
DF03	Kapsajoki	67.9403	25.1931	244	16	9	2011	14	7	Helsinki+ Oulu	Lennartz 3-D lite	Reftek 130	100
DF04	Kiimalaki	97.9886	25.1042	274	16	9	2011	15	22	Oulu	Mark L4a	Reftek 130	100
DF05	Outa-Perttunen	68.0715	25.4131	302	15	9	2011	12	30	Oulu	Mark L4a	Reftek 130	100
DF06	Mietrikkilehto	68.1015	25.6388	314	15	9	2011	9	47	Helsinki	Lennartz 3-D lite	Reftek 130	100
DF07	Suasselkä	68.0262	25.4365	286	16	9	2011	6	34	Helsinki	Lennartz 3-D lite	Reftek 130	100
DF08	Arabiankangas	68.0395	25.6588	326	15	9	2011	7	51	Helsinki	Lennartz 3-D lite	Reftek 130	100
DF09	Salo	67.9091	25.0926	227	14	9	2011	7	10	Oulu	Trillium compact	PR6-24	100
DF10	Tepsänjänkkä	67.9595	25.5397	271	15	9	2011	14	40	Helsinki	Lennartz 3-D lite	Reftek 130	100
DF11	Rajalompolontie	68.0110	25.7622	340	17	9	2011	7	21	Helsinki	Lennartz 3-D lite	Reftek 130	100
DF12	Pokka	68.1596	25.7760	282	13	9	2011	14	36	Oulu	Trillium compact	SeisCom P	100

nent seismic networks detected no natural events from the SPGF area. This can be partly explained by large distances between the fault and existing permanent seismic stations. The second problem for detection of natural seismic events in northern Finland is a huge number of production and development blasts originating from numerous underground and open pit mines. The DAFNE/FINLAND array recorded up to 100 of such blasts per day from northern Sweden, Russia and Finland. Due to this, it was not possible to use a routine LTA/STA analysis for automatic event detection. Therefore, the manual data analysis was used. The continuous data were accessed and reviewed with the Seismic Handler Motif (SHM) program package (http://www.seismic-handler.org/portal).

As a first step, we analysed all the continuous data for the period of September 2011–October 2011 and relocated all local events seen in the DAFNE data. Then, we compiled a dataset of waveforms of explosions from the mines outside the target area and detected typical blasting time intervals for these mines. This dataset was used to exclude such events from further analysis. This was a necessary data processing step, because about 30–50 % of all such explosions are not included in the FENCAT bulletin and other regional bulletins.

The continuous data were filtered using the Butterworth third-order 2–40 Hz band pass filter and waveforms were analysed using SMH software. In total, 1188 events in

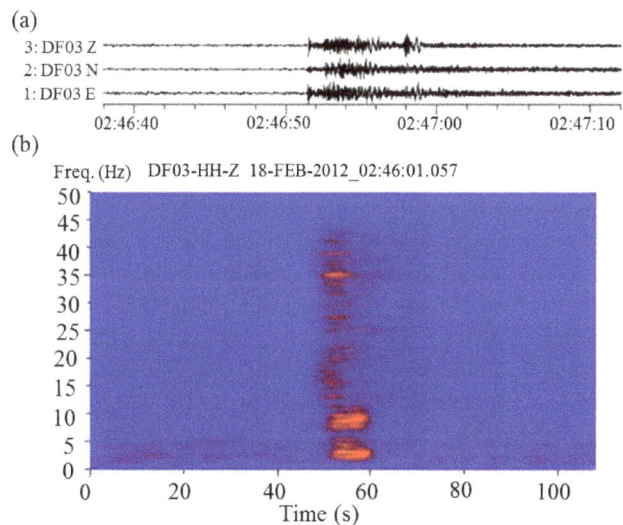

Figure 4. (a) An example of waveforms of production blast in Kittilä Mine recorded by DF03 station. (b) An example of a spectrogram of the same event.

September 2011–October 2011 were analysed and relocated using the DAFNE array data.

For relocation we used the LocSAT seismic event location program (Bratt, 1988) and manually picked first arrivals of P and S waves. The first arrivals of P waves were picked on Z component of the data filtered by the Butterworth third-

(a)

(b)

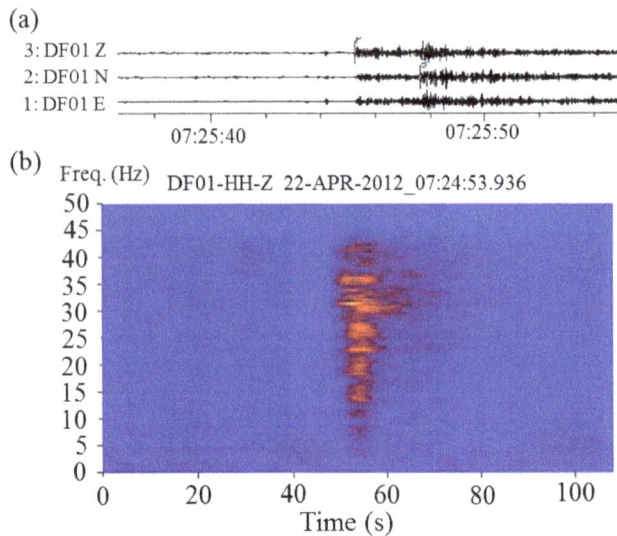

Figure 5. (a) Waveforms of a local earthquake on 22 April 2012, 07:25:41 UT, at 67.8° N, 24.76° E, depth 5.8 km recorded by DF01 station. **(b)** The spectrogram of the same event.

order 2–40 Hz band pass filter. The first arrivals of S waves were picked either on horizontal E or N component filtered by the Butterworth third-order 2–6 Hz band pass filter, depending on the number of traces in which the same signal was seen.

The distance between epicentre and stations of the array is less than 60 km for events, originating from the target area. In this case, the first arrivals of P and S wave correspond to direct Pg and Sg waves, refracting in the upper crust and the LocSAT procedure provides hypocentre coordinates with precision satisfactory for event detection.

In order to test the precision of the location procedure, we used blasts originating from the Kittilä gold mine. These blasts occurred inside known blasting time windows and have similar waveforms. Information about time windows for production and development blasting time was kindly provided by Engineering Superintendent André van Wageningen (personal communication, 2012) of Agnico-Eagle Kittilä Mine.

The hypocentres of selected blasts, obtained after relocation, were concentrated inside the mining area, which is about 5 km long and 2 km wide (www.agnicoeagle.com), with depths varying from 0 to 1 km. This location precision is satisfactory for proper detection of events originating from the fault area. As a result of this preliminary analysis, we distinguished two types of events, originating from our target area, but having different waveforms:

1. blasts originating from the Kittilä mine (Fig. 4);

2. events originating from the SPGF area and its surrounding, which could be of natural origin (Fig. 5).

Table 2. List of seismic events in the area of Suasselkä postglacial fault detected by the DAFNE seismic array.

Year	Month	Day	Hour	Min	Sec	Lat (deg.)	Long (deg.)	Depth (km)
2011	9	28	22	8	59	67.85	24.86	12.5
2011	10	8	22	26	13	67.31	24.67	5.2
2011	11	17	1	18	10	68.01	25.5	0
2011	11	17	1	18	10	68.01	25.51	4.3
2011	12	3	18	5	52	68.45	23.41	1
2011	12	7	22	36	16	67.94	25.51	6.3
2011	12	9	3	47	51	68.47	25.92	7.5
2011	12	10	5	47	18	68.26	23.6	3.1
2011	12	12	12	50	23	67.85	25.21	5.2
2011	12	24	4	52	44	67.25	25.59	4.9
2011	12	24	19	42	1	68.44	25.82	7.5
2011	12	31	3	24	27	67.79	24.58	0
2012	1	7	7	25	40	67.79	25.07	6.2
2012	1	12	0	15	19	67.55	24.12	8.6
2012	1	17	3	22	39	68.12	24.21	17
2012	1	18	21	44	15	68.14	25.93	6.5
2012	1	20	5	36	31	68.14	25.9	7.3
2012	1	23	16	36	47	67.1	25.72	6.5
2012	1	23	23	24	23	67.81	25.08	6.8
2012	2	11	18	44	4	67.99	25.57	0
2012	2	18	0	33	4	68.14	25.93	0
2012	2	19	1	24	20	67.77	24.68	8.5
2012	3	4	14	22	44	67.92	25.4	0
2012	3	4	23	27	31	67.92	25.4	0
2012	3	19	23	49	17	68.23	23.93	2.5
2012	4	9	1	3	41	68.19	24.15	14.2
2012	4	15	15	39	29	67.47	24.46	4.6
2012	4	17	16	0	19	68.19	26.14	1.3
2012	4	18	20	47	32	67.71	27.43	9.5
2012	4	20	11	28	9	68.09	25.75	0.2
2012	4	22	7	25	41	67.8	24.77	5.8
2012	4	24	8	58	5	68.16	23.19	7.8
2012	4	26	16	38	51	67.68	24.01	2.4
2012	4	30	4	15	1	68.95	23.57	0
2012	5	1	9	29	39	67.91	25.4	0
2012	5	8	22	38	34	67.44	24.27	11.1
2012	5	14	22	10	11	67.28	25.8	7.1
2012	5	16	10	55	31	67.39	25.92	0
2012	5	21	14	16	24	68.5	25.78	2.1
2012	5	25	18	27	58	67.67	27.68	4.7

The waveforms of these events were used in further manual analysis of continuous data for the period until 31 May 2012. Totally, the DAFNE network recorded 10 230 local events during September 2011–May 2012. From this amount, we selected 40 events with the waveforms similar to those of the group 2 and located them, using manually picked first arrivals of P and S waves and the LocSAT software. Additional control on discriminating natural events from blasts in Kittilä gold mine was performed using analysis of waveforms spectra (Glitterman and van Eck, 1993; Glitterman and Shapira, 1998). Coordinates of hypocentres of events, identified as natural ones, are presented in Table 2 and their epicentres are shown in Fig. 3. In our study precision of hypocentre coordinate determination by LocSat is usually of the order of 2–5 km and the depth of hypocentres of local earthquakes varies from about 2 to 15 km, with an average error of about 3–4 km. Hypocentres of events, originating from the

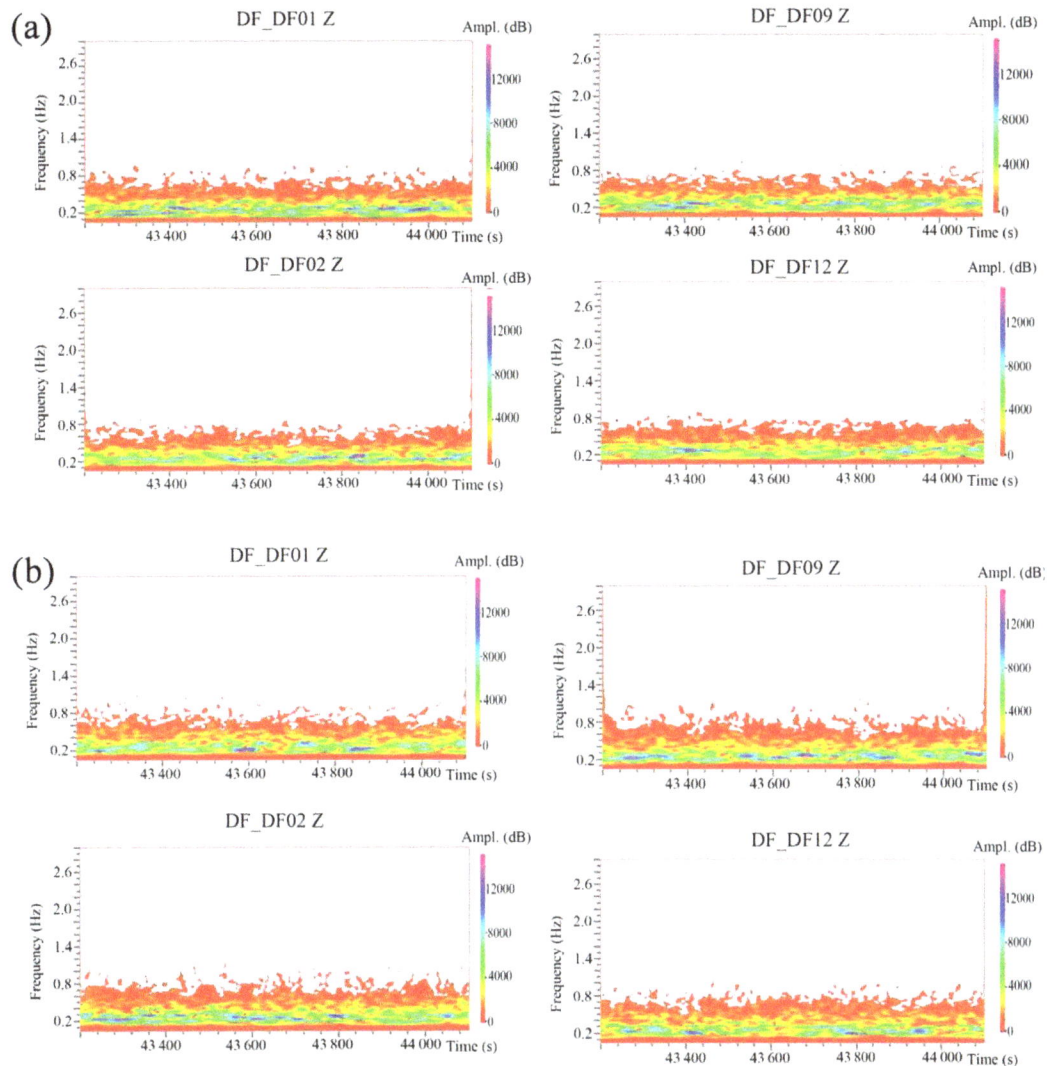

Figure 6. Examples of typical ambient noise spectrograms estimated for the low-frequency band (0.1–3 Hz). **(a)** During weekend and **(b)** during the working week.

SPGF, have depths up to 8.5 km. As can be seen, epicentres of many local earthquakes recorded by DAFNE/FINLAND array show good spatial coincidence with the Suasselkä postglacial fault. This indicates that the fault zone is still seismically active. Generally, the epicentres are shifted to the southeast from the surface position of the fault. This agrees with the previous studies by Paananen (1987) and Kuivamäki et al. (1998), who suggested that the fault dips southeast.

4   Ambient noise analysis

Analysis of empirical Green's functions (EGFs), estimated from ambient seismic noise, has been widely used in order to study seismic velocities in the subsurface (see Shapiro and Campillo, 2004; Shapiro et al., 2005; Campillo, 2006). Poli et al. (2012a, b, 2013) applied analysis of EGFs to retrieve

body waves reflected from the Moho and upper mantle discontinuities and to estimate the 3-D S-wave velocity of the upper crust of northern Finland. In our study we apply the procedure described in Poli et al. (2012b, 2013) in order to estimate EGFs from continuous recordings of vertical component of all stations of the DAFNE array. The functions are then used in order to estimate seismic velocities in the uppermost crust of the SPGF area.

4.1   Analysis of spectrograms of the ambient noise

Some stations of the DAFNE temporary network were installed in the sites with a high level of anthropogenic noise (for example, station DF02 was installed near the Kittilä Gold mine; Fig. 1). Such noise is not propagating to large distances from the noise source, and this may cause a decline in the quality of EGFs. For identification of anthro-

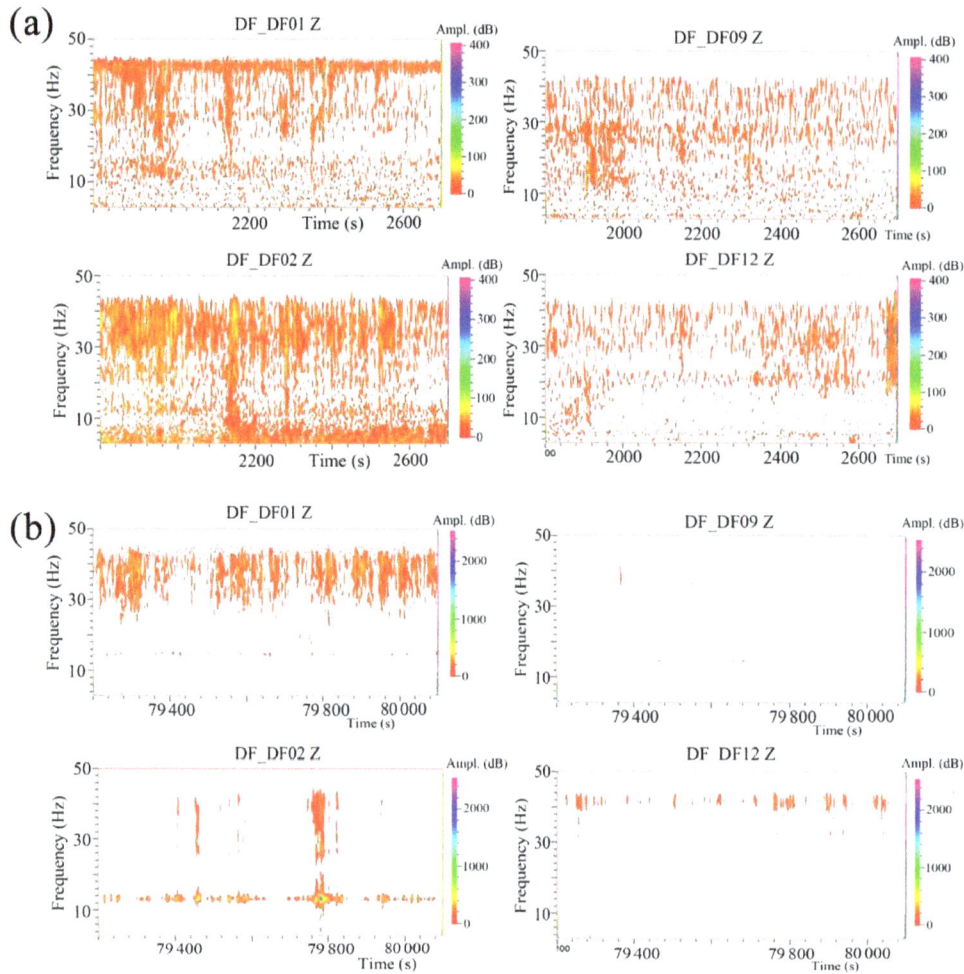

Figure 7. Examples of typical ambient noise spectrograms estimated for the high frequency band (3–50 Hz). **(a)** During the weekend and **(b)** during the working week.

pogenic noise on seismograms and selection of frequency band for filtering of the input data before cross-correlation, we used analysis of spectrograms. These spectrograms were estimated from continuous data recorded by stations closest to the mine (DF01, DF02, and DF09) and by the most distant station from the mine (DF12). We studied high-frequency and low-frequency parts of spectrograms separately and also compared the spectrograms calculated for weekends and for the working week. The examples are shown in Fig. 6a and b, respectively. As can be seen, all the spectrograms are identical in the frequency band of 0.1–3 Hz and characteristics of the noise in this frequency band do not depend on day of the week and location of stations with respect to the mine.

If the high frequencies are considered, one can see stable noise in the frequency band of ∼ 13–14 Hz for workday (Fig. 7b). This noise was recorded by stations in the vicinity of the mine (DF01, DF02, DF09), but it was not registered by the station most distant from the mine (DF12). Therefore, it is anthropogenic noise from the mine. It is worth noting that

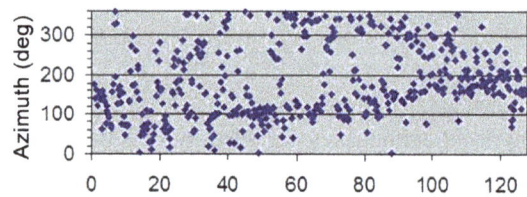

Figure 8. Azimuthal distribution of ambient noise sources during time periods from 1 January to 30 April 2012 and from 10 to 31 December 2012 estimated by beam forming in the time domain. Numbers on the abscissa axis correspond to the relative day numbers (see the text for additional explanation).

station DF12 was installed near the river, and thus the stable amplitude maximum at a frequency of about 42 Hz is most probably the noise of the river (Fig. 6b).

Thus, we applied filtering in the band of 0.1–1 Hz for all input seismograms before computation of EGFs.

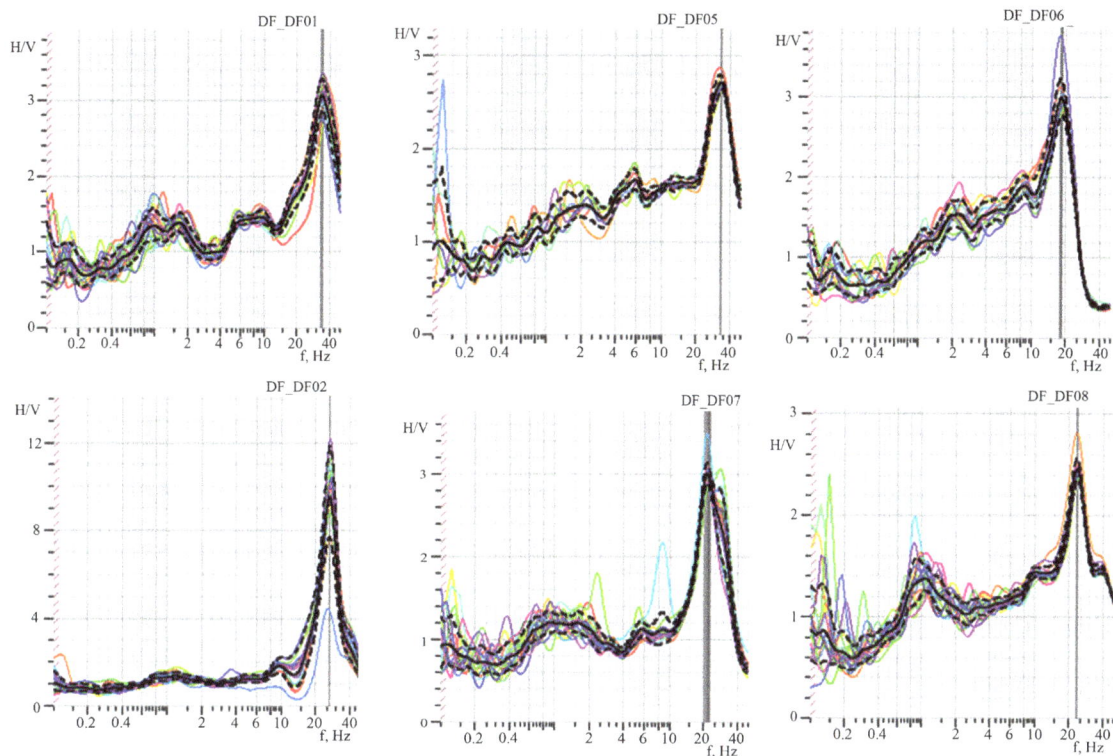

Figure 9. Results of horizontal-to-vertical (H/V) analysis for selected stations of the DAFNE array. The resonance frequency is marked by the grey bar, with width corresponding to error bar.

4.2 Analysis of azimuthal distribution of noise sources

Analysis of azimuthal distribution of ambient noise sources during the data acquisition period is an important part of data preparation for EGFs calculation, as the algorithm of their calculation is strongly dependent on directivity of noise sources. The algorithm is simpler if the noise sources have uniform azimuthal distribution. There are some methods of calculation of azimuthal distribution, such as f-k analysis (Neiddell and Taner, 1971; Douze and Laster, 1979) and beam forming (see Rost and Thomas, 2002; Schweizer et al., 2012). The configuration of the DAFNE array unfortunately does not allow for use of f-k analysis. That is why we analysed the azimuthal noise distribution using beam forming in the time domain. For this we calculated about 5000 EGFs between station DF09 and every other station of the DAFNE array. Six segments per day with a duration of 1 h were selected for each day from 1 January to 30 April 2012 and from 10 to 31. December 2012. During these time periods all the stations of the array were recording data. Surface wave parts of Green's functions were calculated from selected segments, which were band-pass-filtered in the band of 0.1–1 Hz. The result is shown in Fig. 8.

As one can see, during January and February 2012 (days 0–50) the ambient noise was recorded from various directions, with prevailing azimuth of 0–200°. From 1 March

2012 onwards the prevailing azimuths are of approximately 100 and 350–360°, and this tendency is slightly changing during March (days approximately 50–80) and April (days approximately 80–110). In December the prevailing azimuths are approximately 100–210°. The observed changes in azimuthal distribution of noise sources may be caused by reduction of marine microseisms during winter months because the range of 300–360° corresponds to azimuths to the seashore in the north. This distribution is necessary to take into account when interpreting EGF.

4.3 Horizontal-to-vertical (H/V) analysis of the ambient noise

The bedrock in our study area is covered by a thin (up to several dozen metres) layer of Quaternary deposits developed during the late Weichselian glaciation. In our study, we applied H/V analysis of ambient noise (Nakamura, 1989; SESAME H/V User Guidelines, 2005) in order to estimate thickness of the sedimentary layer. It was necessary to do this prior to calculation of velocity models using inversion of EGFs because low velocities in this layer can affect inversion results.

We applied the H/V analysis procedure implemented into Geopsy software (www.geopsy.org) to ambient seismic noise, recorded by selected DAFNE stations during winter and summer months. We used time segments selected from

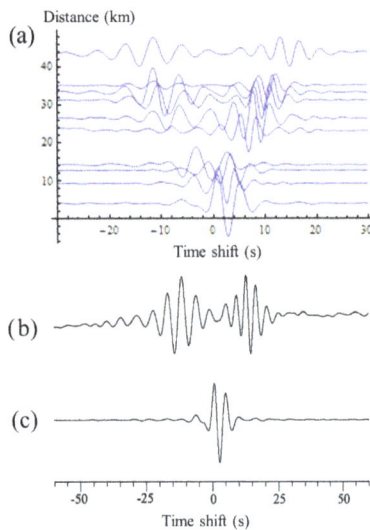

Figure 10. (a) Empirical Green's functions calculated between station DF01 and all other stations of the DAFNE array. **(b)** EGF calculated for stations with the largest interstation distance (DF01 and DF12). **(c)** EGF calculated for stations with the smallest interstation distance (DF01 and DF09).

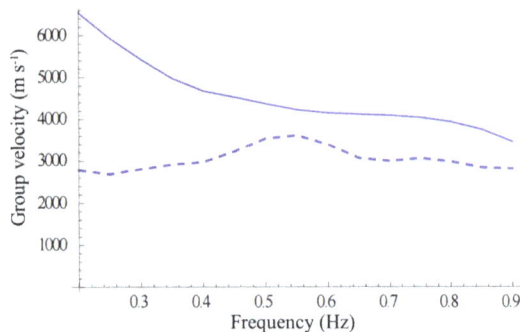

Figure 11. Example of dispersion curves corresponding to the fundamental and first-order higher mode of Rayleigh wave estimated from EGFs. The dashed line denotes the fundamental mode and the solid line denotes the first higher mode.

parts of recordings without anthropogenic noise with duration of 30 min. Results obtained for winter and summer time were identical. Fig. 9 demonstrates an example of the result of H/V analysis for some selected stations.

As can be seen from Fig. 9, the resonance frequency is between 20 and 40 Hz for all the data considered. From petrological data for sedimentary rocks (Gebrande et al., 1982; Kaikkonen, 2007), the S-wave velocities are about 250–350 m s^{-1} and P-wave velocities are about 500–700 m s^{-1} in our study area. Therefore, the maximum thickness of the sedimentary layer in the study area can be estimated as 5 m.

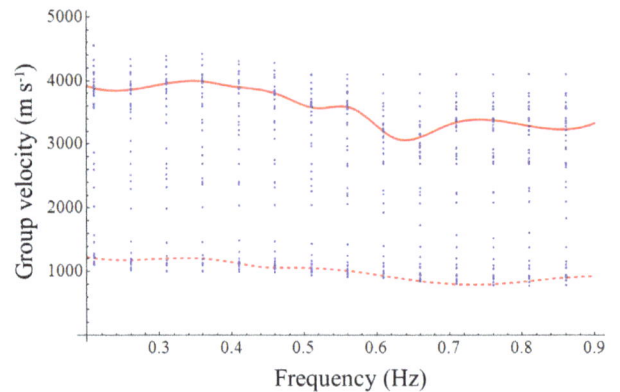

Figure 12. A set of dispersion curves estimated from all EGFs of DAFNE array (blue dots) that demonstrates their bimodal distribution. Averaged dispersion curves calculated for two groups of seismic station pairs (see explanation for Group 1 and Group 2 in the text) are shown by dashed line (Group 1) and solid line (Group 2).

5   Calculating empirical Green's functions and dispersion curves

For evaluation of EGFs we applied pre-processing procedure that includes pre-filtering in the frequency band of 0.1–1 Hz; deconvolution of instrument responses; and removing record parts with earthquakes, quarry blasts and explosions, and a high signal-to-noise ratio. After that, we calculated cross-correlation functions between all pairs of the DAFNE stations. Whitening was not applied. Figure 10a, b and c show Green's functions calculated between station DF01 and all other stations of the DAFNE array.

In Fig. 10 one can see that EGFs are asymmetric for pairs of stations with small interstation distances and symmetric for station pairs with large interstation distances. The asymmetry can be explained by non-uniformed azimuthal distribution of noise sources, different distances between stations and selected frequency band.

For extraction of dispersion curves, we applied narrow band pass filters with widths of 0.125 Hz to surface wave parts of all EGFs. We measured group velocities, but for the selected frequency band they can be approximately taken as phase velocities for dispersion curve evaluation.

Some of the station pairs of the DAFNE array have interstation distances less than the wavelength for some frequencies. That is why the corresponding calculated velocities had large error bars. These velocities were excluded from further processing. Some dispersion curves contained fundamental and the first higher mode of Rayleigh wave (Fig. 11). In our study we did not use the first higher mode and used fundamental mode only.

Figure 12 shows all dispersion curves selected for the analysis. As can be seen, the scatter of dispersion curves is large and distribution of them is generally bimodal. Dispersion curves for intermediate interval with velocities between

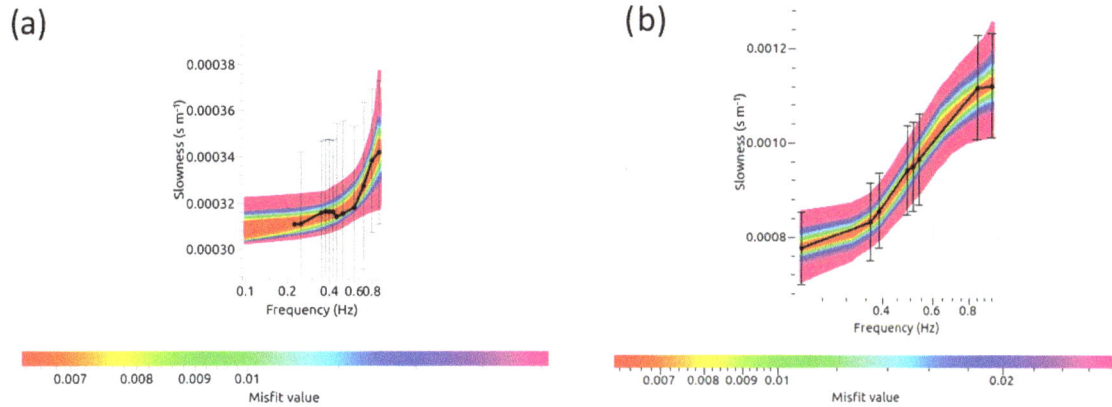

Figure 13. Results of inversion of averaged dispersion curves for station pairs of Group 1 (**a**) and Group 2 (**b**). The plots demonstrate the fit of model dispersion curves to the observed dispersion curve marked by the solid black line with the corresponding error bar. An ensemble of all models that equally fits the data and prior information is shown by colour plots, in which the colour scale indicates the value of misfit function.

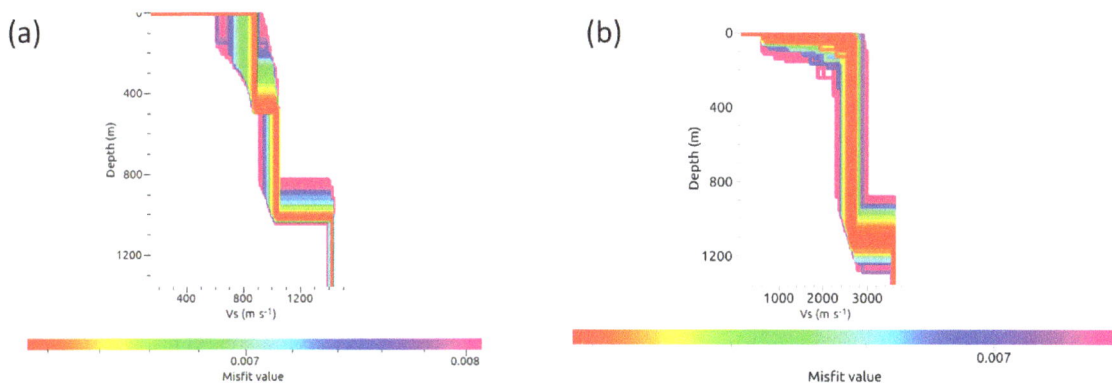

Figure 14. Results of inversion of averaged dispersion curves for station pairs of Group 1 (**a**) and Group 2 (**b**). The plots demonstrate the ensemble of all velocity models that equally fit the data and prior information. The colour scale indicates the value of the misfit function.

2000 and 3000 m s^{-1} correspond to waves propagating both along and across the fault zone, and the number of dispersion curves for this velocity interval is small. Because of this, inversion results for this group of curves would be not stable. In our study we discuss only results for which it was possible to obtain stable and statistically significant inversion results. We separately calculated averaged dispersion curve for EGFs corresponding to two groups of station pairs. The first group (referred hereafter as Group 1) is composed of the pairs in which stations are installed on different sides of the fault or on top of the fault, and the second group (hereafter Group 2) is composed of the pairs of stations installed on the same side of the fault and of the pairs with one of the stations (or both stations) installed on the fault. Figure 12 shows comparison of two different dispersion curves corresponding to Group 1 and Group 2.

In Fig. 12 one can see that dispersion curve for Group 1 indicates significantly smaller velocities than the dispersion curve for Group 2. This result suggests that seismic velocities

inside the fault zone are most probably significantly different from those outside the fault zone. Therefore, we decided to determine this difference using inversion of these two dispersion curves separately.

6 Inversion of dispersion curves

For inversion of dispersion curves we used the Geopsy software (www.geopsy.org). The software uses the neighbourhood global optimization algorithm by Sambridge (1999), modified by Wathelet (2008). As solutions of geophysical inverse problems are generally non-unique, the ideal solution in this method is made of the ensemble of all models that equally fit the data and prior information.

Table 3 presents the starting model and constraints on model parameters we used in inversion. Parameters of the uppermost layer of the model (V_p, V_s and density) were obtained from results of H/V analysis presented in our study and parameters for the other layers were constrained using

Table 3. Parameters of the starting velocity model for inversion of dispersion curves.

Depth (m)	V_p (m s^{-1})	V_s (m s^{-1})	Density (kg m^{-3})
5–15	700	350	1500
200–700	6150	3590	2030.27
800–2300	6350	3710	2096.27
15 000–20 000	6300	3650	2079.77

the P- and S-wave velocity models from previous study by Janik et al. (2009). These models were obtained for seismic controlled-source profiles closest to our study area.

We applied about 500 iterations for calculation of 1-D velocity models with minimum misfit function value. The results of inversion for two dispersion curves corresponding to Group 1 and Group 2 are presented in Fig. 13a and b and Fig. 14a and b. Figure 13a and b show two ensembles of calculated dispersion curves compared to two observed dispersion curves and Fig. 14a and b show two corresponding ensembles of velocity models. The values of the misfit function for each element of solutions ensemble are denoted by the colour scale in both figures. The error bars for the dispersion curve for pairs of stations outside the fault are larger than those for the dispersion curve for pairs of stations across the fault. This can be explained by variations in seismic velocities in the near-fault zone.

Figure 14a shows that 1-D S-wave velocity model for pairs of stations of Group 1 consists of three layers with low velocities (about 900–400 m s^{-1}). The pronounced velocity boundary is located at a depth of about 1000 m. For the pairs of stations of Group 2 the 1-D velocity model consists of two layers with generally higher velocities than in the previous case (from 2500 to 3500 m s^{-1}). As the Green's functions estimated by cross-correlation of recordings of two stations are dependant on the structure between them, this result suggests that an area exists inside the fault zone with seismic velocities significantly lower than those of the bedrock outside the fault zone.

7 Discussion and conclusions

Two major results obtained in our study can be formulated as follows:

1. Suasselkä postglacial fault zone is still seismically active, as shown by distribution of hypocentres of local earthquakes from the fault area detected by the DAFNE array.

2. Analysis and inversion of averaged dispersion curves obtained from EGFs for two groups of seismic stations pairs (e.g. the pairs in which stations are located on opposite sides of the fault and the pairs in which stations

are located outside the fault) revealed significant (about 1000 m s^{-1}) difference between seismic velocities inside the fault zone and outside the fault zone.

Unfortunately, we not able to analyse trapped waves in our study, unlike in the study by Allam et al. (2014). The main reason for this was that the events considered in our study were weak compared to those used for investigation of trapped waves in modern active fault areas. That is why the signal from them was seen clearly at the nearest stations only. At remote stations the signal-to-noise ratio was poor and correlation of phases corresponding to trapped waves was not possible.

As shown in Figs. 2 and 3, regional seismic networks did not register any earthquakes originating from the SPGF area, neither prior to DAFNE experiment nor during the time period considered in Sect. 3. In spite of that, we revealed a number of natural events using the data of the dense DAFNE array installed in the vicinity of the fault. This suggests that the magnitudes of these events are smaller than the magnitude detection threshold for the present configuration of permanent seismic stations. Analysis of the FENCAT catalogue shows that no events originating from northern Finland with local magnitude (M_L) less than 0.7 were detected during the DAFNE data acquisition period. Hence, our study suggests that local magnitudes of events detected in our study are less than 0.7 and the value of 0.5 can be assumed as a conservative estimate. The more precise magnitude estimate was difficult to do in our study, as the events were seen on band-pass-filtered recording only and such a filtering generally distorts the amplitudes of seismic signal.

Concerning the low velocities inside the fault area, these velocities cannot be explained by sediments because results of the H/V analysis presented in Sect. 4.3 suggest that the thickness of sediments is smaller than 5 m in our study area. Therefore, the most plausible explanation is that the low-velocity area is a fractured zone inside the fault that can correspond to the fault damage zone (FDZ). The velocities outside the fault zone are typical S-wave velocities in the felsic rocks of the uppermost crystalline crust documented in a petrophysical study of Finnish bedrock by Kern et al. (1993). The rough estimate of the width of this zone corresponds to a smaller wavelength in the considered frequency band of 0.1–1 Hz and is about 1.5 km. This is in a good agreement with the recent results of Valoroso et al. (2013), who estimated the width of the seismically active L'Aquila fault zone using analysis of high-precision earthquake locations spanning 1 year. They showed that the width of the fault zone varies along strike from 0.3 km, where the fault exhibits the simplest geometry and experienced peaks in the slip distribution, up to 1.5 km at the fault tips, with an increase in the geometrical complexity.

The velocity boundary at a depth of about 1200 m is seen in both velocity models obtained from average dispersion curves for Group 1 and Group 2. The origin of this boundary is not clear, but it cannot correspond to the lower boundary of the FDZ, as seismic velocities obtained for the Group 1 are lower than those for the Group 2 down to the depth of 1300 m. However, we can speculate that the lower boundary of the fault damage zone may correspond to the sub-horizontal area of high reflectivity revealed by Abdi et al. (2015). The authors also demonstrated the complex structure of the fault area down to a depth of 2–3 km, with two sets of segmented and discontinuous dipping reflectors. Slip on small-scale fractures corresponding to these reflectors may be the explanation of small magnitude of events detected by our study.

In summary, our study revealed that the SPGF is still seismically active and that a non-healed mechanically weak fault damage zone (FDZ) exists in the fault area. Most probably, the FDZ was formed after the last major earthquake that occurred after the last glaciation. This suggests that the SPGF has the potential for future reactivation. Our study also confirms that analysis of EGFs estimated by cross-correlation can be an efficient tool in investigating the inner structure of seismogenic faults.

Team list. The DAFNE/FINLAND Working Group consists of the following individuals: Ilmo Kukkonen (University of Helsinki, Department of Physics, Helsinki, Finland), Pekka Heikkinen (University of Helsinki, Institute of Seismology, Helsinki, Finland), Kari Komminaho (University of Helsinki, Institute of Seismology, Helsinki, Finland), Elena Kozlovskaya (Oulu Mining School, University of Oulu/ Geological Survey of Finland, Oulu, Finland), Riitta Hurskainen (Sodankylä Geophysical Observatory, University of Oulu, Oulu, Finland), Tero Raita (Sodankylä Geophysical Observatory, University of Oulu, Oulu, Finland), Hanna Silvennoinen (Sodankylä Geophysical Observatory, University of Oulu, Oulu, Finland).

Competing interests. The authors declare that they have no conflict of interest.

Acknowledgements. The present paper was a part of research projects DAFNE (Drilling into Active Faults in Northern Europe).

The study was partly funded by Posiva Oy and Geological Survey of Finland. The authors are grateful to the technical unit of the SGO for service of the DAFNE station during experiment. We give our particular thanks to Inna Usoskina, who performed tremendous work with manual analysis of the FINLAND/DAFNE data and event relocation for this study. The digital aeromagnetic map of Finland was provided by the Geological Survey of Finland.

Geopsy software (www.geopsy.org) was used for ambient noise analysis and inversion of dispersion curves. We very much appreciate the detailed and constructive comments of Alireza Malehmir, Gregor Hillers and the anonymous reviewer, which help to improve an earlier version of the manuscript.

This work was supported in part by the Federal Agency of Scientific Organizations, project no. AAAA-A16-116052710111-2.

Edited by: T. Nissen-Meyer

References

Ahmadi, O., Juhlin, C., Ask, M., and Lund, B.: Revealing the deeper structure of the end-glacial Pärvie fault system in northern Sweden by seismic reflection profiling, Solid Earth, 6, 621–632, doi:10.5194/se-6-621-2015, 2015.

Allam, A. A., Ben-Zion, Y., and Peng, Z.: Seismic Imaging of a Bimaterial Interface Along the Hayward Fault, CA, with Fault Zone Head Waves and Direct P Arrivals, Pure Appl. Geophys., 171, 2993, doi:10.1007/s00024-014-0784-0, 2014.

Brandes, C., Winsemann, J., Roskosch, J., Meinsen, J., Tanner, D.C., Frechen, M., Steffen, H., and Wu, P.: Activity of the Osning thrust during the late Weichselian: Ice-sheet and lithosphere interactions, Quaternary Sci. Rev., 38, 49–62, doi:10.1016/j.quascirev.2012.01.021, 2012.

Bratt, S. R. and Bache, T. C.: Locating Events with a Sparse Network of Regional Arrays, B. Seismol. Soc. Am., 78, 780–798, 1988.

Campillo, M.: Phase and correlation of "random" seismic fields and the reconstruction of the Green function, Pure Appl. Geophys., 163, 475–502, doi:10.1007/s00024-005-0032-8, 2006.

Chester, F. M. and Logan, J. M.: Composite planar fabric of gouge from the Punchbowl Fault, California, J. Struct. Geol., 9, 621–634, 1986.

Cochran, E., Li, Y-G., Shearer, P. M., Barbot, S., and Vidale, J. E.: Seismic and geodetic evidence for extensive, long-lived fault damage zones, Geology, 37, 315–318, doi:10.1130/G25306A.1, 2009.

Crone, A. J., De Martini, P. M., Machette, M. N., Okumura, K., and Prescott, J. R.: Paleoseismicity of two historically quiescent faults in Australia: Implications for fault behaviour in stable continental regions, B. Seismol. Soc. Am., 93, 1913–1934, 2003.

Davis, G. H. and Reynolds, S. J.: Structural Geology of Rocks and Regions, Wiley, New York, 1996.

Douze, E. J. and Laster, S. J.: Statistics of semblance, Geophysics, 44, 1999–2003, doi:10.1190/1.1440953, 1979.

Gebrande, H., Kern, H., and Rummel, F.: Elasticity and Inelasticity. In: Landolt-Börnstein Tables, Group V, edited by: Angenheister, G., Springer-Verlag, Berlin, 1–233, 1982.

Glitterman, Y. and van Eck, T.: Spectra of quarry blasts and microearthquakes recorded at local distances in Israel, B. Seismol. Soc. Am., 83, 1799–1812, 1993.

Glitterman, Y. and Shapira, A.: Spectral classification methods in monitoring small local events by the Israel seismic network, J. Seismol., 2, 237–256, 1998.

Hillers, G., Campillo, M., Ben-Zion, Y., and Roux, P.: Seismic fault zone trapped noise, J. Geophys. Res.-Sol. Ea., 119, 5786–5799, doi:10.1002/2014JB011217, 2014.

Janik, T., Kozlovskaya, E., Heikkinen, P., Yliniemi, J., and Silvennoinen, H.: Evidence for preservation of crustal root beneath the Proterozoic Lapland-Kola orogen (northern Fennoscandian shield) derived from P- and S- wave velocity models of POLAR and HUKKA wide-angle reflection and refraction profiles and FIRE4 reflection transect, J. Geophys. Res., 114, B06308, doi:10.1029/2008JB005689, 2009.

Juhlin, C. and Lund, B.: Reflection seismic studies over the endglacial Burträsk fault, Skellefteå, Sweden, Solid Earth, 2, 9–16, doi:10.5194/se-2-9-2011, 2011.

Juhlin, C., Dehghannejad, M., Lund, B., Malehmir, A., andPratt, G.: Reflection seismic imaging of the end-glacial Pärvie Fault system, northern Sweden, J. Appl. Geophys., 70, 307–316, 2010.

Kern, H., Walther C., Flüh, E. R., and Marker, M.: Seismic properties of rocks exposed in the POLAR profile region – constraints on the interpretation of the refraction data, Precambrian Res., 64, 169–187, 1993.

Kaikkonen, T.: Kalliopinnan topografian määrittäminen geofysikaalisilla tutkimuksilla hangaskankaan pohjavesialueella Oulussa, MSc Thesis, University of Oulu, 2007 (in Finnish).

Kim, Y.-S., Peacock, D. C. P., and Sanderson, D. J.: Fault damage zones, J. Struct. Geol., 26, 503–517, 2004.

Kujansuu, R.: Recent faults in Lapland, Geologi, 16, 30–36, 1964 (in Finnish).

Kuivamäki, A., Vuorela, P., and Paananen, M.: Indications of postglacial and recent bedrock movements in Finland and Russian Karelia, Geological Survey of Finland, Report YST-99, Espoo, 92 pp., 1998.

Kukkonen, I., Olesen, O., and Ask, M. V. S.: PFDP Working Group. Postglacial faults in Fennoscandia: targets for scientific drilling, Journal of geological society of Sweeden, 132, 71–81, 2010.

Malehmir, A., Andersson, M., Mehta, S., Brodic, B., Munier, R., Place, J., Maries, G., Smith, C., Kamm, J., Bastani, M., Mikko, H., and Lund, B.: Post-glacial reactivation of the Bollnäs fault, central Sweden – a multidisciplinary geophysical investigation, Solid Earth, 7, 509–527, doi:10.5194/se-7-509-2016, 2016.

McKenna, J., Stein, S., and Stein, C. A.: Is the New Madrid seismic zone hotter and weaker than its surroundings? in: Continental Intraplate Earthquakes: Science, Hazard and Policy Issues, edited by: Stein, S. and Mazotti, S., Geological Society of America Special Paper 425, 167–175, doi:10.1130/2007.2425(12), 2007.

Mikko, H., Smith, C. A, Lund, B., Ask, M., and Munier, R.: LiDAR-derived inventory of 25 post-glacial fault scarps in Sweden, J. Geol. Soc. Sweden, 137, 334–338, doi:10.1080/11035897.2015.1036360, 2015.

Nakamura, Y: A method for dynamic characteristics estimation of subsurface using microtremor on the ground surface. Railway Technical Research Institute, Quarterly Reports, 30, 25–33, 1989.

Neiddell, N. and Turhan Taner, M.: Semblance and other coherency measures for multichannel data, Geophysics, 36, 482–497, 1971.

Olesen, O., Blikra, L. H., Braathen, A., Dehls, J. F., Olsen, L., Rise, L., Roberts, D., Riis, F., Faleide, J. I., and Anda, E.: Neotectonic deformation in Norway and its implications: a review, Norw. J. Geol., 84, 3–34, 2004.

Paananen, M.: Geophysical studies of the Venejärvi, Ruostejärvi, Suasselkä and Pasmajärvi postglacial faults in northern Finland, Geological Survey of Finland, Nuclear Waste Disposal Research Report YST-59, 45, 1987.

Palmu, J. P., Ojala, A. E., Ruskeeniemi, T., Sutinen, R., and Mattila, J.: LiDAR DEM detection and classification of postglacial faults and seismically-induced landforms in Finland: a paleoseismic database, Journal of Geological Society of Sweden, 137.4, 344–352, 2015.

Poli, P., Pedersen, H. A., Campillo, M., and the POLENET/LAPNET Working Group: Emergence of body waves from cross-correlation of short period seismic noise, Geophys. J. Int. 188, 549–588, 2012a.

Poli, P., Pedersen, H., Campillo, M., and LAPNET Working Group: Body-wave imaging of Earth's mantle discontinuities from ambient seismic noise, Science, 338, 1063–1065, doi:10.1126/science.1228194, 2012b.

Poli, P., Campillo, M., Pedersen, H., and the POLENET/LAPNET Working Group: Noise directivity and group velocity tomography in a region with small velocity contrasts: the northern Baltic shield application to the northern Baltic Shield, Geophys. J. Int., 192, 413–424, doi:10.1093/gji/ggs034, 2013.

Rost, S. and Thomas, C.: Array seismology: methods and applications, Rev. Geophys., 40, 2-1–2-27, doi:10.1029/2000RG000100, 2002.

Sambridge, M.: Geophysical inversion with a neighbourhood algorithm—I. Searching a parameter space, Geophys. J. Int., 138, 479–494, 1999.

Schweitzer, J., Fyen, J., Mykkeltveit, S., Gibbons, S. J., Pirli, M., Kühn, D., and Kværna, T.: Seismic Arrays, In New Manual of Seismological Observatory Practice(NMSOP-2). GFZ German Research Centre for Geosciences (Deutsches GeoForschungsZentrum GFZ), ISBN 3-9808780-0-7.9., 1–80, 2012.

SESAME H/V User Guidelines: Guidelines for the implementation of the H/V spectral ratio technique on ambient vibrations – measurements, processing and interpretations, SESAME European research project, deliverable D23.12, 2005.

Shapiro, N. M. and Campillo, M.: Emergence of broadband Rayleigh waves from correlations of the ambient seismic noise, Geophys. Res. Lett, 31, L07614, doi:10.1029/2004GL019491, 2004.

Shapiro, N. M, Campillo, M., Stehly, L., and Ritzwoller, M. H.: High-Resolution Surface-Wave tomography from ambient seismic noise, Science, 307, 1615–1618, 2005.

Shipton, Z. K. and Cowie, P. A.: A conceptual model for the origin of fault damage zone structures in high-porosity sandstone, J. Struct. Geol., 25, 333–344, 2003.

Sutinen, R., Hyvönen, E., and Kukkonen, I.: LiDAR detection of paleolandslides in the vicinity of the Suasselkä postglacial fault, Finnish Lapland, Int. J. Appl. Earth Obs., 27, 91–99, 2014.

Stein, S.: Approaches to continental intraplate earthquake issues, in: Continental Intraplate Earthquakes: Science, Hazard and Policy Issues, edited by: Stein, S. and Mazotti, S., Geological Society of America Special Paper, 425, 1–16, doi:10.1130/2007.2425(01), 2007.

Valoroso, L., Chiaraluce, L., Piccinini, D., Di Stefano, R., Schaff, D., and Waldhauser, F.: Radiography of a normal fault system by 64,000 high-precision earthquake locations: The 2009 L'Aquila (central Italy) case study, J. Geophys. Res.-Sol. Ea., 118, 1156–1176, doi:10.1002/jgrb.50130, 2013.

Vidale, J. E. and Li, Y.-G.: Damage to the shallow Landers fault from the nearby Hector Mine earthquake, Nature, 421, 524–526, doi:10.1038/nature01354, 2003.

Wathelet, M.: An improved neighborhood algorithm: Parameter conditions and dynamic scaling, Geophys. Res. Lett., 35, L09301, doi:10.1029/2008GL033256, 2008.

Zigone, D., Ben-Zion, Y., Campillo, M., and Roux, P.: Seismic Tomography of the Southern California Plate Boundary Region from Noise-Based Rayleigh and Love Waves, Pure Appl. Geophys., 172, 1007–1032, doi:10.1007/s00024-014-0872-1, 2015.

11

Active faulting, 3-D geological architecture and Plio-Quaternary structural evolution of extensional basins in the central Apennine chain, Italy

Stefano Gori, Emanuela Falcucci, Chiara Ladina, Simone Marzorati, and Fabrizio Galadini

Istituto Nazionale di Geofisica e Vulcanologia, Rome, Via di Vigna Murata 605, 00143, Italy

Correspondence to: Stefano Gori (stefano.gori@ingv.it)

Abstract. The general "basin and range" Apennine topographic characteristic is generally attributed to the presently active normal fault systems, whose long-term activity (throughout the Quaternary) is supposed to have been responsible for the creation of morphological/structural highs and lows. By coupling field geological survey and geophysical investigations, we reconstructed the 3-D geological model of an inner tectonic basin of the central Apennines, the Subequana Valley, bounded to the northeast by the southern segment of one of the major active and seismogenic normal faults of the Apennines, known as the Middle Aterno Valley–Subequana Valley fault system. Our analyses revealed that, since the late Pliocene, the basin evolved in a double half-graben configuration through a polyphase tectonic development. An early phase, Late Pliocene–Early Pleistocene in age, was controlled by the ENE–WSW-striking and SSE-dipping Avezzano–Bussi fault, that determined the formation of an early depocentre towards the N–NW. Subsequently, the main fault became the NW–SE-striking faults, which drove the formation during the Quaternary of a new fault-related depocentre towards the NE. By considering the available geological information, a similar structural evolution has likely involved three close tectonic basins aligned along the Avezzano–Bussi fault, namely the Fucino Basin, the Subequana Valley, and the Sulmona Basin, and it has been probably experienced by other tectonic basins of the chain. The present work therefore points out the role of pre-existing transverse tectonic structures, inherited by previous tectonic phases, in accommodating the ongoing tectonic deformation and, consequently, in influencing the structural characteristics of the major active normal faults. This has implications in terms of earthquake fault rupture propagation and segmentation. Lastly, the morpho-tectonic setting of the Apennine chain results from the superposition of deformation events whose geological legacy must be considered in a wider evolutionary perspective. Our results testify that a large-scale "basin and range" geomorphological feature – often adopted for morpho-tectonic and kinematic evaluations in active extensional contexts, as in the Apennines – just led by range-bounding active normal faults may be actually simplistic, as it could not be applied everywhere, owing to peculiar complexities of the local tectonic histories.

1 Introduction

The presently active normal fault systems are commonly supposed to be the major ones responsible for the recent and present-day regional morpho-tectonic aspect of the central Apennine chain. Indeed, since the late Pliocene, extension took place through normal fault systems presently occurring at the boundary between intermontane basins and mountain ranges (e.g. Cavinato and De Celles, 1999; Galadini and Messina, 2004). The progressive lowering of the fault hanging walls and the relative uplift of the footwalls created alternating morphological/structural highs and lows, that represent the Quaternary geomorphic leitmotiv of the Apennine chain, producing a typical "basin-and-range" physiography. Nonetheless, extensional deformation displaced an inherited thrust-and-fold belt, whose external thrust fronts were still active when regional extension began to affect the inner sectors of the chain (e.g. Carminati and Doglioni, 2012). This resulted in the overprinting of the extensional deformation on the compressive one. In addition, an ancient morpho-genetic phase, subsequent to the compressive tectonic phase but shortly preceding the onset of extensional deformation, shaped an embryonic central Apennine relief. The related geomorphic signature is represented by the remnants of a low-gradient erosional relict landscape, presently detectable at high elevations along the mountain slopes, carved into the compressively deformed bedrocks and which has been sub-sequently displaced by the extensional faults (Centamore et al., 2003; Galadini et al., 2003).

Therefore, as pointed out by other authors in the past (Valensise and Pantosti, 2001), the "basin and range" phys-iography of the Apennine chain, as being ascribable just to the long-term movements of the presently active extensional faults, can be sometimes illusory. In this perspective, the presence of thrust-top basin sediments, deposited during the compressive tectonic phase, at the margins of some present-day intermontane tectonic basins (e.g. Cosentino et al., 2010, and references therein) suggests that those sectors already represented tectonic lows before extension began, and nor-mal faulting has just contributed to enlarge those depressions.

A further element that complicates the structural setting of the central Apennine chain is the recognition of NE–SW-striking, chain-transverse tectonic structures, recognized by different authors in the past, and whose role in both the Pliocene–Quaternary evolution and the current seismotecton-ics of the belt has been analysed. According to Pizzi and Galadini (2009), NNE–SSW-striking pre-existing structures can act as transfer faults, permitting the propagation of the rupture along adjacent active fault segments, or as structural barriers, halting rupture propagation. Such an opposite role relates to the size of the structure. Specifically, the second case refers to regional basement/crustal oblique pre-existing cross-structures. An example of Apennine chain-transverse fault that can act as barrier to hinder fault rupture propaga-tion has been proposed by Pace et al. (2002), dealing with the source of the M_s 5.8 1984 Sangro Valley earthquake. Ac-cording to Fracassi and Milano (2014), the regional tectonic structure known as the Ortona–Roccamonfina Line allowed soft linkage between active faults across it, as well as the shift of fault dip direction, that is SW-ward and NE-ward, north and south of the Line, respectively.

In the present work, we analyse one of the innermost in-termontane basins of the central Apennines, the NW–SE-striking Subequana Valley, which is bounded to the northeast by the southern segment, called the Subequana Valley fault, of a major active and seismogenic 30–35 km long Middle Aterno Valley–Subequana Valley normal fault system (Fal-cucci et al., 2011), able to rupture during earthquakes with M of up to 6.5–7. Data on the structural framework and evolu-tion of this depression, matched with a 3-D view of the deep geometry of the valley, are gathered to understand the rela-tionship between the activity of the main tectonic structures bounding the basin and, consequently, to decipher the Qua-ternary evolution of the depression. Moreover, we compare

the data obtained in the Subequana Valley with the geolog-ical and geophysical information available for other nearby tectonic depressions, namely the Fucino and the Sulmona basins. Our aim is to shed light on the possible occurrence of a common evolutionary path of the three depressions through the Pliocene–Quaternary. In this perspective, we will investi-gate the role of a regional chain-crossing fault inherited from the compressive tectonic phase in accommodating tectonic deformation through the Pliocene–Quaternary. The analysis of the structural relation between this structure and the active normal faults bounding the Subequana Valley, and the Fu-cino and Sulmona basins can allow us to make inferences on the role of such pre-existing structural features in the evolu-tionary path of the Apennine chain and in its seismotectonic characteristics, in terms of fault segmentation, possibly ap-plicable to many other sectors of the Apennine chain.

To fulfil the objective of the work we adopt a multi-methodological approach that combines geological and geo-physical data. Then, general aspects of the Apennine chain evolution are illustrated, followed by more detailed geologi-cal information concerning the Subequana Valley. The results of our investigation are then illustrated and discussed. We ex-amine the implications in terms of Quaternary structural evo-lution of the basin as well as of a wider portion of the chain, proposing an evolutionary structural model. In the conclud-ing remarks, a summary of the obtained results is provided and general considerations on their meaning in terms of ac-tive tectonic setting of this portion of the belt are made.

2 Geological background

2.1 Tectonic evolution of the central Apennines

Since the Miocene, the tectonic history of the central Apen-nines has involved the occurrence of two kinematically oppo-site tectonic phases. The first phase was compressional, in re-sponse to the Africa, Eurasia, and Adria plates' convergence and subduction of the Ionian lithosphere (e.g. Faccenna et al., 2004; Carminati et al., 2004), and it has been characterized by development of NW–SE-oriented, E–NE-verging thrust-and-fold systems at the expenses of the Meso-Cenozoic car-bonate and siliciclastic marine sequences (e.g. Ciarapica and Passeri, 1998; Centamore et al., 1991; Patacca et al., 1990, 2008, and references therein). This phase built the main structural edifice of the chain, through subsequent episodes of migration of the compressive front, with in-sequence and out-of-sequence thrusting events (e.g. Ghisetti and Vezzani, 1991; Cipollari et al., 1997, 1999; Scrocca, 2006; Cosentino et al., 2010, and references therein).

Since the Pliocene, the compressive front progressively migrated towards the east and the northeast, affecting the Adriatic sectors, while extension began to affect the inner sector of the chain (e.g. Malinverno and Ryan, 1986; Patacca et al., 1990; Doglioni, 1991, 1995), with a progressive migra-

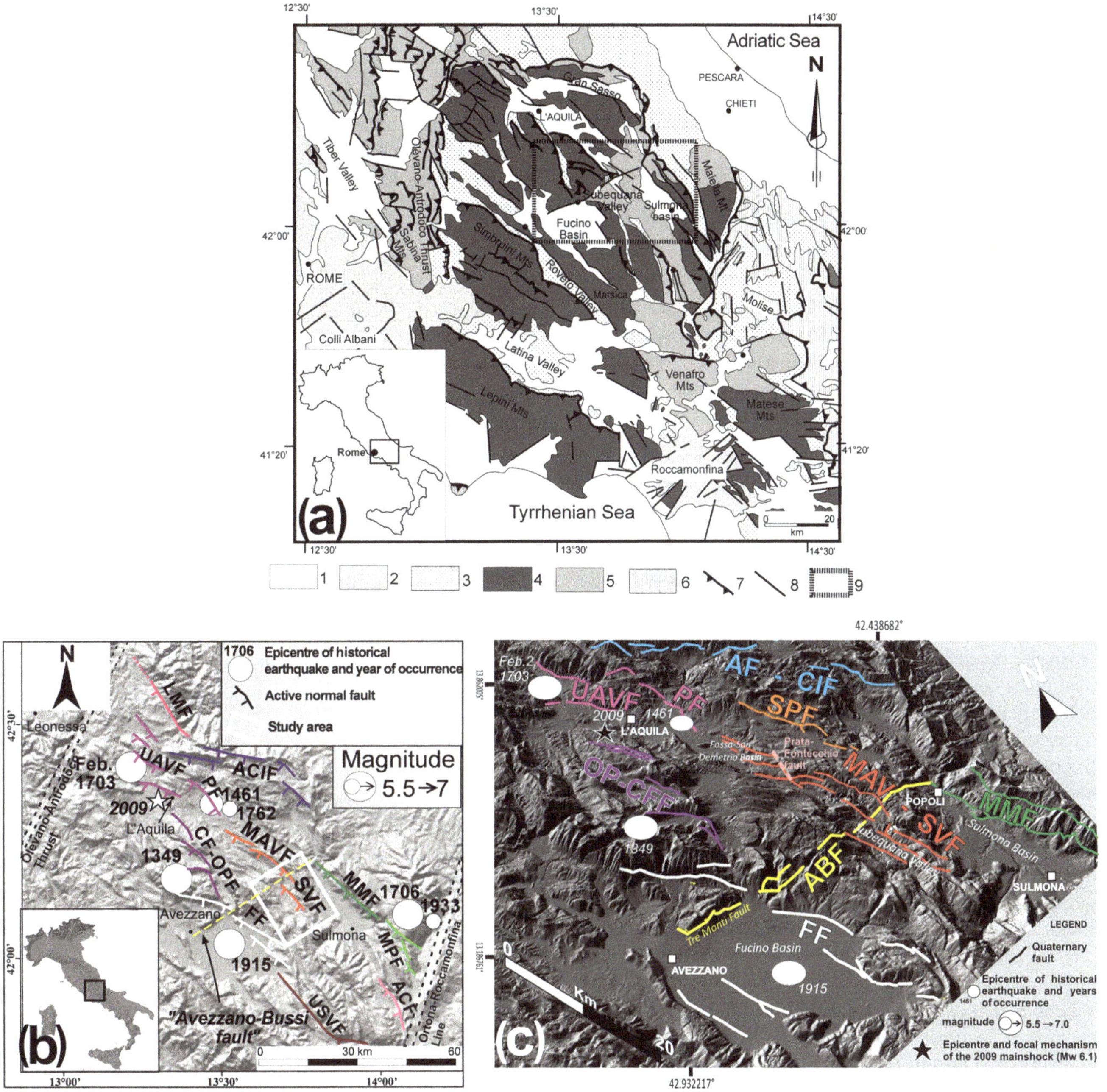

Figure 1. (a) Geological–structural scheme of the central Apennines. (1) Marine and continental clastic deposits (Pliocene–Quaternary); (2) volcanic deposits (Pleistocene); (3) synorogenic hemipelagic and turbiditic sequences (Tortonian–Pliocene); (4) carbonate platform deposits (Triassic–Miocene); (5) slope and pelagic deposits (Lias–Miocene); (6) Molise–Sannio pelagic deposits (Cretaceous–Miocene); (7) main thrust fault; (8) main normal and/or strike–slip fault; (9) study area. **(b)** Seismotectonic framework of the central Apennines (shaded relief); faults (colours indicate splays and segments comprised in the same fault system): MAV-SVF, Middle Aterno Valley–Subequana Valley fault system; ABF, Avezzano–Bussi fault; FF, Fucino fault; MMF, Mt Morrone fault; OP-CFF, Ovindoli–Pezza–Campo Felice fault, SPF, San Pio fault; AF-CIF, Assergi–Campo Imperatore fault system; UAVF, Upper Aterno Valley fault system; PF, Paganica fault. **(c)** Shaded relief in perspective view of the seismotectonic setting of the area under investigation.

tion towards the external sectors of the chain (e.g. Cavinato and De Celles, 1999; Galadini and Messina, 2004; Fubelli et al., 2009, and references therein). The eastward migration of compression–extension pair (e.g. Carminati and Doglioni,

2012) occurred contemporaneously to regional uplift (e.g. D'Agostino et al., 2001; Galadini et al., 2003; Pizzi, 2003). Specifically, uplift seems to have temporally occurred at the compression–extension transition, and a range of evidence indicates that mantle played a role in chain doming and consequent localization of major extension at crest of the topographic bulge (e.g. D'Agostino et al., 2001). Hence, topography of the chain derives from competition of chain uplift and extensional faulting along major normal fault systems, paralleling the axis of the chain. According to D'Agostino et al. (2014), active extensional deformation within the chain would be driven by lateral variation in the gravitational potential energy of the lithosphere. These faults offset the thrust-and-fold-related structures and caused the development of mainly half-graben tectonic depressions, with the main basin-boundary faults on their eastern side, mostly NW–SE-striking and dipping southwestwards. The depressions became traps for continental sedimentary sequences (e.g. Bosi et al., 2003). Many of the extensional faults located in the innermost sector of the chain are presently active (Fig. 1) and responsible for large-magnitude, surface-rupturing seismic events (M_w of up to 7; Galadini and Galli, 2000; Basili et al., 2008; Galli et al., 2008). One of the active normal faults is the NW–SE Paganica fault, activated during the M_w 6.2 2009 L'Aquila earthquake (Falcucci et al., 2009; Boncio et al., 2010; Emergeo, 2010; Chiaraluce et al., 2011; Cinti et al., 2011; Galli et al., 2010; Vittori et al., 2011; Gori et al., 2012; Moro et al., 2013).

Also the extensional faults get younger towards the east pursuing the compressive front (e.g. Cavinato and De Celles, 1999; Fubelli et al., 2009). An early phase of local tectonic extension has been recognized within some of the central Apennine intermontane tectonic basins, such as the Fucino Basin (Galadini and Messina, 2001) and the L'Aquila Basin (Nocentini, 2016), during the late Pliocene–early Quaternary. This phase seems to have not been accommodated along the presently active basin bounding normal faults, but along pre-Quaternary structures. One of these structures is a major fault system that cuts across the central sector of the chain and it is known as the Avezzano–Bussi fault system (hereafter ABF; Fig. 1). As will be shown below, the available literature agrees in attributing to the ABF a significant role in leading to the first phase of formation of the Fucino Basin. Other major chain-transverse faults are also known, among which the so-called Olevano–Antrodoco Line and the Ortona–Roccamonfina Line, which are considered as the northern Apennines–central Apennines and central Apennines–southern Apennines structural boundaries, respectively. These structures, that probably re-activated older faults that separated different Meso-Cenozoic palaeogeographic marine domains, are defined to have accommodated differential migration velocities of thrust fronts during the compressive tectonic phase, also leading to out-of-sequence compressive tectonic events (e.g. Satolli and Calamita, 2008; Cosentino et al., 2010).

2.2 The Subequana Valley

The Subequana Valley is located in the inner sector of the central Apennine chain. The Aterno River currently drains the valley and hydrologically connects the L'Aquila Basin to the Sulmona Basin (Fig. 2a). The Subequana Valley is one of the less known intermontane depressions of the central Apennines as for the Quaternary continental morpho-stratigraphic evolution. Little information about the Quaternary geological evolution can be derived from the studies of Bosi and Bertini (1970) and Miccadei et al. (1997), who identified a thick outcropping lacustrine sequence attributed to the Early Pleistocene by the former authors and to the Middle Pleistocene by the latter (Fig. 2a). In particular, Bosi and Bertini (1970) proposed the presence of an Early Pleistocene single lacustrine basin encompassing the Subequana Valley, the Middle Aterno River valley, and the Fossa–San Demetrio–L'Aquila Basin, from south to north. Miccadei et al. (1997) also identified alluvial fan bodies, attributed to the late Middle Pleistocene and to the Late Pleistocene, embedded within the lacustrine succession (Fig. 2a).

The formation of the Subequana Valley is due to the activity of a ~ 10 km long normal fault, known as the Subequana Valley fault (hereafter SVF), that bounds the depression to the NE, striking NW–SE and dipping towards SW (Miccadei et al., 1997; Fig. 2a). Recent field investigations permitted Falcucci et al. (2011) to provide a detailed mapping of the fault, which is made of two main parallel fault splays affecting the southwestern slope of Mt Urano, some minor synthetic splays and a major antithetic fault on the western side of the valley (Fig. 2a).

The SVF is the southern segment of a 30–35 km long active fault system, the Middle Aterno Valley–Subequana fault (hereafter MAVFSVF; Falcucci et al., 2011, 2015; Fig. 1). Recent palaeoseismological investigations carried out along the MAVF–SVF system documented that the fault system ruptured twice during the late Holocene, with minimum 0.8 m surface offset per event (Falcucci et al., 2011, 2015).

2.3 The Avezzano–Bussi fault system

The ABF is a complex structure striking about ENE–WSW that bounds the Fucino Basin, the Subequana Valley, and the Sulmona Basin to the north (Fig. 1). According to Ghisetti and Vezzani (1997), who first made extensive structural analyses along the fault system, the ABF is made of two main segments, an eastern segment and a western segment. These are characterized by 1000 m dip slip displacement. The horizontal throw was the prevailing one, with about 5 km right lateral offset. The authors defined a polyphase kinematic history of the ABF – including transtensional to extensional kinematics – and hypothesized it to be a tear fault that absorbed differential velocities of thrust fronts' propagation during the compressive tectonic phase.

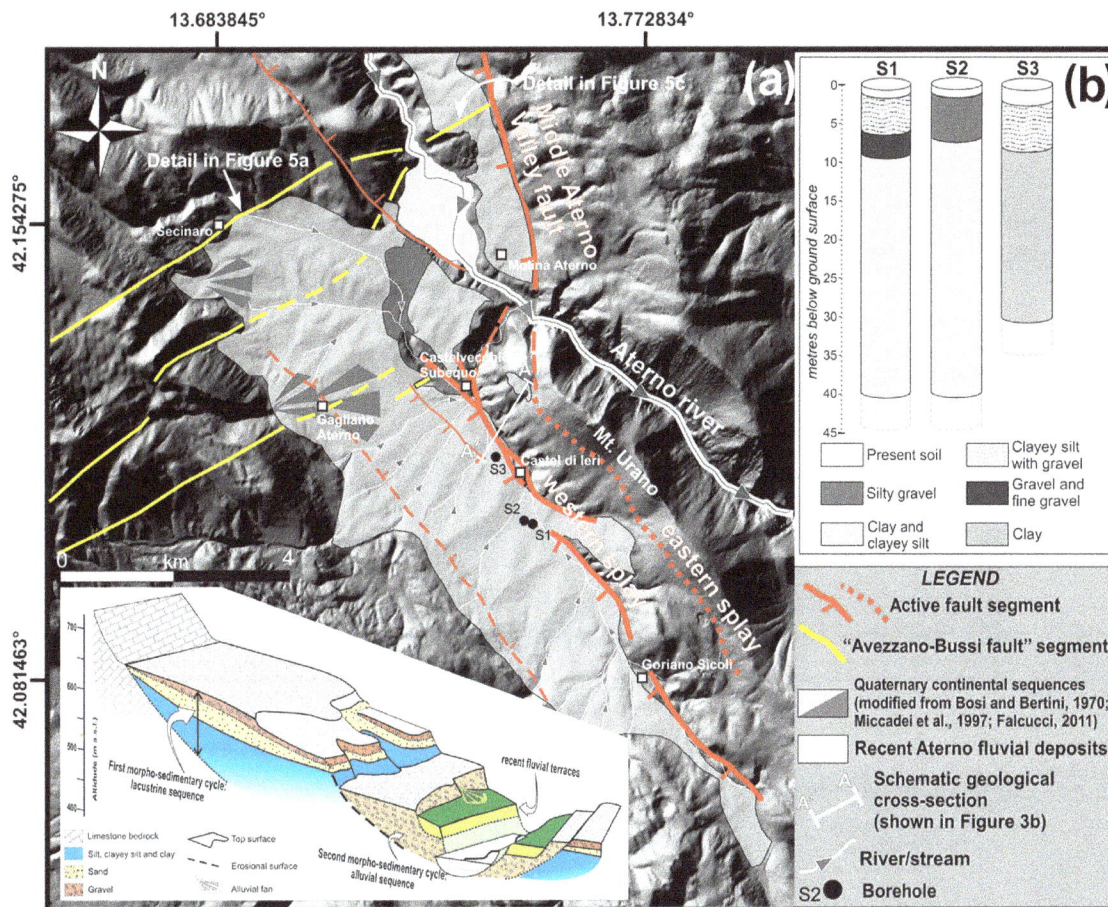

Figure 2. (a) Shaded relief of the Subequana Valley (plan view) on which simplified geological map (modified after Bosi and Bertini, 1970; Miccadei, 1997; Falcucci, 2011) is reported; morpho-stratigraphic reconstruction of the Quaternary succession of the Subequana Valley (Falcucci et al., 2011), inset. **(b)** Stratigraphy of boreholes drilled in the Subequana Valley during the 1970s.

Galadini and Messina (2001) deduced that the ABF has been active during the late Pliocene, controlling the earliest stages of formation of the Fucino Basin. The fault would have been active up to uppermost part of the late Pliocene–beginning of the Early Pleistocene, when the progressive spreading of the Fucino Basin has been driven by the NW–SE-striking and SW-dipping normal fault system on the eastern border of the depression (e.g. Michetti et al., 1996; Galadini et al., 1997; Galadini and Galli, 1999; Gori et al., 2007, 2015a), in relation to regional kinematic changes in the central Apennines tectonics during the Early Pleistocene (Galadini, 1999). The presently active Fucino fault is ~ 35 km long and has been responsible for high magnitude seismic events, e.g. the M_w 7, 1915 Avezzano earthquake (e.g. Galadini et al., 1997). Based on geological and geophysical (reflection seismic lines) observations, Cavinato et al. (2002) confirmed that ENE–WSW faults had a major activity during the Pliocene, and they progressively stopped (or slowed down) during the Quaternary. The NW–SE faults remained active during the Pleistocene and they are currently active.

Overall, this led to a complex 3-D basin architecture, related to the activity of fault systems having almost perpendicular trend. Galadini and Messina (2001) defined that the ENE–WSW-trending Tre Monti fault, a sub-segment of the ABF in the Fucino Basin, strongly reduced its activity since the early Quaternary, being responsible for only few-metres offset of Early Pleistocene slope deposits. According to the mentioned authors, this fault would presently just play the role of minor releasing fault (Destro, 1995), which only accommodates movements along the active NW–SE-trending Fucino fault system.

Galadini and Messina (2001) also made some inferences on the ABF activity in the Subequana Valley. They proposed fault activity during the Pliocene, comparably to the Fucino Basin, that ended before the Early Pleistocene, as early Quaternary land surfaces crosscut the fault. Nonetheless, differently from the Fucino Basin, data about the deep geometry of the Subequana Valley were not available at that time and this prevented the authors from verifying their hypothesis. As will be shown later, our data partly confirm the Galadini and Messina hypothesis but also modify it to some extent.

Lacustrine sequence: silt and sandy silt with overlaying sand and conglomerate

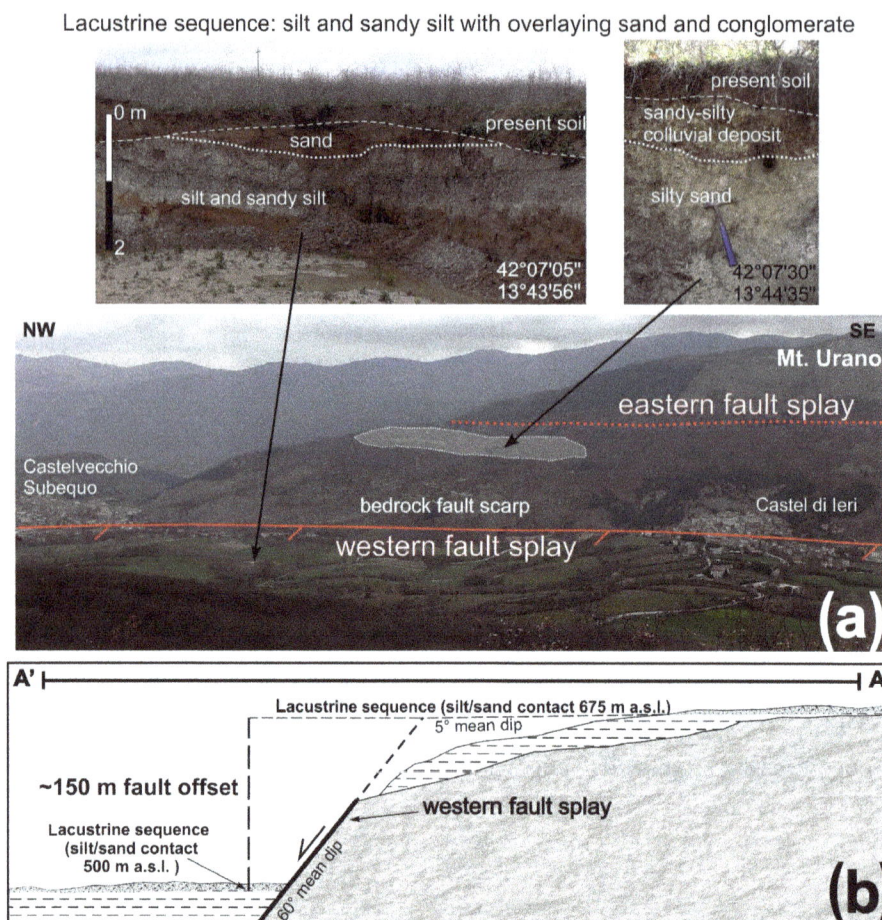

Figure 3. (a) Panoramic view of the southwestern slope of Mt Urano; major normal fault splay of the Subequana Valley fault, red lines; outcrops of the upper lacustrine sequence of the Subequana Valley are pointed by arrows and shown in insets. (b) Schematic geological cross section showing the displacement of the Subequana Valley lacustrine deposits across the western splay of the fault (see Fig. 2 for the section trace).

3   Geological field data

3.1   Morpho-stratigraphic investigation

Extensive field surveys and the analysis of the continental sequence hosted within the Subequana Valley allow us to recognize two main Quaternary morpho-sedimentary cycles (Falcucci, 2011; Gori et al., 2015b; Fig. 2a, inset). The former consists is a lacustrine sequence made of whitish-greyish silt and clayey silt that grades upwards to sand a few metres thick with sparse conglomerate interbedded layers. The sequence crops out all along the valley. Palaeomagnetic analyses made on the lake sediments and U series dating of two tephra layers found within the uppermost portion of the silt sequence mainly provided an Early Pleistocene age, with the sandy and conglomerate unit deposited across and shortly after the Matuyama–Brunhes palaeomagnetic reversal (Falcucci, 2011). This morpho-stratigraphic setting testifies for a mainly Early Pleistocene Subequana Lake basin – in agreement with Bosi and Bertini (1970) – that has been

progressively filled by alluvial deposition sourced from the surrounding reliefs just after the Early–Middle Pleistocene transition.

Afterward, a major erosional phase took place, as testified by an erosional surface deeply cut into the described lacustrine sequence, related to a significant drop of the base level. The second morpho-sedimentary cycle, Middle Pleistocene in age, lies onto this erosional surface, and is represented by fluvial/alluvial deposits cropping out along the northern portion of the valley, of the order of tens of metres thick. These merge to northerly fluvial sediments, that flowed from the Middle Aterno River valley. This suggests that the formerly closed hydrographic system of the Subequana Valley opened at the beginning of the Middle Pleistocene, after the lake basin infill; then, erosion took place and a palaeo-Aterno River system established. The second cycle is presently suspended over the Aterno River thalweg, owing to successive river entrenchment. Comparable morpho-stratigraphic evolution has been observed in many other central Apennines

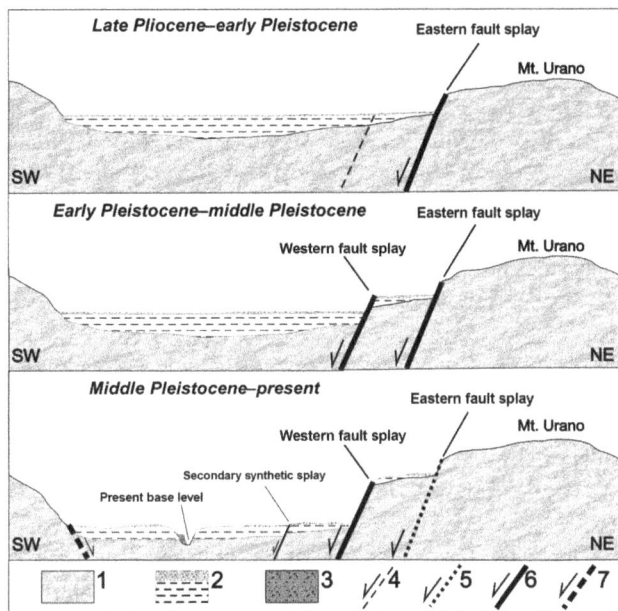

Figure 4. Scheme of the Quaternary structural evolution of the Subequana Valley fault and of Mt Urano southwestern slope. (1) Carbonate bedrock, (2) lacustrine sequence, (3) fluvial-alluvial deposits, (4) incipient normal fault, (5) inactive normal fault or normal fault whose current activity is presently uncertain, (6) active normal fault, and (7) supposed antithetic normal fault.

intermontane basins (e.g. Bosi et al., 2003), most of which hosted lakes during the early phase of formation. These ancient lakes have been subsequently cut by headward erosion of rivers, coming from outside the basins. This processes has been related to the increase of chain uplift which occurred at the Early–Middle Pleistocene transition (e.g. Centamore et al., 2003).

River incision, as well as minor streams, produced exposed thickness of the lacustrine sequence of several tens of metres. We also collected stratigraphic and sedimentological data from boreholes made some 40 years ago for water extraction (Fig. 2b) in the central sector of the basin, near Castel di Ieri. The deepest ones (about 40 m depth from the ground surface) cored the lake silt unit. None of the boreholes reached the bedrock. As will be shown later, this is in agreement with the attitude of the buried basin bottom defined by means of geophysical survey. Moreover, borehole S2 provided evidence of alluvial deposits (gravel and fine gravel) overlying the lake silt unit. This corroborates the stratigraphic setting above described, that is, the presence of the Early Pleistocene lake that has been progressively filled, grading upward into an alluvial environment. The bedrock progressively crops out in places at the northern portion of the basin at the base of the lake sequence (Fig. 2a), thus testifying to a quite articulated geometry of the basin bottom underneath the sedimentary cover.

3.2 Evidence for Quaternary activity of the SVF

The above-described morpho-sedimentary sequence mainly crops out on the hanging wall of the SVF. Nevertheless, remnants of the lake sequence are also seen in the sector comprised between the eastern and western splays of the structure, lying onto the carbonate substratum (Fig. 3a). Here, the sequence is suspended above the basin bottom, from which it is separated by the scarp of the western fault splay (Fig. 3a). This indicates that after the lake basin infill, the sedimentary sequence has been truncated by the western fault splay, which left the deposits hanging on the fault hanging wall. The difference in elevation of the contact between the sand-and-conglomerate body and the underlying silt at the footwall defines a post-Early Pleistocene fault offset of the order of some 150 m across the western fault splay (Fig. 3b).

The following lines of evidence allow us to hypothesize that the fault splay nucleation may have not been synchronous, that is, the eastern splay activated before the western splay or that the activity of the western splay strongly increased after the lake basin infill: (i) the lacustrine sequence occurs solely in the footwall of the western splay (it has never been found in the footwall of the eastern one); (ii) the geometric relation between the lake sediments and the eastern fault scarp seems to indicate that the scarp probably represented the border of the lake; (iii) the sub-horizontal attitude of the stratigraphic contact between the lacustrine unit and the overlying alluvial unit across the western fault suggests that the latter deposited in a palaeo-landscape still in the proximity of the local base level, i.e. the landscape evolved before the formation of the slope (and related topographic gradient) associated with the western fault or the western fault scarp was still too small to significantly influence the alluvial deposition (Fig. 4). The subsequent activation of the western fault or a significant increase of the fault slip would have definitively left the lake sequence hanging over the valley bottom. Such a structural evolution has been observed in other central Apennine active normal fault systems (e.g. Gori et al., 2007, 2014; Falcucci et al., 2015).

3.3 New data from the Avezzano–Bussi fault

As formerly defined by Galadini and Messina (2001), the ABF bounds the Subequana Valley to the northwest. The fault is here represented by discrete parallel splays, that define a roughly 2 km wide fault zone in the bedrock. From a geomorphic point of view, the presence of this tectonic structure is attested by a rather straight trend of the slopes, and by incisions in the carbonate bedrock that are aligned along the ABF; these incisions probably are probably localized along zones of weakness in the limestone rocks determined by fault shearing.

Field survey conducted along the ABF defined that in the sector comprised between the Fucino Basin and the Subequana Valley the tectonic structure is barely visible in the local slope morphology. In agreement with Galadini and Messina (2001), neither evident fault scarps are here visible nor any other morpho-tectonic features that can be related to the recent activity of the ABF. Geomorphic hints of the Quaternary ABF activity seem to end within the Fucino Basin.

The geomorphic and structural evidence of the ABF become again detectable within the Subequana Valley. More in detail, in the area of Secinaro, we found a high angle ENE–WSW-striking fault surface at the base of the slope, aligned along the ABF. Slope-derived breccias are displaced by a few metres and dragged, compatible with dip slip along the fault plane (Figs. 2a, 5a), defining a roughly normal-to-oblique (extensional) sense of motion. Structural observations related to the tectonic structure are shown in Fig. 5a.

The displaced deposits can be related by lithology – that is, angular-to-subangular carbonate clasts in a pink-orange cement – to those deposits widely distributed in the central Apennines, known as the Early Pleistocene "second sedimentary cycle breccias" of Bosi et al. (2003). This chronostratigraphic correlation is corroborated by the fact that they are hanging above the present bottom of local incisions, and their attitude progressively flattens (as moving away from the fault zone) at elevations comparable to the above-described lacustrine sequence. This suggests that the breccias refer to a base level higher than the present one, comparable to that of the Early Pleistocene lake. The slight displacement of the breccias indicates that the activity of minor tectonic features related to the ABF within the Subequana Valley may have lasted until the Early Pleistocene, or even later.

Furthermore, along the northeastern far end of the Subequana Valley, at the overlapping zone between the SVF and the MAVF, another high-angle ENE–WSW-striking fault plane is seen, aligned along the ABF. Here, slope deposits referable to the Middle Pleistocene (Falcucci et al., 2015) are displaced along the fault and affected by minor fault surfaces (Figs. 2a, 5b). The sense of displacement of the Middle Pleistocene slope deposits and structural features, as reconstructed from fault population analysis (Fig. 5b), indicate dominant normal dip-slip kinematics of the fault, but with very small offset. Therefore, together with the above-described faulted breccias, this evidence testifies to a post-Early/Middle Pleistocene activity of secondary faults that can be associated to the ABF within the Subequana Valley.

Moving to the E–NE, having crossed the MAVF, the geomorphic traces of the ABF disappear again along the reliefs that separate the Subequana Valley from the Sulmona Basin. Indeed, comparably to the sector comprised between the Fucino Basin from the Subequana Valley, no fault scarp or any other morpho-tectonic feature that may suggest Quaternary activity of the ABF is here seen. Entering the Sulmona Basin, instead, the northern border of the depression displays a quite rectilinear trend (about ENE–WSW oriented) and a discontinuous and altered bedrock scarp aligned along the ABF is seen at the base of the slopes (Fig. 6a). This geomorphic feature abuts the northern tip of the active normal fault that

bounds the Sulmona Basin to the NE, the Mt Morrone fault, a 23 km long structure active since the Early Pleistocene (e.g. Gori et al., 2011). Other ENE–WSW-striking faults are seen along the northern sector of Mt Morrone. One of these fault splays whose bedrock fault plane is seen in the Sant'Anna Valley, presently accommodates lateral movements of large-scale gravitational mass movements (Gori et al., 2014). Interestingly, sets of ENE–WSW- to NE–SW-striking sub-vertical fault planes are seen in the northernmost portion of the basin (Fig. 6b), just southwest of Popoli, affecting the Early Pleistocene lacustrine sequence (Giaccio et al., 2009; Gori et al., 2011), whose activity is possibly attributable to the Middle Pleistocene owing to stratigraphic similarities with the sequence described by Regattieri et al. (2016). The mainly normal sense of displacement of the lacustrine deposits, the almost vertical and often anastomosed, undulated and wavy attitude and geometry of the majority of the shear planes (lithons of the lake silty deposits are locally distinguished), and the variations of thickness of the same displaced layers across them (Fig. 6b) suggest a major normal kinematics with local normal–oblique component.

4 Geophysical survey in the Subequana Valley

4.1 Methods

In order to achieve information about the geological framework of the Subequana Valley, we combine data from field surveying of the Quaternary stratigraphic sequence filling the valley with data derived from the analysis of unpublished well logs of boreholes made after the 2009 L'Aquila earthquake for emergency housing projects.

To enlighten the deep structure of the Subequana Valley, a campaign of ambient seismic noise measurements was performed between July 2009 and February 2010 (Marzorati et al., 2010; Ladina et al., 2017). Ambient seismic noise is always present and it can provide information about the characteristics of the ground. The analysis of the seismic signals allowed the identification of the resonance frequency of the interface between the bedrock and the continental sedimentary fill. For this purpose, we performed extensive measurements all along the valley. Most of the measurements were performed with good weather condition, as weather can influence the quality of the measurements. However, to mitigate the possible effects of wind and rain, as well as of presence of vegetation on the ground, the sensors were protected by a cap and buried in the soil, where high vegetation could interact with the sensor due to the wind.

Microtremors were recorded through portable seismic stations equipped with Lennartz 5s (www.lennartz-electronic.de) velocimetric sensors and 24 bit Reftek 130 (www.reftek.com) digital acquisition systems (DASs). The stations were connected with a battery of 12 V/12 Ah and GPS antenna to precisely locate each measurement point. The gain was set

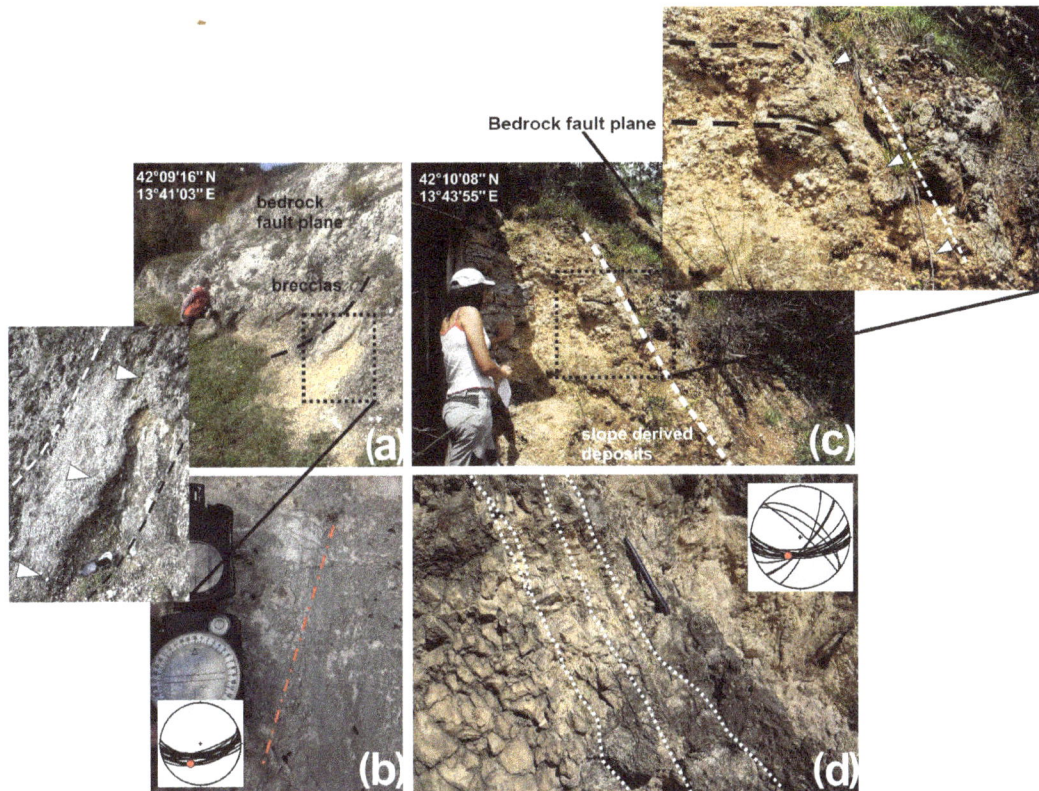

Figure 5. (a) Early Pleistocene breccias dragged along a secondary fault splay parallel to the Avezzano–Bussi fault (see Fig. 2 for location); a detail in inset, where white triangles indicate a fault plane affecting the breccias and dashed lines mark the dragging of the deposits. **(b)** Bedrock fault plane along which the breccias are dragged; slickenlines, red dashed line. on the Schmidt net diagram (inset), the synthetic shear planes detected in the fault zone (black great circles) and the pitch of the slickenlines, derived from statistical analysis of seven measurements (red dots) are plotted. **(c)** Middle Pleistocene slope deposits affected by extensional faults (white dashed line) parallel to the Avezzano–Bussi fault (see Fig. 2 for location); a detail in inset, where black dashed lines mark the attitude of the deposits in the fault zone, and the white dashed line and triangles mark a fault plane affecting the deposits. **(d)** Shear zone related to the fault plane in **(c)**; white dotted lines, synthetic shear planes; the synthetic shear planes detected in the fault zone (black great circles) and the pitch of the slickenlines (red dots) are plotted on the Schmidt net diagram (inset).

high in the digital acquisition to obtain the maximum resolution and the possibility to record the seismic noise vibrations. The acquisitions were set with rate of 100 sps and the length of measurement was at least 30 min each. The three acquisition channels (one for each component) were recorded on memory compact flash.

The ambient seismic noise recordings were manually selected in order to remove the possible traces of signal affected by disturbances and artificial transients (Marzorati et al., 2010). The waveforms of the three components of the signal were windowed in 120 s long time series. Then, a cosine taper 10 % and a bandpass Butterworth four-pole filter between 0.2 and 30 Hz were applied to compute the power spectral density (PSD) of each selected window. Each PSD was smoothed using the Konno and Ohmachi (1998) technique (with smoothing parameter $b = 40$) and used to compute the noise horizontal to vertical spectral ratio – NHVSR (Nogoshi and Igarashi, 1971; Nakamura, 1989, 2000).

4.2 Reconstruction of the deep geometry of the bedrock

Ambient seismic noise measurements provide a 1-D response of the basin at each measure point (Fig. 7a). By interpolating each measure, a 3-D picture of the deep morphology of the bedrock–sedimentary fill interface can be obtained, as illustrated in the following.

Generally, the resonance frequency f_0 of a sedimentary cover is described as the maximum of the ellipticity curve of Rayleigh waves, obtained by the mean of the horizontal-to-vertical (H/V) ratio of spectral amplitudes of noise components (Lachet and Bard, 1994; Tokimatsu, 1997; Bard, 1999; Nakamura, 2000; Fäh et al., 2001; SESAME, 2005). Modelling the resonant frequency by the 1-D quarter-wavelength approximation, the f_0 is given by the following equation:

$$f_0 = \frac{V_S}{4H}, \tag{1}$$

Figure 6. (a) Discontinuous bedrock fault scarp of the Avezzano–Bussi fault along the northern border of the Sulmona Basin. **(b)** Transtensional faults in the Sulmona Basin, aligned along one of the splays of the Avezzano Bussi fault, displacing Early Pleistocene lacustrine sequence.

where V_S is the shear wave velocity and H is the depth of the sediments. This relationship is used to infer thickness of sedimentary cover over a rigid bedrock in the presence of an impedance contrast (Yamanaka et al., 1994; Ibs-von Seht and Wohlenberg, 1999; Nakamura, 2000). From data processing, a set of 65 NHVSR curves resulted as reliable (see Supplement). The f_0 of each site was evaluated averaging all frequencies calculated from analysis windows selected from whole seismic traces (see Supplement). All the curves with no significant peak were considered as flat (SESAME, 2005).

In general, at the edges of the basin, frequencies were between 1 and 5 Hz, with the exception of the area in front of Castelvecchio Subequo, where a series of NHVSR curves were flat, indicating the presence of bedrock close to the ground surface (Fig. S1, Table S1; see Marzorati et al., 2011, and Pagliaroli et al., 2015, for details on the local seismic response in the Castelvecchio Subequo and Castel di Ieri areas). This is in agreement with the above-described geological observations. Also, the northern portion of the basin indicates frequencies around and greater than 1 Hz. The smaller f_0 (between 0.48 and 1 Hz) were founded in the central–

northern part of the basin and they were also present along the longitudinal axis of the basin towards the southeast.

Taking advantage of Eq. (1), it is possible to estimate the depth of the bedrock–sediments interface below each measuring point, using the average values of V_S of the sedimentary layers. Neither geological or geotechnical data are available for the deepest portion of the Subequana Valley. Nonetheless, the described geological setting of the basin and of other surrounding depressions – such as the L'Aquila Basin, the Middle Aterno River valley, and the Sulmona Basin, and the few borehole data available define that the majority of the sediments infilling the Subequana Valley is conceivably represented by the lacustrine sequence, which has geotechnical characteristics similar to those of the mentioned neighbouring basins.

After the 2009 L'Aquila earthquake, a campaign of seismic microzonation was conducted in the epicentral and mesoseismal areas (Gruppo di Lavoro MS-AQ, 2010). A geotechnical model of the lacustrine white carbonate silt was built to estimate the value of V_S in relation to the thickness of the sediments. A set of down-hole (DH) test measurements explored the maximum depth of 50 m. At higher depth, the V_S estimate was defined by measuring the variation of the small shear strain stiffness, G0, and the mean effective stress, p', by resonant column (RC) tests (Lanzo et al., 2011). The V_S value extrapolated to a depth of 200 m reaches 600 m s^{-1}. In the Sulmona Basin, the V_S of deepest lacustrine deposits estimated with joint inversion of seismic array techniques was also about 600 m s^{-1} (Di Giulio et al., 2016).

The piecewise linear function of V_S was interpolated by the following power law (Gruppo di Lavoro MS-AQ, 2010):

$$V_S = 230 H^{0.17}. \tag{2}$$

The V_S model of Eq. (2) provides velocity of deep lacustrine deposits comparable to that estimated for the L'Aquila and Sulmona basins by Lanzo et al. (2011) and Di Giulio et al. (2016), respectively. For the shallow layers (up to 40 m depth), V_S ranges between 300 and 400 m s^{-1}. This shallow V_S agrees with value estimated by means of the multichannel analysis of surface waves (MASW) technique in the Castelvecchio Subequo and Castel di Ieri areas (Salucci, 2010a, b), where the shallow layers are made of silt covered by sand and sandy gravel a few metres thick (Salucci, 2010a, b). Such lithology fits the class of "lacustrine white carbonate silts" as defined by geotechnical model in MZS Working Group (2011), which comprises silt, sand, and gravel layers.

In this perspective, the assumption that the Subequana Valley is mainly filled with the lacustrine sequence can be acceptable, considering the lithologies comprised within the class of "lacustrine white carbonate silts" defined by MZS Working Group (2011). Indeed, if a sand or gravel layer occurs in the deepest portions of the basin, it is lithologically included in the assumed sedimentary class.

Hence, to obtain an estimate of the thickness of the sediments above the bedrock in correspondence with seismic

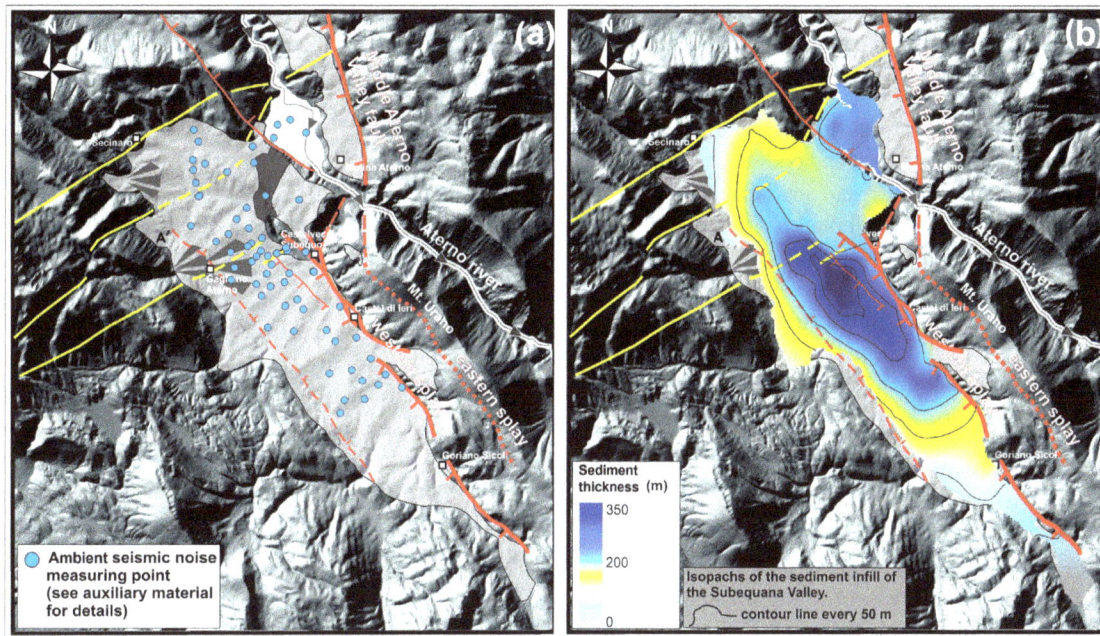

Figure 7. (a) Simplified geological map of the Subequana Valley (legend as in Fig. 2); location of the ambient seismic noise measuring points, blue circles. **(b)** Reconstruction of the deep geometry of the Subequana Valley (legend as in Fig. 2), defined by the sedimentary infill thickness.

noise measurement points, it is possible to derive the following equation, which describes the sediment thickness (H) as a function of the soil resonance frequency (f_0) from Eqs. (1) and (2):

$$H = (0.0174 f_0)^{-\left(\frac{1}{0.83}\right)}. \tag{3}$$

Figure 7b shows the sediment thickness obtained by seismic noise measurements (Fig. S2). The estimation points were interpolated by natural neighbour algorithm, and matched with the geological characteristics at surface, getting a 3-D view of the morphology of the bedrock/sediment infill interface (see Supplement). Specifically, the depth of the interface is obtained by subtracting the sediment thickness from the topographic height of each measurement point. As a result, the obtained reconstruction of the deep basin geometry defines the deepest zone in the central–northern part of the basin, where the estimate of the thickness of the sedimentary layers reaches values of up to 300–350 m (Supplement).

North of the deepest part of the basin, the bedrock gets progressively shallower approaching the ABF shear zone (Fig. 7b). To the east, instead, the bedrock abruptly rises at surface, indicating the presence of a high-angle buried scarp west of Castelvecchio Subequo, which coincides with a secondary synthetic splay of the SVF (Fig. 7b).

5   Discussion

5.1   Morpho-structural evolution of the Subequana Valley and of other basins aligned along the ABF

Geological and geophysical data describe a complex morpho-structural picture of the Subequana Valley. A comparison with the surrounding areas can help to decipher it in evolutionary terms and, consequently, to make hypotheses on the structural evolution of a wider sector of the Apennine chain.

The setting described for the Subequana Valley appears highly comparable to that of the Fucino Basin (Giraudi, 1988; Galadini and Messina, 2001; Cavinato et al., 2002). In detail:

1. The major active faults on the eastern border of both of the depressions active since the late Pliocene–Early Pleistocene are responsible for hundred-metres offset of deposits spanning the Quaternary – as documented in Sect. 3.2 – and determine significant surface offset (up to about 1 m) per event of activation.

2. The two basins are bounded to the north by the ABF that displays evidence of small displacements of early to Middle Pleistocene deposits along some of its strands. In this perspective, Falcucci et al. (2011) have defined that small portions of the ABF are presently used solely as connecting/transfer faults between two main segments of the SVF. Consistently, no evidence of Quaternary activity is seen along the portions of the ABF

in the mountainous sectors between the Subequana Valley and the Fucino Basin, to the west, and between the Subequana Valley and the Sulmona Basin, to the east.

3. Geomorphic/structural features that can be related to the ABF (such as rectilinear slopes and bedrock fault planes aligned along the structure) and evidence of Quaternary deposits displaced by fault planes aligned along the ABF are confined within the Fucino, Subequana Valley, and Sulmona basins, which are bounded by active normal faults. In the areas in between the depressions, instead, no geomorphic imprint of the ABF on the landscape is seen, and it is just geologically detectable for the lateral juxtaposition of different bedrock units.

4. The available geophysical data document an overall half-graben geometry for both the tectonic depressions, with maximum thickness of the infilling deposits – both buried underneath the plains and outcropping – not strictly centred with respect to the presently active faults, as would be expected, but occurring in the northeastern portion of the basins (Fig. 8).

As described earlier in this work, Galadini and Messina (2001) attributed these geological–structural features of the Fucino Basin to a double, superposed half-graben style of extension, firstly controlled by the ABF – mostly during the late Pliocene – and then by the NW–SE-trending and presently active Fucino fault during the whole Quaternary. As a result, the described similarities between the geological and structural characteristics, and between the 3-D deep geometry of the Fucino Basin and Subequana Valley indicate that a very similar evolution for both of the tectonic depression can be plausible.

Furthermore, interestingly, gravimetric investigations made by Di Filippo and Miccadei (1997) depicted a deep geometry of the Sulmona Basin markedly similar to that of the Fucino Basin and the Subequana Valley. Indeed, the authors defined a maximum depth of the depression in the northeastern portion of the basin, i.e. not centred with respect to the major active normal fault of the basin, the above mentioned Mt Morrone active normal fault (Fig. 8). The deep geometry of the Sulmona Basin defined by gravimetric data has been recently confirmed by Di Giulio et al. (2016) through ambient seismic noise measurements – that is, the same technique we adopted to investigate the Subequana Valley deep architecture. The authors defined the frequency of resonance of the bedrock–sedimentary fill interface, and imaged its trend beneath the sedimentary cover.

Therefore, the structural evolution defined for the Fucino Basin can be likely ascribed not only to the Subequana Valley but also to the Sulmona Basin. As for the ABF, in agreement with Galadini and Messina (2001), our observations indicate that, as in the Fucino Basin, it likely played during the Quaternary – and may still play – the role of releasing fault also in the Subequana Valley and the Sulmona Basin, with the dif-

ference that in the Fucino Basin and the Subequana Valley, the active fault systems cut across the ABF, while in the Sulmona Basin, the Mt Morrone fault ends against the ABF. A scheme of the proposed structural evolution of the depression is shown in Fig. 9.

Hence, to sum up, the three basins would have experienced a comparable structural evolution, with an early phase of nucleation during the late Pliocene ruled by the activity of the ABF and, likely, of the former strands of the NW–SE-trending and presently active normal faults. This phase resulted in the formation of early depocentres mostly located in the northern sectors of the present basins, at the intersection of the two almost perpendicular fault systems. Then, since the Early Pleistocene, the NW–SE-trending normal faults became the leading structures, and determined the formation of new depocentres centred on the major active faults, partly superposed on the older ones. Since then, the ABF has acted just as a local release fault (Destro, 1995) of the NW–SE-trending normal faults.

It must be discussed whether the very similar deep asymmetric shape displayed by the three basins could be related to an along-strike non-symmetric slip (i.e. slip that is maximum at the centre of the fault and that symmetrically tapers towards the tips) along the NW–SE faults. Along-strike asymmetric fault slip, indeed, would result in an asymmetric basin geometry, with the maximum thickness of the deposits located where the fault displacement is maximum, i.e. not centred to the fault. This can be ruled out for the following considerations. (1) In the Subequana Valley, the contact between the Quaternary infill and the bedrock reaches the surface at the tips of the SVF, with a certain asymmetry towards the north. However, in the northern part of the Valley, several hundred metres north of the fault tip – that is, where the long-term SVF offset is expected to be very low – a relevant thickness of lake/alluvial deposits is present at the hanging wall of the ABF, lying onto the outcropping bedrock, both on the hanging wall and on the footwall of the SVF. Here, the sedimentary sequence ends against the ABF, near Secinaro (Figs. 2, 5a). This implies that the basin depocentre can be only partly attributed to the Subequana Valley fault movements or only to interaction of the two orthogonal fault systems; otherwise, we should have found early Quaternary deposits only where the downthrown blocks of the two intersecting structures overlap. (2) If basin asymmetry was related just to a non-symmetric slip along the basin bounding active faults, this would imply that similar kinematic behaviour would have characterized all of the three faults, and for a very long time period, likely the whole Quaternary; and this is quite improbable. (3) Asymmetry of the deep geometry of the Sulmona Basin (with the deepest part located in the central-northern portion of the depression), comparable by shape with that of the Subequana Valley and Fucino Basin, does not fit the throw rate estimate in different sector of the Sulmona–Mt Morrone fault. Indeed, fault slip rate estimates, defined through the displacement of Early to Late Pleistocene

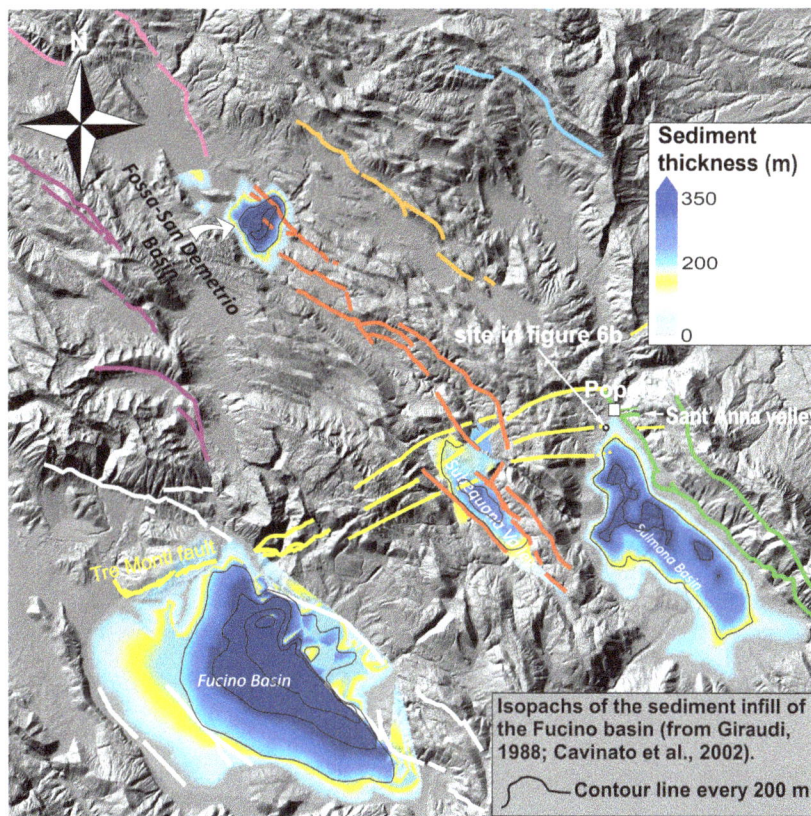

Figure 8. Shaded relief on which the reconstruction of the isopachs of the Fucino Basin, Subequana Valley, and Sulmona Basin sedimentary infill are reported, as well as that of the Fossa–San Demetrio depression, derived by crossing data from Bosi and Bertini (1970), Cesi et al. (2010), Linee Guida MZS (2010), Di Nezza et al. (2010), Di Filippo et al. (2011), Civico et al. (2015), and Pucci et al. (2016); legend of the fault systems as in Fig. 2.

continental deposits (e.g. Gori et al., 2011), defined very similar estimates in the central and southern sectors of the fault; conversely, slip rate values strongly decrease in the northern sector of the fault. This would be not consistent with the deep geometry of the Sulmona Basin. This implies that asymmetry of the Sulmona Basin does not relate solely to the long-term activity of the Mt Morrone fault.

5.2   Tectonic implications at regional and wider scale

In a wider perspective, further sectors of the central Apennine may have undergone a similar structural evolution, with chain-transverse tectonic structures active during the early phases of basin formation and presently acting just as secondary faults. For instance, the available literature suggests that one of these structures may have determined the early nucleation of the Fossa–San Demetrio depression, at the southern sector of the L'Aquila Basin. There, refined gravimetric prospection (Cesi et al., 2010; Di Nezza et al., 2010; Di Filippo et al., 2011), supported by further geophysical investigations (Civico et al., 2015; Pucci et al., 2016), document a gravimetric narrow low, roughly NE–SW oriented, that testifies for a deep bedrock–sedimentary fill interface,

buried by the Quaternary continental sequences. This low shows very steep, quite linear and sharp boundaries towards the north and south (Fig. 8). These boundaries coincide with an abrupt deepening of the carbonate bedrock (which sharply sinks into the basin) and seem to align with local NE–SW-trending cross-basin faults. As proposed by Falcucci et al. (2015), NE–SW-trending faults (such as the Prata–Fontecchio fault) at the northern sector of the active MAVF have been no more active as primary/major faults during the Quaternary, but they locally act only as secondary/transfer faults between the northern major segments of the active MAVF–SVF system. Hence, similarly to the cases of the Fucino Basin, and Subequana Valley and Sulmona basins, it is possible to hypothesize that the early phases of formation of the Fossa–San Demetrio depression have been led by cross-basin structures that have been subsequently cut by the Quaternary NW–SE-trending normal fault occurring in the area (e.g. Bosi and Bertini, 1970; Falcucci et al., 2015). The older structure may presently play only the role of fault segment boundary between the Paganica fault, to the northeast, and the MAVF-SVF system, to the southeast, hindering fault rupture propagation as they are structurally and kinematically separated (Gori et al., 2012; Fig. 8).

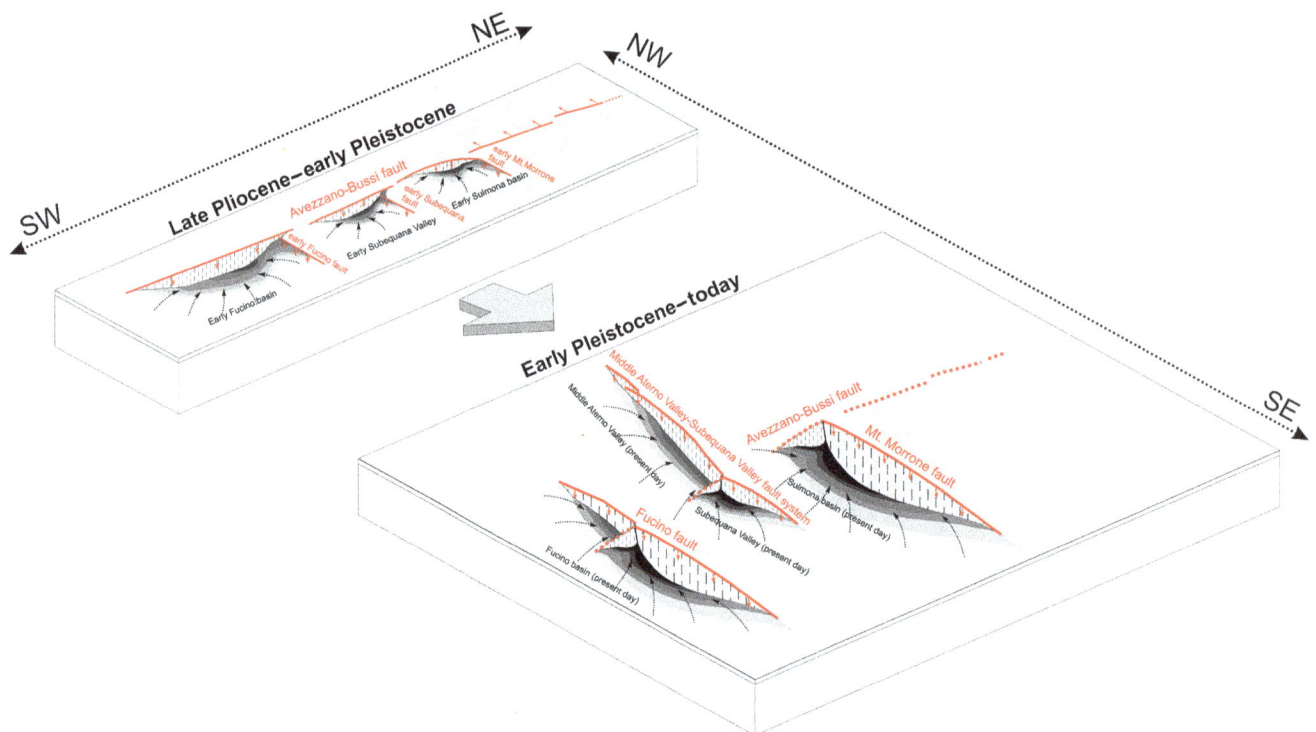

Figure 9. Scheme of the late Pliocene-to-present structural evolution of the Fucino Basin, Subequana Valley, and Sulmona Basin.

The results of our analysis also have important implications on some major seismotectonic aspects of the Apennine chain. The defined secondary role of the ABF in the recent tectonic activity is in agreement with the structural observation made by Roberts and Michetti (2004). Interestingly, the authors defined that the Tre Monti fault is probably not a major fault, owing to a number of structural and geomorphic features exhibited by the structure. Moreover, at a wider scale, the authors proposed an age of 2.5–3.3 Ma for the initiation of NW–SE-trending fault activity in the central Apennines, which is roughly consistent with the age that we propose for the onset of the Fucino, Subequana Valley, and Sulmona faults activity, when the ABF was still leading the formation of the early basins.

Our results, conversely, contrast with the hypothesis of Benedetti et al. (2013) regarding the recent kinematic behaviour of the Fucino fault system. The authors described the ENE–WSW-trending Tre Monti fault, a segment of the ABF, as a 20 km long active fault (actually being just 15 km long, as mapped by the authors) able to rupture primarily. Although the authors defined it as a minor fault, they attributed a high seismic potential to the structure ($M_{\mathrm{w}} > 6.4$ earthquakes originated along the fault, with some 0.7 m offset per event) and a high earthquake probability (probable rupture in the next ~ 0.2 ka). This hypothesis is based on cosmogenic nuclide ^{36}Cl dating of the fault plane exhumation, suggesting that the fault plane has been exposed only for tectonic slip. However, besides being not geometrically coherent with the present-day NE–SW-trending extensional stress regime (as the fault trends parallel to the main extensional axis), they do not discuss any other non-tectonic cause that may contribute to or even determine fault plane exposition, such as landsliding and/or erosion of the debris accumulated at the base of the fault scarp (e.g. Fubelli et a., 2009; Kastelic et al., 2015, 2017). Moreover, the secondary role we defined for the whole ABF implies that the Tre Monti fault is unlikely to nucleate primarily seismic events that strongly. This is in agreement with what Galadini and Messina (2001) already defined for the Tre Monti fault, i.e. the few-metre fault offset of early Quaternary deposits along the fault trace, comparably to what we observed along the ABF in the Subequana Valley. This would be therefore consistent with solely secondary movements of the structure to accommodate slip on the presently NW–SE-trending major active Fucino and Subequana Valley faults. Even if minor seismic events along the ABF cannot be completely ruled out, the 0.7 m slip per event defined by Benedetti et al. (2013) – that is consistent with a $M > 6.5$ earthquake, according to the regressions of Wells and Coppersmith (1994) – can actually incorporate contribution of the above-mentioned non-tectonic processes.

Other extensional tectonic contexts worldwide show interaction between orthogonal (or quasi-orthogonal) fault systems similar to that we describe. For instance, Şengör (1987) described cross-faults in western Turkey, which are controlled by zones of weakness that may be related to older structures and which trend perpendicular to the strike of the major extensional faults, that act as tear faults separating different downthrown blocks. In more detail, the author pointed out the importance of correct identification of different types of cross (or transverse) faults as this can result in incorrect assessments of the tectonic history of a given extensional tectonic setting. This is in agreement with the role of transverse faults in the central Apennines in accommodating tectonic extension, as Pizzi and Galadini (2009) pointed out and we attest for the ABF, and the importance of differentiating transverse structures for correctly assessing the role that they play in accommodating the ongoing tectonic regime as segment boundary or transfer faults. Morley et al. (2004) described oblique alignment of secondary faults and jogs in major faults of Thailand along oblique trends that indicates the structural influence of passive basement transverse fabrics. In particular, they concluded that pre-existing oblique fabrics (faults) can cause the oblique orientation of superposing faults, or can influence the location, geometry, and style of transfer zones, fault linkage, and displacement patterns. In the case we investigated, accordingly, the presence of the transverse ABF influenced the location of transfer faults between major active fault segments, as in the cases of the Fucino and Middle Aterno Valley–Subequana Valley fault systems, and the en-echelon arrangement between the segments.

Wilkins and Shultz (2003) described re-activation as extensional faults of pre-existing faults striking perpendicular to major extensional faults even on Mars, in response to stress changes related to slip change along the border faults.

6 Concluding remarks

Integrated geological and geophysical investigations in the Subequana Valley allowed us to get a 3-D view of the tectonic basin, that we framed in the Quaternary morpho-stratigraphic evolution of the depression.

Our analyses revealed an asymmetric deep geometry of the basin, characterized by maximum depth in the northeastern sector. The comparison with the structural evolution and setting of the nearby Fucino and Sulmona basins suggests that the three depressions experienced a similar double polarity nucleation, with two subsequent phases: an early phase, probably late Pliocene in age, led by the activity the regional chain-crossing, ENE–WSW-trending Avezzano–Bussi fault (ABF), contemporaneously to the early activation of the former strands of the NW–SE-trending normal faults. Afterwards, since the Early Pleistocene, the NW–SE basin bounding extensional faults took on the opening of the depressions, with the ABF just playing the role of release fault to accommodate slip on the major faults since then. A similar evolution may have been experienced by other basins of the Apennine chain.

We observed that the ABF is cut by the Fucino fault and, even more evidently, by the Middle Aterno Valley–Subequana Valley fault system. Conversely, the Sulmona–Mt

Morrone active normal fault ends against the ABF. Hence, following Pizzi and Galadini (2009), the about 40 km long ABF may be a chain-crossing structure at threshold size between acting as transfer fault or as structural barrier. This could have implications in terms of definition of the role of other chain-crossing faults in the seismotectonics of the central Apennines.

As for the cause of the shift in the trend of the extensional deformation – firstly led by chain-crossing faults and then by chain-parallel faults – we can hypothesize that the early phase might be associated to the formation of extensional basins at the bending or stepping of tear faults of the thrust fronts, as releasing bend zones (or pull-apart depressions), following the model proposed by Doglioni (1995). Subsequently, NE–SW-trending Quaternary regional extensional tectonics, related to the E and NE migration of the chain and slab retreat (Carminati and Doglioni, 2012) and, as exposed above, to chain doming caused by mantle dynamics (e.g. D'Agostino et al., 2001; Aoudia et al., 2007), operated accordingly through NW–SE fault systems.

We highlight that where deformation phases quickly follow one another, the tectonic evolution of a region frames within a progressive and continuous process, which implies morpho-structural heritage and interference that must be understood and considered in kinematic assessments of the presently active tectonic regime. In this perspective, a basic morpho-tectonic fault-bounded basin model in which the tectonic depressions are strictly ruled by just the long-term activity of the presently active bounding faults, cannot be applied tout court to the Apennines and, likely, to other extensional settings. In this perspective, the "basin and range" geomorphic setting of the central Apennine chain, where the relief evolution is strictly ruled by the activity of the major normal fault systems during the Quaternary, has to be viewed with great caution, and, in light of this, any tectonic analyses or kinematic evaluations in the central Apennine chain cannot be based on simply assuming a flat landscape preceding the currently active extensional tectonic regime. Instead, studies on the long-term morpho-tectonic evolution in whichever tectonically active region have to take into consideration any possible geological legacy over a time span of interest for neotectonics, which lies at the base of the definition of the term given by Bosi (1992) for the Italian case.

Author contributions. Stefano Gori and Emanuela Falcucci performed geological field investigations, manuscript writing, and tectonic interpretation of the geophysical survey and modelling carried out by Chiara Ladina and Simone Marzorati. General tectonic/structural aspects were discussed and shared with Fabrizio Galadini.

Competing interests. The authors declare that they have no conflict of interest.

Acknowledgements. The work strongly benefited from discussion on the Quaternary stratigraphy and active tectonics with Giandomenico Fubelli, Marco Moro, Lorenzo Santilli, and Michele Saroli, and Gianluca Valensise, who are warmly thanked. We also thank Sandro for providing us with the 1970s' Subequana Valley borehole data. The people living in the Subequana Valley, who were experiencing very hard times during our fieldwork in 2009–2010 after the L'Aquila earthquake, are warmly thanked for all for their interest and spontaneous kindness. Finally, we warmly thank both of the anonymous reviewers, whose comments and suggestions allowed us to significantly improve the overall quality of the paper. We are indebted to them.

Edited by: F. Rossetti

References

Aoudia, A., Ismail-Zadeh, A. T., and Romanelli, F.: Buoyancy-driven deformation and contemporary tectonic stress in the lithosphere beneath Central Italy, Terra Nova, 19, 490–496, 2007.

Bard, P. Y.: Microtremor measurements: a tool for site effect estimation?, in: The Effects of Surface Geology on seismic Motion, edited by: Irikura, K., Kudo, K., Okada, H., and Sasatani, T., Balkema, Rotterdam, 1251–1279, 1999.

Basili, R., Valensise, G., Vannoli P., Burrato, P., Fracassi, U., Mariano, S., Tiberti, M. M., and Boschi, E.: The database of Individual Seismogenic Sources (DISS), version 3: summarizing 20 years of research on Italy's earthquake geology, Tectonophysics, 453, 20–43, 2008.

Benedetti, L., Manighetti, I., Gaudemer, Y., Finkel, R., Malavieille, J., Pou, K., Arnold, M., Aumaître, G., Bourlès, D., and Keddadouche, K.: Earthquake synchrony and clustering on Fucino faults (Central Italy) as revealed from in situ ^{36}Cl exposure dating, J. Geophys. Res., 118, 4948–4974, 2013.

Boncio, P., Pizzi, A., Brozzetti, F., Pomposo, G., Lavecchia, G., Di Naccio, D., and Ferrarini, F.: Coseismic ground deformation of the 6 April 2009 L'Aquila earthquake (central Italy, M_w 6.3), Geophys. Res. Lett., 37, L06308, doi:10.1029/2010GL042807, 2010.

Bosi, C.: Giornate di studio sul tema: La Neotettonica in Italia a dieci anni dalla fine del Progetto Finalizzato Geodinamica, Roma, 2–3 March 1992, Relazione Introduttiva, Il Quaternario, 5, 281–286, 1992.

Bosi, C. and Bertini, T.: Geologia della media valle dell'Aterno, Mem. Soc. Geol. Ital., 9, 719–777, 1970.

Bosi, C., Galadini, F., Giaccio, B., Messina, P., and Sposato, A.: Plio-Quaternary continental deposits in the Latium-Abruzzi Apennines: the correlation of geological events across different intermontane basins, Il Quaternario, 16, 55–76, 2003.

Carminati, E. and Doglioni, C.: Alps vs. Apennines: The paradigm of a tectonically asymmetric Earth, Earth-Sci. Rev., 112, 67–96, 2012.

Carmini, E., Doglioni C., and Scrocca, D.: Alps vs Apennines, Special Volume of the Italian Geological Society for the IGC 32 Florence, 2004.

Cavinato, G. P. and De Celles, P.: Extensional basins in the tectonically bimodal central Apennines fold-thrust belt, Italy: Response to corner flow above a subducting slab in retrograde motion, Geology, 27, 955–958, 1999.

Cavinato, G. P., Carusi, C., Dall'Asta, M., MIccadei, E., and Piacentini, T.: Sedimentary and tectonic evolution of Plio-Pleistocene alluvial and lacustrine deposits of Fucino Basin (Central Italy), Sediment. Geol., 148, 29–59, 2002.

Centamore, E., Cantalamessa, G., Micarelli, A., Potetti, M., Berti, D., Bigi, S., Morelli, C., and Ridolfi, M.: Stratigrafia e analisi di facies dei depositi del Miocene e del Pliocene inferiore dell'avanfossa marchigiano-abruzzese e delle zone limitrofe, Studi Geologici Camerti, 1991/2, 125–131, 1991.

Centamore, E., Dramis, F., Fubelli, G., Molin, P., and Nisio, P.: Elements to correlate marine and continental sedimentary successions in the context of the neotectonic evolution of the central Apennines, Il Quaternario, 16, 77–87, 2003.

Cesi, C., Di Filippo, M., Di Nezza, M., and Ferri, F.: Caratteri gravimetrici della media Valle del Fiume Aterno, Parte I – Geologia e Pericolosità Sismica dell'area Aquilana, Microzonazione Sismica per la ricostruzione dell'area Aquilana, edited by: Naso, G. and Castenetto, S., Regione Abruzzo-Presidenza del Consiglio dei Ministri, Dipartimento della Protezione Civile, L'Aquila, 1, 31–37, 2010.

Chiaraluce, L., Chiarabba, C., De Gori, P., Di Stefano, R., Improta, L., Piccinini, D., Schlagenhauf, A., Traversa, P., Valoroso, L., and Voisin, C.: The April 2009 L'Aquila (central Italy) seismic sequence, Boll. Geofis. Teor. Appl., 52, 367–387, 2011.

Ciarapica, G. and Passeri, L.: Evoluzione paleogeografica degli Appennini, Atti Tic. Sc. Terra, 40, 233–290, 1998.

Cinti, F. R., Pantosti, D., DeMartini, P. M., Pucci, S., Civico, R., Pierdominici, S., and Cucci, L.: Evidence for surface faulting events along the Paganica Fault prior to the 6 April 2009 L'Aquila earthquake (Central Italy), J. Geophys. Res., 116, B07308, doi:10.1029/2010JB007988, 2011.

Cipollari, P., Cosentino, D., and Parotto, M.: Modello cinematico-strutturale dell'Italia centrale, Studi Geologici Camerti, 1995/2, 135–143, 1997.

Cipollari, P., Cosentino, D., Esu, D., Girotti, O., Gliozzi, E., and Praturlon, A.: Thrust-top lacustrine-lagoonal basin development in accretionary wedges: late Messinian (Lago-Mare) episode in the central Apennines (Italy), Palaeogeogr. Palaeocl., 151, 146–166, 1999.

Civico, R., Sapia, V., Di Giulio, G., Villani, F., Pucci, S., Vassallo, M., Baccheschi, P., De Martini, P.M., Amoroso, S., Cantore, L., Di Naccio, D., Smedile, A., Orefice, S., Pinzi, S., Pantosti, D., and Marchetti, M.: Imaging the three-dimensional architecture of the Middle Aterno basin (2009 L'Aquila earthquake, Central Italy) using ground TDEM and seismic noise surveys: preliminary results, in: 6th International INQUA Meeting on Paleoseismology, Active Tectonics and Archaeoseismology, edited by: Blumetti, A. M., Cinti, F., De Martini, P. M., Galadini, F., Guerrieri, L., Michetti, A. M., Pantosti, D., and Vittori, E., 19–24 April 2015, Pescina, Italy, Miscellanea INGV, 27, available at: http://www.ingv.it/editoria/miscellanea/2015/miscellanea27/ (last access: 6 March 2017), 2015.

Cosentino, D., Cipollari, P., Marsili, P., and Scrocca, D.: The Geology of Italy, 2010, in: Electronic edition, edited by: Beltrando, M., Peccerillo, A., Mattei, M., Conticelli, S., and Doglioni, C., Journal of the Virtual Explorer, 36, 1441–8142, 2010.

D'Agostino, N., Jackson, J. A., Dramis, F., and Funiciello, R.: Interactions between mantle upwelling, drainage evolution and active normal faulting: an example from the central Apennines (Italy), Geophys. J. Int., 147, 475–479, 2001.

D'Agostino, N., England, P., Hunstad, I., and Selvaggi, G.: Gravitational potential energy and active deformation in the Apennines, Earth Planet. Sc. Lett., 397, 121–132, 2014.

Destro, N.: Release fault: a variety of cross fault in linked extensional fault systems in the Sergipe-Alagoas Basin, NE Brazil, J. Struct. Geol., 17, 615–629, 1995.

Di Filippo, M. and Miccadei, E.: Studio gravimetrico della conca di Sulmona, Il Quaternario, 10, 489–494, 1997.

Di Filippo, M., Di Nezza, M., and Scarascia Mugnozza, G.: Rilievi gravimetrici per la microzonazione sismica (1 livello), Contributi per l'aggiornamento degli "Indirizzi e criteri per la microzonazione sismica 2008", Ingegneria Sismica, International Journal of Earthquake Engineenring, edited by: Pàtron, E., 2, 18–22, ISSN: 0393-14202011, 2011.

Di Giulio, G., de Nardis, R., Boncio, P., Milana, G., Rosatelli, G., Stoppa, F., and Lavecchia, G.: Seismic response of a deep continental basin including velocity inversion: the Sulmona intramontane basin (Central Apennines, Italy), Geophys. J. Int., 204, 418–439, 2016.

Di Nezza, M., Di Filippo, M., and Ferri, F.: Gravity features of the Middle Aterno Valley, 85° Congresso Nazionale della Società Geologica Italiana, Pisa, 6–8 September 2010.

Doglioni, C.: A proposal of kinematic modelling for W-dipping subductions. Possible applications to the Tyrrhenian–Apennines system, Terra Nova, 3, 423–434, 1991.

Doglioni, C.: Geological remarks on the relationships between extension and convergent geodynamic settings, Tectonophysics, 252, 253–267, 1995.

EMERGEO Working Group: Evidence for surface rupture associated with the M_W 6.3 L'Aquila earthquake sequence of April 2009 (central Italy), Terra Nova, 22, 43–51, 2010.

Faccenna, C., Piromallo, C., Crespo Blanca, A., Jolivet, L., and Rossetti, F.: Lateral slab deformation and the origin of the western Mediterranean arcs, Tectonics, 23, TC1012, doi:10.1029/2002TC001488, 2004.

Fäh, D., Kind, F., and Giardini, D.: A theoretical investigation of average H/V ratios, Geophys. J. Int., 145, 535–549, 2001.

Falcucci, E.: Evoluzione geomorfologica e geologica del Quaternario della conca Subequana e della media valle dell'Aterno, Appennino Abruzzese, PhD thesis, Università degli Studi La Sapienza di Roma, Italy, 139 pp., 2011.

Falcucci, E., Gori, S., Peronace, E., Fubelli, G., Moro, M., Saroli, M., Giaccio, B., Messina, P., Naso, G., Scardia, G., Sposato, A., Voltaggio, M., Galli, P., and Galadini, F.: The Paganica fault and surface coseismic ruptures caused by the 6 April 2009, earthquake (L'Aquila, central Italy), Seismol. Res. Lett., 80, 940–950, 2009.

Falcucci, E., Gori, S., Moro, M., Pisani, A. R., Melini, D., Galadini, F., and Fredi, P.: The 2009 L'Aquila earthquake (Italy): what next in the region? Hints from stress diffusion analysis and normal fault activity, Earth Planet. Sc. Lett., 305, 350–358, 2011.

Falcucci, E., Gori, S., Moro, M., Fubelli, G., Saroli, M., Chiarabba, C., and Galadini, F.: Deep reaching versus vertically restricted Quaternary normal faults: implications on seismic potential assessment in tectonically active regions. Lessons from the middle

Aterno valley fault system, central Italy, Tectonophysics, 651–652, 186–198, 2015.

Fracassi, U. and Milano, G.: A soft linkage between major seismogenic fault systems in the central-southern Apennines (Italy): Evidence from low-magnitude seismicity, Tectonophysics, 636, 18–31, 2014.

Fubelli, G., Gori, S., Falcucci, E., Galadini, F., and Messina, P.: Geomorphic signatures of recent normal fault activity versus geological evidence of inactivity: Case studies from the central Apennines (Italy), Tectonophysics, 476, 252–268, 2009.

Galadini, F.: Pleistocene changes in the Central Apennine fault kinematics: a key to decipher active tectonics, Tectonics, 18, 877–894, 1999.

Galadini, F. and Galli, P.: The Holocene paleoearthquakes on the 1915 Avezzano earthquake faults (central Italy): implications for active tectonics in Central Apennines, Tectonophysics, 308, 143–170, 1999.

Galadini, F. and Galli, P: Active tectonics in the central Apennines (Italy) – input data for seismic hazard assessment, Nat. Hazards, 22, 225–270, 2000.

Galadini, F. and Messina, P.: Plio-Quaternary changes of normal fault architecture in the central Apennines (Italy), Geodin. Acta, 14, 321–344, 2001.

Galadini, F. and Messina, P.: Early-Middle Pleistocene eastward migration of the Abruzzi Apennine (central Italy) extensional domain, J. Geodyn., 37, 57–81, 2004.

Galadini, F., Galli, P., and Giraudi, C.: Paleosismologia della Piana del Fucino (Italia Centrale), Il Quaternario, 10, 27–64, 1997.

Galadini, F., Messina, P., Giaccio, B., and Sposato, A.: Early uplift history of the Abruzzi Apennines (central Italy): available geomorphological constraints, Quatern. Int., 101/102, 125–135, 2003.

Galli, P., Galadini, F., and Pantosti, D.: Twenty years of paleoseismology in Italy, Earth Sci. Rev. 88, 89–117, 2008.

Galli, P., Giaccio, B., and Messina, P.: The 2009 central Italy earthquake seen through 0.5 Myr-long tectonic history of the L'Aquila faults system, Quaternary Sci. Rev., 29, 3768–3789, 2010.

Ghisetti, F. and Vezzani, L.: Thrust belt development in the central Apennines: northward polarity of thrusting and out-of-sequence deformations in the Gran Sasso Chain (Italy), Tectonics 10, 904–919, 1991.

Ghisetti, F. and Vezzani, L.: Interfering paths of deformation and development of arcs in the fold-and-thrust belt of the central Apennines (Italy), Tectonics, 16, 523–536, 1997.

Giaccio, B., Messina, P., Sposato, A., Voltaggio, M., Zanchetta, G., Galadini, F., Gori, S., and Santacroce, R.: Tephra layers from Holocene lake sediments of the Sulmona Basin, central Italy: implications for volcanic activity in Peninsular Italy and tephrostratigraphy in the central Mediterranean area, Quaternary Sci. Rev., 28, 2710–2733, 2009.

Giaccio, B., Castorina, F., Nomade, S., Scardia, G., Voltaggio, M., and Sagnotti, L.: Revised chronology of the Sulmona lacustrine succession, Central Italy, J. Quaternary Sci., 545–551, 2013.

Giraudi, C.: Evoluzione geologica della piana del Fucino (Abruzzo) negli ultimi 30 000 anni, Il Quaternario, 1, 131–159, 1988.

Gori, S., Dramis, F., Galadini, F., and Messina, P.: The use of geomorphological markers in the footwall of active faults for kinematic evaluations: examples from the central Apennines, Ital. J. Geosci., 126, 365–374, 2007.

Gori, S., Giaccio, B., Galadini, F., Falcucci, E., Messina, P., Sposato, A., and Dramis, F.: Active normal faulting along the Mt. Morrone south-western slopes (central Apennines, Italy), Int. J. Earth Sci., 100, 157–171, 2011.

Gori, S., Falcucci, E., Atzori, S., Chini, M., Moro, M., Serpelloni, E., Fubelli, G., Saroli, M., Devoti, R., Stramondo, S., Galadini, F., and Salvi, S.: Constraining primary surface rupture length along the Paganica fault (2009 L'Aquila earthquake) with geological and geodetic (DInSAR and GPS) data, Ital. J. Geosci., 131, 359–372, 2012.

Gori, S., Falcucci, E., Dramis, F., Galadini, F., Galli, P., Giaccio, B., Messina, P., Pizzi, A., Sposato, A., and Cosentino, D.: Deepseated gravitational slope deformation, large-scale rock failure, and active normal faulting along Mt. Morrone (Sulmona basin, Central Italy): Geomorphological and paleoseismological analyses, Geomorphology, 208, 88–101, 2014.

Gori, S., Falcucci, E., and Galadini, F.: Report on the INQUA 6th International Workshop on "Active Tectonics, Paleoseismology and Archaeoseismology" Fucino 2015, 19–24 April 2015, Pescina (AQ), Alpine and Mediterranean Quaternary, 28, 2279–7335, 2015a.

Gori, S., Falcucci, E., Scardia, G., Nomade, S., Guillou, H., Galadini, F., and Fredi, P.: Early capture of a central Apennine (Italy) internal basin as a consequence of enhanced regional uplift at the Early-Middle Pleistocene transition, in: The Plio-Pleistocene continental record in Italy: highlights on Stratigraphy and Neotectonics, edited by: Monegato, G., Gianotti, F., and Forno, M. G., Abstracts Volume AIQUA Congress 2015, February 24–26, Torino, Miscellanea dell'Istituto Nazionale di Geofisica e Vulcanologia (ISSN 2039-6651), 26, 26–27, 2015b.

Gruppo di Lavoro MS-AQ: Microzonazione sismica per la ricostruzione dell'area aquilana, Regione Abruzzo – Dipartimento della Protezione Civile, L'Aquila, 3, available at: http://www.protezionecivile.gov.it/jcms/it/view_pub.wp?contentId=PUB25330 (last access: 6 March 2017), 2010.

Ibs-von Seht, M. and Wohlenberg, J.: Microtremor measurements used to map thickness of soft sediments, B. Seismol. Soc. Am., 89, 250–259, 1999.

Kastelic, V., Burrato, P., Carafa, M., and Basili, R.: Progressive Exposure of the Central Apennine's "Nastrini": a Falling Tectonics Paradigm?, Gruppo Nazionale di Geofisica della Terra Solida, 34° convegno nazionale, 17–19 November 2015, Trieste, 2015.

Kastelic, V., Burrato, P., Carafa, M. M. C., and Basili, R.: Repeated surveys reveal nontectonic exposure of supposedly active normal faults in the central Apennines, Italy, J. Geophys. Res.-Earth, 122, 114–129, doi:10.1002/2016JF003953, 2017.

Konno, K. and Ohmachi, T.: Ground-motion characteristics estimated from spectral ratio between horizontal and vertical components of microtremors, B. Seismol. Soc. Am., 88, 228–241, 1998.

Lachet, C. and Bard, P. Y.: Numerical and theoretical investigations on the possibilities and limitations of Nakamura's technique, J. Phys. Earth, 42, 377–397, 1994.

Ladina, C., Marzorati, S., Falcucci, E., and Gori, S.: Dataset of the seismic noise measurements performed from 2009 to 2012 in Subequana Valley (central Italy), Zenodo, doi:10.5281/zenodo.376573, 2017.

Lanzo, G., Silvestri, F., Costanzo, A., d'Onofrio, A., Martelli, L.,

Pagliaroli, A., Sica, S., and Simonelli, A.: Site response studies and seismic microzoning in the Middle Aterno valley (L'Aquila, Central Italy), B. Earthq. Eng., 9, 1417–1442, 2011.

Malinverno, A. and Ryan, W. B. F.: Extension in the Tyrrhenian sea and shortening in the Apennines as a result of arc migration driven by sinking of lithosphere, Tectonics, 5, 227–245, 1986.

Marzorati, S., Ladina C., Piccarreda D., and Ameri, G.: Campagna di Misure Sismiche nella conca Subequana, Rapporti Tecnici INGV, 156, available at: http://istituto.ingv.it/l-ingv/produzione-scientifica/rapporti-tecnici-ingv/rapporti-tecnici-2010 (last access: 6 March 2017), 2010.

Marzorati, S., Ladina, C., Falcucci, E., Gori, S., Ameri, G., and Galadini, F.: Site effects "on the rock": the case of Castelvecchio Subequo (L'Aquila, Central Italy), B. Earthq. Eng., 9, 841–868, 2011.

Miccadei, E., Barberi, R., and De Caterini, G.: Nuovi dati geologici sui depositi quaternari della Conca Subaequana (Appennino abruzzese), Il Quaternario, 10, 483–486, 1997.

Michetti, A. M., Brunamonte, F., Serva, L., and Vittori, E.: Trench investigations of the 1915 Fucino earthquake fault scarps (Abruzzo, Central Italy): geological evidence of large historical events, J. Geophys. Res., 101, 5921–5936, 1996.

Morley, C. K., Haranyac, C., Phoosongsee, W., Pongwapee, S., Kornsawan, A., and Wonganan, N.: Activation of rift oblique and rift parallel pre-existing fabrics during extension and their effect on deformation style. examples from the rifts of Thailand, J. Struct. Geol., 26, 1803–1829, 2004.

Moro, M., Gori, S., Falcucci, E., Saroli, M., Galadini, F., and Salvi, S.: Historical earthquakes and variable kinematic behaviour of the 2009 L'Aquila seismic event (central Italy) causative fault, revealed by paleoseismological investigations, Tectonophysics, 583, 131–144, 2013.

Nakamura, Y.: A method for dynamic characteristics estimations of subsurface using microtremors on the ground surface, Quarterly Rept. RTRI Japan, 30, 25–33, 1989.

Nakamura, Y.: Clear identification of fundamental idea of Nakamura's technique and its applications, in: Proceedings of the XII World Conf. Earthquake Engineering, New Zealand, 2656 pp., 2000.

Nocentini, M.: Integrated analysis for intermontane basins studies: tectono-stratigraphic and paleoclimatic evolution of the L'Aquila basin, PhD thesis, Università degli Studi Roma Tre, Italy, 121 pp., 2016.

Nogoshi, M. and Igarashi, T.: On the amplitude characteristics of microtremor (part 2), Journal of Seismological Society of Japan, 24, 26–40, 1971 (in Japanese with English abstract).

Pace, B., Boncio, P., and Lavecchia, G.: The 1984 Abruzzo earthquake (Italy): an example of seismogenic process controlled by interaction between differently-oriented sinkinematic faults, Tectonophysics, 350, 237–254, 2002.

Pagliaroli, A., Avalle, A., Falcucci, E., Gori, S., and Galadini, F.: Numerical and experimental evaluation of site effects at ridges characterized by complex geological setting, B. Earthq. Eng., 13, 2841–2865, 2015.

Patacca, E., Sartori, R., and Scandone, P.: Tyrrhenian Basin and Apenninic arcs: kinematic relations since late Tortonian times, Mem. Soc. Geol. It., 45, 425–451, 1990.

Patacca, E., Scandone, P., Di Luzio, E., Cavinato, G. P., and

Parotto, M.: Structural architecture of the central Apennines: interpretation of the CROP 11 seismic profile from the Adriatic coast to the orographic divide, Tectonics, 27, TC3006, doi:10.1029/2005TC001917, 2008.

Pizzi, A.: Plio-Quaternary uplift rates in the outer zone of the central Apennines fold-and-thrust belt, Italy, Quatern. Int., 101–102, 229–237, 2003.

Pizzi, A. and Galadini, F.: Pre-existing cross-structures and active fault segmentation in the northern-central Apennines (Italy), Tectonophysics, 476, 304–319, 2009.

Pucci, S., Villani, F., Civico, R., Di Naccio, D., Sapia, V., Di Giulio, G., Vassallo, M., Ricci, T., Delcher, E., Finizola, A., Baccheschi, P., and Pantosti, D.: Structural complexity and Quaternary evolution of the 2009 L'Aquila earthquake causative fault system (Abruzzi Apennines, Italy): a three-dimensional image supported by deep ERT, ground TDEM and seismic noise surveys, Proceedings of the 7th International INQUA Meeting on Paleoseismology Active Tectonics and Archeoseismology, 30 May–3 June 2016, Crestone, Colorado, USA, 2016.

Regattieri, E., Giaccio, B., Galli, P., Nomade, S., Peronace, E., Messina, P., Sposato, A, Boschi, C., and Gemelli, M.: A multiproxy record of MIS 11e12 deglaciation and glacial MIS 12 instability from the Sulmona basin (central Italy), Quaternary Sci. Rev., 132, 129–145, 2016.

Roberts, G. P. and Michetti, A. M.: Spatial and temporal variations in growth rates along active normal fault systems: an example from The Lazio-Abruzzo Apennines, central Italy, J. Struct. Geol., 26, 339–376, 2004.

Salucci, R.: Studio geologico-geognostico relative al progetto M.A.P. (Moduli Abitativi Provvisori), Comune di Castelvecchio Subequo (AQ), 56 pp., 2010a.

Salucci, R.: Studio geologico-geognostico relative al progetto M.A.P. (Moduli Abitativi Provvisori), Comune di Castel di Ieri (AQ), 49 pp., 2010b.

Satolli, S. and Calamita, F.: Differences and similarities between the central and the southern Apennines (Italy): Examining the Gran Sasso versus the Matese-Frosolone salients using paleomagnetic, geological, and structural data, J. Geophys. Res., 113, B10101, doi:10.1029/2008JB005699, 2008.

Scrocca, D.: Thrust front segmentation induced by differential slab retreat in the Apennines (Italy), Terra Nova, 18, 154–161, 2006.

Şengör, A. M. C.: Cross-faults and differential stretching of hanging walls in regions of low-angle normal faulting: examples from western Turkey, edited by: Coward, M. P., Dewey, J. F., and Hancock, P. L., Continental Extensional Tectonics, Geological Society Special Publication, 28, 575–589, 1987.

SESAME European project: Guidelines for the implementation of the H/V spectral ratio technique on ambient vibrations – Measurements, processing and interpretation, Deliverable D23.12., available at: ftp://ftp.geo.uib.no/pub/seismo/SOFTWARE/SESAME/USER-GUIDELINES/SESAME-HV-User-Guidelines.pdf (last access: 6 March 2017), 2005.

Tokimatsu, K.: Geotechnical site characterization using surface waves, Geot. Geol. Earthquake., 1333–1368, 1997.

Yamanaka, H., Takemura, M., Ishida, H., and Niwa, M.: Characteristics of long period microtremors and their applicability in the exploration of deep sedimentary layers, B. Seismol. Soc. Am., 84, 1831–1841, 1994.

Valensise, G. and Pantosti, D.: The investigation of potential earthquake sources in peninsular Italy: A review, J. Seismol., 5, 287–306, 2001.

Vittori, E., Di Manna, P., Blumetti, A. M., Comerci, V., Guerrieri, L., Esposito, E., Michetti, A. M., Porfido, S., Piccardi, L., Roberts, G. P., Berlusconi, A., Livio, F., Sileo, G., Wilkinson, M., McCaffrey, K. J. W., Phillips, R. J., and Cowie, P. A.: Surface faulting of the 6 April 2009 M_w 6.3 L'Aquila earthquake in Central Italy, B. Seismol. Soc. Am., 101, 1507–1530, 2011.

Wells, D. L. and Coppersmith, K. J.: New empirical relationships among magnitude, rupture length, rupture width, rupture area, and surface displacement, B. Seismol. Soc. Am., 84, 974–1002, 1994.

Wilkins, S. J. and Schultz, R. A.: Cross faults in extensional settings: Stress triggering, displacement localization, and implications for the origin of blunt troughs at Valles Marineris, Mars, J. Geophys. Res., 108, 5056, doi:10.1029/2002JE001968, 2003.

12

Assessing and analysing the impact of land take pressures on arable land

Ece Aksoy[1], **Mirko Gregor**[2], **Christoph Schröder**[1], **Manuel Löhnertz**[2], **and Geertrui Louwagie**[3]

[1]European Topic Centre on Urban, Land and Soil systems (ETC/ULS), University of Malaga, Malaga, Spain
[2]ETC/ULS, space4environment, Niederanven, Luxembourg
[3]European Environment Agency (EEA), Copenhagen, Denmark

Correspondence to: Ece Aksoy (ece.aksoy@uma.es, eceaksoy@hotmail.com)

Abstract. Land, and in particular soil, is a finite and essentially non-renewable resource. Across the European Union, land take, i.e. the increase of settlement area over time, annually consumes more than $1000\,\text{km}^2$ of which half is actually sealed and hence lost under impermeable surfaces. Land take, and in particular soil sealing, has already been identified as one of the major soil threats in the 2006 European Commission Communication "Towards a Thematic Strategy on Soil Protection" and the Soil Thematic Strategy and has been confirmed as such in the report on the implementation of this strategy. The aim of this study is to relate the potential of land for a particular use in a given region with the actual land use. This allows evaluating whether land (especially the soil dimension) is used according to its (theoretical) potential. To this aim, the impact of several land cover flows related to urban development on soils with good, average, and poor production potentials were assessed and mapped. Thus, the amount and quality (potential for agricultural production) of arable land lost between the years 2000 and 2006 was identified. In addition, areas with high productivity potential around urban areas, indicating areas of potential future land use conflicts for Europe, were identified.

1 Introduction

Land use in Europe has changed drastically during the last 50 years, primarily in relation to the betterment of human well-being and economic development, while unfortunately causing serious environmental problems such as urban sprawl, soil scaling, loss of biodiversity, soil erosion, soil degradation, floods, or desertification.

The changes in land use can also be interpreted as changes in the resources, services, and goods that soils offer to us; moreover, the type of land use change varies among different types of regions. Smith et al. (2015) describe the effects of land use changes (increased change of arable to urban) on different ecosystem services that are provided by soil decreased biomass and decreased availability of water for agricultural use (provisioning services); decreased infiltration, storage, and soil-mediated water regulation (regulating services); decreased genetic diversity (supporting service); and decreased natural environment (cultural service).

Land use changes are a worldwide issue and the impacts of land use changes are the subject of several studies. In recent years, several modelling and foresight studies of land use change have emerged with European research projects, such as VOLANTE – Visions of Land Use Transitions in Europe (EU FP7 project), EU-LUPA – European Land Use Patterns (ESPON project), SENSOR – Sustainable Impact Assessment Tools for Environmental, Social and Economic Effects of Multifunctional Land Use in European Regions (EU FP6 integrated project) (Helming et al., 2006), enviroGRIDS (EU FP7 project), the ATEAM EU FP5 project (Advanced Terrestrial Ecosystem Analysis and Modelling to search global climate and land use change impacts on ecosystem vulnerability in Europe (Rounsevell et al., 2006), the EURURALIS project – addressing socio-economic impacts associated with land use changes in the agricultural sector (Klijn et al., 2005), the SEAMLESS project – approach for multi-scale modelling to asses sustainability impacts of agricultural policies (van Ittersum et al., 2008), and the PRELUDE project of the European Economic Area (EEA) on scenarios for future land use changes in Europe (Hoogeven and Ribeiro, 2007). In the following, we give some examples from the literature on the impacts of land use change. Mancosu et al. (2015) develop different land use change scenarios and discusses their impacts on the Black Sea region. Parras-Alcántara et al. (2013) examine the impacts of land use change on soil carbon and nitrogen in a Mediterranean agricultural area. Adugna and Abegaz (2016) discuss the effects of land use changes on the soil properties in Ethiopia. Mohawesh et al. (2015) reveal the effects of land use changes on soil properties in Jordan and results help in understanding the effects of land use changes on land degradation processes and carbon sequestration potential and in formulating sound soil conservation plans. Wasak and Drewnik (2015) studied the land use effects on soil organic carbon sequestration in calcareous Leptosols in the Tatra Mountains, Poland. Muñoz-Rojas et al. (2015) analysed the long time series (1956–2007) impacts of land use and land cover changes on organic carbon stocks in Mediterranean soils. Liu et al. (2014) studied land use and climate changes and their impacts on runoff in the Yarlung Tsangbo River basin, China. Kalema et al. (2015) showed the impacts of land use changes on woodlands in an equatorial African savanna. Lastly, Trabaquini et al. (2015) examined the effects of the land use changes of physical soil properties in the Brazilian savanna environment.

Land take represents an increase in artificial surfaces or settlement areas (for residential, commercial, industrial, or infrastructural purposes, for example) over time, usually at the expense of rural areas. This process can result in an increase in scattered settlements in rural regions or in an expansion of urban areas around an urban nucleus (urban sprawl, which is defined as "the physical pattern of low-density expansion of large urban areas, under market conditions, mainly into the surrounding agricultural areas"; EEA, 2006a). A clear distinction is usually difficult to make (Prokop et al., 2011).

Land take is a widespread phenomenon in Europe. The assessment as part of the EEA indicator land take (CSI 014/LSI 001) identifies extension of artificial land cover as one of the two major flows that consume arable land; the other one is withdrawal of farming, which is supported by European policies (EEA, 2006a). Tóth (2012) analysed the impact of land take on soil productivity using the Joint Research Center (JRC) Cropland Productivity Index map and combined it with Corine Land Cover (CLC) changes and socio-economic data. He concluded that the European Union (EU) experiences a constant decrease in production capacity (Tóth, 2012).

Soils are used to produce a range of biomass products that serve as food, feed, fibre, and fuel. Biomass production can be particularly relevant in biodiversity conservation and climate change mitigation efforts, through supporting elements of green infrastructure and flood regulation (EEA, 2015).

Biomass production is one of the soil functions recognized in the EU (CEC, 2006) and is severely affected by land take. Urbanized land is not mainly used for agriculture, and furthermore, a large proportion of the land taken for urbanization is actually sealed. Soil sealing can be considered as an almost irreversible process, since "de-sealing" is very costly and the formation of new soil takes decades, i.e. 1 cm in 100 years (Scheffer and Schachtschabel, 2002). Accordingly, soil functions are commonly considered as lost when soils are covered with impervious surfaces.

From the agricultural point of view, land take is a soil–land loss for non-agricultural purposes, so that in a way its effect is similar to soil degradation (caused by severe erosion) and might be considered as a complementary process. It is important to recognize why it is of interest to compare different categories of soil biomass productivity affected by land take and how these classes are connected to soil erosion and degradation. Therefore, the aim of this study is to assess and analyse the impacts of several land cover flows related to urban development (referred to as land take) between the years 2000 and 2006 on soils with good, average, and poor biomass production potentials and identify regions with major impact (hotspots) in Europe.

2 Material and methods

2.1 Material

The main input data for this study are

- soil biomass productivity data on arable land (Tóth et al., 2013).

- land cover and use data (CORINE, 2017).

- land cover and changes (CLC changes and derived land cover flows (LCFs) between the years 2000 and 2006) (EEA, 2013c).

The soil biomass productivity map on arable land was produced with the spatially explicit Soil Productivity Model (SoilProd) for Europe by JRC (Tóth et al., 2011) (Fig. 1). This map provides composite cropland productivity index scores, which are expressed on a scale from 1 to 10. Score 1 represents the lowest and 10 the highest biomass production potential. The productivity index is the sum of the inherent soil productivity index and the fertilizer response rate. The former results from an evaluation matrix set up for eight climatic zones, five inherent productivity classes (derived from second level taxonomic soil units), soil attribute information from the soil database (corrected for topographic conditions), and four available water capacity classes. The fertilizer response rate takes account of the management practices applied. More details about the model and the map production process can be found in Tóth (2012) and Tóth et al. (2011).

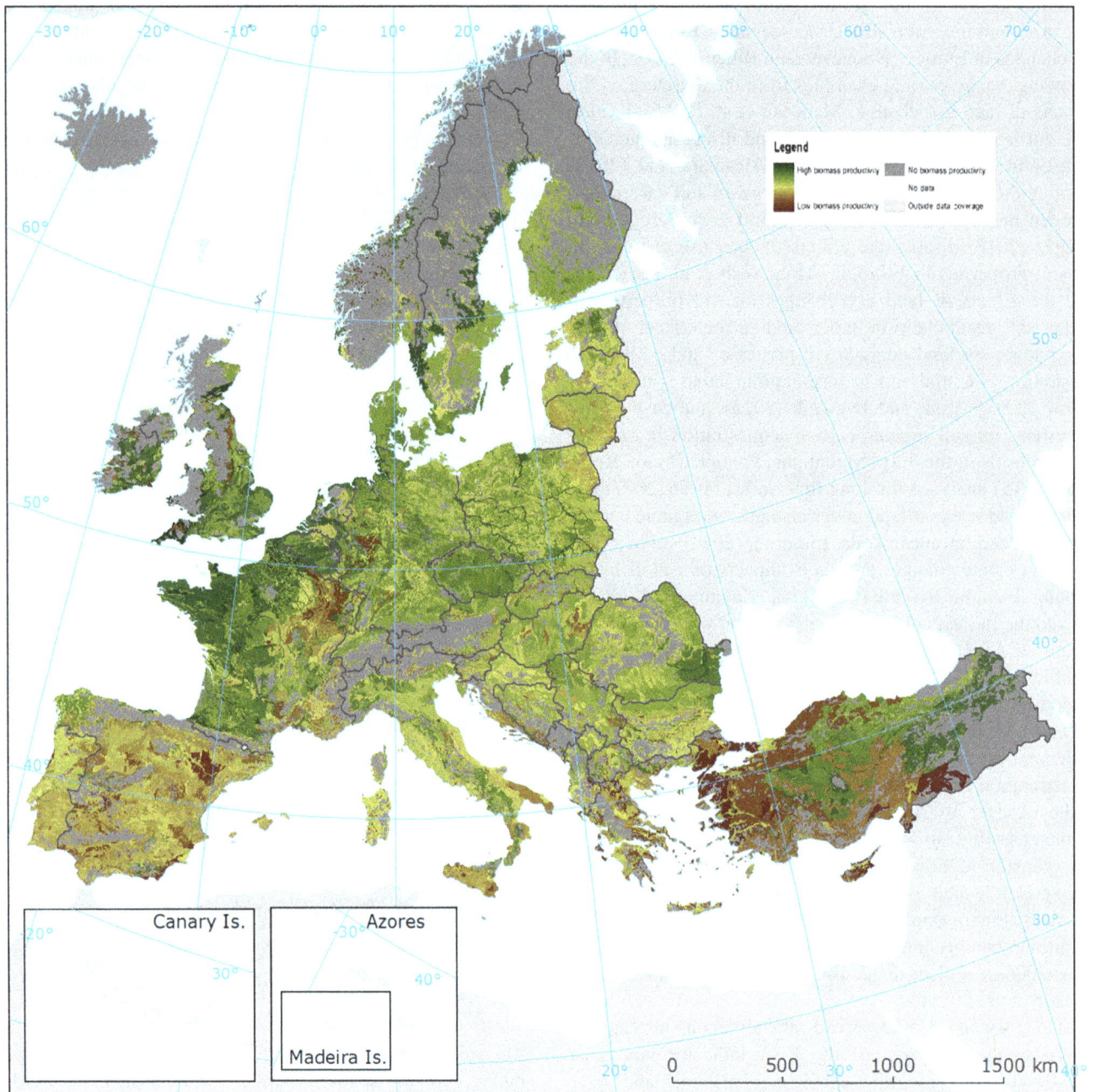

Figure 1. Soil productivity data on arable land (pan-European grid layer) (JRC). Legend shows high biomass productivity (green) to low biomass productivity (brown), no biomass productivity (dark grey), no data (white), and outside data coverage (light grey).

The soil biomass productivity data were provided by JRC, 1 km^2 raster data sets have full coverage of Europe but they are only valid for the corresponding land use types. Therefore, the appropriate CLC classes (based on CLC-Corilis 2000) were identified to build the masks for the extraction of the soil and/or land productivity layers. The CLC classes we used are 2.1 "Arable land" (subclasses 211 "Non-irrigated arable land", 212 "Permanently irrigated land", and 213 "Rice fields") and 2.4 "Heterogeneous agricultural ar-

eas" (subclasses 241 "Annual crops associated with permanent crops" and 242 "Complex cultivation patterns") (Tóth, 2012).

There are nine major LCFs on Level 1 (Land and Ecosystem Accounting, LEAC, 2000–2006) (EEA, 2013a) (Table 1). The combination of the land take flows LCF2 and LCF3 (urban residential sprawl and extension of economic sites and infrastructure) were used for this study. The impact

Table 1. Major land cover flows (LCFs) on Level 1 (EEA, 2013a).

Code	Major type of cover change
LCF1	Urban land management
LCF2	Urban residential sprawl
LCF3	Extension of economic sites and infrastructure
LCF4	Agriculture internal conversions
LCF5	Conversion from forested and natural land to agriculture
LCF6	Withdrawal of farming
LCF7	Forest creation and management
LCF8	Water body creation and management
LCF9	Changes of land cover due to natural and multiple causes

calculation for Greece could not be done because of not having CLC 2006 and LCFs.

The technical assessment of land take on arable land is based on the land cover flows as described below:

- The definition of LCF2 is as follows: urban residential sprawl consists of land uptake by residential buildings altogether with associated services and urban infrastructure (classified in CLC 111 and 112) from non-artificial land (extension over sea may happen). Two subcategories are distinguished, namely urban dense residential sprawl resulting in continuous urban fabric and urban diffuse residential sprawl resulting in discontinuous urban fabric.

- The definition of LCF3 is as follows: sprawl of economic sites and infrastructures consists of land uptake by new economic sites and infrastructures (including sport and leisure facilities) from non-artificial land (extension over sea may happen). This land cover flow includes eight subcategories, namely sprawl of the following infrastructure on non-urban land, i.e. industrial and commercial sites, transport networks, harbours, airports, mines and quarries, dump sites, construction, and sport and leisure facilities (EEA, 2013a).

2.2 Method

The schematic workflow of the study can be seen in Fig. 2. Four main steps were followed to assess the impacts of land take pressures on arable land analysis.

First of all, the soil biomass productivity data were classified into soils with good, average, and poor capacities to provide biomass on arable land (step 1, Fig. 2) with the aim of easier analysis, interpretation, and calculation. This classification is performed based on the value distribution and their statistical parameters. This means that the lower third of all values are classified as poor (class 1), the upper third as good (class 3), and the values in between as average (class 2).

Secondly, a mask was applied to the soil biomass productivity map (step 2, Fig. 2) by using defined CLC classes according to the provisions of Tóth (2012). Then, after the clas-

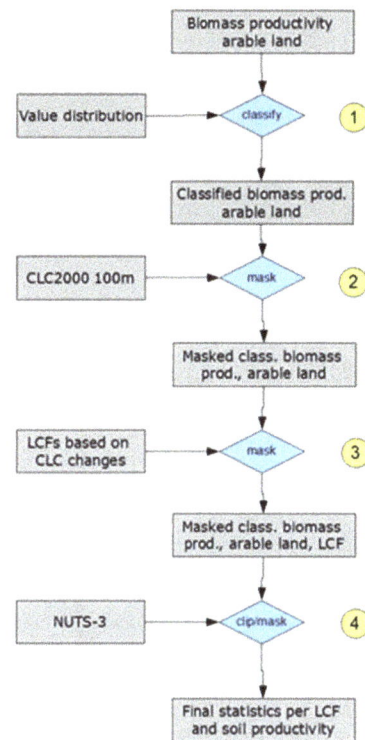

Figure 2. Schematic workflow of the study.

sification and masking processes, the selected LCFs were overlaid onto the masked and classified data to extract the raster cells that contain a land cover change that is relevant for the analysis. This process in fact represents another masking process, as described in Fig. 2 (step 3). Lastly, the raster data were combined with the NUTS-3 reference units to compute the zonal statistics for each of the parameter combinations (impact of a particular LCF or combination of LCFs on a particular soil function potential (Fig. 2, step 4).

The final value of the impact of a particular LCF or combination of LCFs on the capacity of soils to supply a particular soil function is expressed in relation to the share of that specific soil function potential in the NUTS-3 region. This means that the share of, for example, good soils within a NUTS-3 region, is the reference for the calculation, not the entire area of the NUTS-3 region.

Moreover, for interpretation purposes the value ranges can be understood and verbally described regarding their impact (expressed as percentages) as follows (ranked from very low to very high impact; green to red colours in Fig. 5):

- very low impact

- low impact

- intermediate impact

- high impact and

- very high impact.

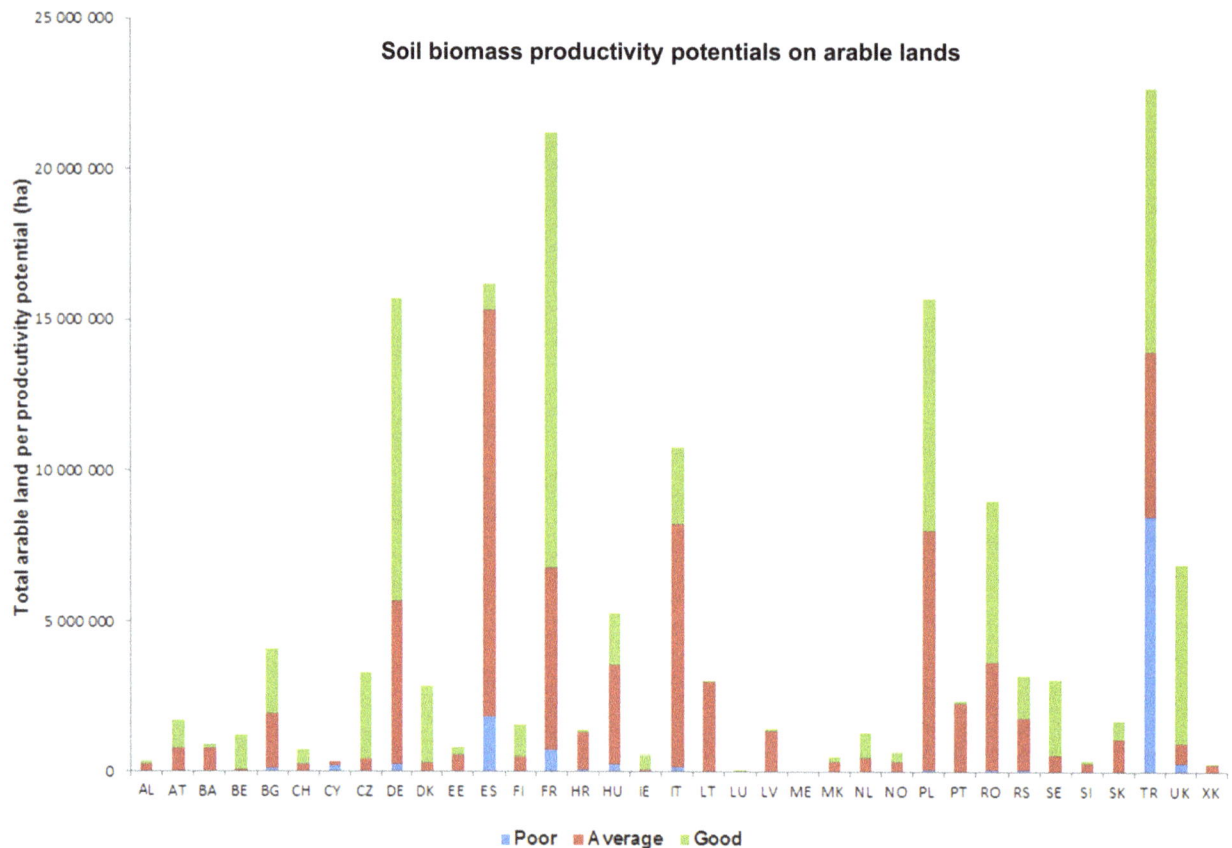

Figure 3. Graphic presentation of soil biomass productivity potentials on arable land for each country.

In addition, the descriptions of the outcomes make reference to relative and absolute impacts. Whereas relative impacts correspond to the percentage values of the impact of a certain LCF on soils of a specific capacity in a NUTS-3 region, the absolute impacts refer to the area (in hectares) that is affected by a particular LCF. Depending on the size of the reference unit (that is, the area of soils of a specific capacity in a NUTS-3 region) high absolute values do not necessarily correspond to high relative values, while low absolute values could well mean high relative values (when the total size of the reference area is very small).

3   Results

According to the results given in Table 2 and Fig. 3, even though the highest share of the total arable land coverage of the whole country is higher than 40 % in the Czech Republic, Germany, Denmark, Hungary, Lithuania, Poland, and Serbia, Turkey, France, Spain, Germany, and Poland each have over 15 000 000 ha of arable land. Moreover, close to half (46.32 %) of the arable lands in the whole study area and half of the countries (18 out of 36 countries) have good productivity potentials. Over 80 % of the arable land in Belgium (BE), the Czech Republic (CZ), Denmark (DK), Ireland (IE),

Sweden (SE), and the United Kingdom (UK) have good productivity potential. Over 80 % of the arable land in Bosnia and Herzegovina (BA), Spain (ES), Croatia (HR), Lithuania (LT), Latvia (LV), Portugal (PT), and Kosovo (XK) have average productivity potential. Only one country, Cyprus, has mostly poor biomass productivity potential on its arable land.

The distribution of the soils according to their potential for biomass production on arable land per NUTS-3 area can be seen in Fig. 4; the proportions are given in relation to the total area of each individual NUTS-3 region. By consequence, the maps nicely illustrate where poor, average, or good soils dominate in Europe and where they are only of minor importance.

Soils that are considered poor for biomass production on arable land mainly dominate in three European regions, (i) Spain, (ii) central and north-eastern France, and (iii) south-eastern Europe (almost all of Turkey and large parts of Greece). Almost all other regions have an intermediate to low share of poor soils for the provision of biomass on arable land. Of the first 20 NUTS-3 regions across Europe, 14 are located in Turkey (Fig. 4). The others are located in the UK, France, and Cyprus (NUTS-3 region boundary corresponds to the entire country). However, most of the men-

Table 2. Statistical distribution of arable lands according to their biomass production potential for each country (bold font shows the major share). Abbreviations of the countries are as follows: AL – Albania, AT – Austria, BA – Bosnia and Herzegovina, BE – Belgium, BG – Bulgaria, CH – Switzerland, CY – Cyprus, CZ – Czech Republic, DE – Germany, DK – Denmark, EE – Estonia, EL – Greece, ES – Spain, FI – Finland, FR – France, HR – Croatia, HU – Hungary, IE – Ireland, IT – Italy, LT – Lithuania, LU – Luxembourg, LV – Latvia, ME – Montenegro, MK – Macedonia, NL – the Netherlands, NO – Norway, PL – Poland, PT – Portugal, RO – Romania, RS – Serbia, SE – Sweden, SI – Slovenia, SK – Slovakia, TR – Turkey, UK – the United Kingdom, and XK – Kosovo.

| Country | Country | Total arable land | Arable land proportion | Soil biomass productivity potential per total arable land | | |
| | | | | Poor | Average | Good |
	$[km^2]$	$[km^2]$	[%]	[%]	[%]	[%]
AL	28 755.06	3729.06	12.97	6.99	**65.48**	27.53
AT	83 947.82	17 291.73	20.6	0.93	43.96	**55.11**
BA	51 399.37	9032.91	17.57	3.2	**84.69**	12.11
BE	30 664.19	12 123.74	39.54	0.25	7.32	**92.43**
BG	110 988.76	41 120.23	37.05	2.81	44.65	**52.54**
CH	41 287.33	7605.61	18.42	1.31	31.71	**66.98**
CY	9249.11	3468.26	37.5	**62**	38	0
CZ	78 869.52	33 054.4	41.91	0.19	13.15	**86.66**
DE	357 737.29	157 211.86	43.95	1.58	34.41	**64.01**
DK	43 174.76	28 487.19	65.98	0.09	11.33	**88.58**
EE	45 335.44	8338.46	18.39	1.39	**65.76**	32.84
EL	131 735.85	28 320.05	21.5	12.47	**74.54**	12.99
ES	505 980.28	161 978.31	32.01	11.51	**83.08**	5.41
FI	337 616.92	15 956.63	4.73	0.7	32.69	**66.61**
FR	638 480.71	212 195.3	33.23	3.58	28.45	**67.97**
HR	56 599.65	13 861.69	24.49	4.85	**91.22**	3.93
HU	93 012.99	52 795.47	56.76	4.9	**62.96**	32.13
IE	69 956.69	6336.24	9.06	7.42	9.04	**83.54**
IT	300 620.28	107 714.14	35.83	1.78	**74.5**	23.72
LT	64 901.2	30 396.19	46.83	0.09	**98.45**	1.46
LU	2595.06	847.7	32.67	13.37	38.21	**48.41**
LV	64 596.24	14 582.72	22.58	1.26	**91.14**	7.6
ME	13 878.81	166.5	1.2	20.16	32.64	**47.2**
MK	25 436.12	5191.04	20.41	3.9	**64.16**	31.95
NL	37 373.99	13 059.17	34.94	1.25	36.62	**62.13**
NO	323 024.51	7033.67	2.18	3.97	46.07	**49.97**
PL	311 942.39	157 056.55	50.35	0.47	**50.49**	49.04
PT	91 969.54	23 630.11	25.69	2.33	**94.53**	3.14
RO	238 364.06	90 033.96	37.77	1.12	39.45	**59.43**
RS	77 313.57	32 165.26	41.6	3.06	**53.01**	43.93
SE	449 563.7	30 977.11	6.89	0.83	17.69	**81.48**
SI	20 273.58	3900.67	19.24	2.41	**79.7**	17.9
SK	49 027.63	17 085.84	34.85	0.38	**64.04**	35.58
TR	780 290.77	226 986.37	29.09	37.49	23.93	**38.58**
UK	244 619.49	68 939.64	28.18	4.68	9.15	**86.17**
XK	11 004.64	2877.41	26.15	0.99	**94.53**	4.48
Grand total	5 821 587.32	1 645 551.2	28.27	8.23	45.45	46.32

tioned regions show very low to intermediate impact of urban expansion; Cyprus shows a high impact though.

Average soils for arable biomass provision are widespread across Europe and can be found in large parts of Spain and Italy, Hungary, Poland, and the southern Baltic countries (Lithuania and Latvia). Regions of Germany, France, Bulgaria, and Greece also possess average soils. Low shares of average soils can be found in Turkey, parts of Greece, Bulgaria and Romania, the Czech Republic, parts of Germany and France, the UK, and Scandinavia. The number of NUTS-3 regions with a high to very high share of average soils for biomass provision on arable land (Fig. 3) is substantially higher compared to those with a high share of poor soils. There are 32 regions that have a majority share, that is, more

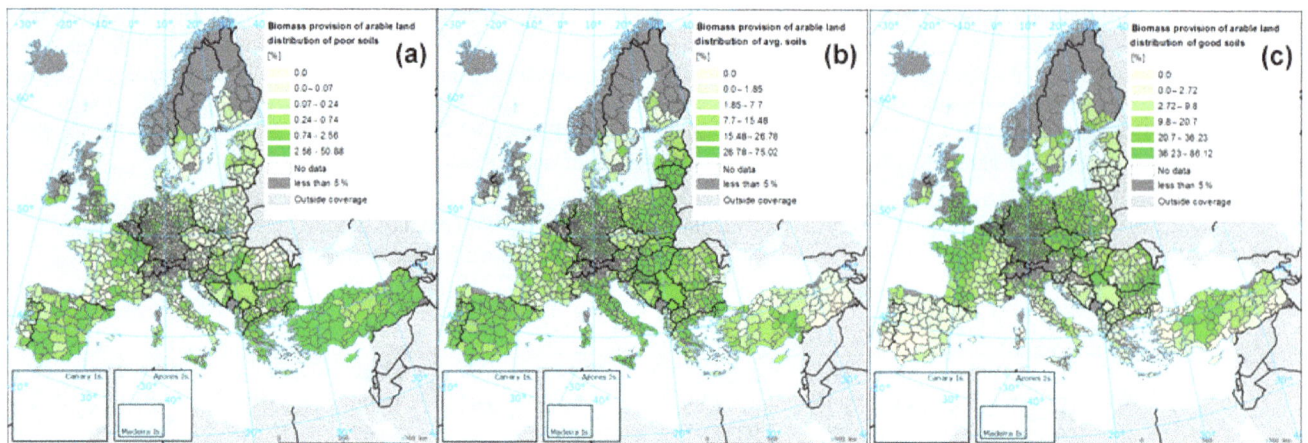

Figure 4. Distribution of soils according to their potential for biomass production on arable land: proportions of poor **(a)**, average **(b)**, and good **(c)** soils (in % of the total NUTS-3 region area). Legend shows the shares (%) from low (light green) to high (dark green), "0" (light yellow), no data (white), less than 5 % (dark grey), and outside data coverage (light grey). "Less than 5 %" means that the total area of arable land is smaller than 5 %. Note that the same colours might represent different percentages as quantiles were used during the map production. Ranges are given between 0 and 50.88 % **(a)**, between 0 and 75.02 % **(b)**, and between 0 and 86.12 % **(c)**.

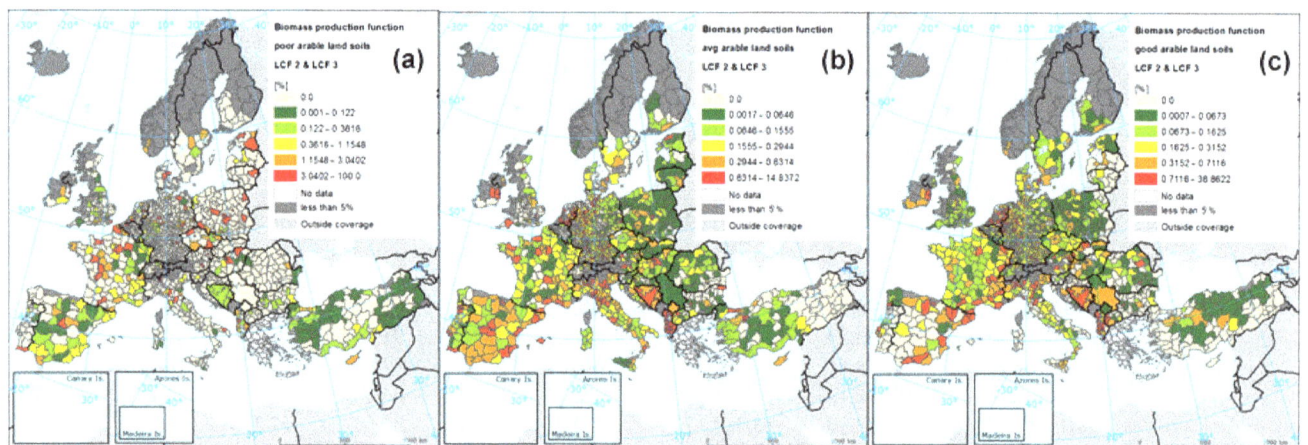

Figure 5. Percentage decline (per NUTS-3 area) of arable land area with poor **(a)**, average **(b)**, and good **(c)** production potentials due to urban residential, commercial, industrial, and infrastructure-related extension (LCF2 and LCF3) between 2000 and 2006. Legend shows the shares (%) from low (dark green) to high (red), "0" (light yellow), no data (white), less than 5 % (dark grey), and outside data coverage (light grey). "Less than 5 %" means that the total area of arable land is smaller than 5 %. Note that the same colours might represent different percentages as quantiles were used during the map production. Ranges are given between 0.001 and 100 % **(a)**, between 0.017 and 14.84 % **(b)**, and between 0.001 and 38.86 % **(c)**.

than 50 %, in the respective NUTS-3 region (only one NUTS-3 region for poor soils), with the highest values of over 70 % in one Spanish (ES418, Valladolid) and two Italian regions (ITH36 and ITH57, Padua and Ravenna, respectively). In general, there is a high share of Italian regions (12 NUTS-3 regions), which are often located in or close to the Po Valley, which used to be one of the most fertile areas in Europe; another remarkable hotspot is Lithuania with 5 regions.

Good soils for the provision of biomass on arable land dominate in large parts of north-western Europe, such as lots of regions in the UK, north-western France, the Benelux countries, Germany, Denmark, Poland, Czech Republic,

Hungary, and Bulgaria. Even some regions in central Turkey have a high share of good soils. Low shares can be found mainly in the western Balkan countries, the Iberian Peninsula, Romania, the Baltic countries and some regions in Finland and Sweden. Compared to the average soils, the number of NUTS-3 regions with a very high share of good soils is even bigger; almost 140 regions have a majority share of good soils, with the upper seven regions exceeding 80 % (four regions in the UK, two in Romania, and one in Germany) (Fig. 4).

The highest land take impacts on the biomass productivity potentials of arable land were found in Albania

Table 3. Statistical distribution of the land take impact on arable land for each country between 2000 and 2006. Abbreviations of the countries are as follows: AL – Albania, AT – Austria, BA – Bosnia and Herzegovina, BE – Belgium, BG – Bulgaria, CH – Switzerland, CY – Cyprus, CZ – the Czech Republic, DE – Germany, DK – Denmark, EE – Estonia, ES – Spain, FI – Finland, FR – France, HR – Croatia, HU – Hungary, IE – Ireland, IT – Italy, LT – Lithuania, LU – Luxembourg, LV – Latvia, ME – Montenegro, MK – Macedonia, NL – the Netherlands, NO – Norway, PL – Poland, PT – Portugal, RO – Romania, RS – Serbia, SE – Sweden, SI – Slovenia, SK – Slovakia, TR – Turkey, UK – the United Kingdom, and XK – Kosovo).

Country	Total arable land (ha)	Total impact on arable land (ha)	Impact on arable land (ha)			Total impacted arable land (%)	Arable land (%) impact on total		
			Poor	Avg	Good		Poor	Avg	Good
AL	372 906	14 795	539	8672	5584	3.97	2.07	3.55	5.44
AT	1 729 173	4137	20	1478	2639	0.24	0.12	0.19	0.28
BA	90 3291	5329	61	4100	1168	0.59	0.21	0.54	1.07
BE	1 212 374	2027	7	111	1909	0.17	0.23	0.13	0.17
BG	4 112 023	1920	145	1289	486	0.05	0.13	0.07	0.02
CH	760 561	784	12	157	615	0.10	0.12	0.07	0.12
CY	346 826	4816	4087	729	0	1.39	1.90	0.55	0.00
CZ	3 305 440	8390	103	900	7387	0.25	1.64	0.21	0.26
DE	15 721 186	47 620	1605	16 053	29 962	0.30	0.65	0.30	0.30
DK	2 848 719	9250	18	1001	8231	0.32	0.69	0.31	0.33
EE	833 846	1522	491	929	102	0.18	4.23	0.17	0.04
ES	16 197 831	71 338	7211	59 786	4341	0.44	0.39	0.44	0.50
FI	1 595 663	1207	0	246	961	0.08	0.00	0.05	0.09
FR	21 219 530	52 096	2919	12 376	36 801	0.25	0.38	0.20	0.26
HR	1 386 169	1409	0	1389	20	0.10	0.00	0.11	0.04
HU	5 279 547	11 382	374	7469	3539	0.22	0.14	0.22	0.21
IE	633 624	4806	193	765	3848	0.76	0.41	1.34	0.73
IT	10 771 414	37 484	179	26 747	10 558	0.35	0.09	0.33	0.41
LT	3 039 619	2522	17	2472	33	0.08	0.64	0.08	0.07
LU	84 770	177	75	37	65	0.21	0.66	0.11	0.16
LV	1 458 272	316	42	243	31	0.02	0.23	0.02	0.03
ME	16 650	1	0	1	0	0.01	0.00	0.02	0.00
MK	51 9104	1330	6	712	612	0.26	0.03	0.21	0.37
NL	1 305 917	18 874	213	6943	11 718	1.45	1.30	1.45	1.44
NO	703 367	557	20	244	293	0.08	0.07	0.08	0.08
PL	15 705 655	14 246	622	6629	6995	0.09	0.85	0.08	0.09
PT	2 363 011	7099	79	6840	180	0.30	0.14	0.31	0.24
RO	9 003 396	5828	59	2178	3591	0.06	0.06	0.06	0.07
RS	3 216 526	2430	0	792	1638	0.08	0.00	0.05	0.12
SE	3 097 711	5728	99	734	4895	0.18	0.38	0.13	0.19
SI	390 067	332	11	280	41	0.09	0.12	0.09	0.06
SK	1 708 584	2660	0	1445	1215	0.16	0.00	0.13	0.20
TR	22 698 637	16 761	7153	4259	5349	0.07	0.08	0.08	0.06
UK	6 893 964	8832	671	1552	6609	0.13	0.21	0.25	0.11
XK	287 741	840	0	832	8	0.29	0.00	0.31	0.06
Total	161 723 114	368 845	27 031	180 390	161 424	0.23	0.2	0.25	0.21

(AL) (3.97 %), the Netherlands (NL) (1.45 %), Cyprus (CY) (1.39 %), and Ireland (IE) (0.76 %) (Table 3 and Fig. 6). However, when expressing the impacts on an absolute (in hectare) rather than on a relative (in percentage) basis, Spain, France, and Germany rank highest (with 71 338, 52 096, and 47 620 ha, respectively). Thus, even though the relative impact may be low in some countries, the absolute impact may be quite high. For example, while the share of land with good

and average productivity potentials is very similar (0.5 and 0.44 %, respectively), the total area of land with good productivity potential is far lower (4341 ha) than that of average productivity potential (59 786 ha) (Table 3). Therefore, it is better to consider the absolute and relative values in parallel.

Figure 5 describes the impact of land take (the combination of LCF2 and LCF3, i.e. residential, commercial, industrial, and infrastructure-related extension) on arable land

Land take impact on arable lands

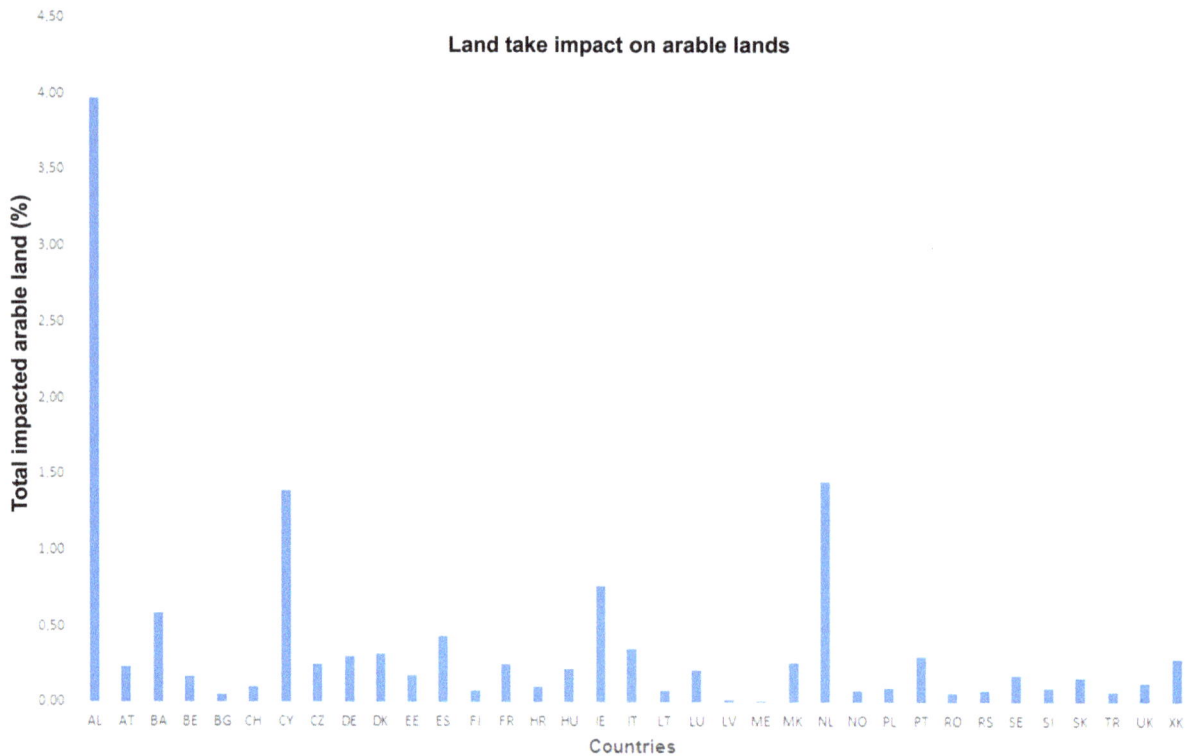

Figure 6. Graphic presentation of land take impact on arable land for each country.

with poor, average, and good potentials for the provision of biomass.

In general, the map illustrates that regions with a very high impact of urban land take on poor soils are scattered across Europe; there is no geographic area with a striking clustering of such regions. However, the south-eastern part of Europe only contains a few NUTS-3 regions with a high impact: Cyprus, Istanbul, and one region in Romania (Galati, RO224). Also, Albania possesses some regions with a high to very high impact of urban land take. Conversely, the NUTS-3 regions with the highest relative impact still possess only low to very low shares of poor soils within the NUTS-3 regions. Most of these regions are located in north-western Europe (the UK, Ireland, and Germany), some isolated ones can be found in south-western France, Italy, and Poland. When looking at absolute impacts (in terms of total area affected) of urban expansion on poor arable soils, four regions, located in southern Europe, stand out. Except for Seville (ES618), all other regions (Cyprus, Istanbul, and Valladolid) also have high to very high relative impacts.

Figure 5 clearly shows that, on the one hand, regions with a high to very high impact of urban expansion activities on average soils are distributed across Europe, but that, on the other hand, some clusters exist. Most striking is Albania, which comprises the two regions with the highest relative impact (AL00B and AL002, Tirana and Durres, with 14.8 and 12.1 %, respectively); followed by the Netherlands, Ger-

many, Italy, and Spain, which also possess a number of regions with a very high impact of urban land take on average soils.

In terms of absolute values, only a few of the previous regions are among the highest ranking. Interestingly, both Albanian regions also possess a large absolute value (2752 and 1997 ha, respectively). However, the region with by far the highest absolute value is the region of Madrid (ES300), with 11 854 ha of average soils lost due to land take, which corresponds to 5.2 % in relative terms. The absolute value of Madrid is more than double that of the second highest region, which is another Spanish region (Toledo), followed by two other Spanish regions (Ciudad Real and Zaragoza). Remarkably, many more Spanish regions follow in the ranking. This implies a very high absolute loss of average soils due to land take, but often with less relevance when it comes to the relative impact (often intermediate, sometimes high values regarding the share of soils with average potential in a particular NUTS-3 area).

Regarding the distribution of regions with a very high impact of land take on good soils, some clusters of regions and/or hotspots exist. One is located in the Netherlands and western Germany, another one in the western Balkans (including Albania), a third one from northern Italy (Umbria and Po valleys) to south-eastern France (Alpes and Provence, Rhone Valley), a fourth one on the Iberian Peninsula, and a fifth one in Ireland. The relative impact ranges from 38.9 %

in Tirana (AL00B) over 34.7 % (NL332, Agglomeratie 's-Gravenhage), 27.7 % (NL327, Het Gooi en Vechtstreek), and 15 % (NL325, Zaanstreek) to several regions between 10.6 and 5 % of an impact.

In terms of absolute values, most of those regions with very high relative impact values do not score very high, though. Only two Albanian regions as well as one Irish region stand out. Otherwise, there are five other regions (next to Tirana, AL00B) that have more than 2000 ha of impacted good soils on arable land. Three of those regions are located in France, one in Turkey, and one in the Czech Republic. In terms of relative impacts, they possess intermediate to high values (between 0.34 and 0.81 %). Interestingly, many of the high-ranked regions possess a share of more than 50 % of good soils; however, there are also some regions with a very low share. One of those regions is again Tirana with a share of 3.4 %; others are AL00A (Shkoder, 5.5 %) and ES523 (Valencia, 4.3 %). The latter two also show very high relative values of the impact of land take on the good soils, that is, of the limited area with good soils available, a high share is affected by land take.

4 Discussion

In general, most of the arable lands have good productivity potentials, both at country level (18 countries out of 36) and when considering the entire coverage of the study area (46.32 %).

However, the European picture is, as expected, very heterogeneous. The urban residential expansion and extension of economic sites and infrastructure activities is spatially distributed across Europe, with very low (green) to very high (red) impact on the biomass productivity of arable land. Several hotspot areas can be identified in which land take clearly affects soils with a capacity to provide biomass.

The highest share of arable land affected by land take was found in Albania (AL) (3.97 %), the Netherlands (NL) (1.45 %), Cyprus (CY) (1.39 %), and Ireland (IE) (0.76 %). However, when the impacted lands are considered in hectare, Spain, France, and Germany are the highest ranked. High and very high impacts on good land can mainly be detected in regions in Ireland, Spain, France, Germany, Italy, and the Balkan countries. Average land is strongly impacted in Albania, the Netherlands, Germany, and Italy. Very high impacts of urban land take on poor soils are scattered across Europe.

When taking the gross domestic product of the outstanding regions into account, there seems to be no direct relation to the economic situation of a region. Both well-developed and less-developed regions experience high to very high impacts of land-take-related land cover flows on the soil productivity.

Several hotspot areas are identified in which land take clearly affects soils with a capacity to provide biomass. The Madrid region is one of the hotspots of urban development in Europe, experiencing a rate of 50 % growth in

the 1990s, compared to 25 % national and 5.4 % EU average rates (EEA, 2006a). The trend attenuated between 2000 and 2006 (around 20 %) but is still present. According to Díaz-Pacheco and García-Palomares (2014) the urban land surface grew at a rate in excess of 4 % per year. Tóth (2012) shows that the urban sprawl of Madrid occurred to a large extent on arable land. According to the EEA report (EEA, 2006a) major drivers are (i) the growing demand for first and second homes caused by economic growth and low interest rates despite a rather modest population growth; (ii) increased mobility; (iii) increasing housing prices, which force more people to move further and further into the city suburbs; and (iv) a weak planning framework. The reasons for land take differ from country to country; nevertheless, these major drivers that were given for the Madrid region might be valid for most of the regions or countries in Europe with the addition of some drivers such as new developments along transportation axes, tourism, and coastline diffusion in general. Moreover, the OECD reports (OECD, 2007) that rapid and partly unplanned development led to urban sprawl in the Madrid region.

Alongside the situation in the Madrid region, the EEA (EEA, 2006a) also presents the example of the occurring urban sprawl along the Spanish and Portuguese coastlines. In these areas, sprawl mainly consists of diffuse settlements adjacent to or disconnected from concentrated urban centres. This residential sprawl is responsible for more than 45 % of coastal zone land transformation into artificial surfaces. In Portugal, 50 % of urban areas are located between Lisbon–Setubal and Porto–Viana do Castelo within 13 km from the shoreline, hence covering only 13 % of the total land area. In Spain economic growth, legislative flexibility, and tourism resulted in an increased number of households and second homes along the coast, in combination with infrastructure and leisure facility development.

Outside the Iberian Peninsula, the Po Valley and the adjacent Emilia-Romagna Plain (ERP) have a long history of urban expansion. The valley has soils that are amongst the most fertile in Europe. Even though the entire region is called "Food Valley", more and more of its agricultural area is irreversibly converted into urban fabric, either for residential or industrial and commercial use, continuing at a rate of 1 ha per day (EC, 2011). The movie "Il suolo minacciato" ("Land under threat") presented during Green Week 2011 uses the example of these two confronting pressures on land to highlight what is currently happening in this region. Malucelli et al. (2014) confirm that while the extent of woodland, grassland, natural areas, and wetlands in the ERP did not change significantly, urban and industrial areas increased to the detriment almost exclusively of cropland. The analysis in the current study highlights that mainly good and average land is affected.

The impacts of land take on regions in southern France are also already described and explained in the EEA report on urban sprawl (EEA, 2006a). This so-called "inverse T" of urban sprawl along the Rhone Valley down to the Mediterranean coast is caused by new developments along transportation axes and coastlines (which are often connected to river valleys).

Another prominent and well-known region of urban sprawl and related land take is the Dublin metropolitan area, which can be recognized on the maps of average and good soils. In the past, population growth and economic development were responsible for the expansion of the metropolitan area further to the outskirts (EEA, 2006a).

In Germany, land take is most prominent in the region comprising the "Ruhrgebiet" (in particular the regions around its core), in parts of southern Germany, but also in eastern Germany, particularly in some regions that are experiencing an improvement in their economic situation (e.g. Leipzig). Prokop et al. (2011) state that despite having defined a target of reducing land take to 30 ha day^{-1} until 2020, the measures taken so far have not been sufficient.

In the Netherlands most regions have experienced and are still experiencing rapid urban expansion along the urban–rural fringes during the past decades, which is still ongoing, although spatial planning policies were seeking to promote compact urban developments (Nabielek et al., 2013). This increase in land take is also documented in Prokop et al. (2011), showing the constant increase in built-up area between the 1960s and 2006 (Fig. 49 in Prokop et al., 2011). A similar picture appears in the Flanders region (Belgium), where the typical ribbon development continues at a rate of 6 ha per day of which 5 ha is due to residential sprawl (Gregor et al., 2015).

Regarding the conversion of arable land to urbanized areas in the central and eastern European countries, it can be assumed that the accession to the EU in 2004 and the related economic development together with benefits from regional development programmes were the leading driving forces to the expansion of residential, but mainly industrial and commercial, areas, primarily at the expense of good and average lands. Very recent statistics on the cohesion funding amount allocated per member state (EC, 2015) confirm that some of the eastern European countries rank amongst the top. For example, Poland is the country with the highest amount allocated, while the Czech Republic and Hungary rank fourth and sixth, respectively.

Without being a member state of the EU, Albania has undergone significant changes with regards to urban expansion and land take. In particular, average and good soils for providing biomass on arable land have been converted into artificial surfaces, according to the most recent assessment of the EEA (2013b) on land take indicator coded CSI 014. This has happened at the expense of grassland and mixed farmland (in total 73 % of the total land uptake), which is of relevance in this context of arable land. Likewise, in Bosnia and Herzegovina 72 % of the total land take occurred on grassland or mixed farmland areas.

5   Conclusion

The potentials and the actual use of the lands were linked, and impact of land take on arable lands with good, average, and poor production potentials were assessed and mapped successfully by this study. According to the results, from 2000 to 2006 0.23 % of the production potential on arable land in the study area was lost as a consequence of land take; over the period 1990–2006, this loss amounted to 0.81 % (EEA, 2015). Especially the arable lands that have good production potential and have been impacted by land take can be considered as lands that have not been used according to their theoretical potentials. This situation of arable lands with high productivity potentials is creating pressure on soils and other ecosystem types, resulting in threats to soil biodiversity, all other soil functions, and ecosystem services. Since Europe is approaching a time when being able to demonstrate good land resource management will be critical for food security and achieving other soil protection and land degradation neutrality targets, this situation is getting attention. Moreover, the countries and hotspot regions that have the highest land take impact on the biomass productivity potentials of arable land, was indicated by this study, can be considered as the areas that also have potential future land use conflicts. Therefore, assessment and mapping of the land take impacts on arable land for the 2006–2012 period is essential to understanding the trends of the countries and to monitor the situation.

Interpreting the assessment results and observing the spatial distribution of the small number of impacts on this scale was quite hard. Even though the impacts were calculated based on pixels, interpreting them on NUTS-3 units was helpful. Also, considering the absolute (in hectares) and relative (as a percentage) values in parallel is important because of the different coverages of the countries. It should be remembered that, even though the relative impact may be low in some countries, the absolute impact may be quite high.

Competing interests. The authors declare that they have no conflict of interest.

Acknowledgements. We are grateful to the two anonymous reviewers for their valuable comments and detailed suggestions to improve this study.

Edited by: Antonio Jordán

References

Adugna, A. and Abegaz, A.: Effects of land use changes on the dynamics of selected soil properties in northeast Wellega, Ethiopia, SOIL, 2, 63–70, https://doi.org/10.5194/soil-2-63-2016, 2016.

CEC (Commission of the European Communities): Communication from the Commission to the Council, the European Parliament, the European Economic and Social Committee and the Committee of the Regions – Thematic strategy for soil protection, Commission of the European Communities. COM, Brussels, Belgium, p. 231, 2006.

CORINE: CORINE Land Cover, available at: http://land.copernicus.eu/pan-european/corine-land-cover, last access: 16 June 2017.

Díaz-Pacheco, J. and García-Palomares, J. C.: Urban Sprawl in the Mediterranean Urban Regions in Europe and the Crisis Effect on the Urban Land Development: Madrid as Study Case, Urban Studies Research, 2014, 807381, https://doi.org/10.1155/2014/807381, 2014.

EC: Green-week events, 25 May 2011, available at: http://ec.europa.eu/environment/archives/greenweek2011/content/film-il-suolo-minacciato-land-under-threat-directed-nicola-dallolio.html (last access: 16 June 2017), 2011.

EC: Financial allocations 2014–2020, available at: http://ec.europa.eu/regional_policy/thefunds/funding/index_en.cfm (last access: 16 June 2017), 2015.

EEA: Urban sprawl in Europe – the ignored challenge- EEA Report No. 10/2006, Copenhagen, Denmark, 2006a.

EEA: Land accounts for Europe 1990–2000, EEA Technical Report No. 11/2006, Copenhagen, Denmark, 2006, available at: https://www.eea.europa.eu/data-and-maps/data/land-cover-flows-based-on-corine-land-cover-changes-database-1990-2000-1 (last access: 16 June 2017), 2006b.

EEA: Changes in European land cover from 2000 to 2006, available at: https://www.eea.europa.eu/data-and-maps/figures/land-cover-2006-and-changes-1 (last access: 16 June 2017), 2013a.

EEA: Land take. Indicator assessment, Data and maps. This report has been generated automagically by the EEA Web content management system on 22 Feb 2017, 12:08 PM, available at: http://www.eea.europa.eu/data-and-maps/indicators/land-take-2/assessment-2 (last access: 22 February 2017), 2013b.

EEA: Dominant land cover flows 2000–2006, available at: https://www.eea.europa.eu/data-and-maps/figures/dominant-land-cover-flow-2000-2006 (last access: 16 June 2017), 2013c.

EEA: SOER 2015, European briefings: Soil, available at: http://www.eea.europa.eu/soer-2015/europe/soil (last access: 16 June 2017), 2015.

ESDAC: ESDAC data, available at: http://esdac.jrc.ec.europa.eu/content/soil-biomass-productivity-maps-grasslands-and-pasture-coplands-and-forest-areas-european, last access: 16 June 2017.

Gregor, M., Löhnertz, M., Philipsen, C., Aksoy, E., Schröder, C., Prokop, G., and Tramberend, P.: The report on land resource efficiency. European Topic Centre on Urban, Land and soil systems (ETC/ULS), Copenhagen, Denmark, 2015.

Helming, K., Konig, B., and Tscherning, K.: SENSOR first annual report (public part), in: SENSOR report series 2006/1, edited by: Helming, K. and Wiggering, H., ZALF, Germany, available at: http://publ.ext.zalf.de/publications/SENSORrs_2006_1_annreport.pdf (last access: 16 June 2017), 2006.

Hoogeven, Y. and Ribeiro, T.: Land Use Scenarios for Europe, Background Report, European Environmental Agency, Kopenhagen, Denmark, 2007.

Kalema, V. N., Witkowski, E. T. F., Erasmus, B. F. N., and Mwavu, E. N.: The Impacts of Changes in Land Use on Woodlands in an Equatorial African Savanna, Land Degrad. Dev., 26, 7 632–641, https://doi.org/10.1002/ldr.2279, 2015.

Klijn, J. A., Vullings, L. A. E., Lammeren, R. J. A., van Meijl, J. C. M., van Rheenen, T., Veldkamp, A., Verburg, P. H., Westhock, H., Eickhout, B., and Tabeau, A. A.: The EURURALIS study: technical document, Alterra Report 1196, Alterra, Wageningen, the Netherlands, 2005.

Liu, Z., Yao, Z., Huang, H., Wu, S., and Liu, G.: Land use and climate changes and their impacts on runoff in the Yarlung Zangbo river basin, China, Land Degrad. Dev., 25, 203–215, https://doi.org/10.1002/ldr.1159, 2014.

Malucelli, F., Certini, G., and Scalenghe, R.: Soil is brown gold in the Emilia-Romagna region, Land Use Policy, 39, 350–357, https://doi.org/10.1016/j.landusepol.2014.01.019, 2014.

Mancosu, E., Gago-Silva, A., Barbosa, A., de Bono, A., Ivanov, E., Lehmann, A., and Fons, J.: Future land-use change scenarios for the Balck Sea catchment, Environ. Sci. Policy, 46, 26–36, https://doi.org/10.1016/j.envsci.2014.02.008, 2015.

Mohawesh, Y., Taimeh, A., and Ziadat, F.: Effects of land use changes and soil conservation intervention on soil properties as indicators for land degradation under a Mediterranean climate, Solid Earth, 6, 857–868, https://doi.org/10.5194/se-6-857-2015, 2015.

Muñoz-Rojas, M., Jordán, A., Zavala, L. M., De la Rosa, D., Abd-Elmabod, S. K., and Anaya-Romero, M.: Impact of Land Use and Land Cover Changes on Organic Carbon Stocks in Mediterranea n Soils (1956–2007), Land Degrad. Dev., 26, 168–179, https://doi.org/10.1002/ldr.2194, 2015.

Nabielek, K., Kronberger-Nabielek, P., and Hamers, D.: The Rural-Urban Fringe in the Netherlands: a Morphological Analysis of Recent Urban Developments, Proceedings REAL CORP 2013, Tagungsband, 20–23 May 2013, Rome, Italy, 2013.

OECD: OECD Territorial Reviews, Madrid, Spain, 2007.

Parras-Alcántara, L., Martín-Carrillo, M., and Lozano-García, B.: Impacts of land use change in soil carbon and nitrogen in a Mediterranean agricultural area (Southern Spain), Solid Earth, 4, 167–177, https://doi.org/10.5194/se-4-167-2013, 2013.

Prokop, G., Jobstmann, H., and Schönbauer, A.: Overview of best practices for limiting soil sealing or mitigating its effects in EU-27. Final report, Study contracted by the EC DG Environment, https://doi.org/10.2779/15146, 2011.

Rounsevell, M. D. A., Reginster, I., Araujo, M. B., Carter, T. R., Dendoncker, N., Ewert, F., House, J. I., Kankaanpaa, S., Leemans, R., Metzger, M. J., Schmidt, C., Smith, P., and Tuck, G.: A coherent set of future land use change scenarios for Europe, Agr. Ecosyst. Environ., 114, 57–68, 2006.

Scheffer, F. and Schachtschabel, P.: Lehrbuch der Bodenkunde, Spektrum, Stuttgart, Germany, 2002.

Smith, P., Cotrufo, M. F., Rumpel, C., Paustian, K., Kuikman, P. J., Elliott, J. A., McDowell, R., Griffiths, R. I., Asakawa, S., Bustamante, M., House, J. I., Sobocká, J., Harper, R., Pan, G., West, P. C., Gerber, J. S., Clark, J. M., Adhya, T., Scholes, R. J., and Scholes, M. C.: Biogeochemical cycles and biodiversity as key drivers of ecosystem services provided by soils, SOIL, 1, 665–685, https://doi.org/10.5194/soil-1-665-2015, 2015.

Tóth, G.: Impact of land take on the land resource base for crop production in the European Union, Sci. Total Environ., 435–436, 202–214, 2012.

Tóth, G., Bódis, K., Ivits, É., Máté, F., and Montanarella, L.: Productivity component of the proposed new European agri-environmental soil quality indicator, in: Land quality and land use information in the European Union, edited by: Tóth, G. and Németh, T., Publication Office of the European Union, Luxembourg, 297–308, 2011.

Tóth, G., Gardi, C., Bódis, K., Ivits, É., Aksoy, E., Jones, A., Jeffrey, S., Petursdottir, T., and Montanarella, L.: Continental-scale assessment of provisioning soil functions in Europe, Ecol. Process., 2, 32, https://doi.org/10.1186/2192-1709-2-32, 2013.

Trabaquini, K., Formaggio, A. R., and Galvão, L. S.: Changes in physical properties of soils with land use time in the Brazilian savanna environment, Land Degrad. Dev., 26, 397–408, https://doi.org/10.1002/ldr.2222, 2015.

Van Ittersum, M. K., Ewert, F., Heckelei, T., Wery, J., Alkan Olsson, J., Andersen, E., Bezlepkina, I., Brouwer, F., Donatelli, M., Flichman, G., Olsson, L., Rizzoli, A. E., van der Wal, T., Wien, J. E., and Wolf, J.: Integrated assessment of agricultural systems – A component-based framework for the European Union (SEAMLESS), Agr. Syst., 96, 150–165, 2008.

Wasak, K. and Drewnik, M.: Land use effects on soil organic carbon sequestration in calcareous Leptosols in former pastureland – a case study from the Tatra Mountains (Poland), Solid Earth, 6, 1103–1115, https://doi.org/10.5194/se-6-1103-2015, 2015.

13

Micromorphological characteristics of sandy forest soils recently impacted by wildfires in Russia

Ekaterina Maksimova[1,2] and Evgeny Abakumov[1,2]

[1]Department of Applied Ecology, Saint Petersburg State University, St Petersburg, Russia
[2]Institute of the Ecology of the Volga Basin, Togliatti, Russia

Correspondence to: Ekaterina Maksimova (doublemax@yandex.ru)

Abstract. Two fire-affected soils were studied using micromorphological methods. The objective of the paper is to assess and compare fire effects on the micropedological organisation of soils in a forest-steppe zone of central Russia (Volga Basin, Togliatti city). Samples were collected in the green zone of Togliatti city. The results showed that both soils were rich in quartz and feldspar. Mica was highly present in soils affected by surface fires, while calcium carbonates were identified in the soils affected by crown fires. The type of plasma is humus–clay, but the soil assemblage is plasma–silt with a prevalence of silt. Angular and subangular grains are the most dominant soil particulates. No evidence of intensive weathering was detected. There was a decrease in the porosity of soils affected by fires as a consequence of soil pores filled with ash and charcoal.

1 Introduction

Fire has an important impact on soil properties as identified by previous works (Certini, 2005, 2013; Guénon et al., 2013; Mataix-Solera et al., 2011; Jain et al., 2012; Bergeron et al., 2013; Dymov and Gabov, 2015; Zharikova, 2015; Maximova and Abakumov, 2015). Soil processes in post-fire soil environments are quite different from those in the soils of natural landscapes or in technogenic ones. In general, changes in morphological organisation and soil mineralogy are well known in soils after fires produced at high temperatures. After the fire, there is an accumulation of ash on topsoil (Pereira et al., 2014), leaching of some nutrients into deeper horizons (Bodi et al., 2014), an over-compaction of the surface, an accumulation of crusts, and a transformation of the soil structure (Mataix et al., 2011).

Micromorphological methods are known as a useful tool for the investigation of soil transformation under natural and human-impacted conditions (Stoops, 2009). The methodology of classical micropedology provides the required information about soil development at the micro level, such as changes in fine earth composition and soil plasma evolution (Kubiëna, 1938, 1970; Stoops and Eswaran, 1986). These methods are widely used for the analysis of soil paleoprocesses (Sedov et al., 2013), soil restoration in post-mining environments (Abakumov et al., 2005), soil elementary process in different environments (Lebedeva et al., 2010; Abakumov et al., 2013) and the specification of soil classification aspects (Kubiëna, 1970). Micromorphological investigations related to the fire effect of soil crusts and fine earth (Greene et al., 1990) and aggregate dynamics in post-fire soil (Mataix-Solera et al., 2002) have been undertaken. However, the micromorphological methods have never been applied to study post-fire soil transformation in the Russian wild-fire environments. Moreover, this paper deals with a comparison of different post-fire scenarios (surface and crown fires) and provides a 5-year monitoring of fire-affected soils and ecosystems as a whole. Little research has been carried out on this topic.

The objective of this work was to characterise the micromorphological indices of microstructure transformation in soils affected by different types of wild fire, as compared to the microstructure of the mature, unaffected soil of pine forests in central Russia.

Table 1. Morphological features and general properties of soils; ± after the mean value means SD. Nd: no data.

Horizon	Depth, cm	Colour	Soil humidity, %	Ignition loss value %	pH	Total organic carbon, $g\,kg^{-1}$	C_{ha}/C_{fa}	Clay content, $g\,kg^{-1}$	Sand content, $g\,kg^{-1}$
				Surface fire					
Apyr	0–3	10 YR 3/2	2.85 ± 0.79	5.68 ± 0.95	8.0 ± 0.06	23.1 ± 0.27	2.14 ± 0.13	20.0	705.0
AY	5–14	10 YR 4/2	1.45 ± 0.19	2.79 ± 0.60	6.2 ± 0.32	12.1 ± 0.50	1.30 ± 0.26	25.0	807.0
AC	14–27	7.5 YR 5/4	1.38 ± 0.36	1.55 ± 1.16	6.0 ± 0.21	7.5 ± 0.48	nd	26.0	835.0
AC	27–36	10 YR 6/4	1.02 ± 0.27	0.80 ± 0.49	5.8 ± 0.21	3.1 ± 0.13	–	19.0	852.0
AC	36–53	10 YR 6/4	0.98 ± 0.42	0.98 ± 0.73	5.3 ± 0.31	2.2 ± 0.05	–	24.0	866.0
C	53–73	7.5 YR 5/4	0.69 ± 0.05	0.81 ± 0.05	5.7 ± 0.21	2.4 ± 0.10	–	13.0	864.0
				Crown fire					
Apyr	0–5	10 YR 3/2	2.37 ± 0.36	5.45 ± 1.41	7.9 ± 0.12	31.9 ± 0.19	1.95 ± 0.34	17.0	720.0
AY	3–10	10 YR 4/2	1.43 ± 0.35	3.01 ± 1.31	5.9 ± 0.38	14.2 ± 0.31	1.18 ± 0.11	20.0	788.0
AC	10–15	7.5 YR 5/2	0.86 ± 0.20	2.91 ± 3.01	5.9 ± 0.25	7.8 ± 0.07	nd	17.0	852.0
AC	15–24	10 YR 7/4	1.11 ± 0.63	0.86 ± 0.29	5.9 ± 0.36	2.6 ± 0.07	–	6.0	862.0
AC	24–44	10 YR 6/3	0.52 ± 0.05	0.66 ± 0.14	5.7 ± 0.12	1.4 ± 0.05	–	4.0	867.0
C	44–64	10 YR 6/3	0.49 ± 0.03	0.63 ± 0.09	5.9 ± 0.25	1.2 ± 0.05	–	9.0	868.0
				Control					
O	0–7	–	5.92 ± 2.27	20.88 ± 5.90	6.5 ± 0.10	nd	1.17 ± 0.18	nd	nd
AY	7–10	10 YR 4/2	1.60 ± 0.47	2.71 ± 1.34	6.3 ± 0.06	19.4 ± 1.35	0.67 ± 0.13	21.0	787.0
AY	10–14	10 YR 6/4	0.78 ± 0.15	1.57 ± 0.82	6.2 ± 0.23	7.8 ± 0.33	nd	18.0	837.0
AC	14–23	7.5 YR 3/2	0.78 ± 0.42	0.66 ± 0.50	6.1 ± 0.23	3.3 ± 0.10	–	13.0	867.0
AC	23–33	2.5 YR 8/6	0.42 ± 0.01	0.57 ± 0.12	5.9 ± 0.20	1.5 ± 0.02	–	4.0	888.0
C	33–50	2.5 YR 8/6	0.33 ± 0.02	0.46 ± 0.01	5.7 ± 0.12	2.1 ± 0.03	–	7.0	891.0
C	50–70	2.5 YR 8/6	0.32 ± 0.03	0.52 ± 0.17	5.8 ± 0.31	1.5 ± 0.05	–	3.0	895.0
				Post hoc test					
Crown control						$p < 0.05$		$p < 0.06$	$p < 0.07$
Surface control						$p < 0.01$		$p < 0.04$	$p < 0.02$
Crown surface						$p < 0.02$		$p < 0.08$	$p < 0.09$

2 Materials and methods

2.1 Study area

The study was conducted in the Samara region situated near the Volga River, in the central part of European Russia (58°39′44.55″ N, 39°17′48.95″ E; 179 m a.s.l.). The extremely hot weather in the summer of 2010 in Russia (especially bad in the entire Russian European area and also Ukraine and eastern Europe) resulted in drought and eventually catastrophic forest fires in vast territories of European and Siberian Russia. The forest fire studied occurred in 2010 and affected more than 8000 ha. Fire severity was very high. The parent material is composed of old (Pleistocene) alluvial-dune landscape.

The affected ecosystem is characterised by a forest-steppe environment with higher pedo-diversity. The vegetation was composed mostly of pine forests (*Pinus sylvestris* L.). There are xerophyte species at dry locations like *Veronica spicata* L., *Sedum acre* L., *Antennaria dioica* L., *Calamagrostis*

epigejos (L.) Roth and *Centaurea marschalliana* Spreng. and hardwoods (*Quercus robur* L., *Betula pendula* Roth, *Populus tremula* L.) in more humid conditions. The herbaceous vegetation consists of rhizomatous and loose-bunch *Gramen* (*Bromus inermis* Leyss., *Elytrigia repens* L., some species of *Poa*, and *Agrostis canina* L.) in post-fire plots.

A variety of Luvisols and Chernozems prevail in the watershed sections (Nosin, 1949; Vasil'eva and Baranova, 2007; Abakumov and Gagarina, 2008; Abakumov et al., 2009; Urusevskaja et al., 2000; Saksonov, 2006), whereas Calcaric Chernozems (southern type) dominate in the south of the region (steppe zone of the Samara region), accompanied by some polypedons of Kastanozems (Nosin, 1949). The soils of investigated area are sandy and sandy loam textured. In the studied area, soils were classified as sandy loam soils on Late Pleistocene alluvial Volga sands – Protoargic Arenosols according to World Reference Base (2015) and to Shishov and Tonkonogov (2004) – and they have weak features of illuvial phenomena without the formation of separate horizons. The

Table 2. Micromorphological characteristics of the soils under investigation.

Horizon	Sampling depth (cm)	Skeleton grains									Type of plasma	Elementary assemblage	Plant residues		Structural separates	Voids	Soil microstructure	Coprolites	
		% character of distribution	size, mm max	min	dominant	size sorting	particle shape	rounding	corrosion	mineral composition			%	decomposition degree				size, mm	description
crown forest fire																			
Apyr	0–15	40–50 uniform	0.07	0.01	0.04	good–medium	subidiomorphic, xeromorphic, regenerate (some minerals in one)	rounded	weak	colourless minerals prevail: quartz, feldspar (orthoclase), carbonates (calcispar)	humus–clay	plasma–silt (little plasma, mostly particles)	10	weak–medium	aggregates	pores	pore structure with separated aggregates		
surface forest fire																			
Apyr	0–14	40–50 uniform	0.07	0.01	0.01	medium–weak	idiomorphic, subidiomorphic, xeromorphic	rounded–not rounded	medium	a lot of specular stone (muscovite), quartz, feldspar (orthoclase)	humus–clay	plasma–silt	5	weak	aggregates	pores	pore structure with separated aggregates	0.4	brown, roundish, harsh with minerals (calcispar)
control																			
O	0–20	40–50 uniform	0.09	0.01	0.02	good–medium	idiomorphic, subidiomorphic, xeromorphic	rounded–not rounded	weak	colourless minerals prevail: quartz, feldspar (orthoclase), specular stone (muscovite)	humus–clay	plasma–silt (little plasma, mostly particles)	30	weak–medium–strong; strong prevails	aggregates	pores	pore structure with separated aggregates		

sand content in these soils is 70.5–86.4 %; their clay content is 0.3–2.6 %.

Three soil pits were sampled in two different fire-affected areas: one at a site affected by a surface forest fire and another affected by a crown fire. A sample was collected in an unburnt area to serve as control. Soil types and vegetation were the same. Soils were sampled as fast as possible after the removal of a state of emergency from the territory in the summer of 2010 and also during the period 2011–2015. Three soil pits were sampled in each studied area. The general characteristics of the soils, including the chemical and physical parameters, were determined by standard methods (Methods of Soil Analysis, 1996). The measurements of these soil properties were performed in triplicate. The normal distribution of the data was verified previously, and analysis of variance (ANOVA) and a post hoc test were conducted using SIGMAPLOT 8.0 software with the aim of comparing differences between plots (site effect). Differences were considered significant at $p < 0.05$. Undisturbed and post-pyrogenic soil samples were collected in 2011 using Kubiëna-type boxes $5 \times 3.5 \times 1.5$ cm in size at the depth of 0–10 cm and taken to the laboratory.

2.2 Micromorphological analysis

Soil samples were air-dried and passed through a 1 mm sieve. Fine sections of soil material were prepared from micro monoliths of soils, sampled in field. Samples were dried and saturated with resin.

Thin sections were investigated with use of a polarisation microscope (Leica DFC 320) in transmitted light and crossed nicols. The following soil micromorphological indexes were investigated: soil microfabric, spatial arrangements of fabric units, soil particle distribution, elements of microstructure and character of organic matter. The terminology used in this paper is published by Stoops (2003), Gagarina (2004) and Gerasimova et al. (2011), where details of the micro organisation of soil were described in detail.

3 Results and discussion

3.1 Soil profile analysis and physico-chemical properties

The soil profile organisation can be described as Apyr (or O in the case of control) – AY – AC – C. The Apyr horizon is a black layer due the deposition of an ash and charcoal at the soil surface, unlike that observed in the control plot (Figs. 1, 2, 3). A wide distribution of coal pieces, a total absence of forest floor remnants and its transformation into ash were diagnosed in 2010 immediately after the fires. At the beginning of the research (2010), thick black horizons were observed at the soil surface, while in the summer of 2011, they were only present as a thin layer at the surface. This testifies to the influence on erosion, as the soil surface was affected by pre-

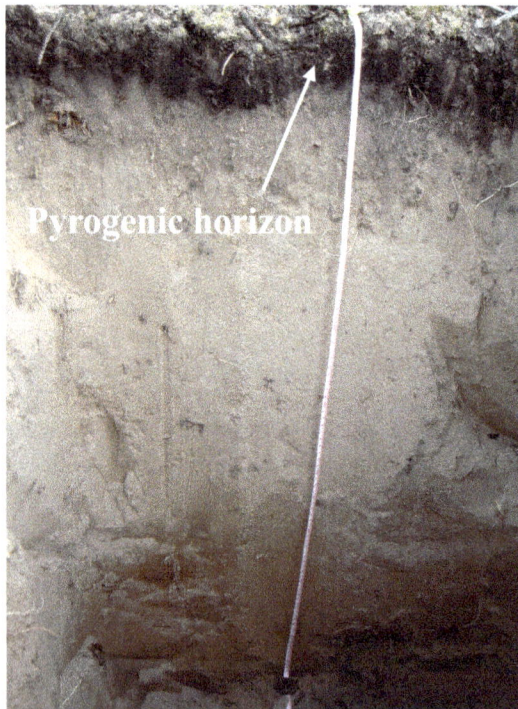

Figure 1. Crown fire-affected soil.

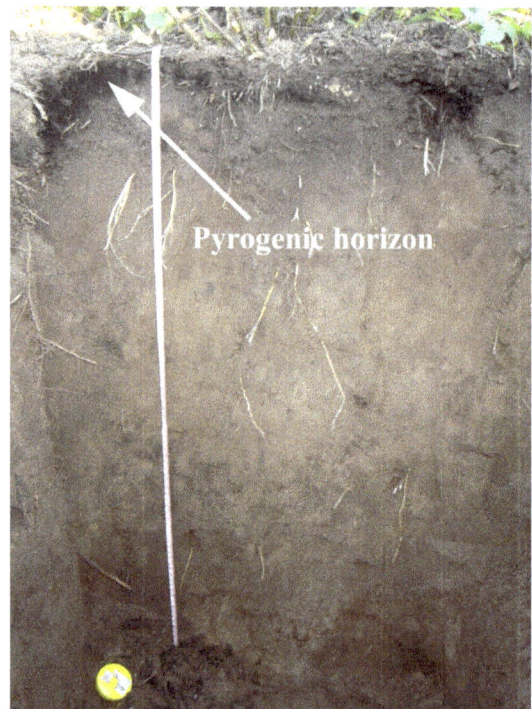

Figure 2. Surface fire-affected soil.

cipitation after the disappearance of forest floor (Robichaud, 2005; Vieira et al., 2014; Delwiche, 2009).

Data on the general characteristics of the soils are given in Table 1.

Organic matter is lost from the surface horizons of the soil, which is related to the destruction of the organic horizons, the mineralisation of root residues and the almost complete absence of fresh plant waste, which could be a material for humification. Humus degradation of the upper horizons was clearly visible by means of the ignition loss value. Ignition loss was more than 20.00 % in the upper layer in the control plot (Table 1), but only 5.45 % in the crown fire and 5.68 % in the surface fire plots. In 2010, the content of organic matter in the ash at the soil surface after the surface fire (2.31 ± 0.27 %) was lower than that after the crown fire (3.19 ± 0.19 %). A similar tendency was observed in the pyrogenically transformed humus-enriched horizons: the content of organic carbon was 1.21 ± 0.50 % after the surface fire and 1.42 ± 0.31 % after the crown fire. Thus, a surface fire, which leads to the complete burning out of the litter and the upper horizons, results in larger losses of organic matter.

The acidity in the upper horizons of the burnt soils decreases significantly, and the burnt litters have an alkaline reaction (pH 7.9–8.0), while the lower horizons have a weakly acidic reaction close to that in the corresponding horizon of the undisturbed forest soil (pH 5.7–5.9). The increase in the pH values of the soils after fires is related to the fact that the water-soluble ash components penetrate into the soil and saturate the soil exchange complex with alkaline-earth el-

ements, which shifts the reaction toward the neutral value. No differences in the pH changes between the effects of the crown and surface fires were observed in the first year of the study. The reaction of the parent rock is similar in all three plots and is characterised as weakly acidic. The partial or sometimes complete mineralisation of organic residues because of pyrolysis resulted in the synchronous input of ash elements onto the soil surface and into the litter, which neutralised the organic acids arriving in the soil solution during the decomposition of the litters. It is therefore obvious that the higher the ash yield (i.e. the more intensive the fire), the more complete and active the neutralisation of the litters.

The group humus composition (C_{ha} to C_{fa} ratio) of the upper soil layers changed as a result of the fires (Table 1). Some authors have noted an increase in humic acid content and a decrease in the carbon-to-nitrogen ratio (Abakumov and Frouz, 2009; Efremova and Efremov, 2006). By contrast, the appearance of the most aggressive fractions presented by fulvic acids was recorded in other studies (Dobrovol'skij, 2002). In our case, the litter of the control plot was characterised by a fulvic–humic type of humus (for the other horizons, humic–fulvic) and an increase in humic acids, which was especially strong after the surface fire, as is characteristic for post-pyrogenic soils. A reduction in the C_{ha} / C_{fa} ratio due to new plant litter was observed in the following years. An increase in humic acid content was also observed in the humus horizon.

A year after the fire, the pH of the burnt litters decreased from 7.9–8.0 (in 2010) to 6.4–6.6 (in 2011), and its absolute

Figure 3. Control soil.

values approached the control level. This is easily explicable: rain and snowmelt waters almost completely removed the soluble ash components for a year; i.e. the alkali elements were removed from the ash at the fire sites.

A post hoc test has shown that significant differences were revealed for the carbon content between surface fire and crown fire ($p < 0.02$), between control and crown fire ($p < 0.05$) and also between control and surface fire plots ($p < 0.01$). As for the silt and clay fraction, there were differences only between control and surface fires ($p < 0.04$). The same situation was characteristic of the sand fraction ($p < 0.02$).

The WRB system (World Reference Base, 2015) does not have any horizons or diagnostic parameters of pyrogenic soils. But the WRB system has pretic horizon as being dark, with a high content of organic matter and phosphorus, low biological activity, and high contents of exchangeable calcium and magnesium, with remnants of charcoal and/or artefacts. A pyrogenic horizon with an abundance of charcoal is formed after wildfires. It can resist degradation when vegetation has not started to recover yet at burnt places and charcoal has not started to be redistributed while erosion and infiltration are in progress. However, black carbon decomposition is controversial, and there are different views on this issue. Some studies argue that black carbon decomposes very slowly (Liu et al., 2008) or is practically non-degradable (González-Pérez et al., 2004), while others show that it is successfully affected by chemical (Cheng et al., 2006, 2008) and microbial (Knicker et al., 2013; Marschner et al., 2008)

oxidation. The assumption of black carbon complete stability in soils is doubtful because its content varies considerably in different soils. This is explained not only by a difference in pyrogenic activity in different natural zones, but also by a difference in humidity (Nguyen and Lehmann, 2009) and temperature (Cheng et al., 2008; Nguyen et al., 2010), by various physical and chemical soil characteristics, by different biological activity, and by land use practices (Czimczik and Masiello, 2007).

3.2 Micromorphological characteristics

Differences in post-pyrogenic and unburnt soils are shown well in the morphological, chemical, physical and biological properties of horizon A. The morphological organisation of solum in burnt soils differs from that in unburnt soils in a number of parameters: wide distribution of charcoal pieces; absence of litter and its transformation into ash that is a mix of mineral soil components, burnt-down plant residues, and small pieces of charcoal (Bodi et al., 2014; Pereira et al., 2014, 2015); and likewise a reduction in the humic horizon's depth. There is soil erosion because of rainfall; this is characterised by a decrease in black surface horizon thickness after several years of investigation as observed in earlier studies (Francos et al., 2016).

Data on soil micromorphological features are presented in Table 2 and Fig. 4. The results showed that the elementary assemblage of crown fire soil is plasma–silt (with a small proportion of plasma and a prevalence of silt particles). In the case of the soil affected by surface fire, the type of assemblage is plasma–silt, which is a result of the accumulation of humus-type organic matter. In both cases the type of the plasma is classified as humus–clay. The structure of all three soils investigated is crumbly, inherited from the previous stages of soil formation. There are no evident features of mass aggregation in the fire-affected soils. The type of the microstructure of all the soils is angular blocky or subangular blocky, which is caused by a low intensity of current weathering in soil mass. The particle shape is subidiomorphic or idiomorphic in cases investigated with weak or medium degree of corrosion.

The mineral composition of the soils can be described as follows: uncoloured minerals, quartz, orthoclase and carbonates (crown fire); quartz and orthoclase with many crystals of muscovite and an absence of carbonates (surface fire); and a predominance of uncoloured minerals with quartz and muscovite in the case of the control plot. The possibility of new mineral formation under the strong heating effect has been reported in previous works (Nobles, 2010; Leon et al., 2014); however, carbonate accumulation does not result from the heating process in the crown fire scenario. In some arrangements, soil microstructure can be classified as skeletal, which is caused by a high content of non-weathered soil particles. No evidence of current weathering (alteration) has been found, but some pore infilling was recognised as a result

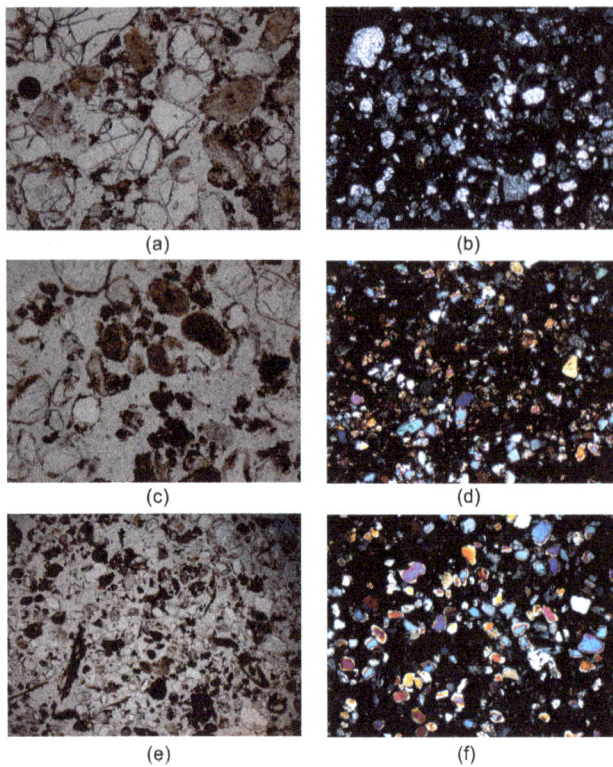

Figure 4. Thin sections. (a, b) Crown fire; (c, d) surface fire; (e, f) nature plot. Left column – transmitted light; right column – crossed nicols.

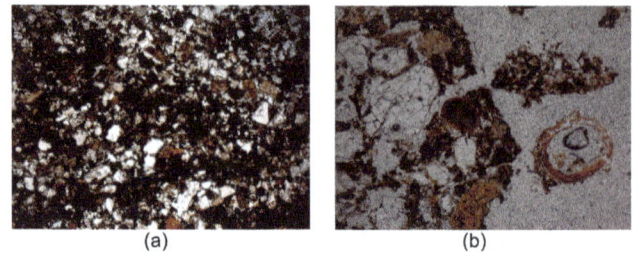

Figure 5. (a) Humons (organic particles with non-humified organic matter), transmitted light; **(b)** organic tissue of non-decomposed remnants.

of ash, charred material and erosion particles accumulating in fire-affected soils (Fig. 5). Non-decomposed organic tissue and residue infill the porous media in these soils. This is the result of an increment of raw forms of organic material in burnt soils. Organic matter under the effect of fire was polymorphic (Fig. 5a, b; Stoops, 1986), while it is monomorphic in the soils of the control plot. The porous media infilling after the fires was described previously by Nobles (2010); however, in this case, it was an accumulation of Mn–Fe-enriched materials. Balfour and Woods (2007) observed similar results in fire-affected soils. So, taking our data into account it is possible to conclude that the infilling of porous media by material of a different composition is typical in burnt soils. Decreasing the porous media area was also described as being the result of an accumulation of ash (Balfour and Woods, 2007), and this explanation of porous media infilling is more appropriate to our case. The investigated soils are characterised by developed system of porous media; this is important regarding the heat penetration into soil and soil resistance to heating. The soils investigated are not as resistant to heating as clay texture ones, where the porous media are not as developed. Pores, cracks and other spaces make it possible for the combustion products to penetrate into the soil and affect the polycyclic aromatic hydrocarbons (PAHs) and other product accumulations.

The type of fire affected the soil organic micromorphology. The quantity of non-humified organic matter in the burnt soils, especially in the samples affected by surface fire, was identified as increased in burnt soils (Fig. 5a). It is also evident that the size of organic residues is higher in fire-affected soils than in mature ones. In mature soils the transformation and humification of organic matter are a gradual process, while new, relatively fresh non-decomposed organic matter has come into the upper soil horizons in fire-affected areas (Gagarina, 2004).

4   Conclusions

The 2010 catastrophic natural fires in the urban forests of Togliatti resulted in the formation of pyrogenically transformed soils, the morphological parameters and the main chemical and physical properties of which significantly differ from those of the undisturbed soils.

The burnt soils differ from the control soil on the macromorphological level only in the upper part of the profile, where the litter is transformed to ash which is identified as a dim grey organomineral mixture. Processes of soil erosion are clearly manifested 1 year after the fire under the effect of precipitation and the illuviation of organic matter to the medium part of the profile, and they will probably continue for several years.

The fires significantly affect the physicochemical and chemical properties of the soils. However, the effect of fires on the properties of the studied soils usually does not spread deeper than 10 cm.

The results of this work showed that mineral composition of all the soils studied consists of quartz and feldspar (orthoclase); in the case of a surface forest fire there was more mica (muscovite), and calcium carbonates appeared in soils affected by crown forest fire. The reasons for this compound accumulation are still not well understood. The type of plasma is humus–clay, but the soil assemblage is plasma–silt, with a prevalence of silt. Angular and subangular grains form the main soil carcass, and no evidence of intensive weathering alteration has been revealed. At the same time, a decrease in the porous media was recognised as the main soil development process after the fires. This is the result of soil porous infilling by ash and charred organic material of different nature: some organic remnants (tissue) come into the porous media after fire, and some transformed, coaled and dark coloured parts also appear in post-fire horizons. Partially decomposed fire-affected particles of soil organic matter accumulate in post-fire soils, especially in porous soil media, which is a result of the soil organic matter accumulation and transformation in post-fire environments.

Competing interests. The authors declare that they have no conflict of interest.

Acknowledgements. The authors thank the director of the Institute of the Ecology of the Volga River Basin of the Russian Academy of Sciences (IEVB RAS), G. S. Rosenberg; the deputy director for science of IEVB RAS, S. V. Saksonov; S. A. Senator for the help in the organisation of work and research support; and Rita Lazareva for her kind assistance with thin section description.

This work was supported by the Russian Foundation for Basic Research, projects 14-04-32132, 15-34-20844, Saint Petersburg University grant No. 1.37.151.2014 and Saint Petersburg State University Internal Grant for the Modernisation of Scientific Equipment No. 1.40.541.2017.

Edited by: A. Jordán

References

Abakumov, E. V., Gagarina, E. I., and Lisitsyna, O. V.: Land reclamation in the Kingisepp area of phosphorite mining, Eurasian Soil Sci., 38, 648–655, 2005.

Abakumov, E. V., Fujitake, N., and Kosaki, T.: Humus and humic acids of Luvisol and Cambisol of Jiguli ridges, Samara Region, Russia, Appl. Environ. Soil Sci., 671359, 2009.

Abakumov, E. V., Gagarina, E. I., Sapega, V. F., and Vlasov, D. Y.: Micromorphological features of the fine earth and skeletal fractions of soils of West Antarctica in the areas of Russian Antarctic stations, Eurasian Soil Sci., 46, 1219–1229, 2013.

Balfour, V. and Woods, S. W.: Does wildfire ash block soil pores? A micromorphological analysis of burned soils, American Geophysical Union, Fall Meeting, abstract no. H43F-1695, 2007.

Bergeron, S. P., Bradley, R. L., Munson, A., and Parsons, W.: Physico-chemical and functional characteristics of soil charcoal produced at five different temperatures, Soil Biol. Biochem., 58, 140–146, 2013.

Bodi, M., Martin, D. A., Santin, C., Balfour, V., Doerr, S. H., Pereira, P., Cerda, A., and Mataix-Solera, J.: Wildland fire ash: production, composition and eco-hydro-geomorphic effects, Earth-Sci. Rev., 130, 103–127, 2014.

Certini, G.: Effects of fire on properties of forest soils: a review, Oecologia, 143, 1–10, 2005.

Certini, G.: Fire as a soil-forming factor, Ambio, 43, 191–195, doi:10.1007/s13280-013-0418-2, 2013.

Cheng, C. H., Lehmann, J., and Engelhard, M. H.: Natural oxidation of black carbon in soils: Changes in molecular form and surface charge along a climosequence, Geochim. Cosmochim. Ac., 72, 1598–1610, 2008.

Cheng, C.-H., Lehmann, J., Thies, J. E., Burton, S. D., and Engelhard, M. H.: Oxidation of black carbon by biotic and abiotic processes, Org. Geochem., 37, 1477–1488, 2006.

Czimczik, C. I. and Masiello, C. A.: Controls on black carbon storage in soils, Global Biogeochem. Cy., 21, GB3005, doi:10.1029/2006gb002798, 2007.

Delwiche, J.: Post-fire soil erosion and how to manage it, JFSP Briefs, Paper 59, 2009.

Dobrovol'skii, G. V.: Degradation and Conservation of Soils, Izd. Mosk. Gos. Univ., Moscow, Moscow, 2002 (in Russian).

Dymov, A. A. and Gabov, D. N.: Pyrogenic alterations of Podzols at the North-east European part of Russia: Morphology, carbon pools, PAH content, Geoderma, 241–242, 230–237, 2015.

Efremova, T. T. and Efremov, S. P.: Pyrogenic transformation of organic matter in soils of forest bogs, Eur. Soil Sci., 39, 1297–1305, 2006.

Francos, M., Pereira, P., Alcaniz, M., Mataix-Solera, J., and Ubeda, X.: Intense rainfall impact on soil properties in fire affected areas, Science Total Environ., 572, 1353–1362, 2016.

Gagarina, E. I.: Micromorphological method of soil investigation, S-Petersburg, 156 pp., 2004.

Gerasimova, M. I., Kovda, I. V., Lebedeva, M. P., and Tursina, T. V.: Micromorphological terms: the state of the art in soil microfabric research, Eurasian Soil Sci., 44, 804–817, 2011.

González-Pérez, J. A., González-Vila, F. J., Almendros, G., and Knicker, H.: The effect of fire on soil organic matter – A review, Environ. Int., 30, 855–870, 2004.

Greene, R. S. B., Chartres, C. J., and Hodgkinson, K. C.: The effects of fire on the soil in a degradet semiarid woodland I. Cryptogam cover and physical and micromorpological properties, Aust. J. Soil Res., 28, 755–777, 1990.

Guénon, R., Vennetier, M., Dupuy, N., Roussos, S., Pailler, A., and Gros, R.: Trends in recovery of Mediterranean soil chemical properties and microbial activities after infrequent and frequent wildfires, Land Degrad. Dev., 24, 115–128, doi:10.1002/ldr.1109, 2013.

Jain, T. B., Pilliod, D. S., Graham, R. T., Lentile, L. B., and Sandquist, J. E.: Index for characterizing post-fire soil environments in temperate coniferous forests, Forests, 3, 445–466, doi:10.3390/f3030445, 2012.

Knicker, H., Hilscher, A., de la Rosa, J. M., González Pérez, J. A., and González-Vila, F. J.: Modification of biomarkers in pyrogenic organic matter during the initial phase of charcoal biodegradation in soils, Geoderma, 197–198, 43–50, 2013.

Kubiëna, W. L.: Micropedology, Ames, Iowa: Collegiate press, 243 pp., 1938.

Kubiëna, W. L.: Micromorphological features of Soil geography, New Jersey: Rutgers University Press, 254 pp., 1970.

Lebedeva, M., Gerasimova, M., and Golovanov, D.: Systematization of the topsoil fabrics in soils of the arid lands in northwest of central Asia, World Congress of Soil Science, Soil Solutions for a Changing World (1–6 August 2010), Brisbane, Australia, Published on DVD, 96, 2010.

León, J., Seeger, M., Badía, D., Peters, P., and Echeverría, M. T.: Thermal shock and splash effects on burned gypseous soils from the Ebro Basin (NE Spain), Solid Earth, 5, 131–140, doi:10.5194/se-5-131-2014, 2014.

Liu, G., Niu, Z., Niekerk, D., Xue, J., and Zheng, L.: Polycyclic aromatic hydrocarbons (PAHs) from coal combustion: emissions, analysis, and toxicology, Rev. Environ. Contam. Toxicol., 192, 1–28, 2008.

Maksimova, E. and Abakumov, E.: Wildfire effects on ash composition and biological properties of soils in forest–steppe ecosystems of Russia, Environ. Earth Sci., 74, 4395–4405, doi:10.1007/s12665-015-4497-1, 2015.

Marschner, B., Brodowski, S., Dreves, A., Gleixner, G., Gude, A., Grootes, P. M., Hamer, U., Heim, A., Jandl, G., Ji, R., Kaiser, K., Kalbitz, K., Kramer, C., Leinweber, P., Rethemeyer, J., Schäffer, A., Schmidt, M. W. I., Schwark, L., and Wiesenberg, G. L. B.: How relevant is recalcitrance for the stabilization of organic matter in soils?, J. Plant Nutr. Soil Sci., 171, 91–110, 2008.

Mataix-Solera, J., Gomez, I., Navarro-Perdenno, J., Guerrero, C., and Moral, R.: Soil organic matter and aggregates affected by wildfire in a *Pinus halepensis* forest in a Mediterranean environment, Int. J. Wildland Fire, 11, 107–114, 2002.

Mataix-Solera, J., Cerdà, A., Arcenegui, V., Jordán, A., and Zavala, L. M.: Fire effects on soil aggregation: a review, Earth-Sci. Rev., 109, 44–60, doi:10.1016/j.earscirev.2011.08.002, 2011.

Methods of Soil Analysis: Soil Science Society of America Inc., Wisconsin, USA, Part 3 Chemical Methods. Wisconsin, USA: Soil Science Society of America Inc., American Society of Agronomy Inc. 677, South Segoe Road, Madison, 1996.

Nguyen, B. T. and Lehmann, J.: Black carbon decomposition under varying water regimes, Org. Geochem., 40, 846–853, 2009.

Nguyen, B. T., Lehmann, J., Hockaday, W. C., Joseph, S., and Masiello, C. A.: Temperature sensitivity of black carbon decomposition and oxidation, Environ. Sci. Technol., 44, 3324–3331, 2010.

Nobles, M. M.: Mineralogical and micromorphological modification in soil affected by slash pile burn, Abstracts of IV International conference on Forest Fire Research (15–18 November 2010), Coimbra, Potugal, 2010.

Nosin, V. A. et al.: Soils of Kujbyshev region, OGIZ, Kujbyshev, 383 pp., 1949.

Pereira, P., Úbeda, X., Mataix-Solera, J., Oliva, M., and Novara, A.: Short-term changes in soil Munsell colour value, organic matter content and soil water repellency after a spring grassland fire in Lithuania, Solid Earth, 5, 209–225, doi:10.5194/se-5-209-2014, 2014.

Pereira, P., Cerdà, A., Úbeda, X., Mataix-Solera, J., Arcenegui, V., and Zavala, L.: Modelling the impacts of wildfire on ash thickness in a short-term period, Land Degrad. Dev., 26, 180–192, 2015.

Robichaud, P. R.: Measurement of post-fire hillslope erosion to evaluate and model rehabilitation treatment effectiveness and recovery, Int. J. Wildland Fire, 14, 475–485, 2005.

Saksonov, S. V.: Samarskaya Luka flora phenomenon, Moscow: Science, 263 pp., 2006.

Sedov, S. N., Sycheva, S., Targulian, V., Pi, T., and Diaz, J.: Last Interglacial paleosols with Argic horizons in upper Austria and Central Russia – pedogenetic and paleoenvironmental inferences from comparison with Holocene analogues, Quaternary Sci. J., 62, 44–58, 2013.

Shishov, L. L. and Tonkonogov, V. D.: Classification and diagnostics of Russian soils, Moscow, Russia: Soil institute of Dokuchayev, 341 pp., 2004.

Stoops, G.: Guidelines for analysis and description of soil and regolith thin section, Published by soil Sci. Soc. Am. Inc. Madison, Wisconsin, USA, 184 pp., 2003.

Stoops, G.: Evaluation of Kubiena's contribution to micropedology. At the Occasion of the Seventieth Anniversary of His book "Micropedology", Eurasian Soil Sci., 42, 693–698, 2009.

Stoops, G. and Eswaran, H.: Soil micromorphology, New York: Van Nostrands Reinhold Company, 345 pp., 1986.

Urusevskaja, I. S., Meshalkina, J. L., and Hohlova, O. S.: Geographic and genetic features of luvisols' humus status, Eur. Soil Sci., 11, 1377–1390, 2000.

Vasil'eva, D. I. and Baranova, M. N.: Natural resources of Samara region, Samara municipal managment institute, Samara, 40 pp., 2007.

Vieira, A., Bento-Gonçalves, A., Lourenço, L., Nunes, A., Meira-Castro, A., and Ferreira-Leite, F.: Soil erosion after forest fires: evaluation of mitigation measures applied to drainage channels in the northwest of Portugal, FLAMMA, 5, 127–129, 2014.

World reference base for soil resources: International soil classification system for naming soils and creating legends for soil maps, 2014, FAO, Rome, 2015.

Zharikova, Y. A.: Post-pyrogenic soil transformation in the forests of the lower Amur region, Bulletin of Altai state agricultural university, 8, 57–61, 2015.

14

Precise age for the Permian–Triassic boundary in South China from high-precision U-Pb geochronology and Bayesian age–depth modeling

Björn Baresel[1], Hugo Bucher[2], Morgane Brosse[2], Fabrice Cordey[3], Kuang Guodun[4], and Urs Schaltegger[1]

[1]Department of Earth Sciences, University of Geneva, Geneva, 1205, Switzerland
[2]Paleontological Institute and Museum, University of Zurich, Zurich, 8006, Switzerland
[3]Laboratoire de Géologie de Lyon, CNRS-UMR 5265, Université Claude Bernard Lyon 1, Villeurbanne, 69622, France
[4]Guangxi Bureau of Geology and Mineral Resources, Nanning, 530023, China

Correspondence to: Björn Baresel (bjoern@heldenepos.de)

Abstract. This study is based on zircon U-Pb ages of 12 volcanic ash layers and volcanogenic sandstones from two deep water sections with conformable and continuous formational Permian–Triassic boundaries (PTBs) in the Nanpanjiang Basin (South China). Our dates of single, thermally annealed and chemically abraded zircons bracket the PTB in Dongpan and Penglaitan and provide the basis for a first proof-of-concept study utilizing a Bayesian chronology model comparing the three sections of Dongpan, Penglaitan and the Global Stratotype Section and Point (GSSP) at Meishan. Our Bayesian modeling demonstrates that the formational boundaries in Dongpan (251.939 ± 0.030 Ma), Penglaitan (251.984 ± 0.031 Ma) and Meishan (251.956 ± 0.035 Ma) are synchronous within analytical uncertainty of ~ 40 ka. It also provides quantitative evidence that the ages of the paleontologically defined boundaries, based on conodont unitary association zones in Meishan and on macrofaunas in Dongpan, are identical and coincide with the age of the formational boundaries. The age model also confirms the extreme condensation around the PTB in Meishan, which distorts the projection of any stratigraphic points or intervals onto other more expanded sections by means of Bayesian age–depth models. Dongpan and Penglaitan possess significantly higher sediment accumulation rates and thus offer a greater potential for high-resolution studies of environmental proxies and correlations around the PTB than Meishan. This study highlights the power of high-resolution radio-isotopic ages that allow a robust intercalibration of patterns of biotic changes and fluctuating environmental proxies and will help recognizing their global, regional or local significance.

1 Introduction

The Permian–Triassic boundary mass extinction (PTBME) is considered the largest mass extinction within the Phanerozoic. About 90 % of all marine species suffered extinction (Raup, 1979; Stanley and Yang, 1994; Erwin et al., 2002; Alroy et al., 2008) and terrestrial plant communities underwent major ecological reorganization (Hochuli et al., 2010). This major caesura in global biodiversity marked the end of the Palaeozoic faunas and the inception of the modern marine and terrestrial ecosystems (e.g., Benton, 2010; Van Valen, 1984). Several kill mechanisms have been proposed, such as global regression (e.g., Erwin 1990; Yin et al., 2014), marine anoxia (e.g., Feng and Algeo, 2014), ocean acidification (e.g., Payne et al., 2010) or a combination thereof. Rapid global warming (e.g., Svensen et al., 2009), high nutrient fluxes from continent into oceans (Winguth and Winguth, 2012) and increased sediment accumulation rates (Algeo and Twitchett, 2010) also came into the play, but their respective relations with the global regression near the PTB and the main extinction peak at the PTB remain unclear. In spite of the rapidly growing amount of data, the detailed timing of available diversity estimates and environmental proxies is still lacking, and the ultimate triggers of the PTBME remain elusive. The most likely cause derives from the temporal coincidence with massive and short-lived volcanism of the Siberian Traps (e.g., Burgess and Bowring, 2015) that injected excessive amounts of volatiles (H_2O, CO_2, SO_2, H_2S) into the atmosphere. Accompanying destabilization of gas hydrates (CH_4) and contact metamorphism of organic carbon-rich sediments (Retallack and Jahren, 2008; Svensen et al., 2009) are likely to have contributed additional volatiles into the atmosphere, thus substantially altering the climate and the chemical composition of the ocean. This presumably close chronological association has led many authors to support a cause–effect relationship between flood basalt volcanism and mass extinctions. Constraining the timing and duration of the PTBME in a precisely and accurately quantified model that combines relative (i.e., biostratigraphy, environmental changes) and sequences of absolute (zircon geochronology) ages is key to reveal the cascading causes and effects connecting rapid environmental perturbations to biological responses.

The South China block provides a few exceptional marine successions with a continuous stratigraphic record across the PTB (e.g., Yin et al., 2014). Among these is the Global Stratotype Section and Point (GSSP) in Meishan D (Yin et al., 2001), where the PTB is defined by the first occurrence (FO) of the Triassic conodont *Hindeodus parvus*. Additionally, these South Chinese sections reflect intense regional volcanic activity during late Permian and Early Triassic times as manifested by many intercalated zircon-bearing ash beds (Burgess et al., 2014; Galfetti et al., 2007; Lehrmann et al., 2015; Shen et al., 2011). High-precision U-Pb zircon geochronology can be applied to these ash beds by assuming that the age of zircon crystallization closely approximates the age of the volcanic eruption and ash deposition. Earliest U-Pb geochronological studies (e.g., Bowring et al., 1998; Mundil et al., 2004; Ovtcharova et al., 2006; Shen et al., 2011) do not reach decamillennial resolution, which is necessary to resolve biotic events, such as extinction or recovery. Recent improvements in the U-Pb dating technique by the development of the chemical abrasion–isotope dilution–thermal ionization mass spectrometry (CA-ID-TIMS; Mattinson, 2005), by the revision of the natural U isotopic composition (Hiess et al., 2012), by the development of data reduction software (Bowring et al., 2011; McLean et al., 2011) and by the calibration of the EARTHTIME ^{202}Pb-^{205}Pb-^{233}U-^{235}U tracer solution (Condon et al., 2015) now provide more accurate weighted mean zircon population dates at the < 80 ka level (external uncertainty) for a PTB age, which allow for more precise calibration between biotic and geologic events during mass extinctions and recoveries. Two of the cases benefiting from this improved technique are the highly condensed GSSP defining the PTB at Meishan (Burgess et al., 2014) and the expanded Early–Middle Triassic boundary in Monggan (Ovtcharova et al., 2015).

The aim of this work is (1) to date the PTB in two sedimentary sections that are continuous with a conformable PTB and with higher sediment accumulation rates over the same duration than in Meishan, using the highly precise and accurate dating technique of CA-ID-TIMS, and (2) to test the age consistency between the PTB as defined paleontologically in Meishan and as recognized by conformable formational boundaries in the deeper water sections of Dongpan and Penglaitan. Our high-precision dates provide a future test for the synchronicity of conodont biozones, chemostratigraphic correlations, and other proxies involved in the study of the PTBME. Moreover, applying Bayesian age modeling (Haslett and Parnell, 2008) based on these high-precision data sets allows us to model variations in sediment accumulation rate and to directly compare other proxy data across different PTB sections, inclusive of the Meishan GSSP.

Our data demonstrate that the PTB, as recognized in our sections by conformable boundaries between late Permian and basal Triassic formations, is synchronous within analytical uncertainty of ca. 40 ka. We also show that Bayesian age models produce reproducible results from different sections, even though U-Pb datasets originate from different laboratories. We construct a coherent age model for the PTB in Dongpan and Penglaitan, which is also in agreement with the PTB age model from Meishan (Burgess et al., 2014). These results further demonstrate that $^{206}Pb / {}^{238}U$ dates produced in two different laboratories using the EARTHTIME tracer solution provide reproducible age information at the 0.05 % level of uncertainty.

2 Geological setting

2.1 Regional context

The newly investigated volcanic ash beds were sampled from two PTB sections: Dongpan in southwestern Guangxi Province and Penglaitan in central Guangxi Province in South China (Fig. 1a, exact sample locations are given in Sect. S1 in the Supplement). Both sections are within the Nanpanjiang Basin (Lehrmann et al., 2015), a late Permian–Early Triassic pull-apart basin in a back arc context located on the present-day southern edge of the South China block. This deep-marine embayment occupied an equatorial position in the eastern paleo-Tethys Ocean (e.g., Golonka and Ford, 2000; Lehrmann et al., 2003; Fig. 1b). The basin was dominated by a mixed carbonate–siliciclastic regime during Permian and Early Triassic times and underwent a major change to a flysch-dominated regime in later Triassic times (e.g., Galfetti et al., 2008; Lehrmann et al., 2007). Decimeter- to meter-thick beds of mixed volcanic and clastic material as well as millimeter- to centimeter-thick volcanic ash beds are locally abundant and especially well preserved in downthrown blocks recording deep water records in low energy environments and, to a lesser degree, on up-thrown blocks recording shallow water to outer platform settings. Volcanic ash beds are, however, usually not preserved in traction-

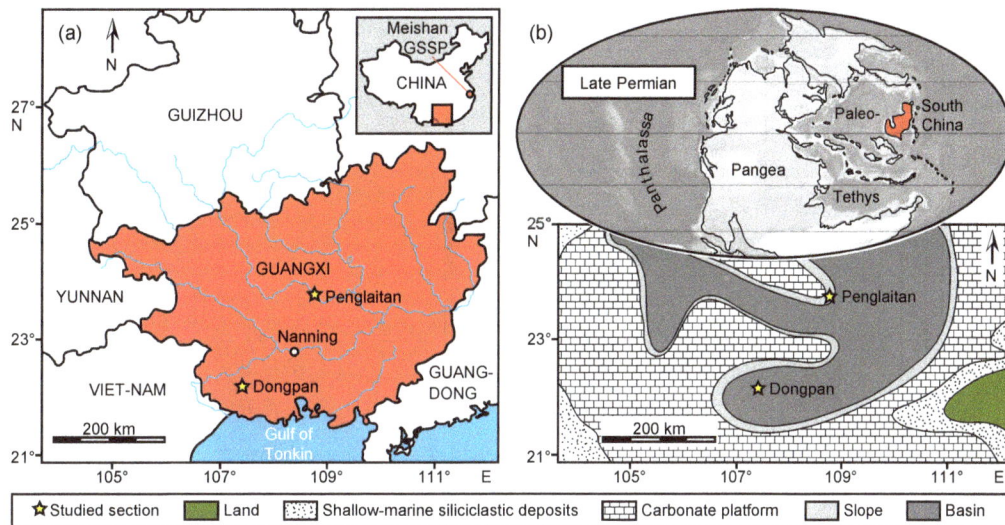

Figure 1. (a) Locality map showing the position of Dongpan, Penglaitan and Meishan Global Stratotype Section and Point (GSSP), China. (b) Late Permian paleogeographic reconstruction after Ziegler et al. (1997), indicating the location of the South China block in the peri-Gondwana region. Beneath the paleogeographic map of the Nanpanjiang Basin in South China showing the position of the investigated Dongpan and Penglaitan section during late Permian times (base map modified after Wang and Jin, 2000).

dominated slope deposits. Genetically related volcanic rocks crop out in the southwestern part of the basin towards Vietnam, suggesting the proximity of a volcanic arc related to the convergence between Indochina and South China (Faure et al., 2016). The volcanism produced by this convergence is the most likely source of the analyzed volcanic ashes (Gao et al., 2013).

In Dongpan and Penglaitan, the PTB is manifested by a sharp and conformable transition from the late Permian Dalong Formation (= Talung Formation) to the basal Triassic Ziyun Formation. Late Permian rocks in these two sections are classically assigned to the Dalong Formation of Changhsingian age. However, we note that there are substantial facies differences between these two late Permian records. The Dalong Formation in Dongpan is composed of thin-bedded siliceous mudstones, numerous ash layers and minor limestone beds (Fig. 2a). This facies association is in agreement with the vast majority of reported occurrences of this formation within the South China block. The Dalong Formation is interpreted as a basinal depositional environment with restricted circulation and an estimated water depth of 200 to 500 m (He et al., 2007; Yin et al., 2007). In Guangxi and Guizhou, the thickness of the typical Dalong Formation is highly variable and ranges from a couple of meters to ca. 60 m. Rocks assigned to the Dalong Formation in Penglaitan markedly diverge from those of the typical Dalong Formation. In Penglaitan, the Dalong Formation reaches an unusual thickness of ca. 650 m and is lithologically much more heterogeneous, with a marked regressive episode in its middle part (Shen et al., 2007). Moreover, in Penglaitan the Dalong Formation contains numerous volcanogenic sandstones distributed within the entire succession, a distinc-

tive feature when compared to other sections. Only the lower part of the "Dalong Formation" in Penglaitan can be unambiguously assigned to this formation. The middle and upper part of this section are notably shallower, showing cross bedding and ripple marks in the uppermost 30 m of the Permian, which are underlain by upper shoreface to foreshore facies deposits containing coal seams and abundant plant fossils (Shen et al., 2007). The uppermost part of the Dalong Formation was deposited in relatively deep water settings that comprise thin-bedded dark-grey limestone intercalated with thick volcanogenic sandstones and thin volcanic ash beds (Fig. 3a). All associated volcanogenic sandstones were deposited by geologically instantaneous turbidites, mainly reflecting the basal part (Bouma A–B sequence) of such gravity flow deposits. Gradual accumulations and sediment mixing are restricted to sands bars occurring in the middle part of the section, in association with coal seams during an intervening regressive episode. Hence, the volcanogenic sandstones from the top of the Dalong Formation in Penglaitan may not suffer from substantial sediment reworking and mixing and do not represent substantial cumulative amounts of time relative to the interlayered shales and thin-bedded limestones. The depositional setting of Penglaitan is interpreted as that of a fault-bounded block successively thrown down and up. Hence, Penglaitan stands in marked contrast with the homogenous deeper water facies of the typical Dalong Formation in other sections.

The conformably overlying Early Triassic rocks have been previously assigned to the Luolou Formation in both Penglaitan and Dongpan (Feng et al., 2007; He et al., 2007; Shen et al., 2012; Zhang et al., 2006). At its type locality and elsewhere in northwestern Guangxi and southern Guizhou, the

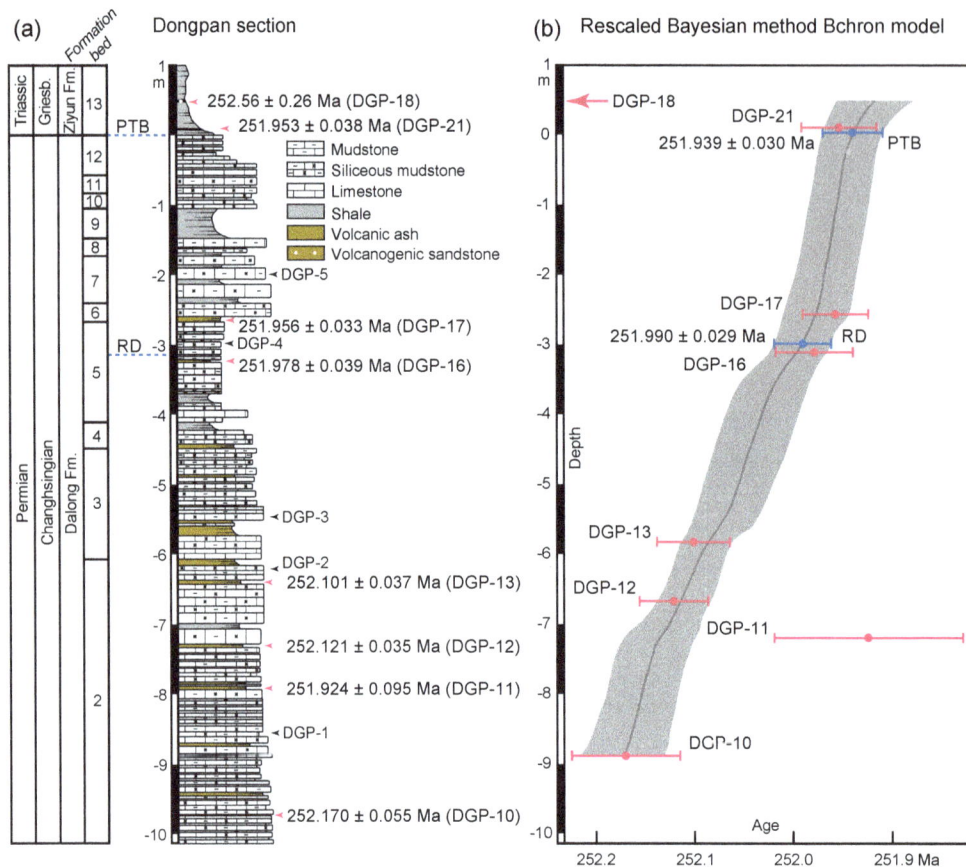

Figure 2. Stratigraphy, geochronology and Bayesian age–depth modeling for the Dongpan section from late Changhsingian to Griesbachian. (a) Weighted mean ^{206}Pb / ^{238}U dates of the volcanic ash beds and volcanogenic sandstones are given in Ma. U-Pb data of DGP-21 are taken from Baresel et al. (2016). Investigated radiolarian samples (DGP-1 to DGP-5) are shown in their stratigraphic positions. (b) The rescaled Bayesian Bchron age–depth model is presented with its median (middle grey line) and its 95 % confidence interval (grey area). Radioisotopic dates together with their uncertainty (red horizontal bars) are presented as ^{206}Pb / ^{238}U weighted mean dates of the dated volcanic ash beds in their stratigraphic positions. Predicted dates for the onset of the radiolarian decline (RD) and the Permian–Triassic boundary (PTB) are calculated with their associated uncertainty using the rescaled Bayesian Bchron age–depth model assuming stratigraphic superposition.

base of the Luolou Formation is invariably represented by shallow water microbial limestone. In contrast, the onset of the Triassic at Dongpan and Penglaitan is represented by ca. 30 m of laminated black shales overlain by several hundred meters of thin-bedded, mechanically laminated, medium- to light-grey limestone. In Dongpan, edgewise conglomerates and breccias are occasionally intercalated within the platy, thin-bedded limestone unit. This succession of facies illustrates a change from basinal to slope depositional environments and is identical to that of the Ziyun Formation at its type locality 3 km east of Ziyun, Guizhou Province (Guizhou Bureau of Geology and Mineral Resources, 1987). Therefore, Early Triassic rocks in Dongpan and Penglaitan are here reassigned to the Ziyun Formation, whose base is of Griesbachian age. In most sections in Guangxi and Guizhou, where latest Permian rocks are represented by the Dalong Formation, these are consistently and conformably overlain by basal black shales of the Early Triassic Ziyun Forma-

tion or the Daye Formation (e.g., Feng et al., 2015). In these downthrown blocks, the effects of the Permian–Triassic global regression were negligible in comparison to those observed in adjacent, up-thrown blocks that recorded pronounced unconformities or condensation.

2.2 The Dongpan section

Numerous litho-, bio- and chemo-stratigraphic studies (e.g., Feng et al., 2007; He et al., 2007; Luo et al., 2008; Zhang et al., 2006) have been published on the Dongpan section during the last two decades. However, the volcanic ash beds of this continuous PTB section have never been dated. The classic lithostratigraphic subdivisions of the Dongpan section (bed 2 to 13; indicated in Fig. 2a) (Meng et al., 2002) can easily be recognized in the field. Based on the conodont alteration index (CAI), Luo et al. (2011) established that the section shows only a low to moderate thermal overprint equivalent to a maximal burial temperature of 120 °C. Our own

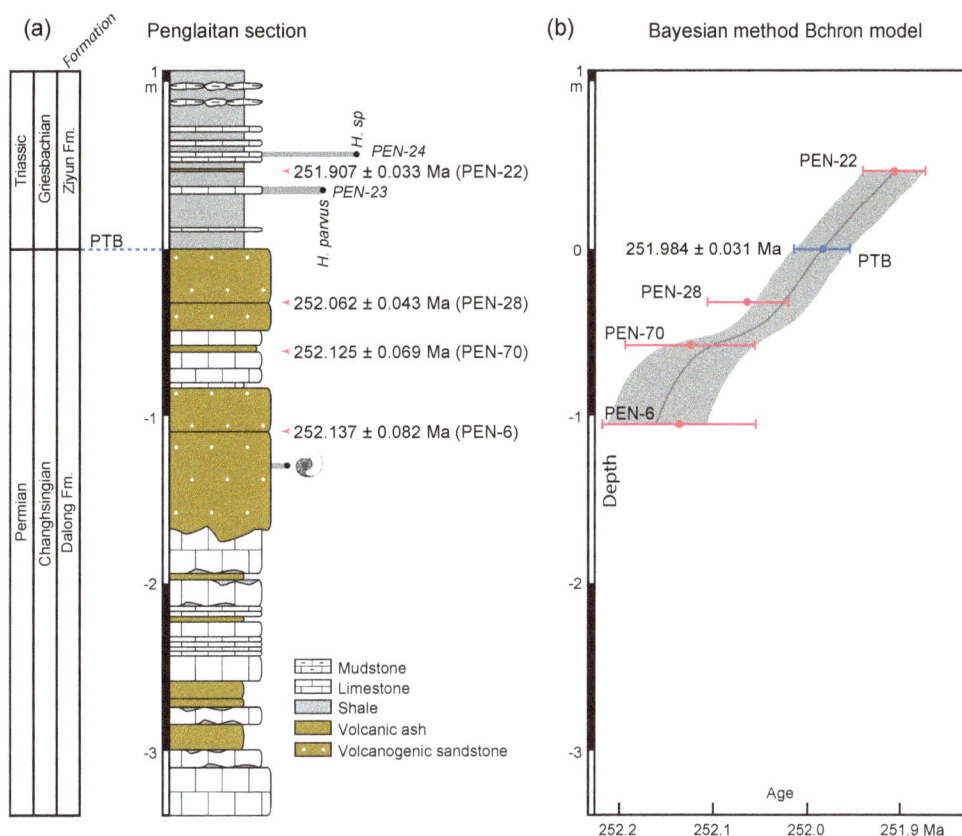

Figure 3. Stratigraphy, geochronology and Bayesian age–depth modeling for the Penglaitan section from late Changhsingian to Griesbachian. **(a)** Weighted mean $^{206}Pb / ^{238}U$ dates of the volcanic ash beds and volcanogenic sandstones are given in Ma. U-Pb data of PEN-28 and PEN-22 are taken from Baresel et al. (2016). Investigated conodont samples (PEN-23 and PEN-24; see also Sect. S5) and first occurrence of Triassic conodonts are shown in their stratigraphic positions. A poorly preserved Permian nautiloid is indicated in its stratigraphic position ca. 1.3 m below the Permian–Triassic boundary (PTB). **(b)** The rescaled Bayesian Bchron age–depth model is presented with its median (middle grey line) and its associated 95 % confidence interval (grey area). Radioisotopic dates together with their uncertainty (red horizontal bars) are presented as $^{206}Pb / ^{238}U$ weighted mean dates of the dated volcanic ash beds in their stratigraphic positions. Predicted date for the PTB is calculated with its associated uncertainty using the rescaled Bayesian Bchron age–depth model assuming stratigraphic superposition.

estimation of the CAI of conodont elements obtained from the same beds points toward values around 3, thus confirming the estimation of Luo et al. (2011).

Beds 2 to 5 consist of thin (dm to cm) siliceous mudstones, mudstones, minor lenticular limestone horizons and numerous intercalated volcanic ash beds. These beds yield radiolarians, foraminifera (Shang et al., 2003), bivalves (Yin, 1985), ammonoids (Zhao et al., 1978), brachiopods (He et al., 2005), ostracods (Yuan et al., 2007), and conodonts (Luo et al., 2008) of Changhsingian age. Chinese authors have provided very detailed studies of radiolarian occurrences from the top of the Dongpan section, documenting about 160 species belonging to 50 genera (Feng et al., 2007; Wu et al., 2010; Zhang et al., 2006). Most of these radiolarians belong to the *Neoalbaillella optima* assemblage zone of late Changhsingian age (Feng and Algeo, 2014), although it is unclear whether some of the Permian taxa reported from the top of the section by previous authors (i.e. above bed 6; Feng

et al., 2007) still belong to this assemblage or to a provisional ultimate Permian biozone (Xia et al., 2004).

We collected five samples with visible radiolarians (DGP-1 to DGP-5; see Fig. 2a) for this study. Our goal was not to duplicate the detailed faunal studies performed at Dongpan by previous authors but essentially to correlate these previous results with our U-Pb ages using own radiolarian data. A selection of well-preserved taxa is illustrated in Sect. S4. We also report the occurrence of morphotypes belonging to the genus *Hegleria*, which was previously reported from the section but not illustrated. Our data confirm that radiolarians of the Dongpan section belong to the *Neoalbaillella optima* assemblage zone.

The conodont fauna obtained from beds 3 and 5 was assigned to the *Neogondolella yini* interval zone by Luo et al. (2008). *Neogondolella yini* is also a characteristic species of the UAZ1 zone, which is the oldest zone of a new high-accuracy zonation around the PTB constructed by means of unitary associations (Brosse et al., 2016). Bed 6 is composed of a yellow fine-grained volcanic ash bed and thin-bedded siliceous mudstone. Beds 7 to 12 contain more frequent mudstone and yield a diverse Permian fauna (Feng et al., 2007; He et al., 2007; Yin et al., 2007). Additionally, He et al. (2007) showed that end-Permian brachiopods underwent a size reduction in the uppermost beds of the Dalong Formation, which they linked with a regressive trend.

The sharp and conformable base (bed 13) of the Early Triassic Ziyun Formation consists of brown-weathering black shales containing a few very thin (mm to cm) volcanic ash beds and volcanogenic sandstones. Previous studies did not recognize how recent weathering superficially altered these black shales. Bed 13 contains abundant bivalves and ammonoids of Griesbachian age (Feng et al., 2007; He et al., 2007), which are also known from other sections where the equivalent black shales are not weathered. Therefore, the formational boundary placed between beds 12 and 13 is reasonably well-constrained in terms of paleontological ages. Even in the absence of any close conodont age control, this boundary has been unanimously acknowledged as the PTB in all previous contributions, thus emphasizing the significance of this formational change.

2.3 The Penglaitan section

The Penglaitan section is well known for its Guadalupian–Lopingian boundary (Capitanian–Wuchiapingian GSSP; Jin et al., 2006; Shen et al., 2007). However, the part of the section that straddles the PTB has not been the focus of any detailed published work. Shen et al. (2007) report Changhsingian *Peltichia zigzag–Paryphella* brachiopod assemblage from a volcanogenic sandstone bed at ~ 28 m below the PTB. In addition, *Palaeofusulina sinensis* is abundant in the uppermost limestone units of the Dalong Formation and conodonts in the topmost part were assigned to the *Clarkina yini* Zone. A poorly preserved Permian nautiloid was recovered from the volcanogenic sandstone 1.3 m below the PTB (Fig. 3a). About 0.3 m above the PTB, concretionary, thin-bedded micritic layers intercalated within the basal black shales of the Ziyun Formation yielded one P1 element of *Hindeodus parvus* (Fig. 3a; see also Sect. S5). Pending the age confirmation of new paleontological data, and in full agreement with Shen et al. (2007), we place the PTB at this sharp but conformable formational boundary.

3 Methods

3.1 CA-ID-TIMS analysis

Sample preparation, chemical processing and U-Pb CA-ID-TIMS zircon analyses were carried out at the University of Geneva. Single zircon grain dates were produced relative to the EARTHTIME ^{202}Pb-^{205}Pb-^{233}U-^{235}U tracer solution (Condon et al., 2015). All uncertainties associated with

weighted mean ^{206}Pb / ^{238}U ages are reported at the 95 % confidence level and given as $\pm x/y/z$, with x as analytical uncertainty, y including tracer calibration uncertainty, and z including ^{238}U decay constant uncertainty. The tracer calibration uncertainty of 0.03 % (2σ) has to be added if the calculated dates are to be compared with other U-Pb laboratories not using the EARTHTIME tracer solution. The ^{238}U decay constant uncertainty of 0.11 % (2σ) should be used if compared with other chronometers such as Ar-Ar. All ^{206}Pb / ^{238}U single-grain ages have been corrected for initial ^{230}Th-^{238}U disequilibrium assuming Th/U$_{magma}$ of 3.00 ± 0.50 (1σ). This should best reflect the Th/U of the whole rock and is identical to the Th/U$_{magma}$ used by Burgess et al. (2014) for the Meishan ash beds in order to provide an unbiased comparison of the Dongpan, Penglaitan and Meishan chronology. Th-corrected ^{206}Pb / ^{238}U dates are on average 80 ka older than the equivalent uncorrected dates when applying this correction, but changes in the Th/U$_{magma}$ have only minor effects on the deposition ages of the Dongpan and Penglaitan volcanic beds. Compared to the Th/U$_{magma}$ of 3.00 ± 0.50 (1σ) used in this study, they would become max. 11 kyr younger with Th/U$_{magma}$ of 2.00 ± 0.50 (1σ) and max. 7 kyr older with Th/U$_{magma}$ of 4.00 ± 0.50 (1σ). All Th-corrected ^{206}Pb / ^{238}U dates are presented as mean ages of selected zircon populations and their associated $\pm 2\sigma$ analytical uncertainties in Figs. 2 and 3, and as single-grain ^{206}Pb / ^{238}U age ranked distribution plots in Fig. 4. The full data table and analytical details are given in Sect. S2.

3.2 Bayesian chronology

In this study we use Bayesian interpolation statistics to establish a probabilistic age model based on our high-precision U-Pb zircon dates of each individual ash bed and its stratigraphic position, as it is incorporated in the free Bchron R software package (Haslett and Parnell, 2008; Parnell et al., 2008) to constrain the chronological sequence and sedimentation history of the investigated sections. By assuming normal distribution of our U-Pb dates within one sample and based on the principle of stratigraphic superposition, which requires that any stratigraphic point must be younger than any point situated below in the stratigraphic sequence, it models the age and its associated 95 % confidence interval for any depth point within the studied sedimentary sequence. The model is based on the assumption of random variability of sediment accumulation rate, yielding a family of dispersed piecewise monotonic sediment accumulation models between each dated stratigraphic horizon. The number of such accumulation models is inferred by a Poisson distribution, and the size of the sediment accumulation rates by a gamma distribution. The strength of this approach is its flexibility that allows changes in sediment accumulation rate from zero (hiatus in sedimentation) to very large values (sedimentation event at high rate). In contrast to standard linear regression models, this approach leads to more realistic

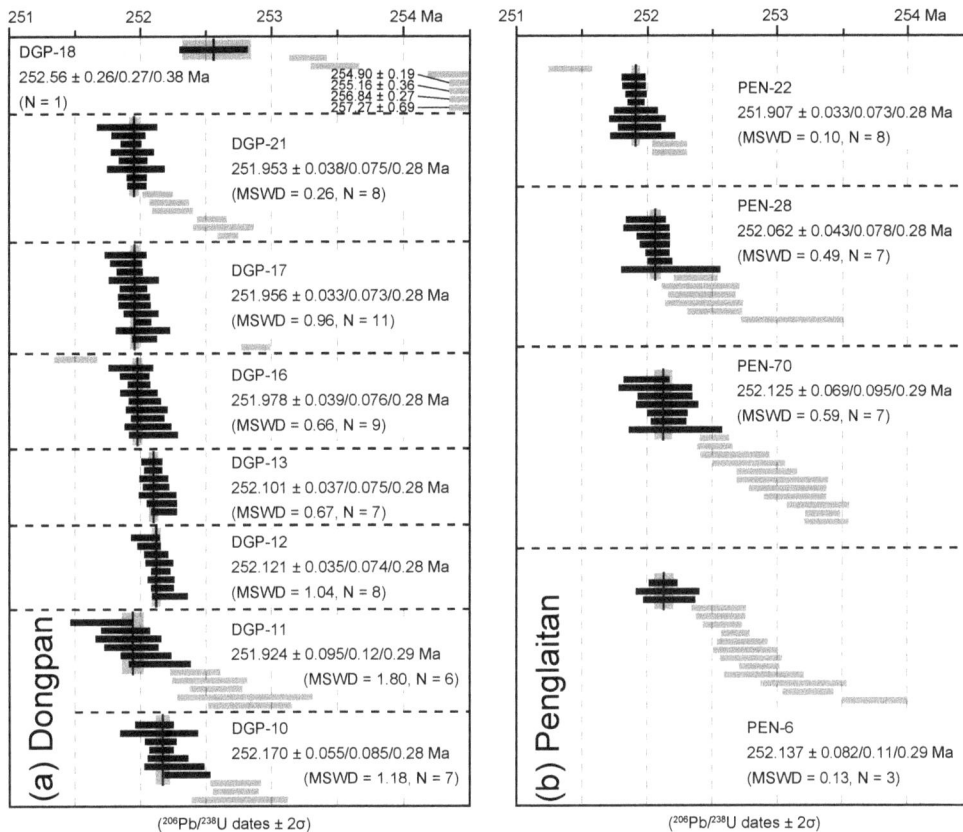

Figure 4. Single-grain zircon analysis and $^{206}Pb/^{238}U$ weighted mean dates for **(a)** Dongpan and **(b)** Penglaitan volcanic ash beds and volcanogenic sandstones. U-Pb data of DGP-21, PEN-28 and PEN-22 are taken from Baresel et al. (2016). Each horizontal bar represents a single zircon grain analysis including its 2σ analytical (internal) uncertainty, whereas grey bars are not included in the weighted mean calculation. Vertical lines represent the weighted mean age, with the associated 2σ uncertainty (in grey). Uncertainty in the weighted mean dates is reported as 2σ internal, 2σ external uncertainty including tracer calibration and 2σ external uncertainty including ^{238}U decay constant uncertainty; MSWD – mean square of weighted deviates.

uncertainty estimates, with increasing uncertainty at growing stratigraphic distance from the dated layers. The model also detects and excludes outliers, which conflict with other evidence from the same sequence in order to produce a coherent and self-consistent chronology; no predefined outlier determination is required from the user. One of the drawbacks of this Bayesian approach is that a change in the sediment accumulation rate is assumed to occur at each dated stratigraphic position, though it is unlikely that the change in sedimentation occurs exactly at the depth of a dated bed. Another drawback is that the sedimentation parameters are shared across the whole sequence. In consequence, Bchron does not allow much opportunity for users to individually influence the chronology behavior.

In this study we use the Bayesian Bchron model as it is part of the Bchron package (http://cran.r-project.org/web/packages/Bchron/index.html). This model outperforms other Bayesian age–depth models, as shown by an extensive comparison conducted on radiocarbon dates from Holocene lake sediments (Parnell et al., 2011). It provides a non-parametric

chronological model according to the compound Poisson–gamma model defined by Haslett and Parnell (2008), requiring the weighted mean $^{206}Pb/^{238}U$ age and the stratigraphic position of the investigated ash beds as input parameters. Since the Bchron model was initially coded for radiocarbon dating with a commonly unknown duration of accumulation for a radiocarbon-dated bed, it also allows for the input thickness of such a horizon to be defined. However, the thickness of the geologically instantaneously deposited volcanic ashes was reduced to zero and the lithostratigraphy has been rescaled in order to remove the thickness of the volcanic horizons and to produce a more accurate age–depth model (Figs. 2a and 3a; see also Sect. S3). The technical details were given in Haslett and Parnell (2008). The Bchron model uses a Markov chain Monte Carlo (Brooks et al., 2011) rejection algorithm which proposes model parameters and accepts or rejects them in order to produce probability distributions of dates for a given depth that match likelihood and do not violate the principle of stratigraphic superposition. In order to create an adequate number of accepted samples, the model was run for 10 000 iterations. The Bchron R scripts of Dongpan, Penglaitan and Meishan are provided in Sect. S3.

4 Samples

In total, 12 volcanic ash beds and volcanogenic sandstones were sampled from the Dalong Formation of late Permian age and from the overlying Ziyun Formation of Early Triassic age at Dongpan and Penglaitan (see Sect. S1). Most of the dated samples exhibit ^{206}Pb / ^{238}U age dispersions that exceed the acceptable scatter from analytical uncertainty and are interpreted as reflecting magmatic residence or a combination of the latter with sedimentary recycling. Only in two cases (DGP-16, PEN-22) do we find single-grain analyses younger than our suggested mean age and interpret them as unresolved Pb loss since they violate the stratigraphic order established by the chronology of the volcanic ash beds.

At Dongpan, six fine- to medium-grained volcanic ash beds (DGP-10, DGP-11, DGP-12, DGP-13, DGP-16 and DGP-17) in the uppermost 10 m of the Dalong Formation, one fine-grained ash bed (DGP-21) just 10 cm above the base of the Ziyun Formation, and one thin-bedded volcanogenic sandstone (DGP-18) 40 cm stratigraphically higher were collected for geochronology. At Penglaitan, the basal part of a 25 cm thick volcanogenic sandstone (PEN-6), one thin-layered volcanic ash bed (PEN-70) and the base of a 30 cm thick volcanogenic sandstone (PEN-28), all together representing the uppermost 1.1 m of the Dalong Formation, were dated. A single fine-grained and extremely thin (2–3 mm) volcanic ash bed (PEN-22) was sampled 50 cm above the base of the Ziyun Formation and thus closely brackets the formational boundary. U-Pb CA-ID-TIMS geochronology following procedures described above and in the Appendix was applied to a number of single crystals of zircon extracted from these volcanic beds. Trace element and Hf isotopic compositions of these dated zircons will be presented elsewhere. Stratigraphic positions of volcanic ash beds at Dongpan and Penglaitan and weighted mean ^{206}Pb / ^{238}U dates of individual zircon grains for the samples below are given in Figs. 2 and 3.

5 Results

The U-Pb isotopic results are presented in Fig. 4 as ^{206}Pb / ^{238}U age ranked plots for each individual sample and in Table S1 (Sect. S2).

5.1 U-Pb age determinations from the Dongpan section

5.1.1 Sample DGP-10

This volcanic ash bed was sampled 9.7 m below the formational boundary. All 10 dated zircons are concordant within analytical error, where the seven youngest grains de-

fine a cluster with a weighted mean ^{206}Pb / ^{238}U age of 252.170 ± 0.055/0.085/0.28 Ma (mean square of weighted deviates (MSWD) = 1.18) for the deposition of DGP-10.

5.1.2 Sample DGP-11

This volcanic ash bed was sampled 7.9 m below the formational boundary. Eleven zircon crystals were analyzed, resulting in scattered ^{206}Pb / ^{238}U dates of 251.662 ± 0.263 to 252.915 ± 0.352 Ma. The six youngest zircons yield a weighted mean ^{206}Pb / ^{238}U age of 251.924 ± 0.095/0.12/0.29 Ma (MSWD = 1.80) that is too young with respect to the stratigraphic sequence defined by over- and underlying ash beds. Therefore, we have to assume that abundant unresolved lead loss affected these zircons, despite application of the same chemical abrasion procedure as for all other samples. It is worth noting that all zircons from DGP-11 were almost completely dissolved after chemical abrasion and show elevated ^{206}Pb / ^{238}U age uncertainties of ∼ 0.30 Ma compared to other volcanic ash beds from Dongpan.

5.1.3 Sample DGP-12

This volcanic ash bed was sampled 7.3 m below the formational boundary. The weighted mean age of 252.121 ± 0.035/0.074/0.28 Ma (MSWD = 1.04) is derived from eight concordant grains representing the youngest zircon population of this ash bed.

5.1.4 Sample DGP-13

This volcanic ash bed was sampled 6.4 m below the formational boundary. Analyses of seven individual zircons yield a statistically significant cluster with a weighted mean ^{206}Pb / ^{238}U age of 251.101 ± 0.037/0.075/0.28 Ma (MSWD = 0.67) representing the youngest zircon population of this ash bed.

5.1.5 Sample DGP-16

This volcanic ash bed was sampled 3.2 m below the formational boundary. Nine zircons yield a weighted mean ^{206}Pb / ^{238}U age of 251.978 ± 0.039/0.076/0.28 Ma (MSWD = 0.66). The youngest grain shows unresolved lead loss and was discarded because it violates the stratigraphic superposition. Incorporating this zircon into the mean age calculation would also lead to a statistically flawed MSWD of 4.80.

5.1.6 Sample DGP-17

This volcanic ash bed was sampled 2.7 m below the formational boundary. A total of 11 zircons define a weighted mean ^{206}Pb / ^{238}U age of 251.956 ± 0.033/0.073/0.28 Ma (MSWD = 0.96). One single zircon displays inheritance with an ^{206}Pb / ^{238}U age of 252.896 ± 0.108 Ma and was consequently excluded from the weighted mean age calculation.

5.1.7   Sample DGP-21

This volcanic ash bed was sampled 0.1 m above the formational boundary. Fourteen zircons were dated, among which the eight youngest yield a cluster with a weighted mean ^{206}Pb / ^{238}U age of $251.953 \pm 0.038/0.075/0.28$ Ma (MSWD $= 0.26$). The six oldest grains display an inherited component as suggested by their scattered ^{206}Pb / ^{238}U dates ranging from 252.145 ± 0.120 to 252.715 ± 0.084 Ma. The U-Pb data of DGP-21 have already been published in a companion study (Baresel et al., 2016) that deals with the stratigraphic correlation of ash beds straddling the PTB in deep- and shallow-marine successions of the Nanpanjiang Basin.

5.1.8   Sample DGP-18

This bed was sampled 0.5 m above the formational boundary. The re-sedimented nature of this volcaniclastic bed is reflected in the ^{206}Pb / ^{238}U zircon ages ranging from 252.559 ± 0.261 to 257.274 ± 0.689 Ma. This sample was excluded from the age–depth model because it clearly violates the stratigraphic superposition.

5.2   U-Pb age determinations from the Penglaitan section

5.2.1   Sample PEN-6

PEN-6 comes from the base of a volcanogenic sandstone. It was sampled 1.1 m below the formational boundary. Fifteen zircon grains were dated. The three youngest grains define a weighted mean ^{206}Pb / ^{238}U age of $251.137 \pm 0.082/0.11/0.29$ Ma (MSWD $= 0.13$). Because zircon dates from this bed spread over almost 2 Myr, recycling of older volcanic material via sedimentary processes appears more likely than via magmatic recycling.

5.2.2   Sample PEN-70

This volcanic ash bed was sampled 0.6 m below the formational boundary. Eighteen zircon grains were analyzed. As in the case of PEN-6, they yield a scatter of ^{206}Pb / ^{238}U dates spanning 1.5 Myr, ranging from 251.994 ± 0.169 to 253.371 ± 0.165 Ma. The weighted mean age of $252.125 \pm 0.069/0.095/0.29$ Ma (MSWD $= 0.59$) for the deposition of this ash bed is calculated by using the seven youngest concordant grains.

5.2.3   Sample PEN-28

This sample was taken 0.3 m below the formational boundary. It is derived from the base of a 30 cm thick volcanogenic sandstone which represents the youngest Permian bed in Penglaitan. Analyses of seven zircon grains

yield a cluster with a weighted mean ^{206}Pb / ^{238}U age of $252.062 \pm 0.043/0.078/0.28$ Ma (MSWD $= 0.49$), reflecting the last crystallization phase of this zircon population. Six older grains, ranging from 252.364 ± 0.156 to 253.090 ± 0.375 Ma, indicate either magmatic or sedimentary recycling. The U-Pb data of PEN-28 have already been published in Baresel et al. (2016).

5.2.4   Sample PEN-22

This 2 mm thick volcanic ash bed was sampled 0.5 m above the formational boundary. Eight zircons yield a weighted mean ^{206}Pb / ^{238}U age of $251.907 \pm 0.033/0.073/0.28$ Ma (MSWD $= 0.10$). One zircon grain shows a significantly younger age suggesting lead loss. Two slightly older grains reflect noticeable pre-eruptive crystallization. Incorporation of these grains into the weighted mean calculation would lead to an excessive MSWD of 3.6 and 1.9, respectively.

However, we noticed that some volcanic ash beds and volcanogenic sandstones in these sections show a large age dispersion of up to 2 Myr, incompatible with recycling of zircon that previously crystallized within the same magmatic system and became recycled into later melt batches, leading to dispersion of dates of a few 100 kyr (e.g., Broderick et al., 2015; Samperton et al., 2015), but pointing to sedimentary reworking. The U-Pb data of PEN-22 have already been published in Baresel et al. (2016).

5.3   Age–depth models

Figure 5 shows a comparison of three different age–depth models based on linear interpolation, cubic spline interpolation and Bayesian statistics, each applied to exactly the same U-Pb dataset of Dongpan (Fig. 5a) and Penglaitan (Fig. 5b). As discussed in the Methods section, the Bayesian Bchron model leads to more realistic uncertainty estimates, producing an increased uncertainty of the model age with increasing distance from the stratigraphic position of a U-Pb dated sample. Due to the well constrained U-Pb dates of Dongpan and Penglaitan, all three age–depth models predict (within uncertainty) similar ages for the PTB in Dongpan (Fig. 5a) and Penglaitan (Fig. 5b). Given that the Bayesian Bchron model evaluates the age probability distribution of each U-Pb date with respect to the other dates of the sequence, it provides a more robust and better constrained chronology, which even results in smaller uncertainties of the predicted model dates compared to the standard linear regression models (as indicated by the smaller uncertainty of the Bchron model age for the PTB in Dongpan and Penglaitan). In contrast to the other models, the Bayesian Bchron model can identify U-Pb dates that violate the principle of stratigraphic superposition, as shown for the Dongpan ash beds DGP-11 (outlier probability of 67 %) and DGP-18 (outlier probability of 100 %). Including them into the age–depth chronology of Dongpan results in unrealistic negative sediment accumulation rates,

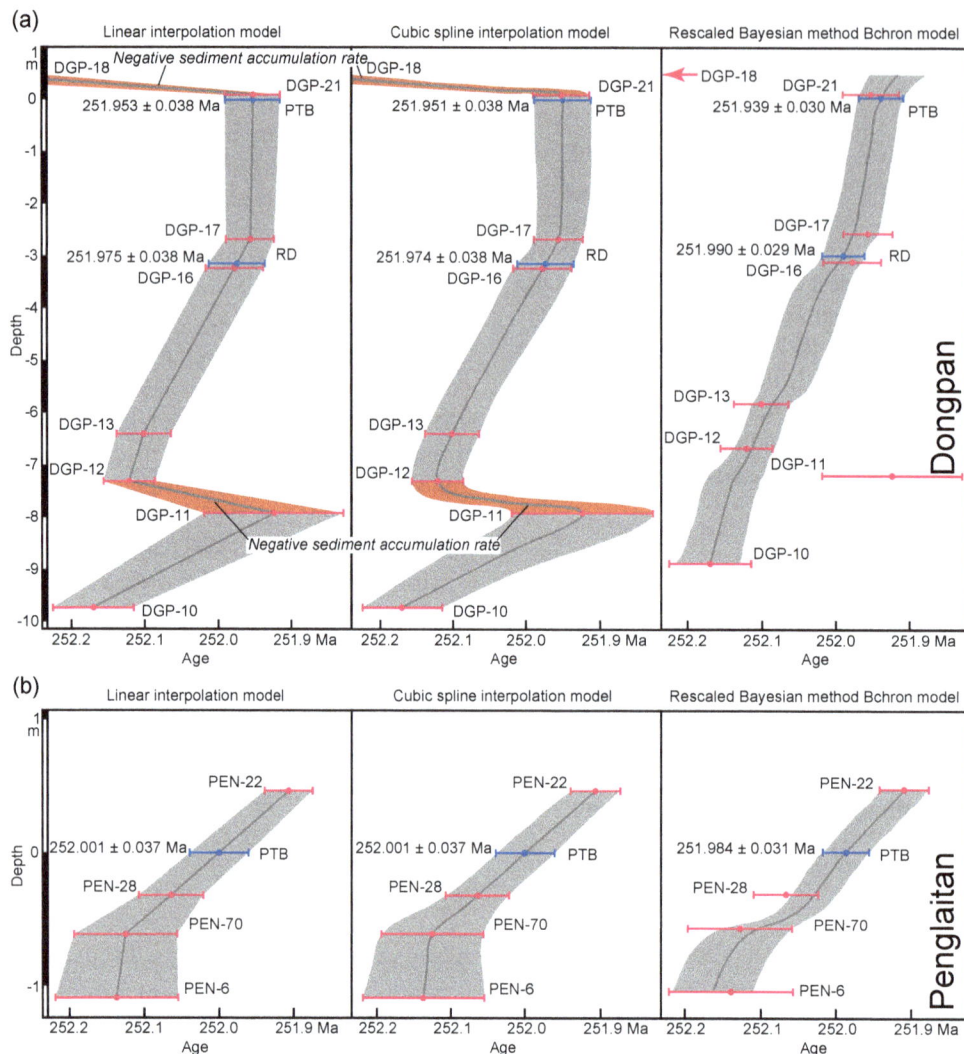

Figure 5. Comparison of the different age–depth models based on linear interpolation, cubic spline fit and Bayesian statistics for **(a)** Dongpan and **(b)** Penglaitan. Each age–depth model is presented with its median (middle grey line) and its associated 95 % confidence interval (grey area). Radioisotopic dates, used in the age–depth models, together with their uncertainty (red horizontal bars) are presented as ^{206}Pb / ^{238}U weighted mean dates of the Dongpan and Penglaitan volcanic ash beds and volcanogenic sandstones in their stratigraphic positions. U-Pb data of DGP-21, PEN-28 and PEN-22 are taken from Baresel et al. (2016). Predicted dates (blue horizontal bars) for the onset of the radiolarian decline (RD) and the Permian–Triassic boundary (PTB) in Dongpan and Penglaitan are calculated with their associated uncertainty using the different age–depth models.

as reflected by the linear and cubic interpolation models for the interval between DGP-11 and DGP-12, and for the interval between DGP-21 and DGP-18 (Fig. 5a).

In all Bchron models, the thickness of the geologically instantaneously deposited volcanic ash beds was virtually removed and the lithostratigraphy has been rescaled in order to create accurate deposition rate models for the investigated sedimentary successions. This approach has only minor effects in the Bchron age–depth model of Dongpan, where the changes in the calculated age of the PTB and the radiolarian decline (RD) are negligible (see Sect. S3), mainly due to the limited thickness (max. 8 cm) of the volcanic hori-

zons in Dongpan. The Bchron model of Penglaitan would be much more affected by such a rescaling if the thickness of the volcanogenic sandstones were also removed, but it is not clear whether each volcanogenic sandstone represents only one "instantaneous" turbidity current event or might reflect a series of several turbidite deposits over a certain time. Hence, also in Penglaitan only the thickness of the volcanic ash beds was removed. However, the relative substantial thickness of "instantaneously" deposited turbiditic volcanogenic sandstone at the top of the Penglaitan section may indeed induce some distortions in the Bchron model. Facies analysis did not reveal any signs of an omission surface at the formational boundary, but the strong contrast in sediment accumulation rates between the "instantaneous" deposition of the last Permian bed and the much slower accumulation of next overlying black shales likely generates a distortion of the Bchron model at the formational boundary. Hence, the Bchron model derived from Dongpan is certainly more reliable than that derived from Penglaitan.

The aim for applying Bayesian age modeling to the dated volcanogenic beds from these two sections was to obtain an age model for the PTB. The age–depth models yield ages of 251.939 ± 0.030 Ma (Dongpan; Figs. 2a and 5a) and of 251.984 ± 0.031 Ma (Penglaitan; Figs. 3b and 5b) for the lithological boundary between the Dalong and Ziyun Formation in both sections. These two ages overlap within uncertainties and thus demonstrate the synchronicity of the PTB in the two sections. Making the reasonable assumption of absence of significant gaps in these two sections, the new U-Pb dates can be used to infer sediment accumulation rates. The age–depth model of Dongpan suggests increased sediment accumulation rates in the uppermost part of the Dalong Formation from bed 6 (DGP-17) upwards. Below bed 6, calculated sediment accumulation rates appear to be relatively constant with 3.6 ± 1.2 cm kyr^{-1}, but above bed 6 they jump to 6.0 ± 2.4 cm kyr^{-1}. In Penglaitan, the sediment accumulation rate of the uppermost Dalong Formation and basalmost Ziyun Formation is significantly lower than in Dongpan with 0.7 ± 0.3 cm kyr^{-1}. Previously published U-Pb zircon geochronology from Penglaitan (Shen et al., 2011), including a weighted mean date of 252.16 ± 0.09 Ma from a volcanogenic sandstone at 26.7 m below the PTB, was not considered in our age model, since substantial improvements in the analytical protocol hamper comparing these dates with our U-Pb results.

6 Comparison of the Dongpan and Penglaitan sections with Meishan GSSP results

6.1 The change of the PTB age through analytical improvement of U-Pb dating

The first geochronological studies in the GSSP at Meishan D have been carried out on a volcanic ash in bed 25, whose base starts 4 cm below the formational PTB, by U-Pb sensitive high-resolution ion microprobe (SHRIMP) analysis of zircons yielding a ^{206}Pb / ^{238}U age of 251.2 ± 3.4 Ma (Claoué-Long et al., 1991) and by ^{40}Ar/^{39}Ar dating of sanidine at 249.91 ± 0.15 Ma (Renne et al., 1995). However, these dates are either not sufficiently precise to allow calibrating magmatic and biological timescales at resolution adequate for both groups of processes or are biased by a systematic age offset between the U-Pb and Ar-Ar systems of ~ 1.0 % (Schoene et al., 2006). In order to properly compare the two systems, all older ^{40}Ar/^{39}Ar data have to be corrected for the revised age of the standard Fish Canyon

sanidine of 28.201 ± 0.046 Ma (Kuiper et al., 2008) and the decay constant uncertainty has to be added to U-Pb and Ar-Ar ages, which would drastically expand the ^{40}Ar/^{39}Ar age error and recalculate the ^{40}Ar/^{39}Ar age of Renne et al. (1995) to 251.6 ± 0.6 Ma. In a first detailed ID-TIMS study, U-Pb ages of mechanically abraded zircons were published by Bowring et al. (1998) for six volcanic ash beds at Meishan, placing the PTB at 251.4 ± 0.3 Ma. Though much more precise than former studies, these ages are mainly based on multi-grain zircon analyses. It was shown by Mundil et al. (2001), by confining data selection to single-crystal analyses of the same horizons, that the multi-grain approach might disguise complexity of zircon population ages which are caused by pervasive lead loss and inheritance. In a second attempt, driven by further improvements in the U-Pb ID-TIMS technique (e.g., chemical abrasion of zircon grains by hydrofluoric acid exposure to remove zircon domains with lead loss; reduced procedural common Pb blanks), the PTB extinction horizon in Meishan and Shangsi (China) was dated at 252.6 ± 0.2 Ma by Mundil et al. (2004). Unlike previous studies, Shen et al. (2011) dated larger number of zircon grains per ash bed in order to overcome inheritance, magmatic residence, and lead loss phenomena of zircon population ages. They determined the duration of the PTB extinction interval (bed 25 to bed 28 in Meishan) at 200 ± 100 kyr starting at 252.28 ± 0.08 Ma in bed 25 together with a sharp negative δ^{13}C excursion. By using the same mineral separates from identical ash beds as in Shen et al. (2011), the extinction period at Meishan was determined by Burgess et al. (2014) between 251.941 ± 0.037 Ma (bed 25) and 251.880 ± 0.031 Ma (bed 28). The differences in age and precision compared to Shen et al. (2011) reflect significant progress of the EARTHTIME community in data acquisition and reduction such as refined tracer calibration, new error propagation algorithms, and the development of the EARTHTIME ^{202}Pb-^{205}Pb-^{233}U-^{235}U tracer solution.

Figure 6 illustrates the three Bayesian age–depth models based on our U-Pb dates from Dongpan and Penglaitan compared to the latest generation of U-Pb ages from Meishan GSSP (Burgess et al., 2014). Such a comparison is possible because all the dates from these three sections were obtained with the same analytical procedures, including identical data reduction procedures, error propagation and Th correction, thus leading to closely comparable precision and accuracy of the ages. This tight temporal framework allows us to perform a quantitative comparison of the Dongpan, Penglaitan and Meishan sections in terms of lithostratigraphy, biostratigraphy and chemostratigraphy via the Bayesian statistics.

6.2 Comparison of lithostratigraphy

All three interpolated ages of the formational boundary in Dongpan (251.939 ± 0.030 Ma), Penglaitan (251.984 ± 0.031 Ma) and Meishan (251.956 ± 0.035 Ma) are in agreement within errors (Fig. 6). They support the

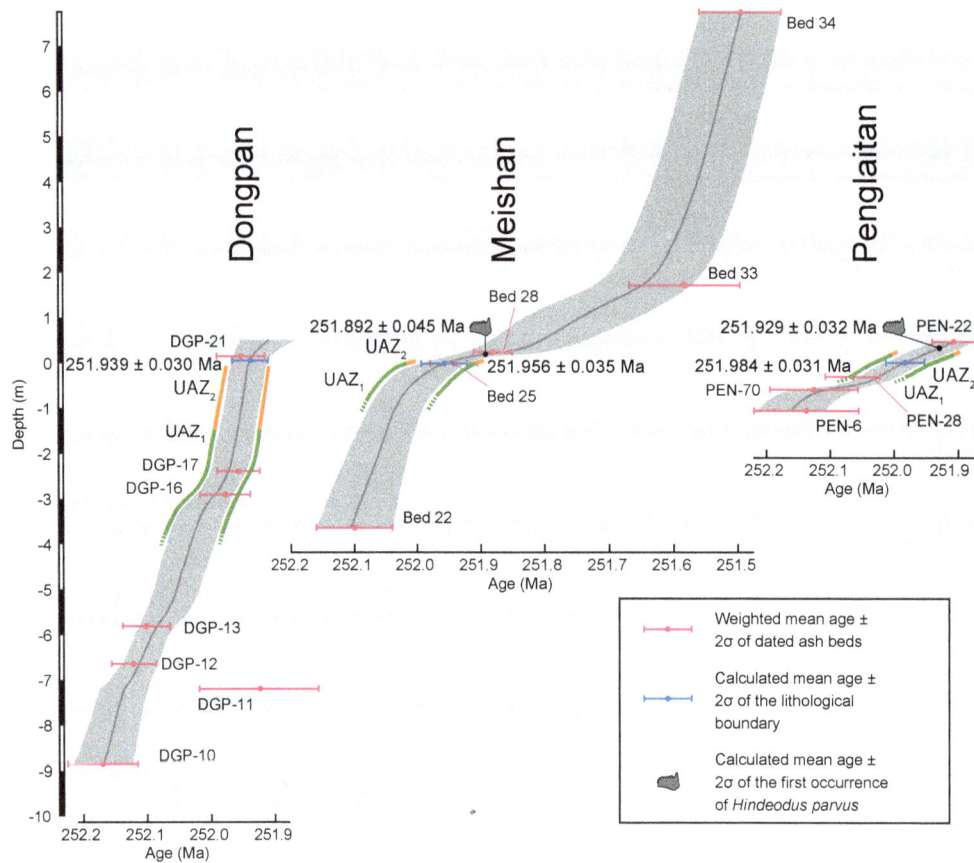

Figure 6. Comparison of Bayesian Bchron age models for Dongpan, Penglaitan and the Meishan GSSP. Predicted dates together with their uncertainty for the lithological boundaries and the first occurrence of the conodont *Hindeodus parvus* at Dongpan and Penglaitan are calculated using U-Pb ages of this study and of Baresel et al. (2016), and the U-Pb ages of Burgess et al. (2014) for Meishan. The durations of the conodont unitary association zones (UAZs) UAZ1 and UAZ2 (Brosse et al., 2016) are inferred from the Bchron age model of Meishan and projected to the model of Dongpan and Penglaitan, respectively.

synchronicity of the conformable boundary between the Dalong Formation and the Ziyun Formation in Dongpan and Penglaitan, and also demonstrate their temporal coincidence with the conformable boundary in Meishan between the Changhsing Formation and Yinkeng Formation. The age model also confirms the extreme condensation around the PTB in Meishan, with a maximal sediment accumulation rate of $0.4 \, \text{cm} \, \text{kyr}^{-1}$ as reported by Burgess et al. (2014) for the 26 cm thick interval between the base of bed 25 and the base of bed 28. In this respect, Dongpan and Penglaitan offer a greater potential for higher-resolution studies of environmental proxies around the PTB with maximal sediment accumulation rates for the same interval of 8.4 and $1.0 \, \text{cm} \, \text{kyr}^{-1}$, respectively. The increased sediment accumulation rate above bed 6 in Dongpan is in agreement with the previously inferred sedimentary fluxes deduced from elemental chemical analyses (Shen et al., 2012). From bed 7 upward, He et al. (2007) showed a clear increase in Al_2O_3 and TiO_2, indicating increased fluxes of terrestrial input into this trough. The accompanying size reduction of

brachiopods (He et al., 2007) led them to infer a regressive trend in the upper part of the Dongpan section. The ecological consequences of any regressive trend or increased clastic input might conceivably impact the diversity of marine species, but distinguishing between increased fluxes and regression remains difficult because both causes may have converging consequences.

6.3   Comparison of biostratigraphy

The PTB is defined by the FO of *H. parvus* in the Meishan GSSP in bed 27c. This definition is complicated by the suggested existence of a hardground within bed 27, which is at the position of the previously defined PTB (Chen et al., 2009). Others have suggested that the FO of *H. parvus* at Meishan is not the timing of the true evolutionary origination of this species (Jiang et al., 2011; Yuan et al., 2015). In Meishan, the FO of *H. parvus* in bed 27c is interpolated at $251.892 \pm 0.045 \, \text{Ma}$ (Fig. 6) and is located 21 cm above the formational boundary which occurs between beds 24 and 25. Temporally coincident, the FO of *H. parvus* in Penglaitan is interpolated at 251.929 ± 0.032 Ma (Fig. 6) and is located 33 cm above the formational boundary. With respect to formational boundaries, the higher stratigraphic position of the FO of *H. parvus* in Penglaitan than in Meishan is to be expected because of the higher sediment accumulation rate. However, in Meishan *H. parvus* first occurs 64 ± 56 kyr and in Penglaitan 53 ± 46 kyr after the formational boundary, indicating perfect synchronicity within our temporal resolution. In Dongpan, the lack of conodont bearing beds around the PTB hampers testing the synchronicity of the FO of *H. parvus* between Dongpan and the two other sections.

Brosse et al. (2016) established a new and robust conodont zonation based on unitary associations around the PTB in South China that includes the Meishan GSSP. This zonation contains six unitary association zones (UAZs), with the PTB falling into the separation interval between UAZ2 (bed 25) and UAZ3 (bed 27a–d). This new zonation also places the UAZ-based PTB in Meishan closer to the conformable boundary between the Changhsing Formation and the Yinkeng Formation than the FO of *H. parvus* (bed 27c) does. Available conodont data from Meishan allow the assignment of bed 24a–e to UAZ1 (UAZ1 might reach further down as indicated by a dashed line in Fig. 6), bed 25 to UAZ2, bed 27a–d to UAZ3 and bed 28 to UAZ4 (Brosse et al., 2016). The stratigraphic thickness comprised between the base of UAZ1 and the top of UAZ4 amounts to 1.22 m. By using the three-section age–depth models, we attempted to project the respective thickness corresponding to the UAZ1–UAZ4 interval in Meishan onto the two other sections. This projection resulted into a pronounced, artificial lengthening of UAZs in Dongpan and Penglaitan. UAZ1 is the penultimate Permian conodont UAZ in Meishan (Brosse et al., 2016). When projected onto the age–depth models of Dongpan and Penglaitan, this UAZ1 is artificially expanded and even crosses the PTB in Penglaitan (Fig. 6). In Penglaitan, the last Permian UAZ2 projects correctly above UAZ1 without overlap but is completely within the Triassic. The cause of these contradictions stems from the irreconcilable conjunction of (i) extreme condensation in Meishan, (ii) high evolutionary rates of conodonts, and (iii) the ~ 40 ka precision of the last generation of U-Pb dates.

In Dongpan, the onset of a protracted radiolarian diversity decline in bed 5 reported by Feng and Algeo (2014) is here interpolated at 251.990 ± 0.029 Ma, occurring 51 ± 42 kyr before the formational boundary (Fig. 2). Excess SiO_2 values of this bed (Shen et al., 2012) suggest a genuine diversity pattern at the local scale, which seems to be unrelated to any substantial change or trend in the local redox conditions (as shown by Co/Al, Cr/Al, Cu/Al and V/Al measurements of He et al., 2007).

6.4 Comparison of chemostratigraphy

Organic carbon isotope chemostratigraphy of Dongpan (Fig. 7) extending from the Permian bed 5 to the Triassic

bed 13 was provided by Zhang et al. (2006) and for Meishan (Fig. 7) extending from the Permian bed 24 to the Triassic bed 29 by Cao et al. (2002). The correlation of these $\delta^{13}C_{org}$ records by Zhang et al. (2006), based on the occurrence of ash beds in both sections, is largely over-interpreted. With the exception of a short negative excursion followed by a more prominent positive excursion between beds 9 and 11, the Permian part of the $\delta^{13}C_{org}$ record in Dongpan is relatively stable and oscillates between -28 and $-27‰$. With the exception of a negative excursion culminating in beds 25 and 26, the Permian part of the $\delta^{13}C_{org}$ record in Meishan shows a sustained positive trend from -30 to $-26‰$. The basal Triassic part of these two records is also incompatible in that they display opposite trends. With the possible exception of the Xinmin section (J. Shen et al., 2013), the $\delta^{13}C_{org}$ record of Dongpan does not correlate with that of any other South Chinese section, but even Xinmin shows a $\sim 3‰$ offset of the base trend in comparison to Dongpan. However, we note that in Meishan an abrupt decline in $\delta^{13}C_{carb}$ occurs in bed 24e at 251.950 ± 0.042 Ma (Burgess et al., 2014) and slightly above in bed 26 in $\delta^{13}C_{org}$ at 251.939 ± 0.032 Ma, which is temporally coincident with the main negative $\delta^{13}C_{org}$ excursion in bed 9 in Dongpan at 251.956 ± 0.030 Ma. The second smaller negative $\delta^{13}C_{org}$ excursion at the PTB in Dongpan at 251.941 ± 0.030 Ma and in Meishan at 251.892 ± 0.045 Ma cannot be distinguished within uncertainty from the main excursion, which hampers the correlation of the $\delta^{13}C_{org}$ records based on U-Pb ages. However, interpreting organic carbon records requires the simultaneous analysis of palynofacies, which are not documented in Dongpan. Shen et al. (2012) also showed that the total organic carbon (TOC) never exceeds 0.2%, thus indicating a generally poor preservation of the organic matter in this section. As shown by Shen et al. (2012), this preservation bias is further supported by coincident peaks in both terrestrial (spore and pollens) and marine (algae and acritarchs) organic material (see Fig. 7). This uneven preservation of the organic matter further hampers the understanding of the $\delta^{13}C_{org}$ signal in Dongpan. More generally, the consistency and lateral reproducibility of the late Permian carbonate and organic carbon isotope records in South China remain equivocal (e.g., S. Z. Shen et al., 2013). These records are probably influenced by the local graben and horst paleotopography that hampered efficient circulation of water masses with the open ocean, thus reflecting more local than global changes.

6.5 Comparison of astrochronology

Sedimentary cycles driven by orbital forcing (100 kyr eccentricity cycles) were inferred by Peng et al. (2008) on the basis of Ce/La fluctuations in Dongpan. These cycles were also used by Feng and Algeo (2014) to calibrate their radiolarian extinction and survival intervals. The duration of these two intervals amounts to ~ 260 kyr (see their Fig. 5). For the same stratigraphic interval, our U-Pb ages (interval

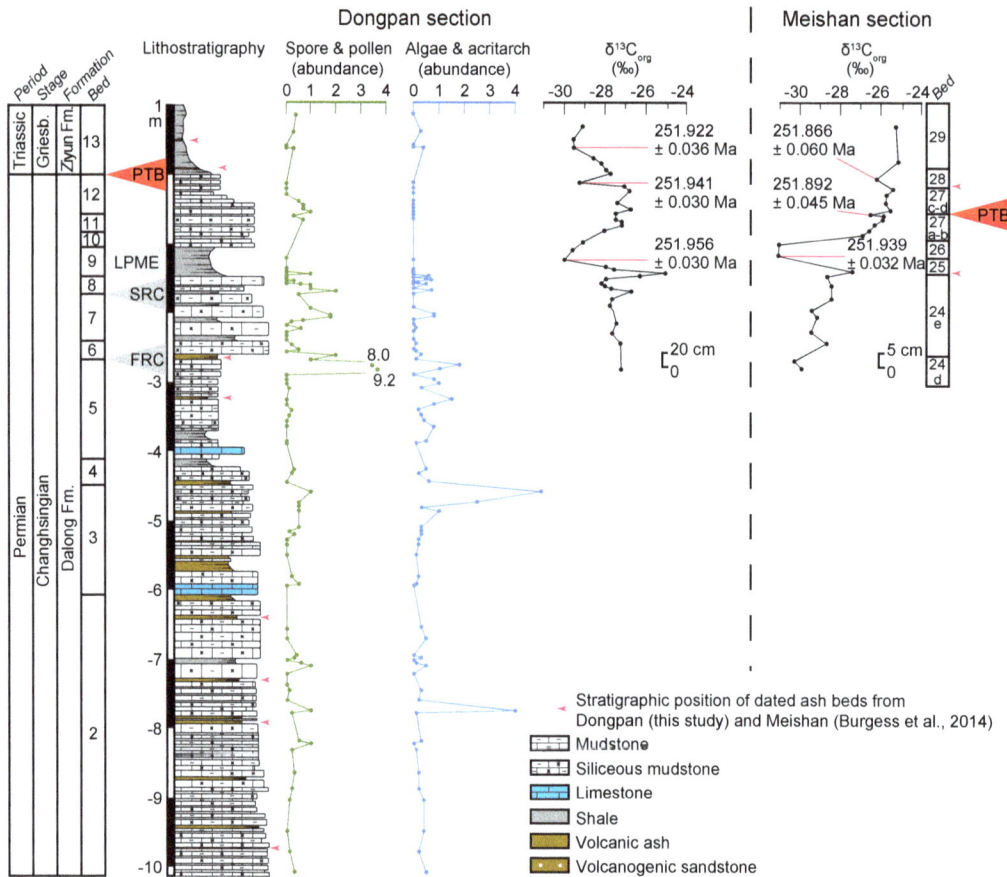

Figure 7. Comparison of the organic carbon isotope chemostratigraphy of Dongpan (Zhang et al., 2006) with that of Meishan (Cao et al., 2002). Dates and their associated uncertainty for the negative carbon isotope excursions in both sections are revealed from the Bayesian Bchron age models of Dongpan and Meishan, respectively. Abundance of spores and pollen, as well as algae and acritarchs (i.e., spores per 100 cm^2 of soil) in Dongpan, is from Shen et al. (2012). Stratigraphic positions of the first radiolarian crisis (FRC) and second radiolarian crisis (SRC) (after Feng et al., 2007), the latest Permian extinction event (LPME) in Dongpan and the Permian–Triassic boundary (PTB) in both sections are indicated as well. Meishan section thickness is not to scale with the Dongpan section.

from DGP-16 to DGP-21) indicate a much shorter duration of max. 75 kyr. It is not clear whether this chemical cyclicity might represent either precession instead of eccentricity cycles or rather a local signal of the sedimentary-chemical system. Huang et al. (2011) produced an astrochronological timescale across the PTB in China and Austria with an estimated duration of 700 kyr for the extinction interval. However, the extinction interval in their study is too long and ranges from the start of the *Neogondolella meishanensis* conodont zone to the base of the *Isarcicella isarcica* zone, defining a prolonged extinction interval stretching from the top of bed 24e to the base of bed 29 in Meishan. Wu et al. (2013) reported Milankovitch cycles from late Permian strata at Meishan and Shangsi, South China, indicating a 7.793 Myr duration for the Lopingian epoch based on 405 kyr orbital eccentricity cycles. Their inferred duration of 83 kyr for the extinction interval in Meishan between the base of bed 25 and the top of bed 28 is in good agreement with the radioisotopically dated duration of 61 ± 48 kyr for the same interval

(Burgess et al., 2014). This is consistent with the study of Li et al. (2016), whose astronomical-cycle tuning of spectral gamma-ray logs constrains the extinction interval in Meishan to less than 40 % of a 100 kyr eccentricity cycle (i.e., < 40 kyr).

7 Conclusions

– The comparison of our high-precision U-Pb zircon data from Dongpan and Penglaitan sections in South China with the data of Burgess et al. (2014) from Meishan D GSSP provides convincing evidence that dates elaborated through the use of the EARTHTIME ^{202}Pb-^{205}Pb-^{233}U-^{235}U tracer solution are comparable at the 0.05 % level or better even if coming from different laboratories. This fact underlines a substantially increased accuracy and precision of U-Pb ID-TIMS dating, now capable of elucidating environmental change and biotic response on a decamillennial scale.

– Applying Bayesian age modeling to sections with such high-precision time information allows for quantitative comparison of disparate information from different sections. Along with the work of Ovtcharova et al. (2015) at the Early–Middle Triassic boundary, these are the two first proof-of-concept studies adopting age–depth modeling to compare coeval sections with different fossil contents, different facies and disparate sediment accumulation rates at highest temporal resolution. We anticipate that this approach will need to become the future standard in the assessment of the geologic timescale.

– Bchron age–depth models demonstrate synchronicity of the conformable lithological boundaries in Dongpan, Penglaitan, and Meishan. These results highlight the temporal reliability of the environmental changes as expressed by formational boundaries in platform-slope and deeper water basinal sections.

– The higher sediment accumulation rates of Dongpan and Penglaitan provide a much better prospect for the high-resolution study of environmental proxies around the PTB than the condensed GSSP section in Meishan. Our age–depth models also reveal that the combination of condensed deposition with high evolutionary rates of conodonts and the ~ 30 ka resolution of the last generation of U-Pb ages makes it impossible to project stratigraphic data points or intervals of Meishan onto expanded PTB sections without distortions. This intrinsic problem of the Meishan GSSP section should stimulate the search of alternative sections with more expanded records.

– The seemingly erratic late Permian carbon isotope record in South China does not allow laterally reproducible intercalibration with the newly obtained U-Pb dates. This stands in sharp contrast with the Early Triassic carbon isotope record, which is of global significance (e.g., Galfetti et al., 2007), thus making the Early Triassic interval the ideal target of future studies that integrate chemostratigraphy, geochronology and astrochronology in a Bayesian age–depth modeling approach.

Competing interests. The authors declare that they have no conflict of interest.

Acknowledgements. The authors acknowledge financial support by the Swiss NSF through projects 200020_137630 (to Urs Schaltegger) and 200021_135446 (to Hugo Bucher). Aymon Baud (Lausanne) is thanked for support in the field. Tom Algeo (Cincinnati) and Shen Shuzhong (Nanjing) are thanked for insightful discussions and for providing Excel tables of carbon isotopic data from published sections. Special thanks go to technical and scientific members of the Geneva and Zurich research groups who helped during all stages of this study. M. Schmitz and S. Burgess are gratefully acknowledged for comments and suggestions that substantially improved the manuscript.

Edited by: M. Heap

References

Algeo, T. J. and Twitchett, R. J.: Anomalous Early Triassic sediment fluxes due to elevated weathering rates and their biological consequences, Geology, 38, 1023–1026, doi:10.1130/G31203.1, 2010.

Alroy, J., Aberhan, M., Bottjer, D. J., Foote, M., Fürsich, F. T., Harries, P. J., Hendy, A. J. W., Holland, S. M., Ivany, L. C., Kiessling, W., Kosnik, M. A., Marshall, C. R., McGowan, A. J., Miller, A. I., Olszewski, T. D., Patzkowsky, M. E., Peters, S. E., Villier, L., Wagner, P. J., Bonuso, N., Borkow, P. S., Brenneis, B., Clapham, M. E., Fall, L. M., Ferguson, C. A., Hanson, V. L., Krug, A. Z., Layou, K. M., Leckey, E. H., Nürnberg, S., Powers, C. M., Sessa, J. A., Simpson, C., Tomasovych, A., and Visaggi, C. C.: Phanerozoic trends in the global diversity of marine invertebrates, Science, 321, 97–100, doi:10.1126/science.1156963, 2008.

Baresel, B., D'Abzac, F. X., Bucher, H., and Schaltegger, U.: High-precision time-space correlation through coupled apatite and zircon tephrochronology: an example from the Permian-Triassic boundary in South China, Geology, 45, 83–86, doi:10.1130/g38181.1, 2016.

Benton, M. J.: The origins of modern biodiversity on land, Philos. T. Roy. Soc. Lond. B, 365, 3667–3679, doi:10.1098/rstb.2010.0269, 2010.

Bowring, J. F., McLean, N. M., and Bowring, S. A.: Engineering cyber infrastructure for U-Pb geochronology: tripoli and U-Pb_Redux, Geochem. Geophys. Geosyst., 12, Q0AA19, doi:10.1029/2010gc003479, 2011.

Bowring, S. A., Erwin, D. H., Jing, G. Y., Martin, M. W., Davidek, K., and Wang, W.: U/Pb zircon geochronology and tempo of the end-Permian mass extinction, Science, 280, 1039–1045, doi:10.1126/science.280.5366.1039, 1998.

Broderick, C., Wotzlaw, J.-F., Frick, D., Gerdes, A., Günther, D., and Schaltegger, U.: Linking the thermal evolution and emplacement history of an upper-crustal pluton to its lower-crustal roots using zircon geochronology and geochemistry (southern Adamello batholith, N. Italy), Contrib. Mineral. Petr., 170, 28, doi:10.1007/s00410-015-1184-x, 2015.

Brooks, S., Gelman, A., Jones, G., and Meng, X.-L. (Eds.): Handbook of Markov Chain Monte Carlo, CRC Press, Boca Raton, Florida, USA, 619 pp., 2011.

Brosse, M., Bucher, H., and Goudemand, N.: Quantitative biochronology of the Permian-Triassic boundary in South China based on conodont Unitary Associations, Earth-Sci. Rev., 155, 153–171, doi:10.1016/j.earscirev.2016.02.003, 2016.

Burgess, S. D. and Bowring, S. A.: High-precision geochronology confirms voluminous magmatism before, during, and after Earth's most severe extinction, Sci. Adv., 1, 1–14, doi:10.1126/sciadv.1500470, 2015.

Burgess, S. D., Bowring, S. A., and Shen, S. Z.: High-precision timeline for Earth's most severe extinction, P. Natl. Acad. Sci. USA, 111, 3316–3321, doi:10.1073/pnas.1317692111, 2014.

Cao, C. Q., Wang, W., and Jin, Y.: Carbon isotope variation across Permian-Triassic boundary at Meishan section in Zhejiang province, China, Bulletin of Sciences (Chinese edition), 47, 302–306, doi:10.1360/02tb9252, 2002.

Chen, J., Beatty, T. W., Henderson, C. M., and Rowe, H.: Conodont biostratigraphy across the Permian-Triassic boundary at the Dawen section, Great Bank of Guizhou, Guizhou Province, South China: implications for the Late Permian extinction and correlation with Meishan, J. Asian Earth Sci., 36, 442–458, doi:10.1016/j.jseaes.2008.08.002, 2009.

Claoué-Long, J. C., Zhang, Z. C., Ma, G. G., and Du, S.H.: The age of the Permian-Triassic boundary, Earth Planet. Sci. Lett., 105, 182–190, doi:10.1016/0012-821x(91)90129-6, 1991.

Condon, D. J., Schoene, B., McLean, N. M., Bowring, S. A., and Parrish, R. R.: Metrology and traceability of U-Pb isotope dilution geochronology (EARTHTIME Tracer Calibration Part I), Geochim. Cosmochim. Ac., 164, 464–480, doi:10.1016/j.gca.2015.05.026, 2015.

Erwin, D. H.: The End-Permian Mass Extinction, Annu. Rev. Ecol. Syst., 21, 69–91, doi:10.1146/annurev.es.21.110190.000441, 1990.

Erwin, D. H., Bowring, S. A., and Jin, Y.-G.: End-Permian mass-extinctions: a review, in: Catastrophic events and mass extinctions: impacts and beyond, edited by: Koeberl, C. and MacLeod, K. G., Geol. S. Am. S., 356, 353–383, doi:10.1130/0-8137-2356-6.363, 2002.

Faure, M., Lin, W., Chu, Y., and Lepvrier, C.: Triassic tectonics of the southern margin of the South China Block, C. R. Geosci., 348, 5–14, doi:10.1016/j.crte.2015.06.012, 2016.

Feng, Q. L. and Algeo, T. J.: Evolution of oceanic redox conditions during the Permo-Triassic transition: Evidence from deepwater radiolarian facies, Earth-Sci. Rev., 137, 34–51, doi:10.1016/j.earscirev.2013.12.003, 2014.

Feng, Q. L., He, W. H., Gu, S. Z., Meng, Y. Y., Jin, Y. X., and Zhang, F.: Radiolarian evolution during the latest Permian in South China, Global Planet. Change, 55, 177–192, doi:10.1016/j.gloplacha.2006.06.012, 2007.

Feng, Z. Z., Bao, Z.-D., Zheng, X.-J., and Wang, Y.: There was no "Great Bank of Guizhou" in the Early Triassic in Guizhou Province, South China, J. Palaeogeogr., 4, 99–108, doi:10.3724/SP.J.1261.2015.00070, 2015.

Galfetti, T., Bucher, H., Ovtcharova, M., Schaltegger, U., Brayard, A., Brühwiler, T., Goudemand, N., Weissert, H., Hochuli, P. A., Cordey, F., and Guodun, K.: Timing of the Early Triassic carbon cycle perturbations inferred from new U–Pb ages and ammonoid biochronozones, Earth Planet. Sc. Lett., 258, 593–604, doi:10.1016/j.epsl.2007.04.023, 2007.

Galfetti, T., Bucher, H., Martini, R., Hochuli, P. A., Weissert, H., Crasquin-Soleau, S., Brayard, A., Goudemand, N., Brühwiler, T., and Guodun, K.: Evolution of Early Triassic outer platform paleoenvironments in the Nanpanjiang Basin (South China) and their significance for the biotic recovery, Sediment. Geol., 204, 36–60, doi:10.1016/j.sedgeo.2007.12.008, 2008.

Gao, Q., Zhang, N., Xia, W., Feng, Q., Chen, Z.-Q., Zheng, J., Griffin, W. L., O'Reilly, S. Y., Pearson, N. J., Wang, G., Wu, S., Zhong, W., and Sun, X.: Origin of volcanic ash beds across the Permian-Triassic boundary, Daxiakou, South China: Petrology and U-Pb age, trace elements and Hf-isotope composition of zircon, Chem. Geol., 360–361, 41–53, doi:10.1016/j.chemgeo.2013.09.020, 2013.

Golonka, J. and Ford, D.: Pangean (Late Carboniferous-Middle Jurassic) paleoenvironment and lithofacies, Palaeogeogr. Palaeocl., 161, 1–34, doi:10.1016/s0031-0182(00)00115-2, 2000.

Guizhou Bureau of Geology and Mineral Resources: Regional geology of Guizhou Province, scale 1 : 500 000, Geological Memoir, Beijing, 1, 700 pp., 1987 (in Chinese, with English summary).

Haslett, J. and Parnell, A.: A simple monotone process with application to radiocarbon-dated depth chronologies,J. Roy. Stat. Soc. C-App., 57, 399–418, doi:10.1111/j.1467-9876.2008.00623.x, 2008.

He, W. H., Shen, S. Z., Feng, Q. L., and Gu, S. Z.: A Late Changhsingian (Late Permian) deepwater brachiopod fauna from the Talung Formation at the Dongpan section, southern Guangxi, South China, J. Paleontol., 79, 927–938, doi:10.1666/0022-3360(2005)079[0927:ALCLPD]2.0.CO;2, 2005.

He, W. H., Shi, G. R., Feng, Q. L., Campi, M. J., Gu, S. Z., Bu, J. J., Peng, Y. Q., and Meng, Y. Y.: Brachiopod miniaturization and its possible causes during the Permian-Triassic crisis in deep water environments, South China, Palaeogeogr. Palaeocl., 252, 145–163, doi:10.1016/j.palaeo.2006.11.040, 2007.

Hiess, J., Condon, D. J., McLean, N., and Noble, S. R.: ^{238}U/^{235}U Systematics in terrestrial uranium-bearing minerals, Science, 335, 1610–1614, doi:10.1126/science.1215507, 2012.

Hochuli, P. A., Hermann, E., Vigran, J. S., Bucher, H., and Weissert, H.: Rapid demise and recovery of plant ecosystems across the end-Permian extinction event, Global Planet. Change, 74, 144–155, doi:10.1016/j.gloplacha.2010.10.004, 2010.

Huang, C., Tong, J., Hinnov, L., and Chen, Z. Q.: Did the great dying of life take 700 k.y.? Evidence from global astronomical correlation of the Permian-Triassic boundary interval, Geology, 39, 779–782, doi:10.1130/g32126.1, 2011.

Jiang, H., Lai, X., Yan, C., Aldridge, R. J., Wignall, P., and Sun, Y.: Revised conodont zonation and conodont evolution across the Permian-Triassic boundary at the Shangsi section, Guangyuan, Sichuan, South China, Global Planet. Change, 77, 102–115, doi:10.1016/j.gloplacha.2011.04.003, 2011.

Jin, Y. G., Shen, S. Z., Henderson, C. M., Wang, X. D., Wang, W., Wang, Y., Cao, C. Q., and Shang, Q. H.: The Global Stratotype Section and Point (GSSP) for the boundary between the Capitanian and Wuchiapingian stage (Permian), Episodes, 29, 253–262, 2006.

Kuiper, K. F., Deino, A., Hilgen, F. J., Krijgsman, W., Renne, P. R., and Wijbrans, J. R.: Synchronizing Rock Clocks of Earth History, Science, 320, 500–504, doi:10.1126/science.1154339, 2008.

Lehrmann, D. J., Payne, J. L., Felix, S. V., Dillett, P. M., Wang, H., Yu, Y., and Wei, J.: Permian-Triassic Boundary Sections from Shallow-Marine Carbonate Platforms of the Nanpanjiang Basin, South China: Implications for Oceanic Conditions Associated with the End-Permian Extinction and Its Aftermath, Palaios, 18, 138–152, doi:10.1669/0883-1351(2003)18<138:pbsfsc>2.0.co;2, 2003.

Lehrmann, D. J., Pei, D., Enos, P., Ellwood, B. B., Zhang, J., Wei, J., Dillett, P., Koenig, J., Steffen, K., Druke, D., Gross, J., Kessel,

B., and Newkirk, T.: Impact of differential tectonic subsidence on isolated carbonate platform evolution: Triassic of the Nanpanjiang basin, south China, Am. Assoc. Petr. Geol. B., 91, 287–320, doi:10.1306/10160606065, 2007.

Lehrmann, D. J., Stepchinski, L., Altiner, D., Orchard, M. J., Montgomery, P., Enos, P., Ellwood, B. B., Bowring, S. A., Ramezani, J., Wang, H., Wei, J., Yu, M., Griffiths, J. D., Minzo, M., Schaall, E. K., Lil, X., Meyerl, K. M., and Payne, J. L.: An integrated biostratigraphy (conodonts and foraminifers) and chronostratigraphy (paleomagnetic reversals, magnetic susceptibility, elemental chemistry, carbon isotopes and geochronology) for the Permian-Upper Triassic strata of Guandao section, Nanpanjiang Basin, south China, J. Asian Earth Sci., 108, 117–135, doi:10.1016/j.jseaes.2015.04.030, 2015.

Li, M., Ogg, J., Zhang, Y., Huang, C., Hinnov, L., Chen, Z.-Q., and Zou, Z.: Astronomical tuning of the end-Permian extinction and the Early Triassic Epoch of South China and Germany, Earth Planet. Sc. Lett., 441, 10–25, doi:10.1016/j.epsl.2016.02.017, 2016.

Luo, G. M., Lai, X. L., Feng, Q. L., Jiang, H. S., Wignall, P., Zhang, K. X., Sun, Y. D., and Wu, J.: End-Permian conodont fauna from Dongpan section: Correlation between the deep- and shallow-water facies. Sci. China Ser. D, 51, 1611–1622, doi:10.1007/s11430-008-0125-1, 2008.

Luo, G. M., Wang, Y., Yang, H., Algeo, T. J., Kump, L. R., Huang, J., and Xie, S. C.: Stepwise and large-magnitude negative shift in $\delta^{13}C_{carb}$ preceded the main marine mass extinction of the Permian-Triassic crisis interval, Palaeogeogr. Palaeocl., 299, 70–82, doi:10.1016/j.palaeo.2010.10.035, 2011.

Mattinson, J. M.: Zircon U-Pb chemical abrasion ("CA-TIMS") method: combined annealing and multi-step partial dissolution analysis for improved precision and accuracy of zircon ages, Chem. Geol., 220, 47–66, doi:10.1016/j.chemgeo.2005.03.011, 2005.

McLean, N. M., Bowring, J. F., and Bowring, S. A.: An algorithm for U-Pb isotope dilution data reduction and uncertainty propagation, Geochem. Geophys. Geosyst., 12, Q0AA18, doi:10.1029/2010gc003478, 2011.

Meng, Y. Y., Zhou, Q., and Li, Y. K.: The characteristics and controlling sedimentary facies and granitoid analysis of the middle part of Pingxing-Dongmeng large fault, Guangxi Geology, 15, 1–4, 2002 (in Chinese).

Mundil, R., Metcalfe, I., Ludwig, K. R., Renne, P. R., Oberli, F., and Nicoll, R. S.: Timing of the Permian-Triassic biotic crisis: Implications from new zircon U/Pb age data (and their limitations), Earth Planet. Sc. Lett., 187, 131–145, doi:10.1016/s0012-821x(01)00274-6, 2001.

Mundil, R., Ludwig, K. R., Metcalfe, I., and Renne, P. R.: Age and Timing of the Permian Mass Extinctions: U/Pb Dating of Closed-System Zircons, Science, 305, 1760–1763, doi:10.1126/science.1101012, 2004.

Ovtcharova, M., Bucher, H., Schaltegger, U., Galfetti, T., Brayard, A., and Guex, J.: New Early to Middle Triassic U-Pb ages from South China: calibration with ammonoid biochronozones and implications for the timing of the Triassic biotic recovery, Earth Planet. Sc. Lett., 243, 463–475, doi:10.1016/j.epsl.2006.01.042, 2006.

Ovtcharova, M., Goudemand, N., Hammer, O., Guodun, K., Cordey, F., Galfetti, T., Schaltegger, U., and Bucher, H.: Developing a strategy for accurate definition of a geological boundary through radio-isotopic and biochronological dating: The Early–Middle Triassic boundary (South China), Earth-Sci. Rev., 146, 65–76, doi:10.1016/j.earscirev.2015.03.006, 2015.

Parnell, A. C., Haslett, J., Allen, J. R. M., Buck, C. E., and Huntley, B.: A flexible approach to assessing synchroneity of past events using Bayesian reconstructions of sedimentation history, Quaternary Sci. Rev., 27, 1872–1885, doi:10.1016/j.quascirev.2008.07.009, 2008.

Parnell, A. C., Buck, C. E., and Doan, T. K.: A review of statistical chronology models for high-resolution, proxy-based Holocene palaeoenvironmental reconstruction, Quaternary Sci. Rev., 30, 2948–2960, doi:10.1016/j.quascirev.2011.07.024, 2011.

Payne, J. L., Turchyn, A. V., Paytan, A., DePaolo, D. J., Lehrmann, D. J., Yu, M. Y., and Wei, J. Y.: Calcium isotope constraints on the end-Permian mass extinction, P. Natl. Acad. Sci. USA, 107, 8543–8548, doi:10.1073/pnas.0914065107, 2010.

Peng, X. F., Feng, Q. L., Li, Z. B., and Meng, Y. Y.: High-resolution cyclostratigraphy of geochemical records from Permo-Triassic boundary section of Dongpan, southwestern Guangxi, South China, Sci. China Ser. D, 51, 187–193, 2008.

Raup, D. M.: Size of the Permo-Triassic bottleneck and its evolutionary implications, Science, 206, 217–218, doi:10.1126/science.206.4415.217, 1979.

Renne, P. R., Black, M. T., Zichao, Z., Richards, M. A., and Basu, A. R.: Synchrony and causal relations between Permian-Triassic Boundary crises and Siberian flood volcanism, Science, 269, 1413–1416, doi:10.1126/science.269.5229.1413, 1995.

Retallack, G. J. and Jahren, A. H.: Methane release from igneous intrusion of coal during Late Permian extinction events, J. Geol., 116, 1–20, doi:10.1086/524120, 2008.

Samperton, K. M., Schoene, B., Cottle, J. M., Keller, C. B., Crowley, J. L., and Schmitz, M. D.: Magma emplacement, differentiation and cooling in the middle crust: Integrated zircon geochronological-geochemical constraints from the Bergell Intrusion, Central Alps, Chem. Geol., 417, 322–340, doi:10.1016/j.chemgeo.2015.10.024, 2015.

Schoene, B., Crowley, J. L., Condon, D. J., Schmitz, M. D., and Bowring, S. A.: Reassessing the uranium decay constants for geochronology using ID-TIMS U-Pb data, Geochim. Cosmochim. Ac., 70, 426–445, doi:10.1016/j.gca.2005.09.007, 2006.

Shang, Q. H., Vachard, D., and Caridroit, M.: Smaller foraminifera from the Late Changhsingian (Latest Permian) of Southern Guangxi and discussion on the Permian-Triassic boundary, Acta Micropalaeontologica Sinica, 20, 377–388, 2003.

Shen, J., Algeo, T. J., Zhou, L., Feng, Q., Yu, J., and Ellwood, B.: Volcanic perturbations of the marine environment in South China preceding the latest Permian mass extinction and their biotic effects, Geobiology, 10, 82–103, doi:10.1111/j.1472-4669.2011.00306.x, 2012.

Shen, J., Algeo, T. J., Hu, Q., Xu, G. Z., Zhou, L., and Feng, Q. L.: Volcanism in South China during the Late Permian and its relationship to marine ecosystem and environmental changes, Global Planet. Change, 105, 121–134, doi:10.1016/j.gloplacha.2012.02.011, 2013.

Shen, S. Z., Wang, Y., Henderson, C. M., Cao, C. Q., and Wang, W.: Biostratigraphy and lithofacies of the Permian System in the Laibin-Heshan area of Guangxi, South China, Palaeoworld, 16, 120–139, doi:10.1016/j.palwor.2007.05.005, 2007.

Shen, S. Z., Crowley, J. L., Wang, Y., Bowring, S. A., Erwin, D. H., Sadler, P. M., Cao, C. Q., Rothman, D. H., Henderson, C. M., Ramezani, J., Zhang, H., Shen, Y., Wang, X. D., Wang, W., Mu, L., Li, W. Z., Tang, Y. G., Liu, X. L., Liu, L. J., Zeng, Y., Jiang, Y. F., and Jin, Y. G.: Calibrating the end-Permian mass extinction, Science, 334, 1367–1372, doi:10.1126/science.1213454, 2011.

Shen, S. Z., Cao, C. Q., Zhang, H., Bowring, S. A., Henderson, C. M., Payne, J. L., Davydov, V. I., Chen, B., Yuan, D. X., Zhang, Y. C., Wang, W., and Zheng, Q. F.: High-resolution $\delta^{13}C_{carb}$ chemostratigraphy from latest Guadalupian through earliest Triassic in South China and Iran, Earth Planet. Sc. Lett., 375, 156–165, doi:10.1016/j.epsl.2013.05.020, 2013.

Stanley, S. M. and Yang, X.: A double mass extinction at the end of the Paleozoic era, Science, 266, 1340–1344, doi:10.1126/science.266.5189.1340, 1994.

Svensen, H., Planke, S., Polozov, A. G., Schmidbauer, N., Corfu, F., Podladchikov, Y. Y., and Jamtveit, B.: Siberian gas venting and the end-Permian environmental crisis, Earth Planet. Sc. Lett., 277, 490–500, doi:10.1016/j.epsl.2008.11.015, 2009.

Van Valen, L.: A resetting of Phanerozoic community evolution, Nature, 307, 50–52, doi:10.1038/307050a0, 1984.

Wang, Y. and Jin, Y.: Permian palaeogeographic evolution of the Jiangnan Basin, South China, Palaeogeogr. Palaeocl., 160, 35–44, doi:10.1016/s0031-0182(00)00043-2, 2000.

Winguth, C. and Winguth, A. M. E.: Simulating Permian-Triassic oceanic anoxia distribution: implications for species extinction and recovery, Geology, 40, 127–130, doi:10.1130/g32453.1, 2012.

Wu, H., Zhang, S., Hinnov, L. A., Jiang, G., Feng, Q., Li, H., and Yang, T.: Time-calibrated Milankovitch cycles for the late Permian, Nat. Commun., 4, 2452, doi:10.1038/ncomms3452, 2013.

Wu, J., Feng, Q. L., Gui, B. W., and Liu, G. C.: Some new radiolarian species and genus from Upper Permian in Guangxi Province, South China, J. Paleontol., 84, 879–894, doi:10.1666/09-057.1, 2010.

Xia, W. C., Zhang, N., Wang, G. Q., and Kakuwa, Y.: Pelagic radiolarian and conodont biozonation in the Permo-Triassic boundary interval and correlation to the Meishan GSSP, Micropaleontology, 50, 27–44, doi:10.1661/0026-2803(2004)050[0027:pracbi]2.0.co;2, 2004.

Yin, H. F.: Bivalves near the Permian-Triassic boundary in South China, J. Paleontol., 59, 572–600, 1985.

Yin, H. F., Zhang, K. X., Tong, J. N., Yang, Z. Y., and Wu, S. B.: The global stratotype section and point (GSSP) of the Permian-Triassic boundary, Episodes, 24, 102–114, 2001.

Yin, H. F., Feng, Q. L., Lai, X. L., Baud, A., and Tong, J. N.: The protracted Permo-Triassic crisis and multi-episode extinction around the Permian-Triassic boundary, Global Planet. Change, 55, 1–20, doi:10.1016/j.gloplacha.2006.06.005, 2007.

Yin, H. F., Jiang, H. S., Xia, W. C. Feng, Q., Zhang, N., and Shen, J.: The end-Permian regression in South China and its implication on mass extinction, Earth-Sci. Rev., 137, 19–33, doi:10.1016/j.earscirev.2013.06.003, 2014.

Yuan, A., Crasquin-Soleau, S., Feng, Q. L., and Gu, S. Z.: Latest Permian deep-water ostracods from southwestern Guangxi, South China, J. Micropalaeontol., 26, 169–191, doi:10.1144/jm.26.2.169, 2007.

Yuan, D., Shen, S., Henderson, C. M., Chen, J., Zhang, H., and Feng, H.: Revised conodont-based integrated high-resolution timescale for the Changhsingian Stage and end-Permian extinction interval at the Meishan sections, South China, Lithos, 204, 220–245, doi:10.1016/j.lithos.2014.03.026, 2015.

Zhang, F., Feng, Q. L., He, W. H., Meng, Y. Y., and Gu, S. Z.: Multidisciplinary stratigraphy across the Permian-Triassic boundary in deep-water environment of Dongpan section, south China, Norw. J. Geol., 86, 125–131, 2006.

Zhao, J. K., Liang, X. L., and Zheng, Z. G.: Late Permian Cephalopods in South China, Science Press, Beijing, China, 1978 (in Chinese).

Ziegler, A. M., Hulver, M. L., and Rowley, D. B.: Permian World Topography and Climate, in: Late glacial and postglacial environmental changes: Quaternary, Carboniferous-Permian, and Proterozoic, edited by: Martini, I. P., Oxford University Press, New York, USA, 343 pp., 1997.

Permissions

The contributors of this book come from diverse backgrounds, making this book a truly international effort. This book will bring forth new frontiers with its revolutionizing research information and detailed analysis of the nascent developments around the world.

We would like to thank all the contributing authors for lending their expertise to make the book truly unique. They have played a crucial role in the development of this book. Without their invaluable contributions this book wouldn't have been possible. They have made vital efforts to compile up to date information on the varied aspects of this subject to make this book a valuable addition to the collection of many professionals and students.

This book was conceptualized with the vision of imparting up-to-date information and advanced data in this field. To ensure the same, a matchless editorial board was set up. Every individual on the board went through rigorous rounds of assessment to prove their worth. After which they invested a large part of their time researching and compiling the most relevant data for our readers.

The editorial board has been involved in producing this book since its inception. They have spent rigorous hours researching and exploring the diverse topics which have resulted in the successful publishing of this book. They have passed on their knowledge of decades through this book. To expedite this challenging task, the publisher supported the team at every step. A small team of assistant editors was also appointed to further simplify the editing procedure and attain best results for the readers.

Apart from the editorial board, the designing team has also invested a significant amount of their time in understanding the subject and creating the most relevant covers. They scrutinized every image to scout for the most suitable representation of the subject and create an appropriate cover for the book.

The publishing team has been an ardent support to the editorial, designing and production team. Their endless efforts to recruit the best for this project, has resulted in the accomplishment of this book. They are a veteran in the field of academics and their pool of knowledge is as vast as their experience in printing. Their expertise and guidance has proved useful at every step. Their uncompromising quality standards have made this book an exceptional effort. Their encouragement from time to time has been an inspiration for everyone.

The publisher and the editorial board hope that this book will prove to be a valuable piece of knowledge for researchers, students, practitioners and scholars across the globe.

List of Contributors

Guillaume Desbois, Nadine Höhne and Janos L. Urai
Structural Geology, Tectonics and Geomechanics, RWTH Aachen University, Lochnerstrasse 4–20, 52056 Aachen, Germany

Pierre Bésuelle and Gioacchino Viggiani
Univ. Grenoble Alpes, CNRS, Grenoble INP, 3SR, 1270 Rue de la Piscine, 38610 Gières, France

Giancarlo Molli
Dipartimento di Scienze della Terra, Università di Pisa, Via S. Maria, 53, Pisa 56126, Italy

Luca Menegon
School of Geography, Earth and Environmental Sciences, Plymouth University, Plymouth, UK

Alessandro Malasoma
TS Lab and Geoservices, Via Vecchia Fiorentina, 10, Cascina 56023, Pisa, Italy

Ashwani Kant Tiwari, Arun Singh and Chandrani Singh
Department of Geology and Geophysics, Indian Institute of Technology Kharagpur, Kharagpur, India

Tuna Eken
Department of Geophysical Engineering, Istanbul Technical University, Istanbul, Turkey

Daniel Schweizer, Philipp Blum, and Christoph Butscher
Karlsruhe Institute of Technology (KIT), Institute for Applied Geosciences (AGW), Kaiserstr. 12, 76131 Karlsruhe, Germany

Robert Delhaye
Geophysics Section, School of Cosmic Physics, Dublin Institute for Advanced Studies (DIAS), 5 Merrion Square, Dublin 2, Ireland
National University of Ireland, Galway, University Road, Galway, Ireland

Volker Rath and Mark R. Muller
Geophysics Section, School of Cosmic Physics, Dublin Institute for Advanced Studies (DIAS), 5 Merrion Square, Dublin 2, Ireland

Alan G. Jones
Geophysics Section, School of Cosmic Physics, Dublin Institute for Advanced Studies (DIAS), 5 Merrion Square, Dublin 2, Ireland
Complete MT Solutions, Ottawa, Canada

Derek Reay
Geological Survey of Northern Ireland (GSNI), Belfast, UK

Katrin M. Wild, Patric Walter, and Florian Amann
Department of Earth Sciences, Geological Institute, ETH Zurich, 8092 Zurich, Switzerland

Yue Li, Shi Jie Wang and Luo Yi Qin
State Key Laboratory of Environmental Geochemistry, Institute of Geochemistry, Chinese Academy of Sciences, Guiyang, Guizhou, 550002, PR China
Puding Comprehensive Karst Research and Experimental Station, Institute of Geochemistry, CAS and Science and Technology Department of Guizhou Province, Puding, Guizhou, 562100, PR China

Xiao Yong Bai
State Key Laboratory of Environmental Geochemistry, Institute of Geochemistry, Chinese Academy of Sciences, Guiyang, Guizhou, 550002, PR China
Puding Comprehensive Karst Research and Experimental Station, Institute of Geochemistry, CAS and Science and Technology Department of Guizhou Province, Puding, Guizhou, 562100, PR China
Institute of Mountain Hazards and Environment, Chinese Academy of Sciences, Chengdu, Sichuan, 610041, PR China

Yi Chao Tian and Guang Jie Luo
State Key Laboratory of Environmental Geochemistry, Institute of Geochemistry, Chinese Academy of Sciences, Guiyang, Guizhou, 550002, PR China
Graduate School of Chinese Academy of Sciences, Beijing 100029, PR China

Zheng-Guo Sun and Hai-Yang Tang
College of Agro-grassland Science, Nanjing Agricultural University, 1 Weigang, Nanjing, Jiangsu 210095, People's Republic of China

Jie Liu
Department of Environmental Science, Hokkaido University, Sapporo 060-0810, Japan

Mathias Ronczka, Kristofer Hellman, Roger Wisén and Torleif Dahlin
Engineering Geology, Lund University, Lund, Sweden

Thomas Günther
Leibniz Institute for Applied Geophysics, Hanover, Germany

Nikita Afonin
Federal Centre for Integrated Arctic Research RAS, Arkhangelsk, Russia

Elena Kozlovskaya
Oulu Mining School, POB-3000, 90014, University of Oulu, Finland
Geological Survey of Finland, P.O. Box 96, 02151, Espoo, Finland

Ilmo Kukkonen
Department of Physics, University of Helsinki, P.O. Box 64, 00014, Helsinki, Finland

Stefano Gori, Emanuela Falcucci, Chiara Ladina, Simone Marzorati and Fabrizio Galadini
Istituto Nazionale di Geofisica e Vulcanologia, Rome, Via di Vigna Murata 605, 00143, Italy

Ece Aksoy and Christoph Schröder
European Topic Centre on Urban, Land and Soil systems (ETC/ULS), University of Malaga, Malaga, Spain

Mirko Gregor and Manuel Löhnertz
ETC/ULS, space4environment, Niederanven, Luxembourg

Geertrui Louwagie
European Environment Agency (EEA), Copenhagen, Denmark

Ekaterina Maksimova and Evgeny Abakumov
Department of Applied Ecology, Saint Petersburg State University, St Petersburg, Russia
Institute of the Ecology of the Volga Basin, Togliatti, Russia

Björn Baresel and Urs Schaltegger
Department of Earth Sciences, University of Geneva, Geneva, 1205, Switzerland

Hugo Bucher and Morgane Brosse
Paleontological Institute and Museum, University of Zurich, Zurich, 8006, Switzerland

Fabrice Cordey
Laboratoire de Géologie de Lyon, CNRS-UMR 5265, Université Claude Bernard Lyon 1, Villeurbanne, 69622, France

Kuang Guodun
Guangxi Bureau of Geology and Mineral Resources, Nanning, 530023, China

Index

www.ingramcontent.com/pod-product-compliance
Lightning Source LLC
Chambersburg PA
CBHW070154240326
41458CB00126B/4835